W9-DGB-913

MASS SPECTROMETRY BASICS

Christopher G. Herbert
Robert A.W. Johnstone

CRC PRESS

Boca Raton London New York Washington, D.C.

Library of Congress Cataloging-in-Publication Data

Herbert, Christopher G.
Mass spectrometry basics / Christopher G. Herbert, Robert A.W. Johnstone.
p. ; cm.
Compilation of articles previously published in the series Back-to-basics.
Includes bibliographical references and index.
ISBN 0-8493-1354-6 (alk. paper)
1. Mass spectrometry. I. Johnstone, R.A.W. (Robert Alexander Walker) II. Title.
[DNLM: 1. Spectrum Analysis, Mass--Collected Works. 2. Spectrum Analysis, Mass--Handbooks. QC 454.M3 H536m 2002]
QP519.9.M3 H47 2002
543′.0873—dc21 2002025935

This book contains information obtained from authentic and highly regarded sources. Reprinted material is quoted with permission, and sources are indicated. A wide variety of references are listed. Reasonable efforts have been made to publish reliable data and information, but the author and the publisher cannot assume responsibility for the validity of all materials or for the consequences of their use.

Neither this book nor any part may be reproduced or transmitted in any form or by any means, electronic or mechanical, including photocopying, microfilming, and recording, or by any information storage or retrieval system, without prior permission in writing from the publisher.

The consent of CRC Press LLC does not extend to copying for general distribution, for promotion, for creating new works, or for resale. Specific permission must be obtained in writing from CRC Press LLC for such copying.

Direct all inquiries to CRC Press LLC, 2000 N.W. Corporate Blvd., Boca Raton, Florida 33431.

Trademark Notice: Product or corporate names may be trademarks or registered trademarks, and are used only for identification and explanation, without intent to infringe.

Visit the CRC Press Web site at www.crcpress.com

© 2003 by CRC Press LLC

No claim to original U.S. Government works
International Standard Book Number 0-8493-1354-6
Library of Congress Card Number 2002025935
Printed in the United States of America 2 3 4 5 6 7 8 9 0
Printed on acid-free paper

LONGWOOD UNIVERSITY LIBRARY
REDFORD AND RACE STREET
FARMVILLE, VA 23909

Dedication

Dedicated to
Christine, Steven, and Fiona

RAWJ

Pauline, Daniel, Joseph, and Emma

CGH

LONGWOOD LIBRARY
1000371673

Preface

This book began as a small series of brief articles designed to assist engineers and salesmen in understanding some aspects of ion chemistry and mass spectrometry. Each article was composed of a very short summary and a longer section that described aspects of the subject matter in somewhat greater detail. To facilitate rapid reading and assimilation, extensive use was made of figures and tables, which contain additional information in their legends or footnotes.

From the outset of writing these articles, it was never intended that they should be exhaustively comprehensive, because this approach would have defeated the aims of the whole exercise, viz., the provision of short, quick explanations of major elements of mass spectrometry and closely allied topics. The traditional, more all-embracing approach to writing about mass spectrometry was left to the many excellent authors, who have provided impressive textbooks. In contrast, the major objective of the *Back-to-Basics* series was the provision of quick explanations of fundamental concepts in mass spectrometry, without overelaboration. As far as possible, descriptions of processes, applications, and underlying science were made with a minimum of text backed up by easily and rapidly understood pictures. Although some major equations of relevance to mass spectrometry have been introduced, the mathematical derivations of these equations were largely omitted. This was not an attempt to dumb down an important discipline. Rather, the intent was to make some of the esoteric aspects of an important area of analysis readily comprehensible to the many people who have to deal with mass spectrometers but who have not been trained specifically in this branch of science and engineering. The series began about ten years ago and encompasses recent and past developments in mass spectrometry. However, since it is the principles of mass spectrometry that are explained and not specific instrumentation, the information content remains as relevant now as it was ten years ago.

The effort to clarify and articulate the principles of spectrometry as simply as possible appears to have struck a welcome response among those who use mass spectrometers. The original series was not advertised, but, once it had been discovered outside the Micromass organization, there were inquiries as to how it could be purchased. From the start, Micromass offered the series free in a ring-binder format and later also provided it free on CD ROM and then on the Internet. Updating of the first CD led to over 600 requests for it on the day after its release had been announced on the Internet. Partly because of this response — but mainly because users or potential users frequently like to have a traditional reference book — it was decided to publish the series in this present book form.

This preface would not be complete without the authors sincerely thanking a number of people and, in particular, Micromass U.K. for its past keen support and for its permission to proceed with publishing this book. A successful business is composed of enthusiastic people, and it would be remiss of the authors not to thank all those many individuals in Micromass who have provided countless pieces of advice and help. Lastly, the authors wish to acknowledge with gratitude the support of Norman Lynaugh, Tony Hickson, and especially John Race for his original perspicacious decision to encourage the series to go ahead and for the ideas he proffered throughout. As he once famously remarked, "There are a lot of poor souls out there who are being asked to run mass spectrometers but who know next to nothing about mass spectrometry." At least we think it was "souls" he said!

Chris Herbert
Bob Johnstone

The Authors

Robert A.W. Johnstone, Ph.D., is a full professor in the Department of Chemistry at the University of Liverpool, Liverpool, U.K. He possesses a D.Sc., a Ph.D., and a B.Sc. in chemistry. He is also a Fellow of The Royal Society of Chemistry (FRSC). An author and contributor to several books, Dr. Johnstone has published over 200 journal papers in natural product chemistry, catalysis in oxidation and reduction, and spectroscopy. His work has resulted in several new synthetic procedures in organic chemistry that are widely used and have provided some 30 full patents. Part of his research with industry led to a Queen's Award for Technological Achievement, an award not usually given to industrial and business enterprises and not normally given to university staff. The award concerned large-scale manufacture of an important ingredient of timber preservative by a new catalytic method, which produces about 3000 t per annum of the product. Apart from his career in universities, he is also a director of three consulting, biotechnological, and animation companies, and he has worked for the Medical Research Council as a research scientist. Professor Johnstone has acted as visiting professor at a number of universities in Europe, China, and Australia. In his private life, Bob has a wife Christine and two children: Steve, a geological scientist in Australia; and Fiona, a Director of Public Health in the National Health Service. He has played rugby and Association football, hockey, and badminton but now accepts more leisurely walking as his main physical exercise. As part of his interests in woodworking and metalwork, he has found time to acquire the status of qualified plumber.

Christopher G. Herbert is the Marketing Materials and Reprographics Manager with Micromass U.K. Ltd., Atlas Park, Simonsway, Manchester, U.K. He possesses a Higher National Diploma in applied physics and is a graduate of The Institute of Physics. He has over 32 years of experience working in the field of mass spectrometry with Micromass U.K. Ltd. His earlier roles in the company include development engineer, test and installation engineer, test and installation manager, technical author (instruction manuals), and webmaster. He is author and coauthor of several hundred publications on mass spectrometry and its applications. In his private life, Chris has a wife Pauline and three children: Daniel, a chartered engineer; Joseph, a research scientist at the Daresbury Laboratory; and Emma, a full-time mother. Chris has a team of racing pigeons competing in races up to 500 miles and is chairman and clock setter for his local pigeon-flying club. In an effort to prove he is as fast as his pigeons, Chris has three times completed the London marathon, twice with his wife, raising over £1000 for the British Heart Foundation and other charities.

The Authors

Introduction

Back-to-Basics was not originally designed to be a book but, rather, a series of articles, each on a topic dealing directly or indirectly with an aspect of mass spectrometry. Each article in the series was intended to be a self-contained, brief (but not superficial) explanation of an area of mass spectrometric instrumentation or theory. The aim was to provide nonexperts with enough information to understand the major fundamental concepts underlying any chosen topic. Because basic facts were being described, the title of the series became *Back-to-Basics*. For example, one of the earliest articles covered the theory and uses of electron ionization in mass spectrometry. To ease understanding, a complex rigorous approach was forsaken in favor of a straightforward account, aided by many clear diagrams. By describing many of the basics of mass spectrometry and not individual instruments per se, the series was not subject to the obsolescence that typically afflicts books composed around a rapidly advancing subject area such as mass spectrometry. The basic articles cover principles, which largely do not change, and therefore the articles remain useful over longer periods of time. Of course, fashions change in mass spectrometry as in anything else. For example, thermospray inlets for mass spectrometers have given way to more efficient ones, but the principles underlying a thermospray device are still valid and are worth retaining on record.

Although each *Back-to-Basics* article is quite brief, a very short summary of each article is provided in Appendix A for readers who want a succinct, informative account that provides a quick working understanding of the topic.

As the number of articles increased, many of them had to be cross-referenced to other entries in the series. Therefore, although the series began as a few articles on isolated topics, as the series built up, the cross-referencing network increased as well. The collected articles were originally offered in ring-binder format but, as the number of articles increased, this format became unwieldy. Although the hardcopy versions of the *Back-to-Basics* series remained available to the hundreds of regular subscribers, it was decided to post the series on the Internet for wider availability. Finally, the articles were assembled in CD format and again made freely available by Micromass U.K.

The result of the *Back-to-Basics* series is an accumulation of some 50 separate but interrelated expositions of mass spectrometric principles and apparatus. Some areas of mass spectrometry, such as ion cyclotron resonance and ion trap instruments, have not been covered except for passing references. This decision has not been due to any bias by the authors or Micromass but simply reflects the large amount of writing that had to be done and the needs of the greatest proportion of users.

For various reasons, it has been decided to close the present series of *Back-to-Basics* and collect the articles in a book that can be placed on a shelf and used as a reference source. It is an unusual book format, having a large number of rather short chapters together with a collection of brief summaries, included as Appendix A. Thus the book will still serve the dual purposes of describing fundamentals and additionally providing a concise synopsis of each article.

In another departure from regular scientific book format, no attempt has been made to include references inside each chapter. This approach was intentional from the start because the basics that were being described were founded on a great deal of knowledge accumulated by very many research workers, and it seemed invidious to single out particular individuals. However, in book form, it was felt that *Back-to-Basics* should include a bibliography of mass spectrometry, which a reader could use to follow up and expand on anything gleaned in the various chapters. To this end, we have included a list of some 100 textbooks that are now available, at least through libraries. Each title in the list gives a good idea of the book's contents, and the books themselves yield many leading references. The book list covers publications from about 1965, when there were few books on mass spectrometry produced annually, to the present, when it is not unusual to find several books published each year. This growth reflects the huge range of applications of mass spectrometry, engendered by its scope and sensitivity in analysis, and by the variety of excellent commercial instruments now available. The list of books is not meant to be exhaustive, nor is it selective. If any book author reads this introduction and finds that his own book is not on the list, please accept our apologies for an unintended oversight.

For readers of this *Back-to-Basics* volume who want to learn about current advances, we have also included a list of journals that deal exclusively with mass spectrometry or contain a significant number of articles on the subject. Another good place to catch up on current work lies in the many large and small conferences or meetings that are held each year all over the world. Readers of this *Back-to-Basics* tome will find that these conferences are good venues for meeting many of the experts in mass spectrometry, where they can be questioned about esoteric aspects of the subject in a friendly and often convivial atmosphere. Who knows, you may even meet the present authors at some meeting and pass on your remarks directly, accompanied by ale, nectar of the gods or Adam's ale!

Contents

Chapter 1

Chemical Ionization (CI)

The Ionization Process

This chapter should be read in conjunction with Chapter 3, "Electron Ionization." In electron ionization (EI), a high vacuum (low pressure), typically 10^{-5} mbar, is maintained in the ion source so that any molecular ions ($M^{\bullet+}$) formed initially from the interaction of an electron beam and molecules (M) do not collide with any other molecules before being expelled from the ion source into the mass spectrometer analyzer (see Chapters 24 through 27, which deal with ion optics).

Decomposition (fragmentation) of a proportion of the molecular ions ($M^{\bullet+}$) to form fragment ions (A^+, B^+, etc.) occurs mostly in the ion source, and the assembly of ions ($M^{\bullet+}$, A^+, B^+, etc.) is injected into the mass analyzer. For chemical ionization (CI), the initial ionization step is the same as in EI, but the subsequent steps are different (Figure 1.1). For CI, the gas pressure in the ion source is typically increased to 10^{-3} mbar (and sometimes even up to atmospheric pressure) by injecting a reagent gas (R in Figure 1.1).

The substance being investigated (M) is present as only a small fraction of the reagent gas pressure. Thus, the electrons in the electron beam mostly interact with the reagent gas to form reagent gas ions ($R^{\bullet+}$) and not $M^{\bullet+}$ ions. At the higher pressures, the initial reagent gas ions almost immediately suffer multiple collisions with neutral reagent gas molecules (R). During this process, new ions (RH^+) are produced (step 2, Figure 1.1); these ions are reagent gas ions.

Because many of these ions are produced, there is a high probability that they will collide with sample molecules (M) and that a proton (H^+) will be exchanged to give protonated molecular ions $[M + H]^+$, as shown in step 3, Figure 1.1. These quasi-molecular ions contain little excess of internal energy following CI and therefore tend not to fragment. Whereas EI spectra contain peaks corresponding to both molecular and fragment ions, CI spectra are much simpler, mostly having only protonated molecular ion peaks. Negative reagent gases give abundant $[M - H]^-$ or $[M + X]^-$ ions.

Example of the Chemical Ionization Process

The ion source, across which an electron beam passes, is filled with methane, the reagent gas. There is a high vacuum around the ion source, so, to maintain a high pressure in the source itself, as many holes as possible must be blocked off or made small. Interaction of methane (CH_4) with electrons (e^-) gives methane molecular ions ($CH_4^{\bullet+}$), as shown in Figure 1.2a.

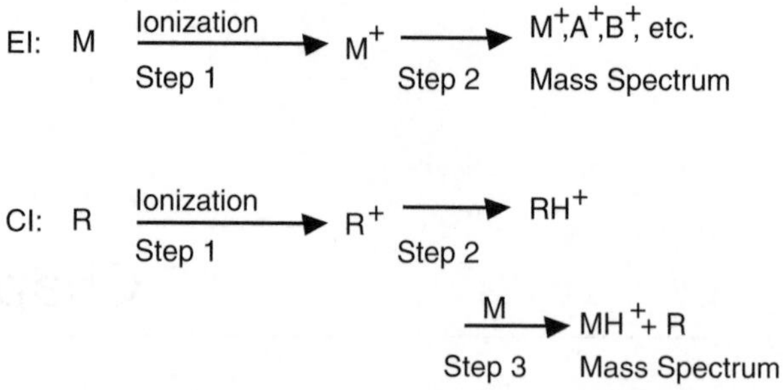

$$\text{EI: } M \xrightarrow[\text{Step 1}]{\text{Ionization}} M^{+} \xrightarrow[\text{Step 2}]{} M^{+}, A^{+}, B^{+}, \text{etc. (Mass Spectrum)}$$

$$\text{CI: } R \xrightarrow[\text{Step 1}]{\text{Ionization}} R^{+} \xrightarrow[\text{Step 2}]{} RH^{+} \xrightarrow[\text{Step 3}]{M} MH^{+} + R \text{ (Mass Spectrum)}$$

Figure 1.1
Comparison of basic EI and CI processes showing different types of molecular ions and the formation of fragment ions in EI.

$$CH_4 + e^- \longrightarrow CH_4^{+} + 2e^- \quad \text{(a)}$$

$$CH_4^{+} + CH_4 \longrightarrow CH_5^{+} + CH_3 \quad \text{(b)}$$

$$CH_5^{+} + M \longrightarrow CH_4 + MH^{+} \quad \text{(c)}$$

Figure 1.2
Formation of reactive ions (CH_5^+) from methane (CH_4) reagent gas and their reaction with sample molecules (M) to form protonated molecular ions $[M + H]^+$.

Newly formed ions ($CH_4^{\bullet +}$) collide several times with neutral molecules (CH_4) to give carbonium ions (CH_5^+) (Figure 1.2b). The substance (M) to be investigated is vaporized into the ion source, where it collides with these carbonium ions (CH_5^+). Proton exchange occurs because organic substances (M) are usually stronger bases than is CH_4, so protonated molecular ions (written as MH^+ or, better, as $[M + H]^+$) are formed (Figure 1.2c). These ions are expelled from the ion source into the mass analyzer of the mass spectrometer (see Chapters 24 through 27 for information on ion optics). Protonated molecular ions are also called quasi-molecular ions.

Little or no fragmentation of $[M + H]^+$ ions occurs, so CI spectra are very simple compared with EI spectra. A comparison of the two is shown in Figure 1.3, where it can be seen that the EI spectrum gives few molecular ions ($M^{\bullet +}$) but many fragment ions (A^+, B^+, etc.), whereas the CI spectrum shows many (abundant) quasi-molecular or protonated ions $[M + H]^+$ and few fragment ions. Thus, CI and EI spectra are complementary and there is often considerable advantage in obtaining both, since EI gives structural information and CI confirms the relative molecular mass (molecular weight).

Other Reagent Gases

The example of methane as a reagent gas can now be extended to other species, some of which are shown in Figure 1.4, along with the principal ions formed. The various reagent gases do not all act in the same proton-exchange manner described above, and some of the major variations are detailed below. However, all of the ionization effected by the reagent gases is characterized by its propensity to give spectra having very few fragment ions.

Negative reagent ions, such as Cl^- (from CH_2Cl_2), $O^{\bullet -}$ (from NO), and OH^- (from NO plus CH_4), react with molecules (M) either by abstracting a proton to give $[M - H]^-$ ions or by

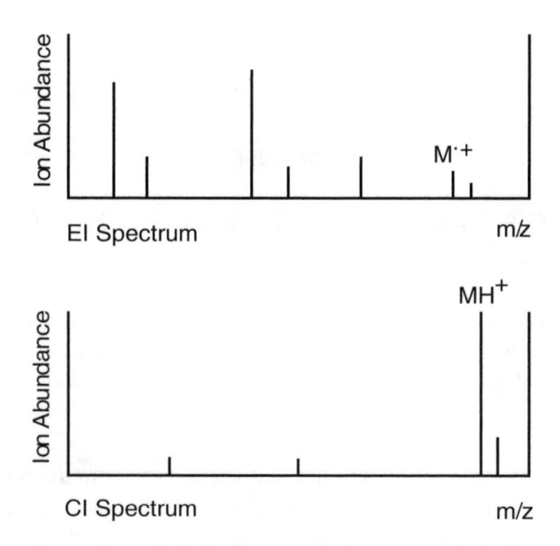

Figure 1.3
Comparison of EI and CI mass spectra illustrating the greater degree of fragmentation in the former and the greater abundance of quasi-molecular ions in the latter.

addition to form $[M + X]^-$ ions. Of these, OH^- is particularly useful because its high proton affinity leads to hydrogen abstraction from most classes of organic compounds (except alkanes) with little fragmentation. As with the EI/CI combinations, it is often convenient for structure elucidation to generate both positive and negative ion data by CI. Instruments are available to measure both positive and negative CI spectra simultaneously. For example, a reagent gas mixture of CH_4 and NO gives two corresponding reagent gases, CH_5^+ and OH^-. Thus, both $[M + H]^+$ and $[M - H]^+$ quasi-molecular ions can be formed from sample molecules (M). By alternately injecting (pulsing) positive ions and negative ions (a process termed PPINICI) into the mass analyzer (usually a quadrupole or ion trap), almost simultaneous positive and negative CI spectra can be measured.

Other Ionization Routes

Molecules of substrate M may not be ionized simply by the proton-transfer mechanism shown in Figure 1.2c. For example, with ammonia reagent gas, either $[M + H]^+$ ions or $[M + NH_4]^+$ ions can be formed, depending on the nature of the substrate (Figure 1.5). Process 5a is proton exchange, but Figure 1.5b shows an example in which the whole of the reagent gas ion attaches itself to the substrate to form a quasi-molecular ion $[M + NH_4]^+$. In Figure 1.5a, the protonated molecular ion has a mass

Reagent gas	**Molecular ion**	**Reactive reagent ion**
H_2	H_2^+	H_3^+
C_4H_{10}	$C_4H_{10}^+$	$C_4H_{11}^+$
NH_3	NH_3^+	NH_4^+
CH_3OH	CH_3OH^+	$CH_3OH_2^+$
NO	NO	NO^+

Figure 1.4
Some types of reagent gases and their reactive ions.

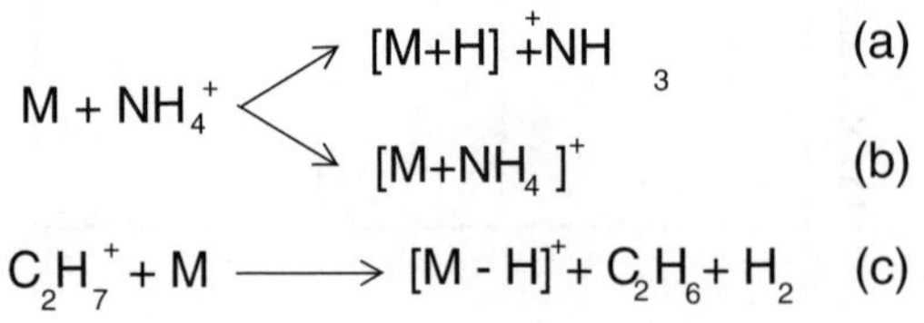

$$M + NH_4^+ \begin{cases} \rightarrow [M+H]^+ + NH_3 & \text{(a)} \\ \rightarrow [M+NH_4]^+ & \text{(b)} \end{cases}$$

$$C_2H_7^+ + M \longrightarrow [M - H]^+ + C_2H_6 + H_2 \quad \text{(c)}$$

Figure 1.5
Typical CI processes in which neutral sample molecules (M) react with NH_4^+ to give either (a) a protonated ion $[M + H]^+$ or (b) an adduct ion $[M + NH_4]^+$; the quasi-molecular ions are respectively 1 and 18 mass units greater than the true mass (M). In process (c), reagent ions ($C_2H_7^+$) abstract hydrogen, giving a quasi-molecular ion that is 1 mass unit less than M.

that is one unit greater than the true relative molecular mass, since the mass of H is one. In Figure 1.5b, the adduct ion is greater than M by 18 mass units: ($N = 14$, $4 \times H = 4$, total $NH_4 = 18$).

If the substrate (M) is more basic than NH_3, then proton transfer occurs, but if it is less basic, then addition of NH_4^+ occurs. Sometimes the basicity of M is such that both reactions occur, and the mass spectrum contains ions corresponding to both $[M + H]^+$ and $[M + NH_4]^+$. Sometimes the reagent gas ions can form quasi-molecular ions in which a proton has been removed from, rather than added to, the molecule (M), as shown in Figure 1.5c. In these cases, the quasi-molecular ions have one mass unit less than the true molecular mass.

Uses of CI

Some substances under EI conditions fragment so readily that either no molecular ions survive or so few survive that it is difficult to be sure that the ones observed do not represent some impurity. Therefore, there is either no molecular mass information or it is uncertain. Under CI conditions, very little fragmentation occurs and, depending on the reagent gas, ions $[M + X]^+$ (X = H, NH_4, NO, etc.) or $[M - H]^+$ or $[M - H]^-$ or $[M + X]^-$ (X = F, Cl, OH, O, etc.) are the abundant quasi-molecular ions, which do give molecular mass information.

Fragmentation under EI conditions yields structural information, but CI yields little or none because it gives few fragment ions. Thus, CI is used mostly for molecular mass information and is frequently used with EI as a complement. Because there is little structural information, in contrast to EI, there are no extensive libraries of CI spectra. CI spectra are apparent also in atmospheric-pressure ionization systems (see Chapters 9 and 11). CI is often called a *soft* ionization method because little excess energy is put into the molecules (M) when they are ionized. Therefore, substances that might not otherwise give mass spectra containing molecular ions will give molecular mass information under CI conditions.

Use of CI/EI in Tandem in GC/MS

As shown above, CI and EI spectra complement each other and they are used frequently in such techniques as gas chromatography/mass spectrometry (GC/MS), where successive mass spectral scans are recorded. Alternate scans can be arranged to be either EI or CI by alternate evacuation of reagent gas from or pressurization with reagent gas into the ion source through which the GC effluent is flowing. With modern pumping systems, this switchover is complete within a few seconds. Figure 1.6 illustrates the EI/CE switching process and the sort of information obtained.

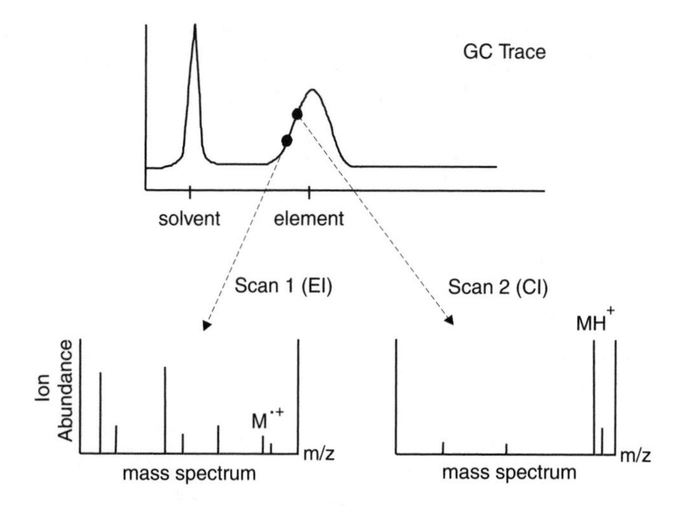

Figure 1.6
Successive spectra taken during elution of a substance from a GC column. The first spectrum obtained by EI shows many fragment ions, while the second, obtained by CI, gives molecular mass information.

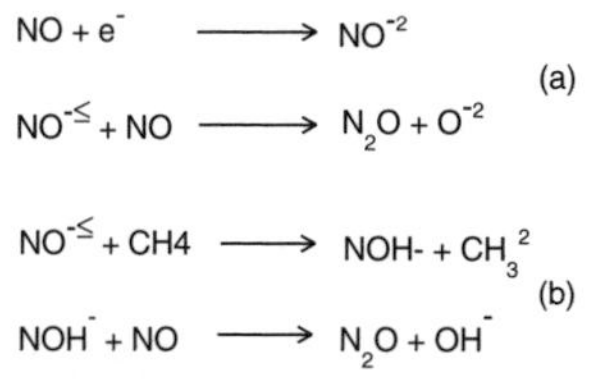

$$NO + e^{-} \longrightarrow NO^{-2}$$
$$NO^{-\leq} + NO \longrightarrow N_2O + O^{-2} \quad (a)$$

$$NO^{-\leq} + CH4 \longrightarrow NOH- + CH_3^{\,2}$$
$$NOH^{-} + NO \longrightarrow N_2O + OH^{-} \quad (b)$$

Figure 1.7
Negative reactant gas ions $O^{\bullet -}$ and OH^- can be produced easily from (a) NO and (b) NO/CH_4, respectively.

Negative Ions

Analogous to the formation of protonated ions, negative ions can be formed by deprotonation (Figure 1.7). Reagent gas ions such as OH^- or $O^{\bullet-}$ are strongly basic and capable of abstracting a proton from sample molecules M to give $[M - H]^-$ ions, one mass unit less than the true molecular mass. Negative-ion CI is a useful sensitive technique for substances having a high electron affinity, such as halo compounds and polycyclic aromatic hydrocarbons.

Conclusion

At higher pressures ($\approx 10^{-3}$ mbar) reagent gases such as methane, butane, ammonia, etc. form reagent ions (CH_5^+, $C_4H_{11}^+$, NH_4^+, etc.). These reagent gas ions react with sample molecules (M) to give stable $[M + X]^+$ ions (X = H, NH_4, etc.) or $[M - H]^+$ ions. These ions have relative molecular masses different from that of the molecule (M) and are called quasi-molecular ions. Negative ions can be obtained from negative reagent gases such as Cl^- or OH^-. The CI technique is particularly useful for providing molecular mass information because little or no fragmentation occurs, unlike with EI. Ion sources for CI usually operate under both CI and EI conditions, with fast switching between modes so that alternate CI, EI spectra can be recorded.

Chapter 2

Laser Desorption Ionization (LDI)

The Ionization Process

A molecule naturally possesses rotational, vibrational, and electronic energy. If it is a liquid or a gas, it will also have kinetic energy. Under many everyday circumstances, if a molecule or group of molecules in a solid have their internal energy increased (e.g., by heat or radiation) over a relatively long period of time (which may be only a few microseconds), the molecules can equilibrate the energy individually and together so that the excess energy is dissipated to the surroundings without causing any change in molecular structure. Beyond a certain threshold of too much energy in too short a time, the energy cannot be dissipated fast enough, so the substance melts and then vaporizes as internal energy of vibration and rotation is turned into translational energy (kinetic energy or energy of motion). Simultaneous electronic excitation may be sufficient such that electrons are ejected from molecules to give ions. Thus, putting much energy into a molecular system in a very short space of time can cause melting, vaporization, possible destruction of material and, importantly for mass spectrometry, ionization (Figure 2.1).

A laser is a device that can deliver a large density of energy into a small space. The actual energy released by a laser is normally relatively small, but, because the laser beam is focused into a very tiny area, the energy delivered per unit area is very large. An analogy can be drawn by considering sunlight, which will not normally cause an object to heat so that it burns. However, if the sunlight is focused into a small area by means of a lens, it becomes easy to set an object on fire or to vaporize it. Thus, a low total output of light radiation concentrated into a tiny area actually gives a high density or flux of radiation (we could even say a high light pressure). This situation is typical of a laser. As an example, a Nd-YAG laser operating at 266 nm can deliver a power output of about 10 W, somewhat like a sidelight on a motor car. However, this energy is delivered into an area of about 10^{-7} cm^2, so that the power focused onto the small irradiated area is about $10/10^{-7}$ = 10^8 W/cm^2 = 10^5 kW/cm^2 — the same effect as focusing the heat energy from 100,000 1-bar electric fires onto the end of your finger!

A molecular system exposed to a laser pulse (or beam) has its internal energy greatly increased in a very short space of time, leading to melting (with increased rotational and vibrational electronic energy), vaporization (desorption; increased kinetic or translational energy), some ionization (electronic excitation energy leading to ejection of an electron), and possibly some decomposition (increase in total energy sufficient to cause bond breaking). If enough energy is deposited into a sample in a very short space of time, it has no time to dissipate the energy to its surroundings, and it is simply blasted away from the target area because of a large gain in kinetic energy. (The material

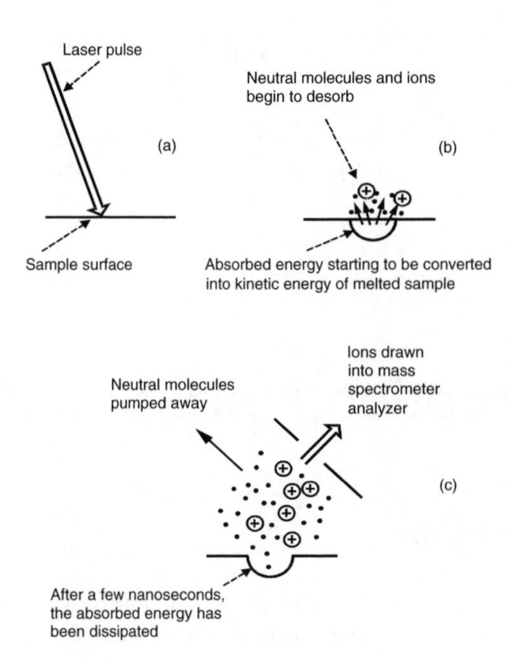

Figure 2.1
A laser pulse strikes the surface of a sample (a), depositing energy, which leads to melting and vaporization of neutral molecules and ions from a small, confined area (b). A few nanoseconds after the pulse, the vaporized material is either pumped away or, if it is ionic, it is drawn into the analyzer of a mass spectrometer (c).

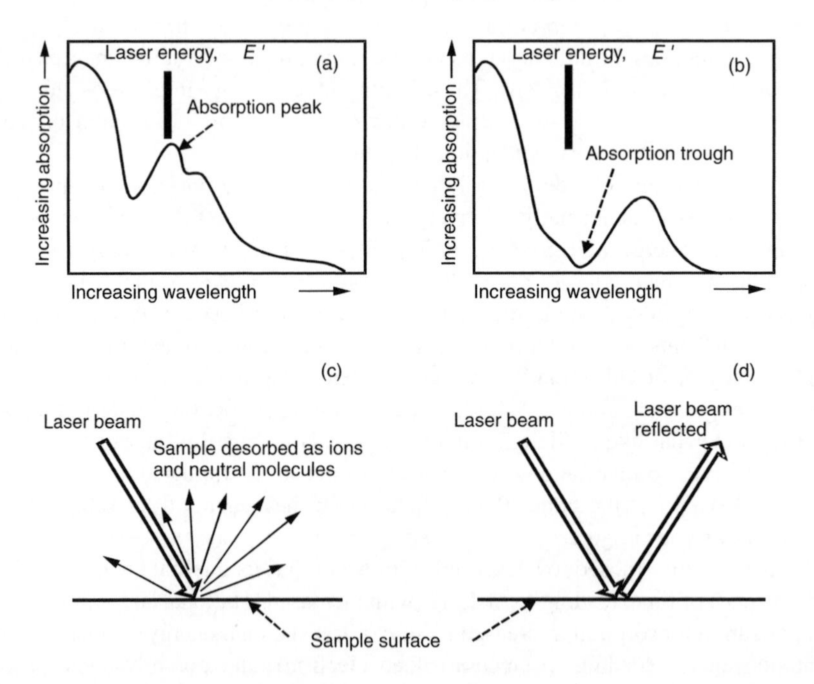

Figure 2.2
In (a), a pulse of laser light of a specific wavelength of energy, E', strikes the surface of a specimen that has a light absorption spectrum with an absorption peak near the laser wavelength. The energy is absorbed, leading to the ablation of neutral molecules and ions (c). In (b), the laser strikes the surface of a specimen that does not have a corresponding absorption peak in its absorption spectrum. The energy is not absorbed but is simply reflected or scattered (d), depending on whether the surface is smooth or rough.

is said to be ablated.) Laser desorption ionization is the process of beaming laser light, continuously or in pulses, onto a small area of a sample specimen so as to desorb ions, which can then be examined by a mass spectrometer.

With continuous lasers (for example an argon ion laser), the energy delivered is usually much less than from pulsed ones, and the focusing is not so acute. Thus, the irradiated area of the sample is more like 10^{-4} cm^2 rather than 10^{-7} cm^2, and the energy input is much less, about 100 kW/cm^2 rather than the 100,000 kW/cm^2 described earlier.

Other Considerations on Laser Desorption Ionization

Consider a laser emitting radiation of energy, E'. For a substance to absorb the energy, it must have an absorption spectrum (ultraviolet, visible, or infrared) that matches the energy. Figure 2.2 shows two cases: one (a) in which a substance can absorb the energy, E', and one (b) in which it cannot absorb this energy. In the second case, since energy cannot be absorbed, the laser radiation is reflected and none of its energy is absorbed. In the first case, much or all of the available energy can be absorbed and must then be dissipated somehow by the system. This dissipation leads to the effects itemized above. It follows that the capacity of a laser to desorb or ionize a substance will depend on three factors:

1. The actual wavelength (energy, E') of the laser light
2. The power of the laser
3. The absorption spectrum of the substance being irradiated.

When the first and third factors match most closely and much power is available (large light flux), much of the laser energy can be absorbed by the substance being examined; when the first and third factors mismatch, whatever the power, little or none of the laser energy is absorbed. Therefore, for any one laser wavelength, there will be a range of responses for different substances. For this reason, it is often advantageous to use a tunable laser so that various wavelengths of irradiation can be selected to suit the substance being examined.

Use of a Matrix

There is another way of dealing with the variability of ionization during laser irradiation. Suppose there is a sample substance (a matrix material) having an absorption band that closely matches the energy of the laser radiation. Upon irradiation, this material will be rapidly increased in energy and will desorb and ionize quickly, as described above. Now suppose that this matrix material is mixed with a substance to be examined with the matrix material. In this case, at least some of the energy absorbed by the matrix can be passed on to the sample substance, causing it to desorb and ionize (Figure 2.3).

This technique depends on the laser energy matching an absorption band in the matrix, and a match with the sample substance is unnecessary. This general method is called matrix-assisted laser desorption ionization (MALDI). Commonly, sinapic acid (3,5-dimethoxy-4-hydroxycinnamic acid) or nicotinic acid is used as a matrix material for examining organic and other compounds. The ions produced are usually protonated molecules, $[M + H]^+$, with few fragment ions. Because most of the ablated material is the acidic matrix, a high proportion of the co-ablated sample molecules are protonated. The MALDI method gives good yields of ions.

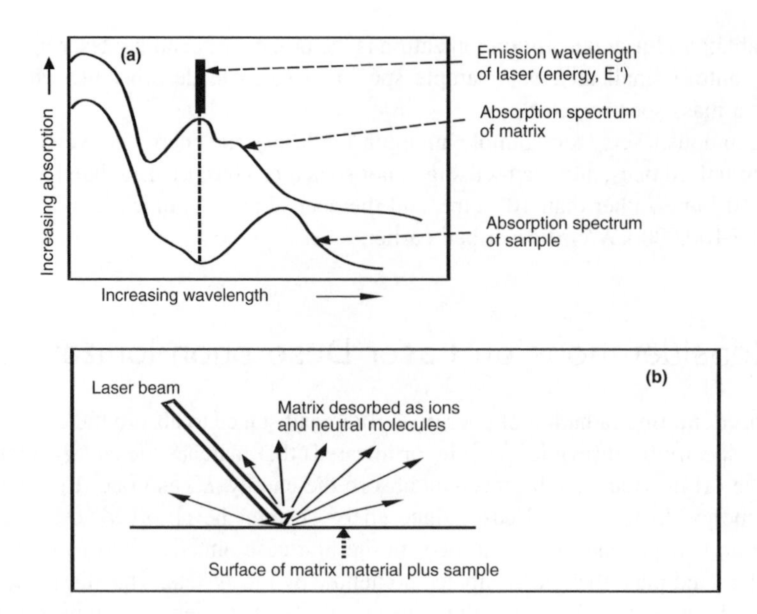

Figure 2.3
In a MALDI experiment, the sample is mixed or dissolved in a matrix material that has an absorption spectrum matching the laser wavelength of energy, E'. The sample may not have a matching absorption peak (a), but this is not important because the matrix material absorbs the radiation, some of which is passed on to the dissolved sample. Neutral molecules and ions from both sample and matrix material are desorbed (b).

Types of Laser

In theory, any laser can be used to effect desorption and ionization provided that it supplies enough energy of the right wavelength in a short space of time to a sample substance. In practice, for practical reasons, the lasers that are used tend to be restricted to a few types. The laser radiation can be pulsed or continuous (continuous wave). Typically, laser energies corresponding to the ultraviolet or near-visible region of the electromagnetic spectrum (e.g., 266 or 355 nm) or the far infrared (about 20 mm) are used. The lasers are often tunable over a range of energies.

The so-called peak power delivered by a pulsed laser is often far greater than that for a continuous one. Whereas many substances absorb radiation in the ultraviolet and infrared regions of the electromagnetic spectrum, relatively few substances are colored. Therefore, a laser that emits only visible light will not be as generally useful as one that emits in the ultraviolet or infrared ends of the spectrum. Further, with a visible-band laser, colored substances absorb more or less energy depending on the color. Thus two identical polymer samples, one dyed red and one blue, would desorb and ionize with very different efficiencies.

Secondary Ionization

Much of the energy deposited in a sample by a laser pulse or beam ablates as neutral material and not ions. Ordinarily, the neutral substances are simply pumped away, and the ions are analyzed by the mass spectrometer. To increase the number of ions formed, there is often a second ion source to produce ions from the neutral materials, thereby enhancing the total ion yield. This secondary or additional mode of ionization can be effected by electrons (electron ionization, EI), reagent gases (chemical ionization, CI), a plasma torch, or even a second laser pulse. The additional ionization is often organized as a pulse (electrons, reagent gas, or laser) that follows very shortly after the

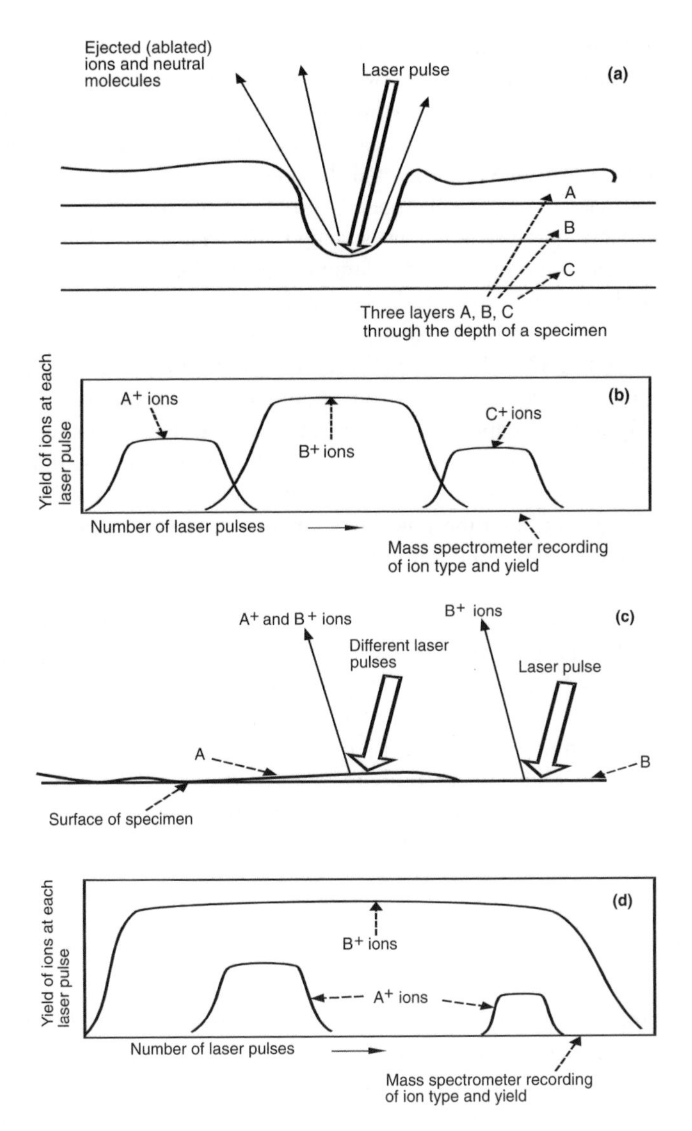

Figure 2.4
A laser pulse strikes the surface of a specimen (a), removing material from the first layer, A. The mass spectrometer records the formation of A^+ ions (b). As the laser pulses ablate more material, eventually layer B is reached, at which stage A^+ ions begin to decrease in abundance and B^+ ions appear instead. The process is repeated when the B/C boundary is reached so that B^+ ions disappear from the spectrum and C^+ ions appear instead. This method is useful for depth profiling through a specimen, very little of which is needed. In (c), less power is used and the laser beam is directed at different spots across a specimen. Where there is no surface contamination, only B^+ ions appear, but, where there is surface impurity, ions A^+ from the impurity also appear in the spectrum (d).

initial laser desorption. Of course, with MALDI this second ionization step is unnecessary, and, in the case of the plasma torch, the ablated material is swept into the plasma flame for ionization.

Uses of Lasers

Laser desorption methods are particularly useful for substances of high mass such as natural and synthetic polymers. Glycosides, proteins, large peptides, enzymes, paints, ceramics, bone, and large polymers are all amenable to laser desorption mass spectrometry, with the sample being examined either alone or as part of a prepared matrix. Because of the large masses involved, for pulsed laser desorption, the method is frequently used with time-of-flight or ion-trap instruments, which need pulses of ions. For MALDI, sample preparation can be crucial, since the number of ions produced varies greatly with both the type of matrix material and the presence of impurities. Fragment ions are few, but the true molecular mass can be misinterpreted because of the formation of adduct ions between the matrix material and the substance under investigation; these adduct ions have greater mass than the true molecular mass. Some impurities, as with common ionic detergents, can suppress ion formation.

The laser approach without a matrix can be employed in two main ways. Since the intensity and spot size of the laser pulse or beam can be adjusted, the energy deposited into a sample ranges from a very large amount confined to a small area of sample to much less spread over a larger area. Thus, in one mode, the laser can be used to penetrate down through a sample, each pulse making the previously ablated depression deeper and deeper. This approach is depth profiling, which is useful for examining variations in the composition of a sample with depth (Figure 2.4a). For example, gold plating on ceramic would show only gold ions for the first laser shots until a hole had been drilled right through the gold layer; there would then appear ions such as sodium and silicon that are characteristic of the ceramic material, and the gold ions would mostly disappear.

By using a laser with less power and the beam spread over a larger area, it is possible to sample a surface. In this approach, after each laser shot, the laser is directed onto a new area of surface, a technique known as surface profiling (Figure 2.4c). At the low power used, only the top few nanometers of surface are removed, and the method is suited to investigate surface contamination. The normal surface yields characteristic ions but, where there are impurities on the surface, additional ions appear.

Laser desorption is commonly used for pyrolysis/mass spectrometry, in which small samples are heated very rapidly to high temperatures to vaporize them before they are ionized. In this application of lasers, very small samples are used, and the intention is not simply to vaporize intact molecules but also to cause characteristic degradation.

Conclusion

Lasers can be used in either pulsed or continuous mode to desorb material from a sample, which can then be examined as such or mixed or dissolved in a matrix. The desorbed (ablated) material contains few or sometimes even no ions, and a second ionization step is frequently needed to improve the yield of ions. The most common methods of providing the second ionization use MALDI to give protonated molecular ions or a plasma torch to give atomic ions for isotope ratio measurement. By adjusting the laser's focus and power, laser desorption can be used for either depth or surface profiling.

Chapter 3

Electron Ionization (EI)

The Ionization Process

When an electron (negative charge) is accelerated through an electric field, it gains kinetic energy. After acceleration through 70 V, the electron has an energy of 70 eV (electronvolts). These energetic electrons can interact with uncharged, neutral molecules (M) by passing close to or even through the molecule, so as to eject an electron from the molecule (Equation 3.1).

$$e^- + M \rightarrow M^{\bullet +} + 2e^- \tag{3.1}$$

Thus two electrons exit the reaction zone, leaving a positively charged species ($M^{\bullet +}$) called an ion (in this case, a molecular ion). Strictly, $M^{\bullet +}$ is a radical-cation. This electron/molecule interaction (or collision) was once called electron impact (also EI), although no impact actually occurs.

At 70eV, in a high vacuum, the interaction between electrons and molecules leaves some ions ($M^{\bullet +}$) with so much extra energy that they break up (fragment) to give ions of smaller mass (A^+, B^+, etc.; Equation 3.2).

$$e^- + M \rightarrow M^{\bullet +} + 2e^- + A^+, B^+, \text{etc.} \tag{3.2}$$

This fragmentation is characteristic for a given substance, similar to a fingerprint, and is referred to as a mass spectrum.

For a limited range of substances, negative radical anions ($M^{\bullet -}$) can be formed rather than positive ions (Equation 3.3). Negative radical anions can be produced in abundance by methods other than electron ionization. However, since most EI mass spectrometry is concerned with positive ions, only they are discussed here.

$$e^- + M \rightarrow M^{\bullet -} \tag{3.3}$$

Mass-to-Charge Ratio (m/z)

The mass of an electron is very small compared with the total mass of the molecule. Consequently, the relative molecular mass of a molecule (M_r) is almost the same as that of the derived molecular ion ($M^{\bullet +}$). For practical purposes in mass spectrometry, $M_r = M_r^{\bullet +}$, and is written, $M^{\bullet +}$.

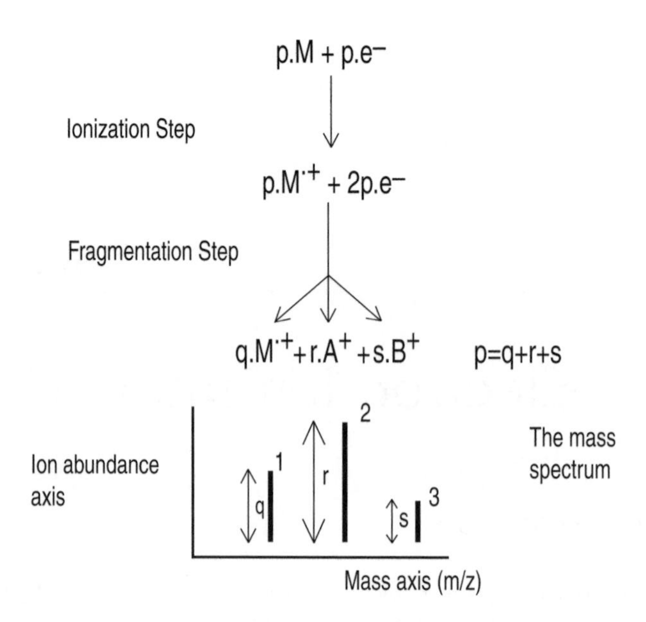

Figure 3.1
The formation of a simple EI mass spectrum from a number (p) of molecules (M) interacting with electrons (e^-). Peak 1 represents $M^{\bullet+}$, the molecular ion, the ion of greatest mass (abundance q). Peaks 2, 3 represent A^+, B^+, two fragment ions (abundances r, s). Peak 2 is also the largest and, therefore, the base peak.

An ion of a given mass (m) having a single positive charge (z = 1) has a mass-to-charge ratio of m/z = m/1 = m. Thus, the mass-to-charge ratio is conveniently equal to the mass of the ion, so $M^{\bullet+}/z = M_r^{\bullet+} = M_r$. Most ions produced by EI have a single charge and, as mass spectrometers utilize the charge to measure ion mass, it is fortunate that usually z = 1. Similarly, the fragment ions (A^+, B^+, etc.) have single positive charges (Figure 3.1).

The Mass Spectrum

An EI mass spectrum is a chart on two axes relating the mass of an ion (m or, strictly, m/z) to its abundance. During the ionization and fragmentation steps in the ion source (where electron/molecule reaction occurs), different numbers of ions (M^+, A^+, B^+, etc.) are produced and subsequently measured by the mass analyzer. The numbers of individual ions are referred to as ion abundance. Thus, a mass spectrum records mass (or m/z) on the *x*-axis and corresponding ion abundance on the *y*-axis as a series of peaks. The peak corresponding to the ion of greatest abundance is called the *base peak,* which can correspond to the molecular ion or to any one of the fragment ions (Figure 3.1).

The mass spectrum is characteristic for different substances and can be used like a fingerprint to identify a substance, either by comparison with an already known spectrum or through skilled interpretation of the spectrum itself (Figure 3.2).

The Ion Source

After ions have been formed by EI, they are examined for mass and abundance by the analyzer part of the mass spectrometer, which can incorporate magnetic sectors, electric sectors, quadrupoles, time-of-flight tubes, and so on. The region in which the ions are first formed is called

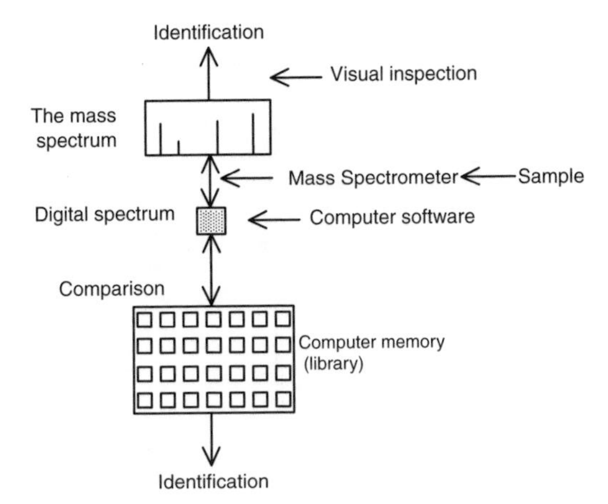

Figure 3.2
Visual and computer-assisted identification of the structure of a sample analyzed by a mass spectrometer.

an ion source. The actual construction of (EI) ion sources varies considerably, but all operate on the same principles.

The source can be regarded as a box that is more or less gas-tight, depending on type and usage (Figure 3.3). Electrons are produced at one side of the box (source) from a heated filament and are further energized by accelerating them through a potential of 70 V. The moving electrons are directed across the box and constrained into a beam by a magnetic field. This electron beam interacts with sample molecules that have been vaporized into the source to give ions, as described above. The ions are extracted from the source by an electric field and passed into the analyzer as an ion beam. The presence and intensity of the electron beam are confirmed by allowing it to impinge on a trap plate on the side of the ion source opposite to the heated filament.

Electron ionization occurs when an electron beam crosses an ion source (box) and interacts with sample molecules that have been vaporized into the source. Where the electrons and sample molecules interact, ions are formed, representing intact sample molecular ions and also fragments produced from them. These molecular and fragment ions compose the mass spectrum, which is a correlation of ion mass and its abundance. EI spectra of tens of thousands of substances have been recorded and form the basis of spectral libraries, available either in book form or stored in computer memory banks.

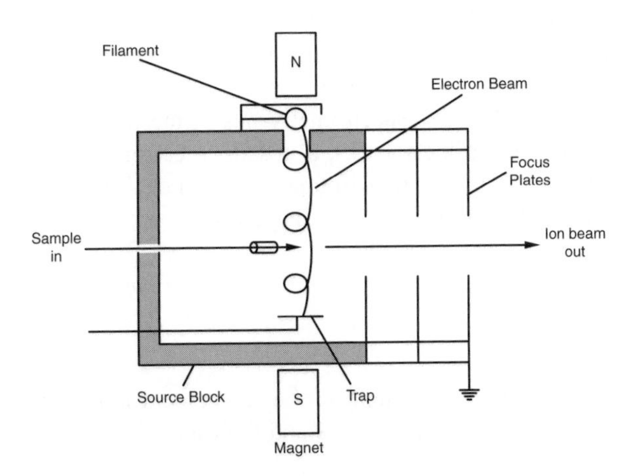

Figure 3.3
Simple ion source, showing the housing (block) with electron beam for EI.

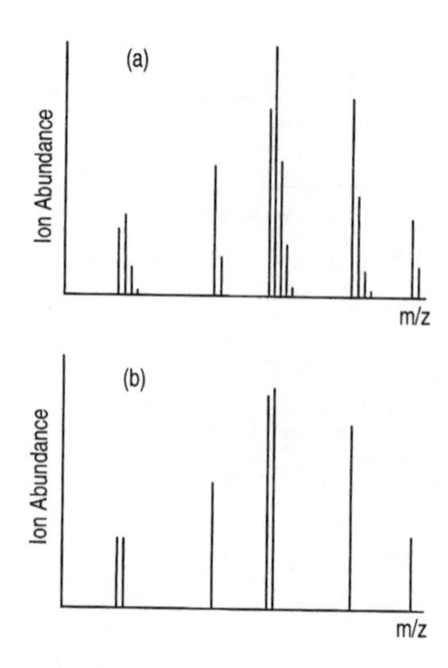

Figure 3.4
Mass spectrum of a carbon compound with (a) and without (b) the ^{13}C isotopes.

Isotopes

A word of warning here. As described above, a simple compound such as methane (natural gas), which has a molecular formula of CH_4, would be expected to give a mass of just 16 (C = 12; H = 1; C + 4H = 16). Only a peak at mass 16 in the spectrum would be expected. However, carbon also occurs in a slightly heavier variety (an isotope) in which C = 13. Therefore, the mass spectrum will show two peaks, one at mass 16 and one at 17 ($^{12}CH_4$ = 16; $^{13}CH_4$ = 17). Because the two carbon isotopes exist in different amounts in nature, the peak heights at 16, 17 will not be equal. Figure 3.4a shows the mass spectrum for a compound with the isotope contributions included, and Figure 3.4b shows the same spectrum in which the ^{13}C isotope peaks have been removed. Clearly, one spectrum looks much more complex than the other. Other common elements such as hydrogen, oxygen, nitrogen, sulfur and so on also have isotopes, thereby further complicating the appearance of a mass spectrum.

Many elements have two or more isotopes, and the presence of these correspondingly gives a spectrum having many peaks. On the other hand, the patterns of these isotopes in a mass spectrum often prove invaluable in identifying which elements are present, with the patterns serving as fingerprints for certain elements.

A further important use of EI mass spectrometry lies in measuring isotope ratios, which can be used in estimating the ages of artifacts, rocks, or fossils. Electron ionization affects the isotopes of any one element equally, so that the true isotope ratio is not distorted by the ionization step. Further information on isotopes can be found in Chapter 46.

Conclusion

Molecules interact with electrons to give molecular and fragment ions, which are mass analyzed. A mass spectrum relates the masses of these ions and their abundances.

Chapter 4

Fast-Atom Bombardment (FAB) and Liquid-Phase Secondary Ion Mass Spectrometry (LSIMS) Ionization

Introduction

It has long been known that bombardment of a solid surface by a primary beam of ions causes the emission of secondary ions characteristic of the solid target. Examination of the spectrum of desorbed (secondary) ions led to the technique of secondary ion mass spectrometry (SIMS). This process was useful for some solids, particularly metals, but achieved no prominence in organic mass spectrometry because of surface charging (organics are usually insulators), the fleeting nature of the production of ions (if at all), and the extensive radiation damage caused to the target substance.

A big step forward came with the discovery that bombardment of a liquid target surface by a beam of fast atoms caused continuous desorption of ions that were characteristic of the liquid. Where this liquid consisted of a sample substance dissolved in a solvent of low volatility (a matrix), both positive and negative molecular or quasi-molecular ions characteristic of the sample were produced. The process quickly became known by the acronym FAB (fast-atom bombardment) and for its then-fabulous results on substances that had hitherto proved intractable. Later, it was found that a primary incident beam of fast ions could be used instead, and a more generally descriptive term, LSIMS (liquid secondary ion mass spectrometry) has come into use. However, note that purists still regard and refer to both FAB and LSIMS as simply facets of the original SIMS. In practice, any of the acronyms can be used, but FAB and LSIMS are more descriptive when referring to the primary atom or ion beam.

When the liquid target is not a static pool but, rather, a continuous stream of liquid, the added description of "dynamic" is used. Thus, dynamic FAB and LSIMS refer to bombardment of a continuously renewed (flowing) liquid target.

Atom or Ion Beams

A "gun" is used to direct a beam of fast-moving atoms or ions onto the liquid target (matrix). Figure 4.1 shows details of the operation of an atom gun. An inert gas is normally used for bombardment because it does not produce unwanted secondary species in the primary beam and avoids contaminating the gun and mass spectrometer. Helium, argon, and xenon have been used commonly, but the higher mass atoms are preferred for maximum yield of secondary ions.

In the gun, inert gas atoms are ionized to give positive ions that are immediately accelerated by an electric potential to give a high-velocity beam of ions. As these ions collide with other inert gas atoms, charge exchange occurs such that many of the fast-moving ions become fast-moving atoms (Figure 4.1). Any residual ions are removed by an electric potential on a deflector plate, leaving a beam of fast-moving atoms that exits from the gun. The beam is somewhat divergent, so the gun is situated near the target.

Instead of the fast-atom beam, a primary ion-beam gun can be used in just the same way. Generally, such an ion gun emits a stream of cesium ions (Cs^+), which are cheaper to use than xenon but still have large mass (atomic masses: Cs, 139; Xe, 131). Although ion guns produce no fragment ions in the primary beam, they can contaminate the mass spectrometer by deposition with continued use.

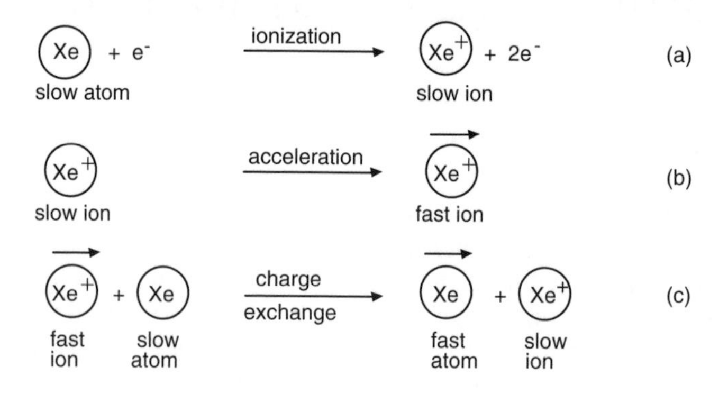

Figure 4.1
(a) Xe atoms are ionized to Xe using electrons. These ions are relatively slow and move in all directions. (b) Xe ions are accelerated through a high electric potential so that they attain a high speed in one direction. (c) Charge exchange between fast Xe ions and slow Xe atoms gives a beam of fast Xe atoms and slow ions. The latter are removed by electric deflector plates, leaving just a beam of fast atoms.

The Ionization Process

When the incident beam of fast-moving atoms or ions impinges onto the liquid target surface, major events occur within the first few nanometers, viz., momentum transfer, general degradation, and ionization.

Momentum Transfer

The momentum of a fast-moving atom or ion is dissipated by collision with the closely packed molecules of the liquid target. As each collision occurs, some of the initial momentum is transferred to substrate molecules, causing them in turn to move faster and strike other molecules. The result is a cascade effect that ejects some of the substrate molecules from the surface of the liquid (Figure 4.2). The process can be likened to throwing a heavy stone into a pool of water — some

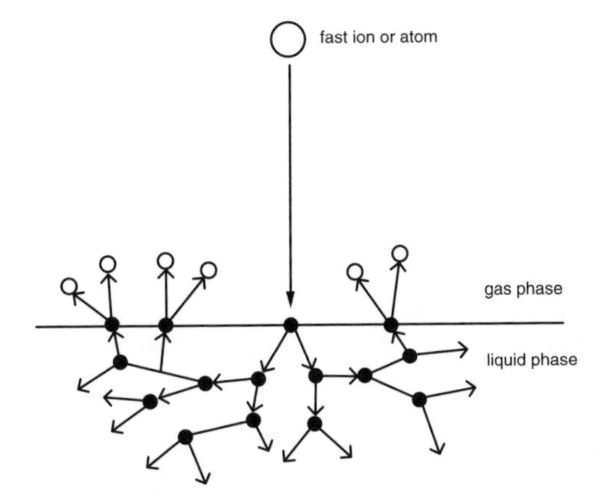

Figure 4.2
A typical cascade process. A fast atom or ion collides with surface molecules, sharing its momentum and causing the struck molecules to move faster. The resulting fast-moving particles then strike others, setting up a cascade of collisions until all the initial momentum has been redistributed. The dots (•) indicate collision points. Ions or atoms (o) leave the surface.

of the water splashes upwards. Clearly, heavier and faster-moving atoms or ions will cause more particles to be ejected from the surface through momentum transfer than will slower-moving, lighter atoms or ions. This is the reason for preferring xenon or cesium.

If the liquid that is being bombarded contains ions, then some of these will be ejected from the liquid and can be measured by the mass spectrometer. This is an important but not the only means by which ions appear in a FAB or LSIMS spectrum. Momentum transfer of preformed ions in solution can be used to enhance ion yield, as by addition of acid to an amine to give an ammonium species (Figure 4.3).

Ionization by Electron Transfer

The close encounter of a fast-moving atom or ion with a neutral molecule can lead to charge exchange in which an electron moves from one particle to another (Figure 4.4). In this way, positive and negative ions are formed separately from any preexisting ions that might be present. Momentum transfer again leads to the newly formed ions being ejected from the liquid.

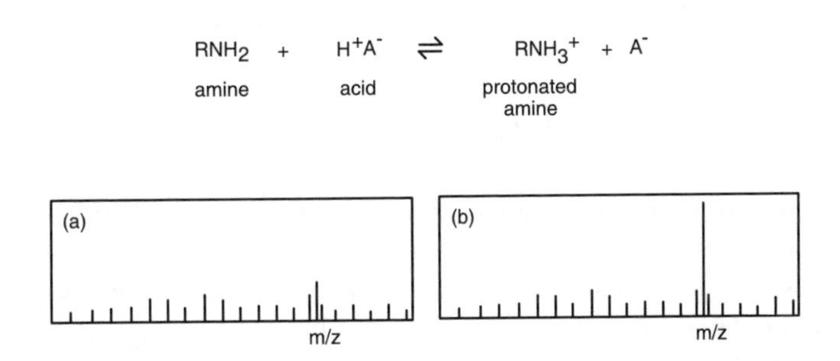

Figure 4.3
An example of enhanced ion production. The chemical equilibrium exists in a solution of an amine (RNH_2). With little or no acid present, the equilibrium lies well to the left, and there are few preformed protonated amine molecules (ions, RNH_3^+); the FAB mass spectrum (a) is typical. With more or stronger acid, the equilibrium shifts to the right, producing more protonated amine molecules. Thus, addition of acid to a solution of an amine subjected to FAB usually causes a large increase in the number of protonated amine species recorded (spectrum b).

$$M + A^{+} \longrightarrow M^{+} + A \quad (a)$$

$$M + A \longrightarrow M^{+} + A^{-} \quad (b)$$

$$M + A \longrightarrow M^{-} + A^{+} \quad (c)$$

$$M(H) + M^{\cdot +} \longrightarrow M(-H)^{\cdot} + MH^{+} \quad (d)$$

Figure 4.4
Collision of a fast-moving ion (A^+) or atom (A) with neutral molecules (M) can lead to an electron being stripped from the molecule by charge exchange to give an ion (M^+) or can lead to an electron being deposited in the molecule to give M– (processes a, b, c). The initially formed ion (M^+) can remove a proton from another molecule, M(H), to give protonated molecular ions, $[M + H]^+$ (process d).

Random Fragmentation

As well as the two specific effects discussed previously (momentum transfer and ionization by electron transfer), bombardment of an organic liquid (solution) with a stream of atoms or ions having very high kinetic energies leads to fairly random bond cleavages to produce radicals and small neutral substances that, in turn, can be further fragmented or can recombine to form other neutrals. All of these bits of fragmented or synthesized molecules can be ionized also. Therefore, an FAB or LSIMS spectrum will have a background of peaks at almost every m/z value and of fairly uniform height. The appearance is rather like viewing grassland by lying down and looking along or through the grass. Indeed, the FAB background is sometimes called "grass."

A major advantage of using a liquid target in SIMS lies in the fact that these randomly produced fragments appear first of all in the surface layers and then diffuse more or less rapidly into the bulk of the liquid, i.e., they are greatly diluted as new surface is constantly formed. On the other hand, sample molecules are distributed throughout the liquid at the start, and, as they disappear from the surface during bombardment, they are continually replaced. Thus, the liquid (matrix) reservoir disperses and dilutes the fragments that would otherwise give a large background of ions and provides a flow of new sample (and solvent) molecules to the surface so that molecular and quasi-molecular ions are continuously replenished.

Despite the high kinetic energy of the bombarding atoms or ions and transfer of some of this energy during collision, the newly formed ions collide with other molecules and, before being ejected from the surface of the liquid, lose most of any excess internal vibrational and rotational energy. Therefore, these ions do not fragment, so any sample substance dissolved in the liquid shows up as molecular or quasi-molecular positive or negative ions. The ionization process is mild, with no additional heat needed for vaporization. Therefore thermally labile molecules like peptides are readily amenable to FAB or LSIMS, giving good molecular mass information.

Properties of the Solvent (Matrix)

Liquids examined by FAB or LSIMS are moved on the end of a probe until the liquid becomes situated in the atom or ion beam. Because of the high-vacuum conditions existing in a mass spectrometer ion source, there would be little point in trying to examine a solution of a sample substance dissolved in one of the common solvents used in chemistry (water, ethanol, chloroform, etc.). Such solvents would evaporate extremely quickly, probably as a burst, upon introduction into the ion source. Instead, it is necessary to use a high-boiling liquid as solvent (matrix). A low-temperature probe has been described, which does utilize low-boiling solvents. Finally, upon bombardment, the solvent itself forms ions that appear as background in a mass spectrum. Very often, protonated clusters of solvent ions can be observed (Figure 4.5).

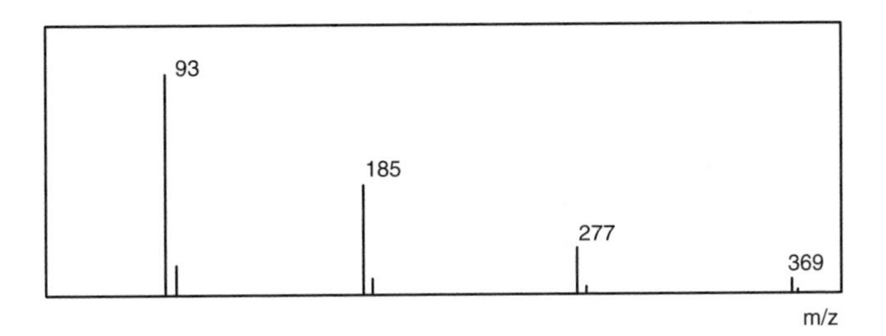

Figure 4.5
A typical FAB mass spectrum of glycerol alone, showing a protonated molecular ion at m/z 93 accompanied by decreasing numbers of protonated cluster ions (m/z, 1 + nx92; n = 2, 3, 4, …).

TABLE 4.1
Some Commonly Used Solvents for FAB or LSIMS

Solvent	Protonated molecular (m/z) ions
Glycerol	93
Thioglycerol	109
3-NOBA[a]	154
NOP[b]	252
Triethanolamine	150
Diethanolamine	106
Polyethylene glycol (mixtures)	- - - -[c]

[a] 3-Nitrobenzyl alcohol
[b] n-Octyl-3-nitrophenyl ether
[c] Wide mass range depending on the glycol used.

In addition to low volatility, the chosen liquid should be a good all-around solvent. Since no one liquid is likely to have the required solvency characteristics, several are in use (Table 4.1). If a mass spectrum cannot be obtained in one solvent, it is useful to try one or more others before deciding that an FAB spectrum cannot be obtained.

The Mass Spectrum

FAB or LSIMS leads to ions being formed from (a) the sample substance, (b) the matrix or solvent (including clusters), and (c) general radiation-induced fragmentation. A typical example of a substance (M; molecular mass, 1000) dissolved in glycerol is shown in Figure 4.6. Protonated molecular ions at m/z 1001 can be observed, together with cluster ions from the solvent (m/z 92, 185, 277, …) and a background of randomly fragmented pieces of ionized solvent and substrate in mainly small abundance. This background is sometimes referred to as "grass." It usually does not prove to be a problem because the abundance of ions making up the "grass" is fairly uniform.

There can be a problem when the sample under investigation itself gives few molecular ions, making it impossible or difficult to distinguish them against the background. At high molecular mass of the sample, the momentum from the bombarding atoms becomes less effective in ejecting molecular ions, which therefore are not abundant and are not easy to discern against the background.

In general, FAB and LSIMS will give excellent molecular mass information in the range (approximately) of m/z 100–2000. Above this value, the abundance of molecular ions tends to diminish until, in the region of m/z 4000–5000, they become either nonexistent or very difficult to

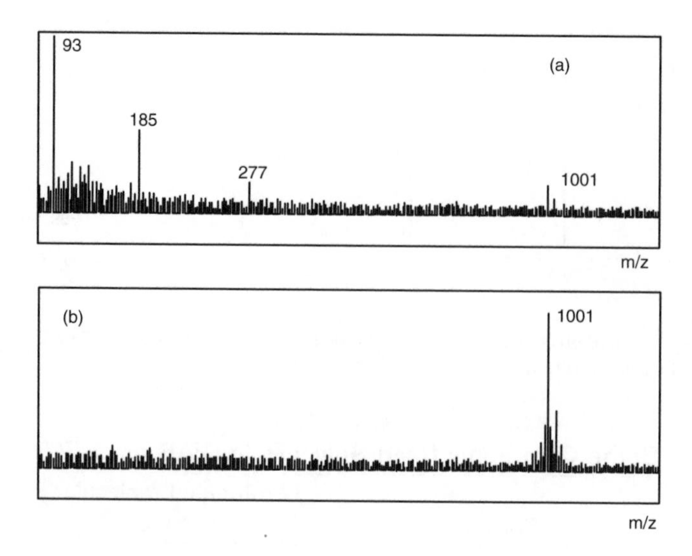

Figure 4.6
(a) A typical FAB mass spectrum in glycerol with protonated sample molecular ions at m/z 1001, protonated glycerol at m/z 93, protonated glycerol clusters at m/z 185, 277, … , and general background ions. (b) The spectrum illustrates the different appearance of recording and expanding from above m/z 300 to m/z 1100. Note the fairly uniform appearances of the background peaks and the absence of solvent cluster ion peaks.

discern against the background. Because background ions above about m/z 200–300 tend to have similar and small abundances, FAB or LSIMS mass spectra are often recorded above, say, m/z 200; this practice also eliminates most of the solvent cluster ion peaks (Figure 4.6b). Alternatively, computer-aided background subtraction can be used to enhance the visibility of molecular ion peaks.

Conclusion

By using a beam of fast atoms or ions incident onto a nonvolatile liquid containing a sample substance, good molecular or quasi-molecular positive and/or negative ion peaks can be observed up to about 4000–5000 Da. Ionization is mild, and, since it is normally carried out at 25–35°C, it can be used for thermally labile substances such as peptides and sugars.

Chapter 5

Field Ionization (FI) and Field Desorption (FD)

Introduction

The main difference between field ionization (FI) and field desorption ionization (FD) lies in the manner in which the sample is examined. For FI, the substance under investigation is heated in a vacuum so as to volatilize it onto an ionization surface. In FD, the substance to be examined is placed directly onto the surface before ionization is implemented. FI is quite satisfactory for volatile, thermally stable compounds, but FD is needed for nonvolatile and/or thermally labile substances. Therefore, most FI sources are arranged to function also as FD sources, and the technique is known as FI/FD mass spectrometry.

Field Ionization

If an electric voltage (potential) is placed across an arrangement such as that shown in Figure 5.1, the lines of equipotential in the resulting electric field crowd in around the needle tip. In this region the field is more intense than, say, near the plate electrode where there are fewer lines of equipotential per unit area. When the needle tip is very fine and the applied potential is large, then very intense electric fields can be generated at the surface (point) of the tip. It is in these high-field regions that ionization occurs.

Unless extremely high potentials are to be used, the intense electric fields must be formed by making the radius of curvature of the needle tip as small as possible. Field strength (F) is given by Equation 5.1 in which r is the radius of curvature and k is a geometrical factor; for a sphere, $k = 1$, but for other shapes, $k < 1$. Thus, if $V = 5000$ V and $r = 10^{-6}$ m, then, for a sphere, $F = 5 \times 10^9$ V/m; with a larger curvature of, say, 10^{-4} m (0.1 mm), a potential of 500,000 V would have to be applied to generate the same field. In practice, it is easier to produce and apply 5000 V rather than 500,000 V.

$$F = V/kr \qquad (5.1)$$

When a neutral molecule settles onto an electrode bearing a positive charge, the electrons in the molecule are attracted to the electrode surface and the nuclei are repelled (Figure 5.2), viz., the electric field in the molecule is distorted. If the electric field is sufficiently intense, this distortion in the molecular field reduces the energy barrier against an electron leaving the molecule (ionization). A process known

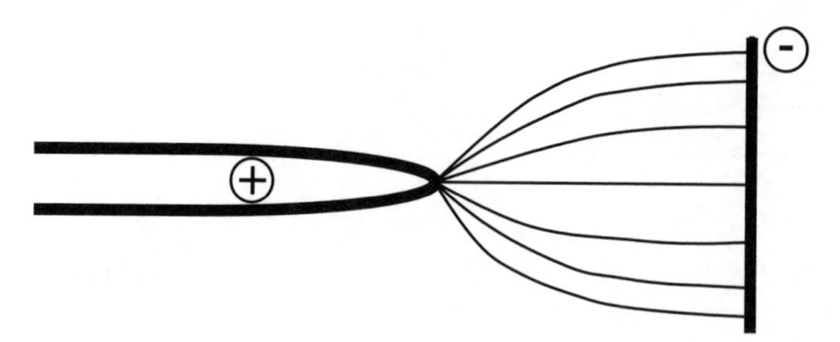

Figure 5.1
An electric potential placed across a needle and a flat (plate) electrode. The lines of equipotential in the resulting electric field are focused around the tip of the needle, where the electric field becomes very large.

as quantum tunneling occurs by which one of the molecular electrons finds itself on the electrode side of the barrier and is promptly neutralized by the positive charges in the electrode. The molecule (M) has then been turned into a positive ion. Because positive charges repel each other, the newly formed positive ion on a positive electrode is repelled by the electrode and flies off into the vacuum of the ion source toward the negative counter electrode (Figure 5.3). A slit in the counter electrode (or a grid electrode) allows the ion to pass into the mass analyzer part of a mass spectrometer.

The electrical reverse of the above arrangement produces negative ions. Thus, a negative needle tip places an electron on the molecule (M) to give a negative ion ($M^{\bullet -}$), which is repelled toward a positive counter electrode.

Design of the Needle Tip Electrode

If there were only one such tip electrode, the yield of ions would be very limited (small surface area and small numbers of ions formed per unit time). To increase ion yield, it is better to use

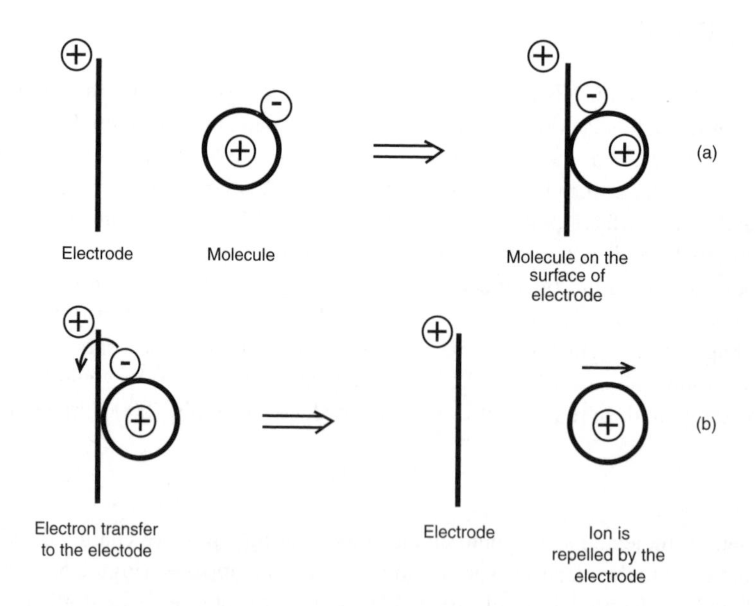

Figure 5.2
In (a), a molecule alights onto a positive electrode surface, its electrons being attracted to the surface and its nuclei repelled. In (b), an electron has tunneled through a barrier onto the electrode, leaving a positive ion that is repelled by and moves away from the positive electrode.

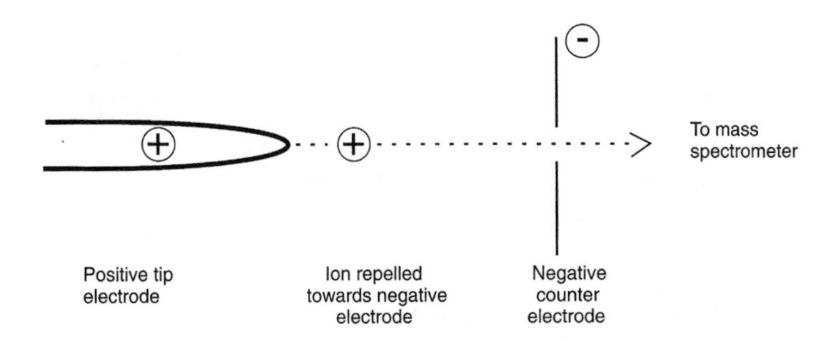

Figure 5.3
A positive ion formed at a positive electrode tip is repelled and travels toward the negative counter electrode, which has a slit in it so that the ion can pass into the mass spectrometer.

many tips or microneedles. This might be achieved by using a sharp edge like a razor blade. Indeed, this was one of the first types of ionization sources to be used, because even the sharpest razor blade, on a molecular scale, is very rough and has many small tips on its surface (Figure 5.4). Such an edge is better than a single needle, but it still does not provide an efficient ion source.

Eventually, methods were found for growing microneedles or "whiskers" on the surface of a thin wire. For example, by maintaining a narrow wire at a high temperature in the vapor of benzonitrile, decomposition of the nitrile on the hot wire produces fine, electrically conducting growths or whiskers (microneedles) having tips of very small radius of curvature (Figure 5.5). Application of a high electric potential to such a wire produces many ionization points and a high yield of ions.

These thin wires are supported on a special carrier that can be inserted into the ion source of the mass spectrometer after first growing the whiskers in a separate apparatus. Although the wires are very fragile, they last for some time and are easily renewed. They are often referred to as emitter electrodes (ion emitters).

Practical Considerations of Field Ionization/Field Desorption

Although there has been some controversy concerning the processes involved in field ionization mass spectrometry, the general principles appear to be understood. Firstly, the ionization process itself produces little excess of vibrational and rotational energy in the ions, and, consequently, fragmentation is limited or nonexistent. This ionization process is one of the mild or soft methods available for producing excellent molecular mass information. The initially formed ions are either simple radical cations or radical anions ($M^{\bullet-}$).

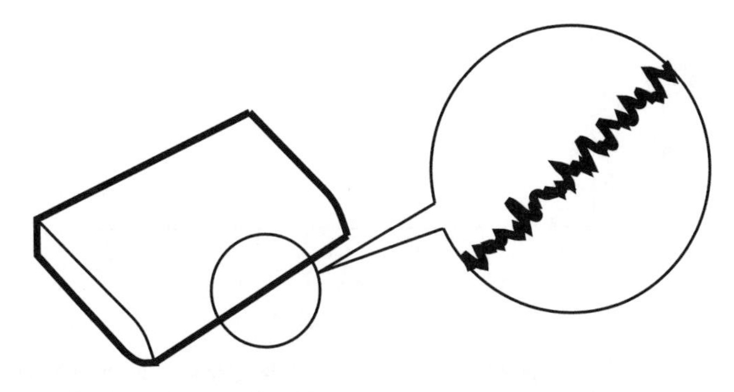

Figure 5.4
Magnification of a sharp edge showing the many tips and valleys on a molecular scale.

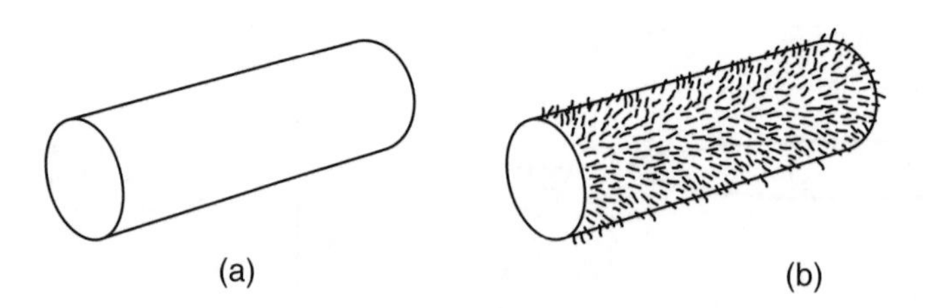

Figure 5.5
A narrow wire (a) heated in the vapor of an organic compound such as benzonitrile causes decomposition of the nitrile and the formation of whiskery growths on the surface of the wire (b). The sizes of the growths are exaggerated for purposes of illustration and are, in fact, very small in relation to the diameter of the wire.

However, in both FI and FD, there are other neutral molecules on or close to the surface of the emitter and, in this region, ion/molecule reactions between an initial ion and a neutral (M(H)) can produce protonated molecular ions ($[M + H]^+$), as seen in Equation 5.2.

$$M(H) + M^{\bullet+} \rightarrow M(-H^{\bullet}) + [M + H]^+ \tag{5.2}$$

For simple FI, the substance to be mass measured is volatilized by heating it close to the emitter so that its vapor can condense onto the surface of the electrode. In this form, an FI source can be used with gas chromatography, the GC effluent being passed over the emitter. However, for nonvolatile and/or thermally labile substances, a different approach must be used.

For nonvolatile or thermally labile samples, a solution of the substance to be examined is applied to the emitter electrode by means of a microsyringe outside the ion source. After evaporation of the solvent, the emitter is put into the ion source and the ionizing voltage is applied. By this means, thermally labile substances, such as peptides, sugars, nucleosides, and so on, can be examined easily and provide excellent molecular mass information. Although still FI, this last ionization is referred to specifically as field desorption (FD). A comparison of FI and FD spectra of D-glucose is shown in Figure 5.6.

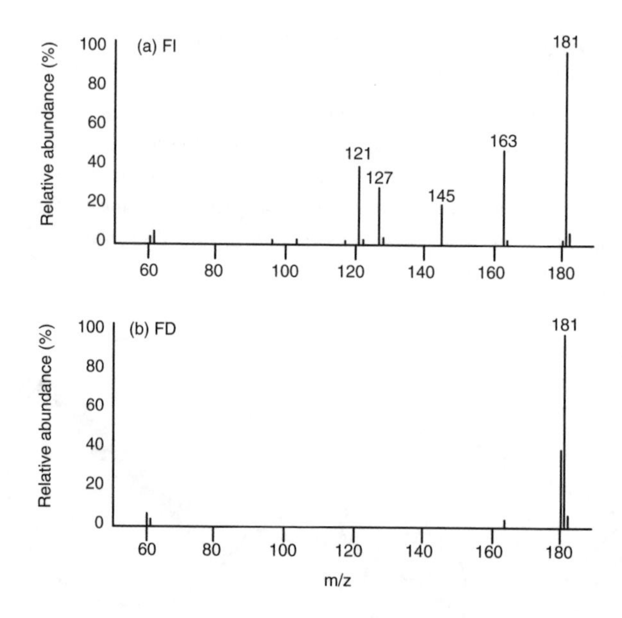

Figure 5.6
A comparison of the (a) FI and (b) FD spectra of D-glucose. Note the greater fragmentation in FI (heat applied for volatilization) and the appearance of ions at m/z 180 in the FD spectrum as well as the $[M + H]^+$ ions at m/z 181 in both spectra.

By intentionally adding inorganic salts to the solution used for FD, cationated molecular ions can be produced in abundance. Equation 5.3 illustrates how addition of NaCl can give rise to $[M + Na]^+$ ions.

$$M + Na^+ + Cl^- \rightarrow [M + Na]^+ + Cl^- \tag{5.3}$$

Sometimes, in FD, the emitter electrode is heated gently either directly by an electrode current or indirectly by a radiant heat source to aid desorption of ions from its surface.

Types of Compounds Examined by FI/FD

Newer developments in ionization methods have tended to overshadow FI, particularly in view of the fragile nature of the emitters and the need for a separate apparatus in which to form the microneedles. In contrast, FD still offers advantages, being able to ionize a wide range of mass spectrometrically difficult substances (peptides, sugars, polymers, inorganic and organic salts, organometallics).

Because there is little fragmentation on FD, it is necessary to activate the molecular or quasi-molecular ions if molecular structural information is needed. This can be done by any of the methods used in tandem MS as, for example, collisional activation (see Chapters 20 through 23 for more information on tandem MS and collisional activation).

Conclusion

FI and FD are mild or "soft" methods of ionization that produce abundant molecular or quasi-molecular positive or negative ions from a very wide range of substances. In the FD mode, it is particularly useful for high-molecular-mass and/or thermally labile substances such as polymers, peptides, and carbohydrates.

Chapter 6

Coronas, Plasmas, and Arcs

Background

Charged particles from the sun, usually protons, encounter the earth's magnetic field, spiral down toward the negatively charged earth, and meet the atmosphere above the magnetic north and south poles (Figure 6.1). The charged particles from the sun are moving at high speed and begin to collide with molecules of oxygen, nitrogen, and other gases in the upper atmosphere. These high-energy collisions cause electrons in the gas molecules to be excited into higher energy orbitals to form excited atoms or ions (Figure 6.2). Additionally, nitrogen and oxygen ions are formed if an electron is ejected from a molecule by the energy of collision. During this process, the remaining electrons in the ions may also be excited. Other gases are present in the atmosphere, such as carbon dioxide, and argon, and these also form excited molecules and ions. When the electrons in the excited atoms or ions return to their original (ground) state, light is emitted. The light can be green, red, or other colors, depending on which molecules have been excited (Figure 6.3).

Such events account for the appearance of the very pretty, mysterious displays of lights in the sky in the polar regions of the northern and southern hemispheres, viz., the aurora borealis and aurora australis. (Sometimes the light can be observed in areas more distant from the poles and, in the north, they are called the "northern lights.") These natural phenomena are manifestations of electric discharge physics, namely the passage of charged particles through a gas. Excitation of atoms or molecules in an electric field by electrons and recombination of ions and electrons causes the formation of excited species, which emit light. Other examples of the discharge can be seen in lightning flashes, the common yellow sodium street lights, fluorescent lighting, and arc welding.

In the laboratory, it has been found that similar effects can be produced if a voltage is applied between two electrodes immersed in a gas. The nature of the laboratory or instrumental discharge depends critically on the type of gas used, the gas pressure, and the magnitude of the applied voltage. The actual electrical and gas pressure conditions determine whether or not the discharge is called a corona, a plasma, or an arc.

Although the discharges attract interest because of the emitted light, the major components of the discharges are ions and electrons. Such electrons can be utilized in mass spectrometry to enhance ionization of sample molecules, and the ions themselves can be used to gain information about the sample (m/z values and abundances of ions). This latter use of discharges is the subject of this chapter. The effects of electrons in these discharges can be greatly enhanced by the application of an external high-frequency electromagnetic field, which leads to the plasma discharge reaching very high temperatures, as in plasma torches (see Chapter 14 for more information on plasma torches).

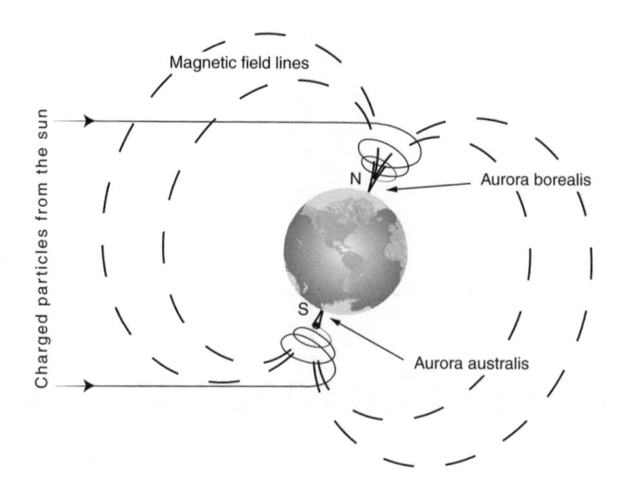

Figure 6.1
Schematic representation of the movement of positively charged particles (mostly protons) from the sun entering the earth's magnetic and electric fields. Under the influence of the magnetic component, the particles spiral down toward the north and south poles of the negatively charged earth, clockwise at one pole and counterclockwise at the other. When the incoming charged particles collide with molecules of oxygen, nitrogen, argon, water, and other gas molecules in the upper atmosphere, the resultant excitation leads to emission of light, known as the aurora borealis in the northern hemisphere and the aurora australis in the southern.

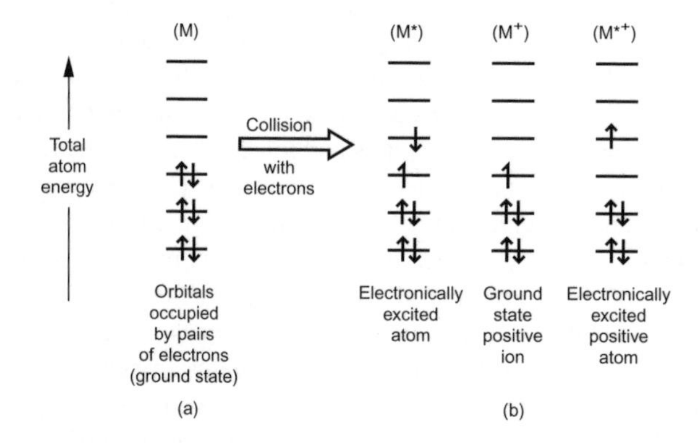

Figure 6.2
(a) Occupied and unoccupied orbital energy levels in an atom. Some of these are occupied by pairs of electrons (the electron spins in the bonding levels are indicated by the arrows), and some are empty in the normal ground state of the atom (M). After collision with a high-energy (fast-moving) electron, one or more electrons are promoted to higher unoccupied orbitals. (b) One electron is shown as having been promoted to the next higher orbital energy to give an electronically excited atom (M^*). Alternatively, the promoted electron may be ejected from the atom altogether, leaving a ground-state positive ion (M^+). Finally, an ion can be formed by loss of an electron but, at the same time, another electron may be promoted to a higher orbital to give an electronically excited ion (M^{*+}).

Electric Discharges in a Gas

In this discussion, only inert gases such as argon or neon are used as examples because they are monatomic, which simplifies description of the excitation. The introduction of larger molecules into a discharge is discussed in later chapters concerning examination of samples by mass spectrometry.

If a gas such as argon is held in a glass envelope that has two electrodes set into it (Figure 6.4), application of an electric potential across the electrodes leads to changes in the gas as the electrons flow from the cathode (negative electrode) to the anode (positive electrode). This passage of electrons

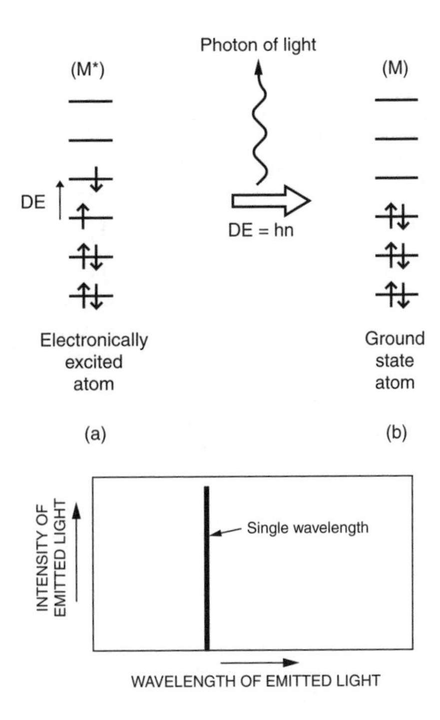

Figure 6.3
(a) The orbital levels of an electronically excited atom (or molecule) show one electron in a higher orbital than it would normally occupy. (b) The atom can return to its normal ground state by emitting a photon of light of frequency υ, for which $\Delta E = h\upsilon$ is the energy change in the atom as the photon is emitted and the electron drops to the more stable state. Since υ is determined by ΔE, the wavelength (λ) of the light is found by dividing the speed of light (c) by the frequency υ, viz., $\lambda = c/\upsilon$. If λ falls in the visible region of the electromagnetic spectrum (approximately 400–700 nm), the emitted radiation appears as colored light of a single wavelength. The colors of the emitted light depend on the nature of the gas atoms or molecules that are excited.

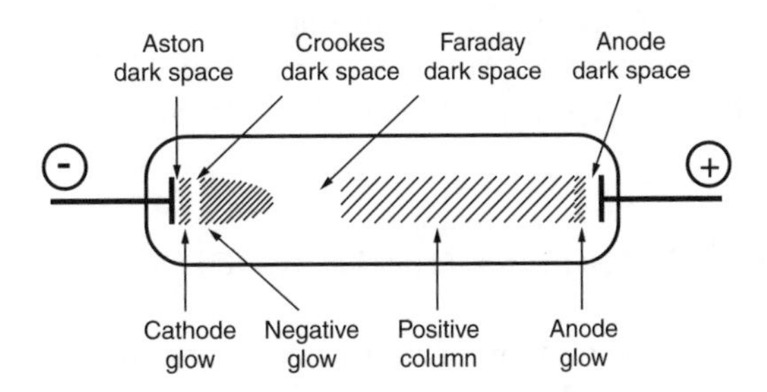

Figure 6.4
A typical representation of light-emitting regions and dark spaces in a gas discharge between two electrodes, one electrically negative (the cathode), the other positive (the anode). The glowing regions are known as the cathode glow, the negative glow, the positive column, and the anode glow. The regions emitting no light are named after earlier pioneers of gas discharges, viz., the *Aston*, *Crookes*, and *Faraday* dark spaces. Actually, the dark spaces do emit some light, but they appear dark in contrast to the much brighter glow regions. The discharge takes up a form controlled by the shapes of the electrodes, and, in this case, it is a cylindrical column stretching from one disc-like electrode to another.

is highly dependent on the type of gas present, the pressure of the gas, and the voltage applied. The most obvious demonstration of the flow of electrons in the gas arises from the emitted light (Figure 6.4).

As described above, this light comes partly from the formation of excited atoms and ions in the gas (Figure 6.3) and partly from their recombination with electrons (Figure 6.5). Unlike atom excitation, which mostly gives rise to light being emitted at one or two fixed wavelengths, the recombi-

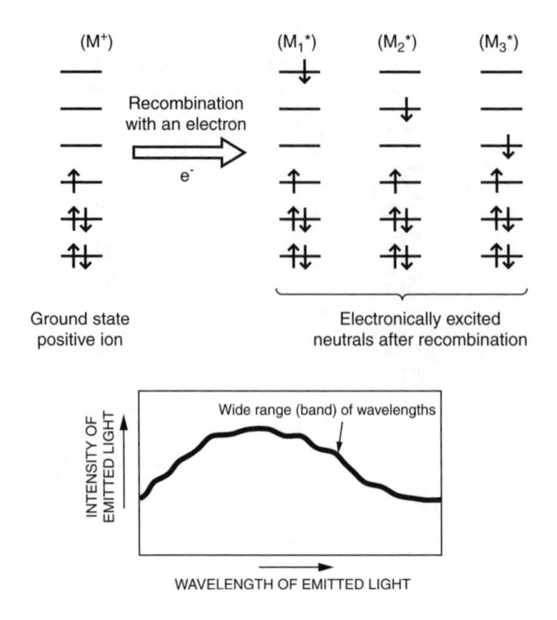

Figure 6.5
When an electron recombines with a positive ion, the incoming electron will attach at any of the vacant atomic orbitals, as illustrated by the three typical states (M_1^*, M_2^*, M_3^*). Other more excited states can be formed as the electron distribution in the newly formed neutral is disturbed (shaken up) by the added energy of the incoming electron. However, most of these states are unstable with respect to the ground-state atom, and the disturbed electrons drop down to more-stable orbitals until the ground state is reached. This increase in stability can only be achieved if energy is lost from the system. Thus, each time an electron drops to a lower orbital, a quantum of light is emitted with a wavelength that depends on the spacing between orbitals. Since many of these excited states are formed upon recombination and decay of the initial excited states, the spectrum of emitted light covers a wide range of wavelengths, viz., it is broadband emission. The light emitted from a gas discharge is a superposition of the line spectra arising from directly excited neutral atoms (Figure 6.3) and broadband spectra from electron/ion recombination. Some typical colors of emitted light are given in Table 6.1.

nation process leads to a "band" spectrum of emitted light in which there is a wide spread of very closely spaced wavelengths (Figure 6.5). A fuller description of the discharge process follows as a typical example, in which a voltage is applied to the electrodes and is then gradually increased.

Light and Dark Regions in the Discharge

The light regions in the discharge result from electron collisions with neutral atoms in the gas and from recombination of electrons and positive ions to give atoms.

Aston Dark Space and the Cathode Glow

The energy of an electron is controlled by its velocity, which is proportional to the applied voltage. Electrons emitted from the cathode are accelerated by the electric field. As they cross the Aston dark space, the electrons gather speed and collide with atoms of gas. Across the Aston dark stage the electrons have insufficient energy to excite gas atoms; any colliding electron and atom pair simply bounce away from each other without any overall transfer of energy other than kinetic. This sort of collision is said to be elastic if no energy is interchanged. Because there is insufficient energy transfer to cause electronic excitation in the neutral atom, this region of the discharge is dark. As the speed of an electron reaches a certain threshold level and collision occurs with an atom, the electric field

TABLE 6.1.
Typical Colors of Light Emitted in Different Regions of the Glow Discharge

Gas used	Cathode glow	Negative glow	Positive glow
Nitrogen	pink	blue	red
Oxygen	red	yellow/white	pale yellow
Air	pink	dark blue	pink
Neon	yellow	orange	red
Argon	pink	dark blue	dark red
Sodium	pink	white	yellow
Mercury	green	green	green

of the electron and the electric field of the electrons in the atom interact with each other. The transferred energy causes one or more electrons in the atom to be promoted to a higher energy level (the atom is energetically excited). The excited state is unstable and quickly loses energy to return to the ground state by emitting a photon ($h\upsilon$). If frequency υ lies in the visible region of the electromagnetic spectrum, then the phenomenon is manifested by the production of light which is the basis of the cathode glow shown in Figure 6.4. The color of the emitted light depends on υ, which in turn depends on the nature of the gas atoms that are excited (Figures 6.3, 6.5). Light is also emitted as a broadband spectrum following electron/ion recombinations (Figure 6.5). The typical excitation colors shown in Table 6.1 are caused by a superposition of line spectra emitted from excited atoms and broadband emission from electron/ion recombination processes.

Because the colliding electron loses energy (slows down) after an inelastic collision, it no longer has sufficient energy to excite any more gas atoms, so the cathode glow appears as a fairly well-defined band (Figure 6.4). The front of this band indicates the place where the colliding electrons have sufficient energy to cause emission from the gas atoms, and the back (furthest from the cathode) marks the region where the electrons have lost energy in collisions and can no longer excite the gas. This last point is also the beginning of the Crookes's dark space.

Crookes's Dark Space and the Negative Glow

After leaving the cathode glow, some of the electrons originally emitted from the cathode have slowed down, but others have suffered few collisions and are traveling considerably faster (have more energy). In the Crookes's dark space there is an assemblage of electrons of various energies being accelerated by the external electric field between the electrodes. In the dark space, all of the electrons are accelerated and two major processes (A, B) occur.

In process A, slow electrons are accelerated until they have sufficient energy to again excite gas atoms; this is the start of the negative glow, just like the process in the cathode glow (Figure 6.4).

In process B, fast electrons are accelerated to even higher speeds (higher energy), eventually being able to remove an electron from an atom entirely, viz., the atom is ionized (Figure 6.2b). In this sort of inelastic collision between an atom and an electron, two electrons leave the collision site (Figure 6.6). The positive ion that is produced will be considered later. However, the two electrons leaving the collision site are accelerated and eventually are able either to cause an atom to emit light (process A) or to cause ionization (process B). Thus the negative glow is a region in which a cascade of electrons is produced by ionizing collisions, which are in addition to the original flux of electrons coming from the cathode. This cascade of electrons leads to many inelastic collisions with gas atoms and therefore to the emission of many photons of light. Thus, the negative glow is much brighter than the cathode glow and spreads over a larger region of space. Also, because of the initial large

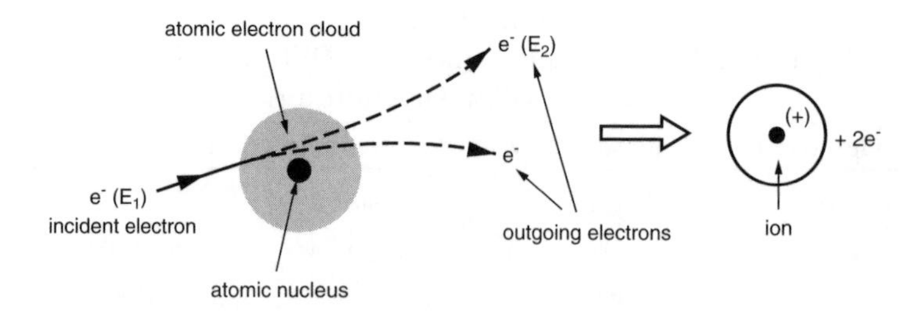

Figure 6.6
After acceleration through a voltage (V), an electron has a new velocity (u) and energy (e _ V), which is equal to the kinetic energy gained by the electron ($mu^2/2$). This is the energy (E_1) of the incoming electron shown in the diagram ($E_1 = eV = mu_1^2/2$). After collision with the atom, two electrons leave the reaction site, the original electron having lost some of its energy (now, $E_2 = mu_2^2/2$). The other newly ejected electron has energy equal to the difference between the ionization energy (I) of the atom and the energy lost by the incident electron ($E_2 = E_1 - I$). Since energy has been given up, $E_2 < E_1$ and $u_2 < u_1$, viz., the exiting electron leaves the collision slower than it moved as an incoming electron. Recall that in the collision process between the incident electron and the atom, the electron is so small and the electronic space surrounding the nucleus of the atom is so large that there is no collision as such. Rather, the incident electron, with its associated wavelike electric field, perturbs the electrons in the atom as it passes through the vast open spaces of the atom's electron cloud. If the perturbation is sufficiently large, one of the atom's electrons will be ejected completely along with the departing (originally incident) electron.

spread of energies as electrons leave the Crookes's dark space and begin causing the negative glow, the back end of the glow is not sharp as with the cathode glow but gradually fades into the following Faraday dark space. The spread of energies of electrons going into the negative glow means that the emitted light results from a range of excitation and recombination possibilities and is usually a different color than that seen from the cathode glow. Although this region of the negative glow is known as a corona, this term is now usually applied when inhomogeneous electric fields are used in gases at or near atmospheric pressure, in which the field is sufficient to maintain a discharge but is unable to produce much of a glow, or it produces none at all.

Faraday Dark Space and the Positive Column

After leaving the negative glow, electrons have insufficient energy for either exciting or ionizing effects, but they begin to be accelerated again. This is the Faraday dark space. Note that in the region from the cathode to the start of the negative glow, the electric field resulting from the applied voltage on the electrodes is high and changes rapidly for reasons discussed below. However, from the end of the negative glow to the anode, the electric-field gradient is small and almost constant; electrons leaving the negative glow are only slowly accelerated as the electrons move toward the anode. In this region, inelastic collisions are less frequent, but ionizing collisions still occur, and there are also some collisions leading to emission of light. Thus the positive column emits light less strongly than does the negative glow, and often the light is a different color, too (Table 6.1).

The colors shown in Table 6.1 are only approximate. Sometimes mixtures of colors are seen as gas pressure or applied voltages change or if impurities are present.

The positive column is a region in which atoms, electrons, and ions are all present together in similar numbers, and it is referred to as a plasma. Again, as with the corona discharge, in mass spectrometry, plasmas are usually operated in gases at or near atmospheric pressure.

Anode Dark Space and Anode Glow

Positive ions formed near the positive electrode (anode) are repelled by it and move into the positive column. Electrons that reach proximity to the anode are accelerated somewhat because the electric-field gradient increases slightly. The more-energetic electrons cause more emission of light near the anode than from the main body of the positive column, so this end region appears brighter than the main body of the positive column. This is the anode glow. Sometimes this glow is not very marked. Also in this region, inelastic collisions with atoms lead to the formation of more positive ions, which are repelled from the anode and into the positive column. After these inelastic collisions with gas atoms, the electrons have lost energy but continue to the anode and are discharged. The last region of the discharge contains electrons with too little energy to cause excitation, and no light is emitted; this is the anode dark space.

Electric-Field Gradients across the Glow Discharge

In Figure 6.4, the two electrodes are marked as cathode and anode, arising from the application of an external voltage between them. Before any discharge occurs, the electric-field gradient between the electrodes is uniform and is simply the applied voltage divided by the their separation distance, as shown in Figure 6.7.

When the discharge has been set up, there is a movement of electrons from cathode to anode and a corresponding movement of positive ions from the anode to cathode. These transfers of electrons and ions to each electrode must balance to maintain electrical neutrality in the circuit. Thus, the number of positive ions discharging at the cathode must equal the number of electrons discharging at the anode. This occurs, but the actual drift velocities of electrons and ions toward the respective electrodes are not equal.

As the electrons move from cathode to anode, they undergo elastic and inelastic collisions with gas atoms. The paths of the electrons are not along straight lines between the electrodes because of the collisions. In effect, the movement of each electron consists of short steps between

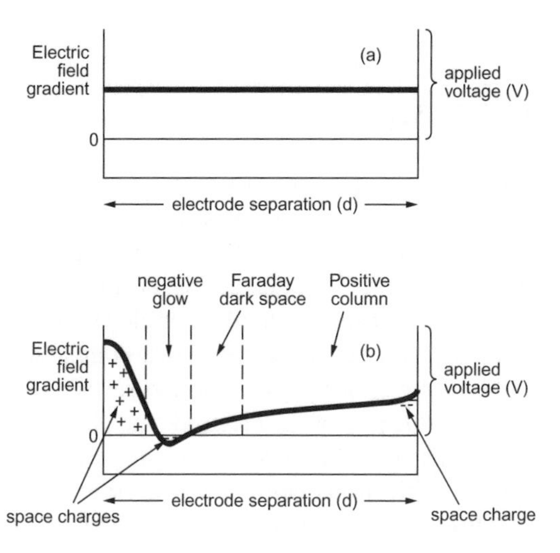

Figure 6.7
(a) Before any discharge occurs, the voltage (*V*) has been applied across two electrodes, distance (*d*) apart. The electric field has a constant gradient (= V/d). (b) The discharge has been set up. Because of space-charge effects, in which there is a preponderance of positive ions or negative electrons, the electric field is no longer uniform. Between the cathode and the start of the negative glow, there is a large fall in potential through the positive space-charge region, leading to a large field gradient. In this region electrons are strongly accelerated. Just in front of the negative glow, an excess of electrons leads to a small negative space charge as the gas density of electrons begins to rise to be greater than the gas density of positive ions. Beyond the negative glow, there is a much smaller change in the field gradient because there are almost equal numbers of ions and electrons. Only near the anode is there another small (electron) space-charge effect, which accelerates ions away from the anode.

collisions, some of which will even cause the electron to move backward, although the overall electric field reverses any such backward recoil. Thus there is a random walk as the electrons gradually make their meandering way across the discharge region. In a perfect vacuum, the velocity u of each electron would be given by the formula, $mu^2 = 2eV$, in which m is the mass of an electron, e is the electronic charge, and V is the applied potential. The time t taken to cross the distance d between the electrodes is then simply $t = d/u$. However, because of the meandering path caused by frequent collisions, the average speed (the drift velocity) is much smaller, and the time taken to reach the anode is much longer than would be the case in the absence of any neutral gas molecules (in a vacuum). This sort of movement of particles in gases and liquids is common and is normally referred to in terms of particle mobility.

The mobility of a positive ion is about 100 times less than that of an electron which means that positive ions move more slowly in the anode-to-cathode direction than electrons move in the opposite direction. Since only the same numbers of ions and electrons can be discharged in unit time, the result of the difference in mobilities is that a large excess of positive ions gathers near the cathode while a much smaller excess of electrons gathers near the anode. These clouds of ions and electrons around the electrodes constitute space charges that affect the electric-field gradient, which is large near the cathode but much smaller near the anode (Figure 6.7). The positive space charge has the overall effect of moving the anode closer to the cathode.

Most of the voltage difference applied to the electrodes falls across a narrow region close to the cathode (Figure 6.7). Thus the electrons, generated at the cathode, are rapidly accelerated into the cathode and negative glow regions but are only slowly accelerated along the positive column. From the negative glow to near the anode dark space, the numbers of ions and electrons are very similar and, in this region, the electric-field gradient is only small, making acceleration of electrons and ions very small. In the region close to the anode, there is a small increase in the electric-field gradient that accelerates the electrons toward the anode and ions away from it. The full profile of the electric field is illustrated in Figure 6.7.

Self-Sustaining Discharge

Once the glow discharge has begun, a number of processes are set in motion to maintain it. Before discharge begins, the cathode emits few electrons unless it is heated (see Chapter 7 for information on thermal ionization) or unless light is shone on it (photoelectric effect). However, once light is being emitted from the discharge glow, the light falling on the cathode induces a photovoltaic emission of more electrons, thereby enhancing the flow of electrons from the electrode.

In addition to this source of extra electrons, there is a bombardment of the cathode by the incoming positive ions. As the positive ions plunge into the surface of the electrode, their kinetic energies are transferred to the constituents of the electrode metal. This momentum transfer causes the emission of secondary electrons and other species, which again improves the flow of electrons. The glow discharge leads to more electrons being released from the cathode than would be the case otherwise, and the total current flow through the discharge increases. The glow becomes self-sustaining as long as an electric potential exists across the electrodes. Thus the starting voltage needed to set up a glow discharge can be reduced once the discharge is underway. A description of the current/voltage changes in a typical discharge is given below.

Sputtering

Ions impacting onto the cathode during a discharge cause secondary electrons and other charged and neutral species from the electrode material to be ejected. Some of these other particles derived from the cathodic material itself migrate (diffuse) to the walls of the discharge tube and form a deposit there. This effect of transferring material from an electrode to other parts of the discharge system is called sputtering.

Effect of Electrode Separation on Discharge

At any one gas pressure, the separation between the electrodes determines the appearance of the discharge. At low pressures of about 1 torr, the appearance of the discharge is similar to that shown in Figure 6.4. If the electrodes are moved farther apart, a greater voltage becomes necessary to maintain the discharge. The positive column increases in length, but there is no effect on the cathode regions because the space charge maintains the high electric-field gradient near the electrode. The positive column therefore simply increases in length since the positive ions and electrons within it have further to migrate.

If the electrodes are moved closer together, the positive column begins to shorten as it moves through the Faraday dark space because the ions and electrons within it have a shorter distance through which to diffuse. Near the cathode, however, the electric-field gradient becomes steeper and electrons from the cathode are accelerated more quickly. Thus atom excitation through collision with electrons occurs nearer and nearer to the cathode, and the cathode glow moves down toward the electrode.

Arcs

As the voltage across the discharge is increased, the glow discharge gets brighter and the current rises as more and more electrons are released through ionization and through bombardment of the cathode by more ions. The negative glow is then almost on top of the cathode; the separation between it and the cathode itself is much less than a millimeter. The positive column or plasma glow increases as the plasma spreads to occupy almost all of the space between the electrodes. At some point, the cathode glow suddenly becomes a bright spot on the cathode, and the voltage falls as the current flow increases again. This is when an arc is struck. There is a very bright narrow column of hot gas between the electrodes. The fairly sudden increase in the flow of electrons as the arc is struck probably arises from three sources. One is an increase in the numbers of secondary electrons emitted from the cathode under increased bombardment from the larger number of positive ions being produced. Another is increased thermal emission of ions as the cathode heats up, and a third is field emission (see Chapter 5 for information on field ionization). As the negative glow approaches ever nearer (10^{-6} m) to the cathode, the electric-field gradient between the cathode and the glow becomes very high, reaching 10^8 to 10^9 V/m for an applied potential of 100 V. This electric-field condition is in the region of that required for field ionization.

A typical temperature in an arc is about 2000 K. An arc can be struck in other ways, as in welding or arc lights. For such uses, two electrodes are first touched together (very low electrical resistance), and a relatively small potential is applied. Because the resistance is small, a large current passes between the electrodes and a rapid sequence of events, as described above for the glow discharge, ends with an arc's being formed. If the electrodes are then drawn apart, there is an increase in resistance and the current density settles. The high current density in the arc causes rapid heating of gas molecules and the emission of large amounts of light. Arcs are usually struck in gases at atmospheric pressure. The intensity of the emitted light is used for the bright arc lights in theaters, but a welder without the protection of a dark glass shield exposed to the intensity of light and the wavelengths it covers — including the ultraviolet — would suffer eye damage. Material sputtered from the electrodes is used in the weld.

The arc discharge is commonly used to volatilize and ionize thermally intractable inorganic materials such as bone or pottery so that a mass spectrum of the constituent elements can be obtained.

Effect of Gas Pressure on Discharge

The appearance of the glow discharge at about 1 torr is shown in Figure 6.4. If the gas pressure is reduced, the space charge is reduced and electrons emerging from the cathode have further to travel to collide with gas atoms. Thus the cathode glow moves away slightly from the cathode, but the negative glow moves strongly toward the Faraday dark space because the cascade of electrons formed by a collisional process has to travel further to meet a sufficient number of gas atoms. At the same time, the positive column shortens, appearing to disappear into the anode region and becoming weaker. If the gas pressure continues to be reduced, there will be too few electron/atom collisions to maintain a cascade of electrons, and the discharge stops (goes out) unless the voltage is increased.

If the gas pressure is increased, the opposite effects occur. The negative glow, the cathode glow, and the dark spaces move toward the cathode, and the positive column gets longer. The lengthening effect of the positive column essentially brings the anode nearer to the cathode. At about 100–200 torr, the negative glow moves almost up to the surface of the cathode, followed by the Faraday dark space. The positive column not only lengthens, but its light emission begins to weaken too and it becomes fainter and narrower.

Effect of Gas Flow on Discharge

With a discharge tube totally enveloping the discharge gas, there is a faster drift of electrons than ions to the walls of the tube, which become negatively charged. Positive ions and sputtered materials are attracted there, reducing the flow of current in the discharge. The buildup of a deposit can eventually lead to most electrons and ions moving to the walls of the tube rather than to the electrodes, and the discharge stops. This situation occurs in the common fluorescent tubes used for lighting. In scientific apparatus, in which coronas and plasmas are struck, it is more usual to have a continuous flow of gas through the discharge region to help prevent a buildup of deposits.

Effect of Electrode Shapes on Discharge

Particularly in mass spectrometry, where discharges are used to enhance or produce ions from sample materials, mostly coronas, plasmas, and arcs are used. The gas pressure is normally atmospheric, and the electrodes are arranged to give nonuniform electric fields. Usually, coronas and plasmas are struck between electrodes that are not of similar shapes, complicating any description of the discharge because the resulting electric-field gradients are not uniform between the electrodes.

In atmospheric-pressure chemical ionization (APCI), a nebulized stream of droplets leaves the sample inlet tube and travels toward the entrance to the mass analyzer. During this passage, ions are produced, but the yield is rather small. By introducing electrodes across the flow of sample material at atmospheric pressure, a discharge can be started, which is essentially of a corona type. In this discharge, electrons and positive ions are formed, and they interact with neutral sample molecules flowing through the discharge. Collisions between electrons and neutral sample molecules produce more sample ions. The newly produced ions collide frequently with other neutral molecules present to give thermalized protonated ions like those produced by normal chemical ionization. Thus the yield of protonated ions from the standard APCI process is greatly increased.

(Chemical ionization and atmospheric-pressure ionization are covered in Chapters 1 and 9, respectively.) The corona discharge is relatively gentle in that, at atmospheric pressure, it leads to more sample molecules being ionized without causing much fragmentation.

In inductively coupled plasmas, sample is introduced into a plasma struck in a flowing gas, frequently argon. The plasma itself is normally formed between the walls of two concentric cylinders so the electric field has a nonuniform gradient. By applying a high-frequency electromagnetic field across the plasma, ions and electrons are made to swing backward and forward as they attempt to follow the changing alternating field. This effect speeds up the ions and particularly the electrons until they have energies equivalent to several thousand degrees. Under these conditions, any sample molecules in this plasma are rapidly degraded to atoms and ions of their constituent elements (see Chapters 14 through 17 for more information on plasma torches). Unlike coronas, these inductively coupled plasmas are very destructive of samples so the original structures are lost and only ions of the constituent elements are observed. This property makes the plasma torch a valuable alternative to thermionic emission as an ion source for isotope ratio analysis.

With arcs, intense bombardment by ions and electrons and the heat produced at the electrodes cause sample molecules to be vaporized and broken down into their constituent elements. These sources are used particularly for analysis or isotope studies when the samples involved are inorganic, nonvolatile, and thermally very stable.

Effect of Voltage Changes on Glow Discharge Characteristics

The glow discharge characteristics described above are typical of those found in a gas at reduced pressures, but, as discussed, changes in gas pressure, the type of gas, and the voltage applied all have important effects on the flow of electrons and ions and thus on the nature of the discharge. This section describes in greater detail the development of a discharge from a weak, non-self-sustaining state through to the corona/plasma/arc self-sustaining conditions.

The graph of Figure 6.8 illustrates the effect of increasing voltage on the electric current between two electrodes immersed in a gas. The circuit is completed by an external resistance, used to limit the current flow. As shown in Figure 6.8, the discharge can be considered in regions, which are described below.

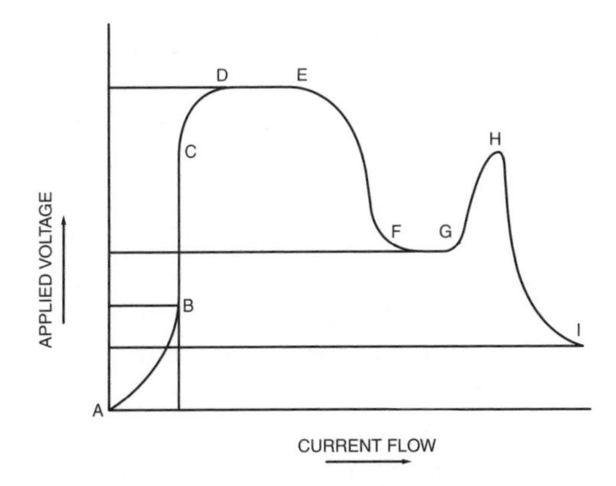

Figure 6.8
The graph shows the variation in current with changes in voltage between two electrodes placed in a gas. These variations in current flow are caused by changes in the flux of electrons passing between the electrodes inside the discharge chamber. An external resistance is used also to limit the total current that can pass. The variations are discussed in sections in the main text, labelled A to H on the diagram. Of particular note are the regions C–E (a corona discharge) and F–H (the start of the arc-forming process). The glow discharge and plasma occur mainly in sections E–F.

Region A–B–C (Non-Self-Sustaining Discharge)

This is the most difficult part of setting up a discharge because the discharge gases used are largely insulating and, theoretically, there are no electrons to start a current flow between the electrodes. However, there are sources of electrons and ions, some natural and some artificial:

Background Cosmic Radiation

Cosmic radiation consists of high-speed charged particles, some of which interact with gas atoms in the discharge tube to produce electrons and ions. As might be imagined, this flow of electrons is spasmodic and not continuous, but under the right conditions of applied voltage and distance between the electrodes it could be enough to initiate a continuous discharge.

Thermal Emission

Application of an electric field between two metal electrodes causes a few ions and electrons to be desorbed and is surface or thermal emission (see Chapter 7 for more information on thermal ionization). Unless the electrodes are heated strongly, the number of electrons emitted is very small, but, even at normal temperatures, this emission does add to the small number of electrons caused by cosmic radiation and is continuous.

Photoelectric Effect

If photons of light of a suitable wavelength (usually ultraviolet or x-rays) impinge on a metal surface, electrons are emitted. This effect is photoelectric (or photovoltaic) and can be used to start a flow of electrons in a discharge tube.

Piezoelectric Spark

By use of a piezoelectric device, as in a gas lighter, a small spark containing electrons and ions can be produced. If the spark is introduced into the gas in a discharge tube, it will provide the extra initial electrons and ions needed to start a continuous discharge. A plasma torch is frequently lit (started) in this fashion.

Given that some electrons and ions are present in the discharge gas from any of the previously described processes (cosmic radiation, thermal emission, photoelectric effect, piezoelectric spark), the applied voltage causes the charged species to drift toward the respective positive and negative electrodes, thereby constituting a small current flow. There is also another process that is important — some ions and electrons recombine to form neutral gas atoms again. Therefore, the electric current is the difference between the rate at which electrons and ions are produced and drift to the electrodes and the rate at which they disappear through electron/ion recombination. (Sometimes this occurs at the walls of the discharge vessel.) As the voltage increases, electrons and ions drift to the electrodes more rapidly and the current rises, seen in region A–B in Figure 6.8. At first, the relationship between the current flowing and the voltage applied is approximately in accord with Ohm's law. However, as the voltage is increased, the current begins to rise less in accord with Ohm's law until, at an applied electric field of about 10 V/m, there is no further increase in current because insufficient numbers of electrons and ions are formed to offset the drift to the electrodes and recombination. At this point, increasing the voltage does not increase the current, which is said to be saturated. This situation is represented by the straight section, B–C. The steady current is given by the equation, $I = \sqrt{(q/r)}$, in which q and r are, respectively, the rates at which electrons and ions are formed and then removed by recombination. This part of the discharge is not self-sustaining because stopping the initial production of electrons and ions ($q = 0$) leads to a shutdown in the total current flow (the discharge stops).

In the region A–B–C, there is no light emitted from the discharge. At the electric-field strengths used in this region, the discharge relies on the formation of ions and electrons by cosmic radiation, which is a spasmodic process, or by thermal emission; the discharge is spasmodic and non-self-sustaining. Irradiating the cathode with UV light improves the flow of electrons, as does heating the electrode. If the electric-field strength is made sufficiently high, even a spasmodic formation of electrons and ions may be enough to initiate a self-sustaining discharge (see below).

Region C–D–E

This region is often referred to as the Townsend breakdown region, in which — with little or no further change in voltage — the current can rise by several orders of magnitude, e.g., from 10^{-12} to 10^{-5} A. There is usually a spark produced during the initiation of this process. The current flow is controlled by the size of the resistance in the external voltage circuit.

The drift of electrons toward the anode under the influence of an electric field is not a simple process. Electrons are accelerated by the electric field and gain kinetic energy ($mu^2 = 2eV$). As the electrons gain speed, they collide with gas atoms and lose some of their kinetic energy through elastic collisions. The collision processes are such that some electrons are knocked sideways, some are even reversed in direction, and some are scarcely affected. Thus, it is necessary to think in terms of a spread in electron energies as the electrons drift toward the anode. In a complete vacuum, electrons would race in straight lines from the cathode to the anode at high speed, but in the discharge gas, the multiple collisions slow them considerably. For example, in a vacuum, a fall of 100 V would accelerate electrons to a speed of 4×10^7 cm/sec, but their drift velocities in a discharge gas are only about 10^3 to 10^4 cm/sec, about a thousand times slower. The distribution of electron energies in a drift situation is approximately of the Boltzmann type, as shown in Figure 6.9.

The electrons have a range of kinetic energies and are therefore at different temperatures. Depending on the strength of the applied electric field, some electrons in the swarm will have

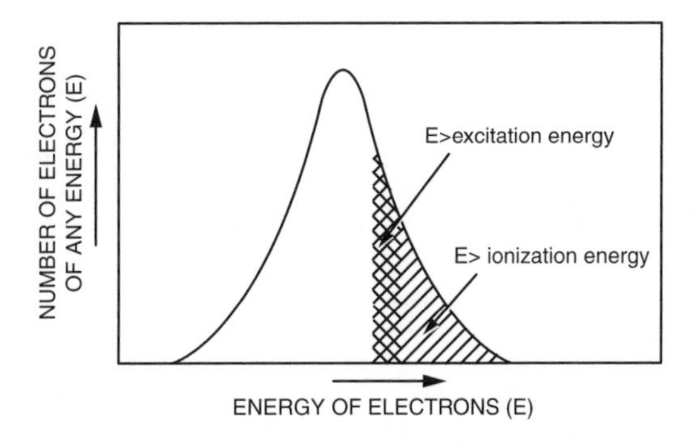

Figure 6.9
An idealized distribution of electron energies in an electron swarm drifting through a discharge gas. In this illustration, most electrons have not been accelerated sufficiently to cause any effect on neutral atoms upon collision. These encounters between an atom and an electron are elastic. A narrow band of electrons has the right energy to excite electrons in an atom upon collision (E > excitation energy). In such an inelastic collision, the incident electron loses kinetic energy and an electron in the impacted atom is raised to a higher orbital level (Figure 6.8). When this electron drops back to a lower energy orbital, a photon of light is emitted; the atom remains neutral throughout. Above an electron energy exceeding the ionization energy of a gas atom (E > ionization energy), an inelastic collision leads to ejection of an electron in the atom altogether, thereby leaving a positive ion. The extra electron produced in the collision causes an increase in the current flow in the discharge.

sufficient energy to cause ionization of neutral gas atoms — their energies are greater than the ionization energy of the discharge gas, and inelastic collisions between these energetic electrons and gas atoms form positive ions and more electrons (Figure 6.6). Thus, in this C–D region, as the external voltage is increased, more and more electrons gain sufficient kinetic energy to ionize gas atoms. Since one incident electron in collision with a neutral gas atom leads to the production of two electrons (plus one positive ion) leaving the collision site, these inelastic collisions start a cascade process, whereby more and more electrons are formed as the swarm drifts down the discharge tube. The current flow, once initiated, becomes self-sustaining in that as many new electrons are produced as reach the anode and are discharged.

At first, this ionization process only takes place near the anode, where some electrons will have gained sufficient energy to cause ionization. However, as the discharge builds up (section D–E), the electric-field gradient near the cathode becomes greater and greater due to the ionic space charge. Because of the steeper field gradient, electrons are accelerated more and more near the cathode and, therefore, ionization through collision begins earlier and earlier in the space between the electrodes. The cascade process — by which electrons produce more electrons by ionization of gas atoms, and then these electrons produce even more electrons — gives rise to a sharp increase in current with no increase in voltage.

Region E–F–G

In the region (D–E), the resistance in the external circuit across the electrodes can be reduced because the self-sustaining process produces sufficient electrons and ions to maintain the discharge. Near E, the discharge gas begins to glow slightly because some electrons have gained just enough energy to cause excitation of atoms but not enough to cause ionization (Figure 6.9). This band of electron energies is relatively narrow so there are few exciting collisions and, therefore, the glow is initially fairly faint. This faint glow is sometimes referred to as the subnormal glow. At this point, other processes begin to produce more and more electrons.

The extra sources of electrons that become important are known as secondary ionization processes and are caused by:

1. Bombardment of the cathode by incoming positive ions, which causes release of electrons
2. Irradiation of the cathode by the glow that starts (photoelectric release of electrons)
3. Impact of excited atoms onto the cathode (similar to the electron release caused by ions)

As these extra sources of electrons become more important, the glow increases and, along region F–G, it becomes steady and similar to that shown in Figure 6.4. (It is known as the normal glow.)

Region G–H–I

In the previous region (F–G), the glow covers only part of the cathode at F but the whole of the cathode at G. At this last point, the discharge has run out of efficient ways of generating electrons, and, for the current density to be maintained, more and more of the cathode is covered by the glow. The current density (= current flowing/area of cathode covered by the discharge = j) remains constant throughout the region F–G. If the voltage is now increased again, the current rises slowly, but for a large increase in electron flow to be obtained other ways of generating electrons have to be initiated. The glow begins to cover not only the cathode itself but also its supports and even the walls of the discharge tube. (This region is the abnormal glow.) Eventually, new methods of producing electrons begin to be effective (H in Figure 6.8). At this point, the voltage can be reduced because an arc strikes between the electrodes. The earlier glow becomes concentrated into a very bright spot on the cathode surface, and the positive column (plasma) glows brightly. The electric current flow increases by several orders of magnitude.

The new processes that cause this arcing to take place can be summarized as follows:

1. Positive ions bombarding the cathode produce more and more electrons and cause the electrode to heat. The heating causes thermal (surface) emission of electrons. For this process to be important, the temperature of the electrode needs to be high, but this high temperature can lead to it melting if the melting point is too low or if the heat generated cannot be dissipated rapidly. Thus, for an arc struck with carbon electrodes, the high melting point of carbon leads to the electrode becoming very hot and emitting a good supply of electrons, without melting, which is referred to as a hot discharge. Other arc discharges are cold. For low-melting-point electrode materials to produce an arc without melting, such as copper or mercury, an alternative process to thermal emission is necessary, which is discussed next.
2. In the arc discharge, the cathode and negative glows are so close to the cathode itself that their distances are of the order of 10^{-7} m. This closeness makes the electric-field strength very high (field strength = voltage/distance) at about 10^8 to 10^9 V/m. If there are small imperfections (points, edges, corners having very small radii of curvature) on the surfaces of the electrodes, these field strengths will increase again, making field ionization a new source of electrons (see Chapter 5 for more information on field ionization). The arc from a cold discharge tends to wander over the cathode surface, much as lightning tends not to strike the same place twice, whereas the arc from a hot discharge tends to remain anchored to one spot on the cathode surface.

In both low- and high-temperature arcs, the discharge begins as a very bright spot on the cathode. The area of the spot depends on the current flowing, but the current density can reach 10^{10} A/m^2. In the area of the spot, the temperature can reach about 2000 K for a hot discharge. At the anode, the temperature discharge is spread over the whole electrode, and the temperature of the latter is accordingly lower.

Overall Process

For a given gas at a given pressure lying between two metal electrodes separated by a given distance, the application of voltage to the electrodes initiates a series of events that is described as an electrical discharge. At low voltages, the discharge current is very small and is non-self-sustaining. At the breakdown region, a cascade of electrons is produced by collision processes between electrons and neutral gas atoms. This region is the corona/plasma region, where the discharge becomes self-sustaining. As the voltage continues to be increased, atoms also become excited sufficiently to emit light (a glow appears). With increasing voltage, the glow moves nearer to the cathode, and the positive column (the plasma) increases in length and glows more strongly, with one end approaching the cathode and the other almost touching the anode. At even higher voltages, more ionization processes begin, and the current flow becomes very high, a bright spot appearing on the cathode. At this stage an arc has been struck.

The various stages of this process depend critically on the type of gas, its pressure, and the configuration of the electrodes. (Their distance apart and their shapes control the size and shape of the applied electric field.) By controlling the various parameters, the discharge can be made to operate as a corona, a plasma, or an arc at atmospheric pressure. All three discharges can be used as ion sources in mass spectrometry.

Conclusion

Under suitable conditions of pressure and other factors, application of a suitable voltage between electrodes immersed in a gas causes a discharge of electric current through the gas. Different stages of the discharge are described as coronas, plasmas, and arcs, which differ according the conditions of gas pressure and the voltages under which they are produced. As well as emitting light, the various discharges contain both ions and electrons and can be used as ion sources in mass spectrometry.

Chapter 7

Thermal Ionization (TI), Surface Emission of Ions

Introduction

It has been known for many years that strongly heating a metal wire in a vacuum causes emission of electrons from the metal surface. This effect is important for thermionic devices used to control or amplify electrical current, but this aspect of surface emission is not considered here. Rather, the discussion here focuses on the effect of heating a sample substance to a high temperature on a metal wire or ribbon.

Placing a sample of any substance onto the surface of a metal (a filament) and then heating it strongly by passing an electric current through the filament in a vacuum causes positive ions and neutral species to desorb from its surface. Because of the high temperatures involved, only certain elements are useful for construction of filaments. Typically platinum, rhenium, tungsten, and tantalum are used because they are metallic and can be heated to temperatures of about 1000°C to over 2000°C without melting. A further important criterion for the filaments is that they should not readily react chemically with surrounding gas or with sample materials placed on them. Hot filaments are used in a high vacuum to facilitate sample evaporation and the subsequent manipulation of emitted ions and neutrals. This method automatically reduces interaction of the filaments with air or background vapors to a low level.

The use of mass spectrometry to measure the masses and abundances of any emitted positive ions is particularly useful in obtaining precise isotope ratios. Few negative ions are produced, and then only through secondary processes that can be suppressed electrically. Negative ions are not considered here. Samples examined by surface emission are almost always inorganic species because, at the high temperatures involved, any organic material is seriously degraded (thermolized) and will react with the filaments. At temperatures of 1000 to 2000°C, most inorganic substances yield positive ions without reacting with the filaments typically used in this procedure.

Positive ions are obtained from a sample by placing it in contact with the filament, which can be done by directing a gas or vapor over the hot filament but usually the sample is placed directly onto a cold filament, which is then inserted into the instrument and heated. The positive ions are accelerated from the filament by a negative electrode and then passed into a mass analyzer, where their m/z values are measured (Figure 7.1). The use of a suppressor grid in the ion source assembly reduces background ion effects to a very low level. Many types of mass analyzer could be used, but since very high resolutions are normally not needed and the masses involved are quite low, the mass analyzer can be a simple quadrupole.

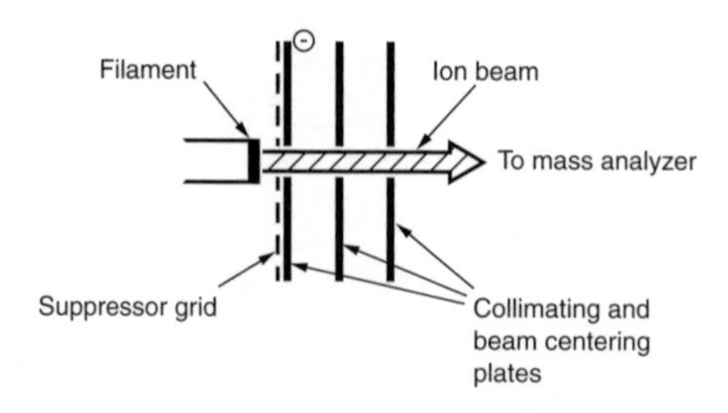

Figure 7.1
A typical filament assembly. Positive ions from the filament are accelerated by a high negative potential of about 1000 V placed on the first collimating plate. The other plates are used for further collimation and centering of the ion beam, which is directed into a suitable mass analyzer. Some positive ions strike the edge of the first collimating slit and produce secondary negative ions and electrons, which would be accelerated back onto the filament without a suppressor grid. This back-bombardment of the filament would lead to the formation of extraneous positive ions. To minimize this process, a suppressor grid (at a potential of about –300 V with respect to the filament) is included to deflect any backscattered ions.

The ion current resulting from collection of the mass-separated ions provides a measure of the numbers of ions at each m/z value (the ion abundances). Note that for this ionization method, all ions have only a single positive charge, z = 1, so that m/z = m, which means that masses are obtained directly from the measured m/z values. Thus, after the thermal ionization process, m/z values and abundances of ions are measured. The accurate measurement of relative ion abundances provides highly accurate isotope ratios. This aspect is developed more fully below.

High Filament Temperatures

The high temperatures necessary to produce ions rapidly vaporize (evaporate) and thermally destroy organic substances. Consequently, this surface ionization technique is generally not used to investigate them. Inorganic substances are generally much more stable thermally but also much less volatile. Although an inorganic sample can be changed upon heating — consider, for example, the formation of calcium oxide from calcium carbonate — the inorganic or metal (elemental) parts of such samples are not destroyed. For example, if a sample of a cesium salt were to be examined, the anionic portion of the sample might well be changed on the hot filament, but the cesium atoms themselves would remain and would still be desorbed as $Cs^{\bullet+}$. Sometimes, the desorbed ions appear as oxide or other species, as with $GdO^{\bullet+}$.

A further consequence of the high temperatures is that much of the sample is simply evaporated without producing isolated positive ions. There is a competition between formation of positive ions and the evaporation of neutral particles. Since the mass spectrometer examines only isolated charged species, it is important for maximum sensitivity that the ratio of positive ions to neutrals be as large as possible. Equation 7.1 governing this ratio is given here.

$$n^+/n^0 = Ae^{-(I-\phi)/kT} \tag{7.1}$$

In Equation 7.1, n^+/n^0 is the ratio of the number of positive ions to the number of neutrals evaporated at the same time from a hot surface at temperature T (*K*), where *k* is the Boltzmann constant and *A* is another constant (often taken to be 0.5; see below). By inserting a value for *k* and adjusting Equation 7.1 to common units (electronvolts) and putting *A* = 0.5, the simpler Equation 7.2 is obtained.

$$n^+/n^0 = 0.5e^{11,600(\phi-I)/T} \tag{7.2}$$

Element	Ionization Energy (eV)
Aluminium	5.98
Calcium	6.11
Carbon	11.26
Caesium	3.89
Copper	7.72
Gold	9.22
Lanthanum	5.61
Lead	7.42
Lithium	5.39
Rubidium	4.18
Strontium	5.69
Thorium	6.95
Uranium	6.08

Figure 7.2
The table lists first ionization energies (electronvolts) for some commonly examined elements. Because only singly charged ions are produced by surface emission from a heated filament, only first ionization energies are given, viz., those for M^+ and not for higher ionization states in which more than one electron has been removed. Note that most elemental ionization energies fall in the range of about 3–12 eV.

The expression $I - \phi$ in Equations 7.1 and 7.2 is the difference between the ionization energy (I, sometimes known as the ionization potential) of the element or neutral from which the ions are formed and the work function (ϕ or sometimes W) of the metal from which the filament is made. The ionization energy and the work function control the amount of energy needed to remove an electron from, respectively, a neutral atom of the sample and the material from which the filament is constructed. The difference between I and ϕ governs the ease with which positive ions can be formed from sample molecules lying on the filament. Both I and ϕ are positive and are frequently reported in units of electronvolts (eV). Their importance is discussed in greater detail below. Some typical values for ionization energies and work functions are given in Figures 7.2 and 7.3.

Thermochemistry of Surface Emission

Adsorption of a neutral (n^0) onto a metal surface leads to a heat of adsorption of Q_a, as the electrons and nuclei of the neutral and metal attract or repel each other. Partial positive and negative charges are induced on each with the formation of a dipolar field (Figure 7.4).

Similarly, adsorption of ions (n^+) onto a metal surface leads to a heat of adsorption of Q_i. Generally, Q_i is about 2–3 eV and is greater than Q_a, which itself is about 1 eV. The difference between Q_i and Q_a is the energy required to ionize neutrals (n^0) on a metal surface so as to give ions (n^+) or vice versa. This difference, $Q_i - Q_a$, can be equal to, greater than, or less than the difference, $I - \phi$, between the ionization energy (I) of the neutral and the ease with which a metal can donate or accept an electron (the work function, ϕ). Where $Q_i - Q_a > I - \phi$, the adsorbed

Element	Work function (eV)	Melting point (K)
Platinum	6.2	2028
Rhenium	4.8	3440
Tantalum	4.2	3120
Tungsten	4.5	3640

Figure 7.3
Values of the average work function (ϕ, electronvolts) for the commonly used filament metals. The melting points of the metals are also shown to give some guidance as to the maximum temperature at which they can be used. Normally, the practical maximum would lie a few hundred degrees below the melting point to prevent sagging of the filament.

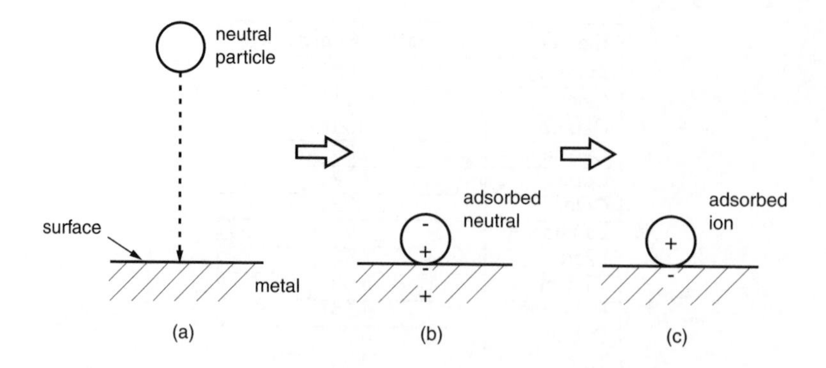

Figure 7.4
Schematic diagram showing the development of a dipolar field and ionization on the surface of a metal filament. (a) As a neutral atom or molecule approaches the surface of the metal, the negative electrons and positive nuclei of the neutral and metal attract each other, causing dipoles to be set up in each. (b) When the neutral particle reaches the surface, it is attracted there by the dipolar field with an energy Q_a. (c) If the values of I and ϕ are opposite, an electron can leave the neutral completely and produce an ion on the surface, and the heat of adsorption becomes Q_i. Similarly, an ion alighting on the surface can produce a neutral, depending on the values of I and ϕ. On a hot filament the relative numbers of ions and neutrals that desorb are given by Equation 7.1,which includes the difference, $I - \phi$, and the temperature, T.

particle will be an ion, whether it was originally a neutral particle or an ion that approached the metal surface. Similarly, if $Q_i - Q_a < I - \phi$, the adsorbed particle will be a neutral species, regardless if it was an ion or a neutral that approached the metal surface. If energy is now added to the system by strongly heating the filament, desorption of ions and neutrals occurs. Clearly, the numbers of desorbing neutral particles and ions must depend on the size and sign of the difference in $(I - \phi)$ and on the added energy, which is controlled by the absolute temperature, T. As shown in Figure 7.5, the critical value for desorption of ions and neutrals at any given temperature is governed by the relation, $Q_a = Q_i - (I - \phi)$. This criterion is used in the derivation of Equation 7.1.

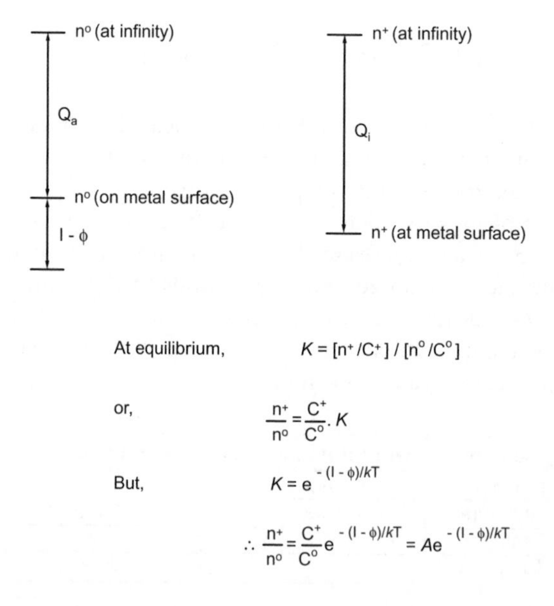

Figure 7.5
On bringing a neutral (n^0) from infinity to the surface of a filament metal, it adsorbs with an energy Q_a (Figure 7.4). Similarly, bringing an ion (n^+) from infinity leads to a heat of adsorption of Q_i. If the difference between Q_a and Q_i is equal to $I - \phi$, the neutral particle can desorb again as a neutral or desorb as an ion, depending on the value of $I - \phi$. Similar arguments apply to an adsorbed ion (n^+), which can desorb again as an ion or desorb as a neutral.

On the surface of a heated filament metal, whether ions or neutrals are adsorbed initially, an equilibrium will be set up between them, with equilibrium constant K (shown in Figure 7.5). For the equilibrium constant, surface concentrations of desorbing neutrals (n^0) and ions (n^+) must be used. The surface concentration of ions is the proportion of ions actually desorbing to the total number on the surface (C^+), viz., the concentration of desorbing ions = $[n^+/C^+]$. Similarly, the concentration of desorbing neutral particles = $[n^0/C^0]$ (Figure 7.5). In the well-known thermodynamic equation that governs an equilibrium process, $\ln K = -\Delta G/RT$, the gas constant R can be replaced by its equivalent (the Boltzmann constant, k) and the total free energy change (ΔG) by the critical energy change ($I - \phi$). From these substitutions, the expression, $K = \exp[-(I - \phi)/kT]$, is obtained. Combining this expression with the expression for K shown in Figure 7.5, Equation 7.1 is revealed.

Therefore, the ratio of the number of ions to the number of neutrals desorbing from a heated filament depends not only on the absolute temperature but also on the actual surface coverage of ions and neutrals on the filament (C^+, C^0) and crucially on the difference between the ionization energy and work function terms, I and ϕ. This effect is explored in greater detail in the following illustrations.

Both of the terms I and ϕ have positive values. Examination of Equation 7.1 or 7.2 reveals that, for $\phi > I$, then $\phi - I$ is positive, and the proportion of positive ions to neutrals diminishes with increasing temperature. For example, with a sample of cesium (ionization energy, 3.89 eV) on a tungsten filament (work function, 4.5 eV) at 1000 K, the ratio of $n^+/n^0 = 591$. Thus, for every cesium atom vaporized, some 600 atoms of Cs^+ ions are produced. At 2000 K, the ratio of n^+/n^0 becomes 17, so only about 20 ions of cesium are evaporated for every Cs atom (Figure 7.6a). For $\phi < I$, as with lead ($I = 7.42$ eV) on tantalum ($\phi = 4.2$ eV), the corresponding figures are 3×10^{-16} at 1000 K and 1×10^{-8} at 2000 K (Figure 7.6b).

Clearly, the lower the ionization energy with respect to the work function, the greater is the proportion of ions to neutrals produced and the more sensitive the method. For this reason, the filaments used in analyses are those whose work functions provide the best yields of ions. The evaporated neutrals are lost to the vacuum system. With continued evaporation of ions and neutrals, eventually no more material remains on the filament and the ion current falls to zero.

Changing the Work Function (Activators)

For an element of ionization energy I, Equation 7.2 shows that at any given temperature, the work function of the surface from which particles are emitted is clearly crucial to the proportion of ions produced in relation to the number of neutrals. As $I - \phi$ changes from negative to positive, the ion yield becomes progressively smaller. Figure 7.3 indicates that platinum would be the filament metal of choice in most applications because it has the biggest work function of the four metals commonly used (Pt, Re, W, or Ta). However, platinum also has the lowest melting point, and to reach the high temperatures needed to effect suitable evaporation rates, it may be necessary to use a metal such as tantalum or rhenium, for which the work functions are smaller. Thus, there is a trade-off between work function and temperature in maximizing ion yield.

For difficult cases, this dilemma can be solved by using activators on the surface of the filaments. The activators commonly used are colloidal or very finely dispersed (high surface area) silicon dioxide or carbon. These substances are much more electronegative than the filament metal and produce a dipolar field (Figure 7.7). This field induces a positive image charge in the filament surface, thereby making removal of electrons more difficult and increasing its effective work function. Since ϕ is increased, the difference $I - \phi$ must change and, therefore, the ion yield ratio n^+/n^0. Activators are used to improve ion yield when examining metals of high ionization energy, as with uranium, lead, or plutonium on tungsten filaments.

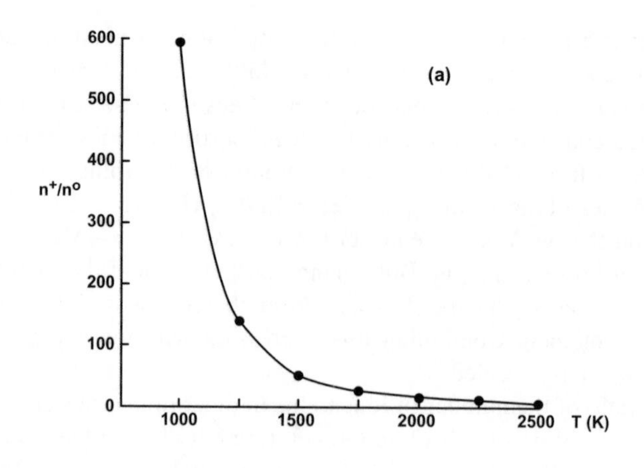

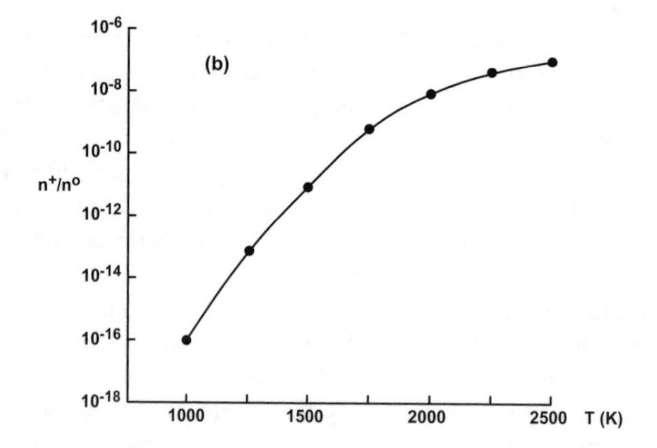

Figure 7.6
Graphs showing the influence of work function and ionization energy on the efficiency of ionization. Using Equation 7.1, the ratio n^+/n^0 was calculated for uranium (I = 6.08 eV) on either (a) a platinum filament (ϕ = 6.2 eV) or (b) a rhenium filament (ϕ = 4.8 eV) at different temperatures. For platinum (a), a good yield of ions is obtained, but the ratio n^+/n^0 falls with increasing temperature. For rhenium (b), the relative ion yield is small but increases with increasing temperature. The best ion yields are given by the use of platinum with uranium, for which $I - \phi$ is negative by about 0.1 eV.

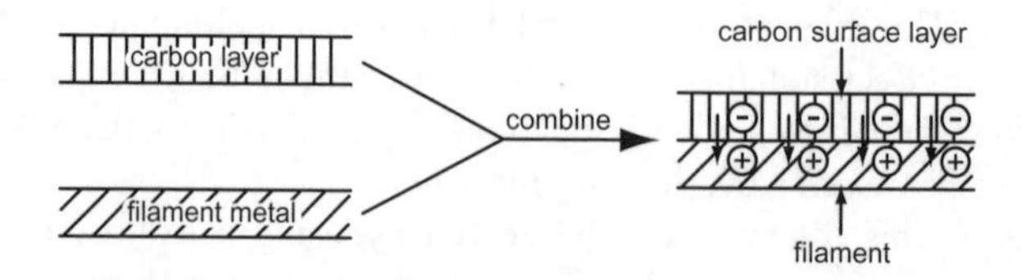

Figure 7.7
Schematic diagram showing how placing a thin layer of highly dispersed carbon onto the surface of a metal filament leads to an induced dipolar field having positive and negative image charges. The positive side is always on the metal, which is much less electronegative than carbon. This positive charge makes it much more difficult to remove electrons from the metal surface. The higher the value of a work function, the more difficult it is to remove an electron. Effectively, the layer of carbon increases the work function of the filament metal. Very finely divided silicon dioxide can be used in place of carbon.

Amount of Sample

The rate of evaporation of ions from a heated surface is given by Equation 7.3, in which Q_i is the energy of adsorption of ions on the filament surface (usually about 2–3 eV) and C_i is the surface density of ions on the surface (a complete monolayer of ions on a filament surface would have a surface density of about 10^{15} ions/cm^{-2}).

$$n^+ = C_i e^{-Q_i/kT} \quad (7.3)$$

Similarly, the rate of evaporation of neutral species from a filament surface is given by Equation 7.4, in which C_0 is the surface density of atoms on the surface (a complete monolayer of atoms would have a surface density of about 10^{15} atoms/cm^{-2}).

$$n^0 = C_0 e^{-(Q_i - \phi + I)/kT} \quad (7.4)$$

Dividing Equation 7.3 by 7.4 yields Equation 7.1, in which $A = C_i/C_0$.

As ions and neutrals evaporate from a heated filament surface, the amount of sample decreases and the surface densities (C_i, C_0) must decrease. Therefore, Equation 7.1 covers two effects. The first was discussed above and concerns the changing value for the ratio n^+/n^0 as the temperature of the filament is varied, and the other concerns the change in the total number of ions desorbing as the sample is used up. The two separate effects are shown in Figure 7.8a,b. Combining the two effects (Figure 7.8c) reveals that if the temperature is increased to maintain the flow of ions, which drops naturally as the sample is used up (time), then eventually the flow of ions and neutrals becomes zero whatever the temperature of the filament because the sample has disappeared from the filament surface.

Measurement of Ratios of Isotopic Abundances

For any one ion type (e.g., Cs^+), measurement of its abundance in a sample requires the sample to be evaporated over a period of time. The total yield of ions is obtained by integrating the area under the ion-yield curve (Figure 7.8c).

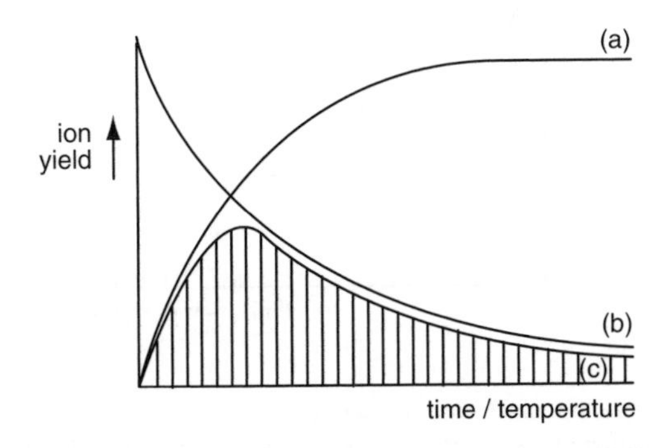

Figure 7.8
(a) The effect of increasing temperature on the ion yield, which increases as the temperature rises. (b) The effect of decreasing surface coverage of the filament surface as ions and neutrals evaporate; as the surface densities of ions and neutrals decrease, the ion yield falls off with time. (c) An example of the shape of a curve resulting from the two effects. As the temperature of the surface is increased to improve ion yield, the surface is depleted of sample more and more rapidly until no sample remains and therefore no ion current. The area under curve (c) represents the total ion yield.

Generally, ratios of isotopic abundances need to be obtained and not individual total ion yields. Experimentally, for two isotopes M_1 and M_2, obtaining the ratios entails the simultaneous measurement of their abundances as given by the ion current for the two masses arriving at the ion collector. For two isotopes, the ion yields are given by Equations 7.5 and 7.6, which are obtained simply from Equation 7.1 by inserting the relevant values for C_i, C_0, and Q. Not only are C_i and C_0 different (because the relative amounts of isotopes are different), but they vary with time, as discussed above. Dividing Equations 7.5 by 7.6 gives Equation 7.7, from which it is clear that at any given temperature, since the ratio of C_1 to C_2 changes with time, the ratio of ion yields for isotopes M_1 and M_2 must change with time.

$$n_1^+/n_1^0 = C_1 e^{-(Q_1 - \phi - I_1)/kT} \tag{7.5}$$

$$n_2^+/n_2^0 = C_2 e^{-(Q_2 - \phi - I_2)/kT} \tag{7.6}$$

$$(n_1+/n_1^0)/(n_2^+/n_2^0) = (C_1/C_2)e^{-(\Delta Q + \Delta I)/kT} \tag{7.7}$$

Figure 7.9 shows a schematic representation of this effect, in which the ratio of the two isotopes changes with time. To obtain an accurate estimate of the ratio of ion abundances, it is better if the relative ion yields decrease linearly (Figure 7.9) which can be achieved by adjusting the filament temperature continuously to obtain the desired linear response. An almost constant response for the isotope ratio can be obtained by slow evaporation of the sample, viz., by keeping the filament temperature as low as is consistent with sufficient sensitivity of detection (Figure 7.9).

The previous discussion demonstrates that measurement of precise isotope ratios requires a substantial amount of operator experience, particularly with samples that have not been examined previously. A choice of filament metal must be made, the preparation of the sample on the filament surface is important (particularly when activators are used), and the rate of evaporation (and therefore temperature control) may be crucial. Despite these challenges, this method of surface ionization is a useful technique for measuring precise isotope ratios for multiple isotopes. Other chapters in this book discuss practical details and applications.

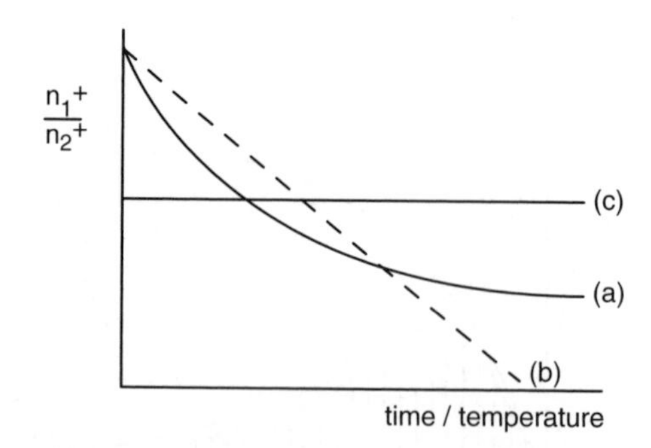

Figure 7.9

Schematic illustrations of the effect of temperature and surface density (time) on the ratio of two isotopes. (a) shows that, generally, there is a fractionation of the two isotopes as time and temperature change; the ratio of the two isotopes changes throughout the experiment and makes difficult an assessment of their precise ratio in the original sample. (b) illustrates the effect of gradually changing the temperature of the filament to keep the ratio of ion yields linear, which simplifies the task of estimating the ratio in the original sample. The best method is one in which the rate of evaporation is low enough that the ratio of the isotopes is virtually constant; this ratio then relates exactly to the ratio in the original sample.

Conclusion

Precise measurement of isotope ratios can be obtained by comparing the yields of isotopic ions desorbing from a sample placed on a strongly heated filament that is generally made from platinum, tantalum, rhenium, or tungsten.

Chapter 8

Electrospray Ionization (ESI)

Introduction

For a more detailed description of the ionization process inherent in electrospray, please see Chapter 9, which discusses atmospheric pressure ionization (API). The reader also should compare electrospray with thermospray (see Chapter 11).

In many applications in mass spectrometry, the sample to be analyzed is present as a solution in a solvent that could be organic (as with methanol or acetonitrile) or aqueous (as with body fluids). The solution could also be an effluent from a liquid chromatography (LC) column. In any case, a solution must flow into the front end of a mass spectrometer (MS), but, before it can provide a mass spectrum, the bulk of the solvent must be removed without losing the sample (solute). If the solvent were not removed, then its vaporization as it entered the vacuum of the mass spectrometer would produce a large increase in pressure and stop the instrument from working. At the same time that this excess of solvent is removed, the dissolved sample must be retained so that its mass spectrum can be measured, viz., there must be differentiation between solvent and solute (sample) molecules. There are several means of effecting this differentiation between carrier solvent and the solute of interest, and electrospray is just one of them. However, there is an additional important consideration in electrospray. Unlike the other methods of introducing a liquid into a mass spectrometer, electrospray frequently produces multicharged ions that make accurate measurement of large masses easier and gives this inlet/ion source a considerable advantage in areas such as peptide and protein research (see below).

One of the first successful techniques for selectively removing solvent from a solution without losing the dissolved solute was to add the solution dropwise to a moving continuous belt. The drops of solution on the belt were heated sufficiently to evaporate the solvent, and the residual solute on the belt was carried into a normal EI (electron ionization) or CI (chemical ionization) ion source, where it was heated more strongly so that it in turn volatilized and could be ionized. However, the moving-belt system had some mechanical problems and could be temperamental. The more recent, less-mechanical inlets such as electrospray have displaced it. The electrospray inlet should be compared with the atmospheric-pressure chemical ionization (APCI) inlet, which is described in Chapter 9.

Differential Solvent Removal

A sample for which a mass spectrum is required may well be dissolved in an organic or aqueous solvent. For example, in searching for drugs in blood plasma, the plasma itself may be investigated (aqueous) or its active components may be first extracted into an organic solvent such as dichloromethane. Alternatively, the sample can first be separated into its components by passage through a liquid chromatographic instrument (see Chapter 37); upon emerging from the column, the sample of interest is present as a solution in the solvents used in the chromatography. In either case, the sample to be examined is in solution and cannot be put straight into a mass spectrometer without first removing most of the solvent and without, of course, removing the dissolved sample also!

Electrospray is one method for effecting this differential solvent removal. The solution is passed along a short length of stainless steel capillary tube, to the end of which is applied a high positive or negative electric potential, typically 3–5 kV (Figure 8.1). When the solution reaches the end of the tube, the powerful electric field causes it to be almost instantaneously vaporized (nebulized) into a jet or spray of very small droplets of solution in solvent vapor. Spraying efficiency can be increased by flowing a gas past the end of the charged capillary tube. Before entering the mass spectrometer proper, this mist of droplets flows through an evaporation chamber that can be heated slightly to prevent condensation. As the droplets move through this region, solvent evaporates rapidly from the surfaces and the droplets get smaller and smaller. In addition to producing the spray, this method of rapid vaporization leaves no time for equilibrium to be attained, and a substantial proportion of the droplets have an excessive positive or negative electrical charge on their surfaces. Thus as the droplets get smaller, the electrical surface charge density increases until the natural repulsion between like charges causes the release of ions and neutral molecules. The end of the capillary tube is aimed at a small hole (target) at the opposite end of this evaporation region. After vaporizing from the surface of a droplet, solvent molecules of low molecular mass quickly and conveniently diffuse away from the line-of-sight trajectory to the inlet target. A Z-spray ion source operates slightly differently (see Chapter 10).

Sample molecular ions and cluster ions have much greater molecular mass (and therefore momentum) than those of the solvent and tend to carry straight on toward the target at the end of the inlet region (Figure 8.1). To assist evaporation of the droplets and the breaking up of unwanted cluster ions, a drying gas (nitrogen) flows along and past the end of the capillary (Figure 8.1). If the gas is arranged to flow between the counter electrode and the nozzle, it is

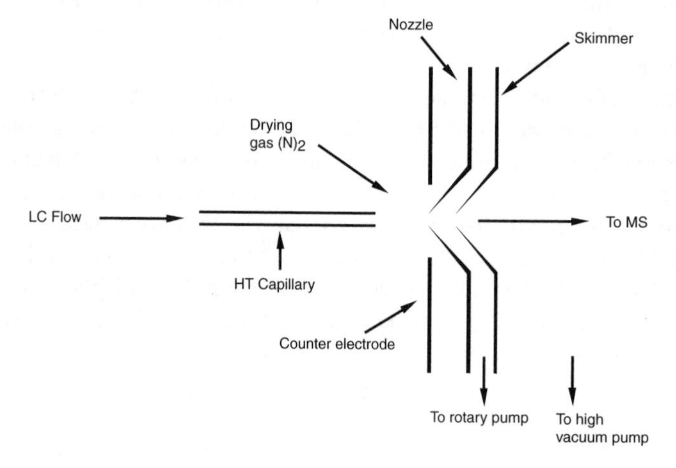

Figure 8.1
Schematic diagram of an electrospray inlet/ion source. A spray produced from the high electrical voltage (HT) on the capillary moves toward a hole in the electrical counter electrode. After removal of much solvent, sample ions continue under their momentum through the hole and then through the nozzle and skimmer, where most remaining solvent is removed.

sometimes referred to as a curtain gas. At the target hole, the heavier ions pass through, but most of the lighter solvent molecules have by this time diffused away and thus do not pass through. In effect, the device is a momentum separator between solvent and solute (sample) molecules. After passing through this hole, the ions pass through two evacuated regions via a nozzle and a skimmer (Figure 8.1). These conically shaped holes refine the separation of sample ions from solvent ions, still mainly on the basis of momentum but also by an extraction and focusing effect of electrical potentials applied to the nozzle and skimmer. Finally, sample ions pass into the analyzer of the mass spectrometer, where their mass-to-charge (m/z) ratios are measured in the usual way. The mass analyzer can be of any type.

The result of the above process means that sample molecules dissolved in a solvent have been extracted from the solvent and turned into ions. Therefore, the system is both an inlet and an ion source, and a separate ion source is not necessary.

The ions passing into the mass spectrometer analyzer from electrospray have little excess internal energy and therefore not enough energy to fragment. Many of the ions are of the form $[M + X]^+$ or $[M - H]^-$, with X representing hydrogen or some other element, such as sodium or potassium. While these quasi-molecular ions are an excellent source of accurate molecular mass information — which may be all that is required — they give little or no information concerning the actual molecular structure of the substance being investigated, which is provided by fragment ions. This situation is entirely analogous to the problem with simple chemical ionization, and similar solutions are available. To give the quasi-molecular ions the extra energy needed to induce fragmentation, they can be passed through a collision gas and the resulting spectra analyzed for metastable ions (MS/MS methods). An alternative arrangement uses the potential difference between the electrodes (cone voltage) to accelerate the ions. If the voltage difference is increased, collisions between the faster moving ions and neutral molecules lead to fragmentation, as in CI. If the cone voltage is reduced, the ions slow and the resulting collisions have insufficient energy for fragmentation.

Multicharged ions

Another type of ion is formed almost uniquely by the electrospray inlet/ion source which makes this technique so valuable for examining substances such as proteins that have large relative molecular mass. Measurement of m/z ratios usually gives a direct measure of mass for most mass spectrometry because z = 1 and so m/z = m/1 = m. Values of z greater than one are unusual. However, for electrospray, values of z greater than one (often much greater), are quite commonplace. For example, instead of the $[M + H]^+$ ions common in simple CI, ions in electrospray can be $[M + n{\cdot}H]^{n+}$ where n can be anything from 1 to about 30.

Thus the m/z value for such ions is $[M + n{\cdot}1]/n$, if the mass of hydrogen is taken to be one. As a particular example, suppose M = 10,000. Under straightforward CI conditions, $[M + H]^+$ ions will give an m/z value of 10,001/1 = 10,001, a mass that is difficult to measure with any accuracy. In electrospray, the sample substance can be associated with, for example, 20 hydrogens. Now the ion has a mass-to-change ratio of $[M + 20{\cdot}H]^{20+}$ and therefore m/z = 10,020/20 = 501. This mass is easy to measure accurately with a wide range of instruments.

Normally, a range of values for n is found, each molecule (M) giving a series of multicharged ions. For example, a series

$$[M + nH]^{n+}, [M + (n + 1)H]^{(n + 1)+}, \ldots , [M + (n + a)H]^{(n + a)+}, \ldots$$

might be observed, each successive quasi-molecular ion having one more hydrogen and one more electrical charge than the preceding one (Figure 8.2). The only difficulty lies in knowing the value of n! Fortunately, this value is relatively easy to extract from the mass spectrum (Figure 8.3).

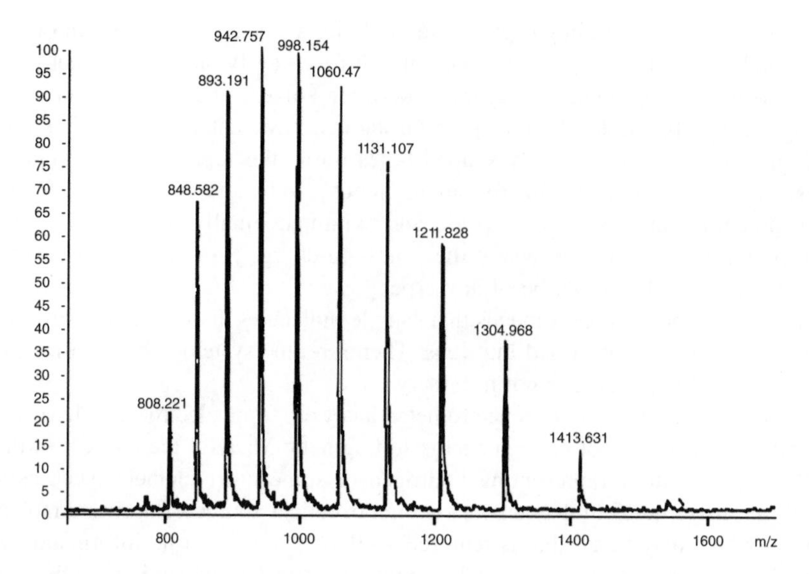

Figure 8.2
Quasi-molecular ions, $[M + nH]^+$, from a protein (myoglobin) of molecular mass 16,951.5 Da. In this case, n ranges from 21 (giving a measured mass of 808.221) to 12 (corresponding to a measured mass of 1413.631). The peaks with measured masses in between these correspond to the other values of n between 12 and 21. By taking successive pairs of measured masses, the relative molecular mass of the myoglobin can be calculated very accurately, as shown in Figure 8.4.

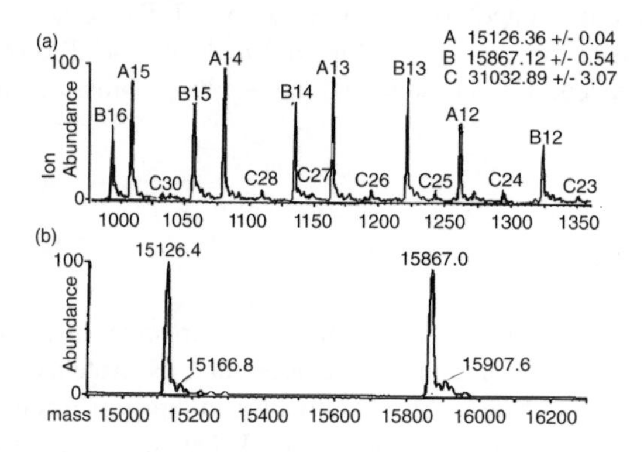

Figure 8.3
Calculation of true mass (M) from measured mass-to-charge ratios $m_1, m_2, \ldots, m_r$.

The derivation has been fully automated to such an extent that the complex of quasi-molecular ions ($m_1, m_2, \ldots, m_r$) measured as shown in Figure 8.3 can now be presented simply by the data system as a transformed spectrum with one molecular ion, M. Such an example is illustrated in Figure 8.4, where the electrospray mass spectrum of hemoglobin, containing nine main measured masses, transforms into a much simpler looking spectrum having two main molecular ions (M_1 and M_2) arising from its alpha and beta chains; close inspection also reveals the molecular ion of the gamma chain.

Uses of Electrospray

Although simple solutions can be examined by these electrospray techniques, often for a single substance dissolved in a solvent, straightforward evaporation of the solvent outside the mass spectrometer with separate insertion of the sample is sufficient. This situation is not true for all substances. Peptides, proteins, nucleotides, sugars, carbohydrates, mass organometallics, and many

Let the unknown mass be M and let the number of unknown charges be n, corresponding to the addition of n protons (n.H+):

For two successive measured mass-to-charge ratios m^1 and m^2 two equations can be written,

$m^1 = (M+n)/n$ (1)

$m^2 = (M+n+1)/(n+1)$ (2)

assuming a value of 1 for the atomic mass of hydrogen.

Equations (1) and (2) are simultaneous and can be solved easily to give M and n.

For a set of successive measured mass-to-charge ratios m^1, m^2.....mr any successive pair can be chosen to calculate M. The best value for M will be obtained by averaging all the values individually calculated from all successive pairs. Thus, if m1 and m^2 gave M' and m^2 and m^3 gave M", a best value for M would be M = (M'+M")/2. Clearly, by averaging a lot of pairs a more accurate value for M can be found.

Figure 8.4

Positive-ion electrospray mass spectrum of human hemoglobin: (a) as initially obtained with all the measured masses, and (b) after calculation of true mass, as in Figure 8.3. The spectrum transforms into two main peaks representing the main alpha and beta chains of hemoglobin with accurate masses as given. This transformation is fully automated. The letters A, B, C refer to the three chains of hemoglobin. Thus, A13 means the alpha chain with 13 protons added.

polar substances, once isolated from solution as solids, are then quite difficult to vaporize in a standard EI or CI ion source without thermally damaging them. For such substances it is best to leave them in solution and obtain their electrospray mass spectra. Electrospray ionization is convenient for most classes of compound, and it is wrong to think of the technique as just an interface for coupling LC to MS. It can be used equally well for single substances by first dissolving them in a suitable solvent and then passing the solution into the electrospray inlet. The ability of electrospray to measure accurately the mass of very large molecules makes it extremely valuable for a wide variety of applications, particularly biochemical and medical.

For mixtures the picture is different. Unless the mixture is to be examined by MS/MS methods, usually it will be necessary to separate it into its individual components. This separation is most often done by gas or liquid chromatography. In the latter, small quantities of emerging mixture components dissolved in elution solvent would be laborious to deal with if each component had to be first isolated by evaporation of solvent before its introduction into the mass spectrometer. In such circumstances, the direct introduction, removal of solvent, and ionization provided by electrospray is a boon and puts LC/MS on a level with GC/MS for mixture analysis. Further, GC is normally concerned with volatile, relatively low-molecular-weight compounds and is of little or no use for the many polar, water soluble, high-molecular-mass substances such as the peptides, proteins, carbohydrates, nucleotides, and similar substances found in biological systems. LC/MS with an electrospray interface is frequently used in biochemical research and medical analysis.

General Comments

Electrospray alone is a reasonably sensitive technique for use with many classes of compounds. Spectacular, unprecedented results have been obtained with accurate mass measurement of high-molecular-mass proteins, but those results should not be allowed to obscure the fact that electrospray is equally well suitable for much smaller molecules and for use as a routine atmospheric-pressure inlet/ionization system for a mass spectrometer. As an atmospheric-pressure ionization method, it is useful to compare it with atmospheric-pressure chemical ionization. Both ES and APCI work well with solutions of thermally sensitive polar molecules, such as peptides, proteins, sugars, and oligomuclestides. Whereas ES gives multicharged ions, $[M + nH]^{n+}$, APCI gives ions similar to those found in chemical ionization, $[M + H]^+$ or $[M + X]^+$.

Conclusion

Nebulizing a solution by a strong electric field produces a spray of small charged droplets from which the solvent can be removed, leaving sample ions to pass straight into the analyzer region of a mass spectrometer. By producing multicharged ions, electrospray is extremely useful for accurate mass measurement, particularly for thermally labile, high-molecular-mass substances.

Chapter 9

Atmospheric-Pressure Ionization (API)

Background

Electron ionization and then also chemical ionization were the principal methods for producing ions for mass spectrometry. Both techniques need to operate in the gas phase, so any sample of interest must be vaporized into the ion source. While vaporization of a sample is satisfactory for many substances, there are also large numbers of them that cannot be vaporized without thermal decomposition taking place. For example, sugars, peptides, proteins, nucleosides, and so on fall into this category. In the same way, these sorts of substances will not pass through a gas chromatographic column. Although methods have been developed for derivatizing nonvolatile and thermally labile substances to make them volatile, they necessitate extra steps and, even so, some substances remain labile or nonvolatile (e.g., proteins). High-performance liquid chromatography (HPLC) as a complement to gas chromatography removed these analytical problems, since no volatilization was needed, but the problem remained of getting these less-volatile substances into a mass spectrometer.

The advent of atmospheric-pressure ionization (API) provided a method of ionizing labile and nonvolatile substances so that they could be examined by mass spectrometry. API has become strongly linked to HPLC as a basis for ionizing the eluant on its way into the mass spectrometer, although it is also used as a stand-alone inlet for introduction of samples. API is important in thermospray, plasmaspray, and electrospray ionization (see Chapters 8 and 11).

The Ionization Process

Ion Evaporation

When a stream of liquid (solvent) containing a sample (substrate) of interest is sprayed from the end of a narrow tube (nebulized) into a larger chamber, a mist of small droplets is produced, which drift along the chamber (Figure 9.1). A proportion of the droplets carry an excess of positive or negative electric charge. The droplets have a large surface-area-to-volume ratio, and solvent evaporates quickly so the droplets get smaller and the electric charge density increases. At some point, the mutual repulsion between like charges becomes so great that it exceeds the

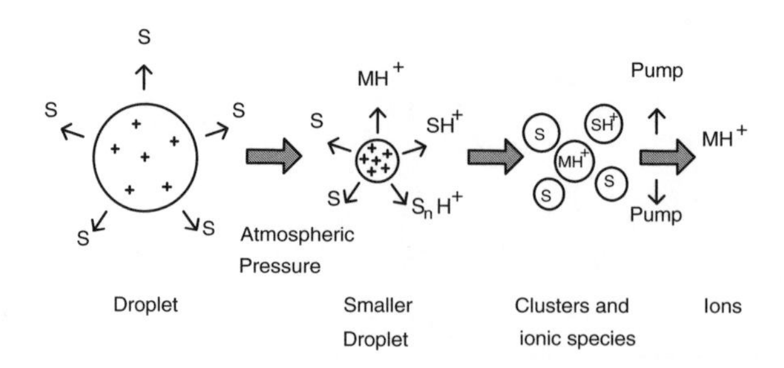

Figure 9.1
After being formed as a spray, many of the droplets contain some excess positive (or negative) electric charge. Solvent (S) evaporates from the droplets to form smaller ones until, eventually, ions (MH^+, SH^+) from the sample M and solvent begin to evaporate to leave even smaller drops and clusters (S_nH^+; n = 1, 2, 3, etc.). Later, collisions between ions and molecules (CI) leave MH^+ ions that proceed into the mass analyzer. Negative ions are formed similarly.

forces of surface tension, and ions begin to leave the droplet as well as the neutral solvent molecules. This process is called ion evaporation. The ions can leave the droplets alone or associated with solvent molecules in clusters (Figure 9.1). As the mixture of ions and ion clusters approaches the mass spectrometer, further evaporation occurs, and another mode of ionization becomes important.

Chemical Ionization (CI)

Further explanation of this CI process can be found in Chapter 1. Briefly, CI results from collision between sample molecules and specially produced reagent gas ions such that ions are formed from sample molecules by various processes, one of the most important of which is the transfer of a proton (H^+, Figure 9.2).

Since ions and neutral molecules are formed close together in an API source, many ion/molecule collisions occur as in CI, and so the ion evaporation process also has impressed upon it the characteristics of CI. Therefore, API is usually thought to involve a mix of ion evaporation and chemical ionization.

The mix of ions, formed essentially at or near ambient temperatures, is passed through a nozzle (or skimmer) into the mass spectrometer for mass analysis. Since the ions are formed in the vapor phase without having undergone significant heating, many thermally labile and normally nonvolatile substances can be examined in this way.

Drying Gas

To aid the evaporation of the droplets, a flow of a gas (often nitrogen) is directed across them. This drying gas helps to reduce the number of cluster ions (Figure 9.3).

$$SH^+ + M \longrightarrow S + MH^+$$

Figure 9.2
An example of proton (H^+) transfer from a protonated solvent molecule (SH^+) or cluster to form a quasi-molecular ion (MH^+) of the substrate (M).

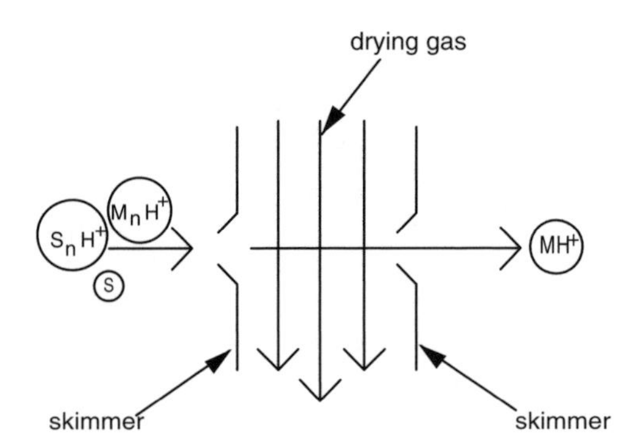

Figure 9.3
Use of a drying gas to remove more solvent (S) and reduce the number of cluster ions, (S_nH^+) or M_nH^+ or $S_mM_nH^+$ (m, n are integers).

The Number of Ions

Under the above conditions, the yield of ions is quite small because most droplets are either neutral or not highly charged electrically. Therefore, the sensitivity of mass spectrometry using simple API is low. Several means of improving the yield of ions have been developed:

- The liquid (solvent) that is nebulized should be polar (e.g., water, acetic acid).
- The liquid (solvent) should already contain ions by inclusion of an electrolyte.
- Additional ionization is effected by including radioactive substances or plasma or glow discharges in the evaporation chamber or by electrical charging of the nebulizer. Such techniques are also discussed in Chapters 8 and 11.

Conclusion

Evaporation from a spray of charged droplets produced from a stream of liquid yields ions that can be analyzed in a mass spectrometer. Thermally labile and normally nonvolatile substances such as sugars, peptides, and proteins can be examined successfully.

Chapter 10

Z-Spray Combined Inlet/Ion Source

Introduction

A solution of an analyte in a solvent can be sprayed (nebulized) from an electrically charged narrow tube to give small electrically charged droplets that desorb solvent molecules to leave ions of the analyte. This atmospheric-pressure ionization is known in various forms, the one most relevant to this section being called electrospray. For additional detail, see Chapters 8, 9, and 11.

As an adaptation of electrospray, Z-spray is a cleaner and more efficient method of generating and separating analyte ions from solvent and buffer agents. In conventional electrospray sources, droplets issue from the end of a narrow inlet tube as a cone-shaped spray. The low-molecular-mass solvent molecules tend to diffuse away toward the edges of the cone, while the high-molecular-mass analyte ions continue along the axis of the cone until they enter the mass spectrometer analyzer through a small orifice (the skimmer). The narrow solution inlet, the cone axis, and the position of the orifice lie along one straight line-of-sight trajectory, as seen in Figure 10.1a. Ions produced in a conventional electrospray source travel along an approximately linear trajectory from formation to entering the analyzer. However, the ions that pass through the skimmer are accompanied by small quantities of neutral materials, and some of these neutral materials strike the edges of the skimmer and are deposited there, where they accumulate and gradually block the skimmer hole.

The Z-spray inlet/ionization source sends the ions on a different trajectory that resembles a flattened Z-shape (Figure 10.1b), hence the name Z-spray. The shape of the trajectory is controlled by the presence of a final skimmer set off to one side of the spray instead of being in-line. This configuration facilitates the transport of neutral species to the vacuum pumps, thus greatly reducing the buildup of deposits and blockages.

The Initial Spray

Two situations need to be considered depending on the type of inlet tube used: either conventional (normal) narrow tubes or very-small-diameter nanotubes.

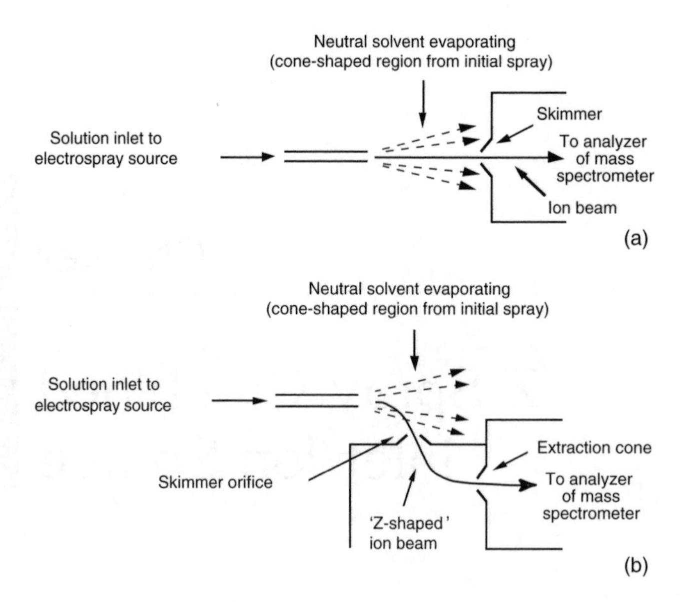

Figure 10.1

(a) The trajectory of analyte ions from a conventional electrospray inlet/ion source is essentially a straight line between the inlet tube carrying the solution of interest and a skimmer orifice placed a short distance away. The neutral solvent molecules are shown diffusing away from the main ion beam over a roughly cone-shaped space. (b) The trajectory of analyte ions from a Z-spray inlet/ion source follows a flattened Z-shape in going from the inlet tube to the final skimmer. Again, neutral solvent molecules are shown diffusing away from the main ion beam over a roughly cone-shaped space. However, unlike the situation in (a), the ion beam first passes through an initial skimmer orifice placed at right angles to the direction of the spray. General gas flow and electrical potentials in the source cause the ion beam to bend toward the skimmer orifice and, having passed through it, to bend once again to pass through the extraction cone.

Nanotube Sprays

Nanotubes may be simply a short section of a capillary tube that holds a small quantity of the solution of interest (Figure 10.2a). Alternatively, Figure 10.2b shows this inlet as the exit from a liquid chromatography apparatus, which is equipped with very narrow "nanocolumns."

For the chromatographic column, flow of solution from the narrow inlet tube into the ionization/desolvation region is measured in terms of only a few microliters per minute. Under these circumstances, spraying becomes very easy by application of a high electrical potential of about 3–4 kV to the end of the nanotube. Similarly, spraying from any narrow capillary is also possible. The ions formed as part of the spraying process follow Z-shaped trajectories, as discussed below.

Normal Inlet Tube Sprays

A common liquid chromatography column is somewhat larger in diameter than a nanocolumn. Consequently, the flow of solution along such a column is measured in terms of one or two milliliters per minute, and spraying requires the aid of a gas flowing concentrically around the end of the inlet tube (Figure 10.2c). An electrical potential is still applied to the end of this tube to ensure adequate electrical charging of the droplets.

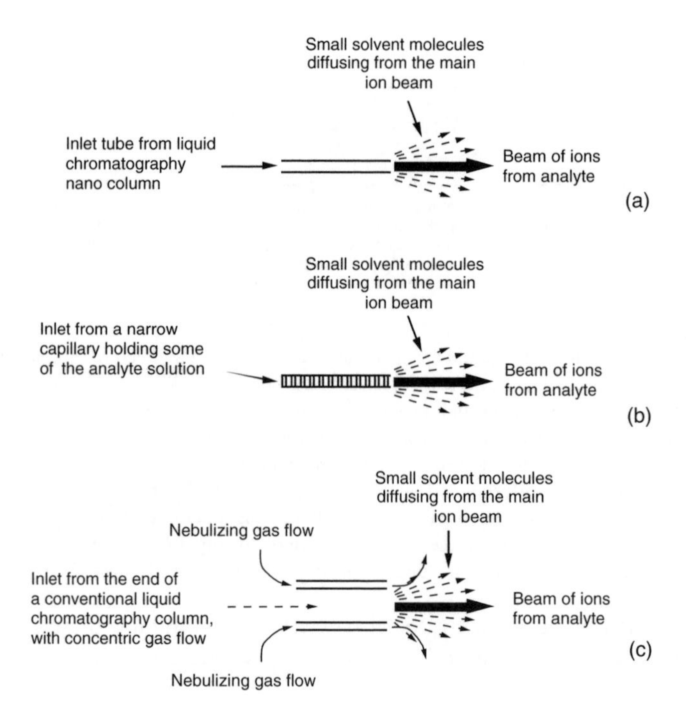

Figure 10.2

Initial trajectories for three kinds of inlet are illustrated. In (a), a high electrical potential of about 3 to 4 kV causes a spray of droplets to be ejected from the end of the inlet from a liquid chromatography nanocolumn as a solution flows to the end of the column. In (b), the same electrical potential applied to a stationary analyte solution held in a very narrow capillary again causes a spray of droplets from the end of the inlet tube, even though no formal liquid flow exists in (a). As the electrically charged droplets move away from the end of the inlet tube, rapid evaporation of small, neutral solvent molecules eventually leaves behind a beam of analyte ions in background gas. Although this ion beam is shown as tightly collimated, in fact it also gradually disperses, but not as rapidly as for the solvent because of the higher molecular masses of analyte molecules compared with solvent molecules. In (c), the inlet from a wider "normal" liquid chromatography column is shown. To obtain a satisfactory spray in this configuration, it is necessary to add a gas flow concentric with the inlet tube. This type of nebulization is analogous to the mechanism used in a common spray can, e.g., for paint or hair lacquer (but without the high potential!).

Trajectories of Ions and Neutrals

Once the spray has formed, electrically charged droplets tend to continue traveling in a straight line according to their initial momenta gained in the electric field. However, neutral solvent molecules are not affected by the electric field, and as they evaporate from the charged droplets, they diffuse, eventually to be drawn into the pumping system. When much of the solvent has evaporated from the droplets, analyte ions also begin to desorb. Analyte ions and any neutral analyte molecules will normally have much greater molecular masses and momenta than the solvent molecules, so they tend not to diffuse from the main axis of the spray as readily as the smaller solvent molecules. If there are any buffers in the original solution sprayed, some is entrained with the analyte ions and molecules. After a short distance from the end of the inlet tube, the initial spray changes from a mix of charged droplets to mainly a beam of analyte ions and neutrals in background gas (usually air) (Figure 10.3).

For conventional electrospray, there is a "line of sight" from the end of the inlet tube to a small hole (the skimmer), through which many of the ions pass on their way to the mass

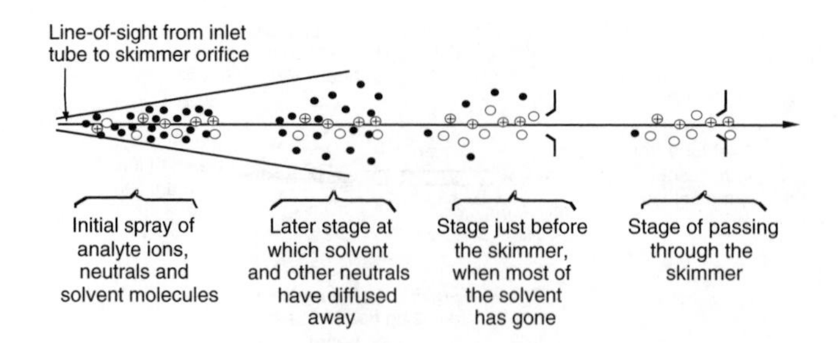

Figure 10.3
In a conventional electrospray inlet, the initial spray of electrically charged and neutral droplets forms a cone-shaped region. As low-molecular-mass solvent evaporates (small black circles), it diffuses from the line-of-sight trajectory to the skimmer; neutral analyte molecules and analyte ions tend not to diffuse so quickly because of their generally much greater molecular mass, allowing them to continue traveling close to the line-of-sight trajectory. As the analyte ions and neutrals near the skimmer, few solvent molecules remain, so mostly analyte ions and neutrals pass through the skimmer. However, because of diffusion and mutual ion repulsion, the ion beam is not closely defined, and some of it strikes the edges of the orifice instead of passing through, causing a buildup of material, which eventually blocks the orifice. The situation is exacerbated if the original solution flowing from the inlet tube contains nonvolatile buffering agents, which cause much faster blocking of the skimmer thus causing need for more frequent cleaning.

analyzer, accompanied by some neutral species. After the skimmer, there is a second stage of removal of residual solvent and neutrals before the ions proceed into the mass analyzer proper (Figure 10.3).

The problem with this linear arrangement is that the species that do not quite make it through the skimmer hole. If these ions and neutrals strike the edges of the skimmer, some will stick there. Gradually, there is a buildup of unwanted material around the skimmer orifice, which is made smaller and smaller and can finally be blocked altogether. Clearly, as the hole diameter gets smaller, fewer and fewer ions will pass through it and the sensitivity of the instrument diminishes. Even before the skimmer is totally blocked, a point is reached when there is no alternative to cleaning the inlet to remove the built-up deposits before the instrument can be used again.

A Z-spray source gets around this problem. Accordingly, a first skimmer orifice is moved from a line-of-sight position to one at right angles to the initial spray direction (Figure 10.4). Now, as the ions form in the background gas, they follow the gas flow toward the vacuum region of the mass spectrometer. Some vapor solvent is also drawn down into the skimmer orifice. More solvent diffuses from the gas stream, which then bends again through a second skimmer (the extraction cone). Mostly ions and background gas molecules (plus some residual solvent molecules) pass through the second skimmer and on to the mass analyzer. There is a drying gas flowing around the entrance to the skimmer to remove more solvent from any residual droplets (Figure 10.4).

Advantages

With the Z-spray design, there is almost no buildup of products on the skimmer orifice, so instrumental sensitivity and performance remain constant over long periods of time. In addition, this arrangement is inherently a better ion collector than the line-of-sight mode and gives a useful gain in instrument sensitivity. The open arrangement resulting from the design gives better access to the inlet tube, facilitating its manipulation, which is particularly important in the placing of nanotubes. Finally, collisionally activated decomposition (CAD) of ions can still be effected by increasing the ion extraction voltage (cone voltage; see Chapter 8).

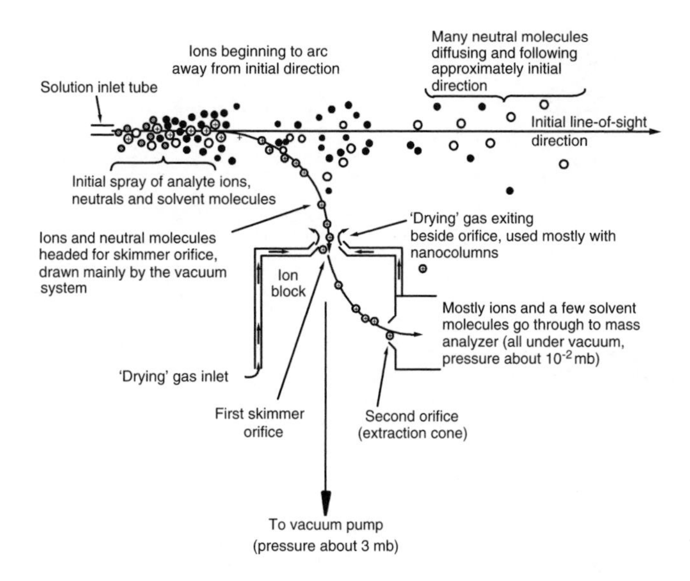

Figure 10.4
Solution issuing from the end of the inlet tube, held at an electrical potential of 3–4 kV, forms a spray of droplets at atmospheric pressure. Solvent evaporates from these droplets. Under the influence of the general gas flow toward the vacuum pumps, ions and neutral molecules move in an arc through the first skimmer orifice, as shown. After this opening, a split between ions and neutral molecules is effected. Most remaining solvent and other neutrals flow on toward the first-stage vacuum pump. But an electric-field gradient causes the ions to flow in an arc toward a second skimmer, often called the "extraction cone," and on to the mass analyzer. A few neutrals diffuse through this second skimmer, as well, because of the differences in pressures on either side of it. Note the overall flattened Z-shape of the ion trajectory.

Conclusion

The Z-spray inlet causes ions and neutrals to follow different paths after they have been formed from the electrically charged spray produced from a narrow inlet tube. The ions can be drawn into a mass analyzer after most of the solvent has evaporated away. The inlet derives its name from the Z-shaped trajectory taken by the ions, which ensures that there is little buildup of products on the narrow skimmer entrance into the mass spectrometer analyzer region. Consequently, in contrast to a conventional electrospray source, the skimmer does not need to be cleaned frequently and the sensitivity and performance of the instrument remain constant for long periods of time.

Chapter 11

Thermospray and Plasmaspray Interfaces

Introduction

For a more detailed description of the ionization process inherent in thermospray and plasmaspray please consult Chapter 9, "Atmospheric Pressure Ionization." The reader should also compare thermospray with electrospray (see Chapter 8).

In many applications in mass spectrometry (MS), the sample to be analyzed is present as a solution in a solvent, such as methanol or acetonitrile, or an aqueous one, as with body fluids. The solution may be an effluent from a liquid chromatography (LC) column. In any case, a solution flows into the "front end" of a mass spectrometer, but before it can provide a mass spectrum, the bulk of the solvent must be removed without losing the sample (solute). If the solvent is not removed, then its vaporization as it enters the ion source would produce a large increase in pressure and stop the spectrometer from working. At the same time that the solvent is removed, the dissolved sample must be retained so that its mass spectrum can be measured. There are several means of effecting this differentiation between carrier solvent and the solute of interest, and thermospray is just one of them. Plasmaspray is a variant of thermospray in which the basic method of solvent removal is the same, but the number of ions obtained is enhanced (see below).

One of the first successful techniques for selectively removing solvent from a solution without losing the dissolved solute was to add the solution dropwise to a moving continuous belt. The drops of solution on the belt were heated sufficiently to evaporate the solvent, and the residual solute on the belt was carried into a normal EI (electron ionization) or CI (chemical ionization) ion source, where it was heated more strongly so that it in turn volatilized and could be ionized. The moving belt system had some mechanical problems and could be temperamental. It can still be found in some laboratories, but the more recent, less-mechanical inlets such as thermospray and electrospray have replaced it. Thermospray alone gives poor ion yields, but thermospray with the help of an electrical discharge (plasmaspray) gives excellent ion yields. Thermospray alone is now obsolete, having given way to plasmaspray (atmospheric-pressure chemical ionization, APCI) and electrospray.

Differential Solvent Removal

A sample for which a mass spectrum is required may well be dissolved in an organic or aqueous solvent. For example, in searching for drugs in blood plasma, the plasma itself can be investigated (aqueous) or its active components can be extracted into an organic solvent such as dichloromethane. Alternatively, the sample can be separated into its components by passage through a liquid chromatographic instrument (see Chapter 37); upon emerging from the column the sample of interest is present as a solution in the solvents used in the chromatography. In either case, the sample to be examined is in solution and cannot be put straight into a mass spectrometer without first removing most of the solvent and without, of course, removing the dissolved sample also!

Thermospray is one method for effecting this differential solvent removal. The solution is passed along a short length of stainless steel capillary tube, the end of which is heated strongly by electrical resistive heating (Figure 11.1). When the solution reaches the hot zone, it is almost instantaneously vaporized and leaves the tube as a supersonic jet or spray of very small droplets of solution in solvent vapor. Before entering the mass spectrometer proper, this mist of droplets flows along a tube, the walls of which are heated slightly to prevent condensation. As the droplets move through this region, solvent evaporates rapidly from their surfaces and the droplets get smaller and smaller. In addition to producing the spray, this method of rapid vaporization leaves no time for equilibrium to be attained, and a substantial proportion of the droplets have an excess of positive or negative electrical charge on their surfaces. Thus, as the droplets get smaller the electrical surface charge density increases until the natural repulsion between like charges causes ions as well as neutral molecules to be released from the surfaces (Figure 11.2).

The end of the capillary tube is aimed somewhat like a gun at a target at the opposite end of the desolvation region. After vaporizing from the surface of a droplet, the solvent molecules, which have low molecular mass, diffuse from the line of fire and are pumped off. However, the sample molecular ions and cluster ions have much greater molecular mass than those of the solvent and, being heavier, tend to carry straight on to the target at the end of the inlet (Figure 11.1). The ions and residual solvent pass through a small hole in the target, after which vacuum pumps reduce the gas pressure and remove more solvent. Desolvated $[M + H]^+$ ions pass to the mass analyzer.

At the target, clusters are broken up and sample molecular ions, accompanied by some remaining solvent ions, are extracted by an electrical potential through a small hole into the mass spectrometer analyzer (Figure 11.1), where their mass-to-charge (m/z) ratios are measured in the usual way. The mass spectrometer may be of any type.

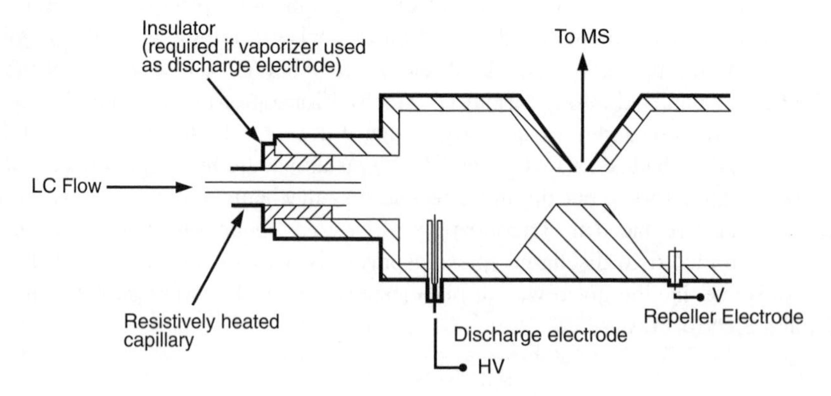

Figure 11.1.
Schematic diagram of a thermospray ion source. This source, of current design, also incorporates (a) a discharge electrode so that the source can be operated in plasmaspray mode and (b) a repeller electrode to induce fragmentation. The vaporizer itself can be used as a discharge electrode.

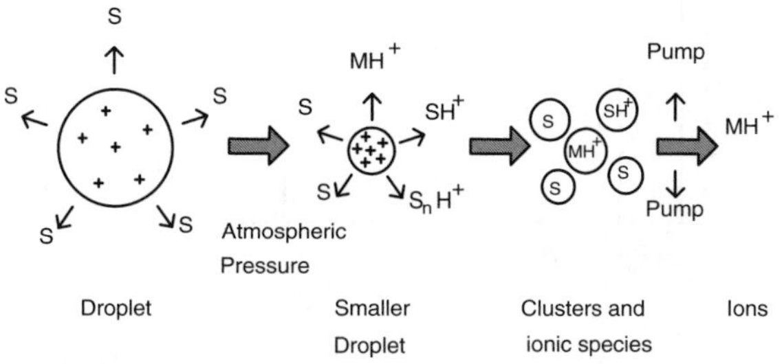

Figure 11.2.
After being formed as a spray, many of the droplets contain some excess positive (or negative) electric charge. Solvent (S) evaporates from the droplets to form smaller ones until, eventually, ions (MH^+, SH^+) from the sample M and solvent begins to evaporate to leave even smaller drops and clusters (S_nH^+; n = 1, 2, 3, etc.). Later, collisions between ions and molecules (CI) leave $[M + H]^+$ ions, which proceed on into the mass analyzer. Ion yield can be enhanced by including a volatile ionic compound (e.g., ammonium acetate) in the initial solution before it reaches the spraying zone.

The thermospray process extracts sample molecules from a solvent and turns them into ions. Therefore, the system is both an inlet and an ion source, so a separate ion source is not necessary.

Ion Yield

The thermospray inlet/ion source does not produce a good percentage yield of ions from the original sample, even with added salts (Figure 11.2). Often the original sample is present in very tiny amounts in the solution going into the thermospray, and the poor ion yield makes the thermospray/mass spectrometer a relatively insensitive combination when compared with the sensitivity attainable by even quite a modest mass spectrometer alone. Various attempts have been made to increase the ion yield. One popular method is described here.

By placing a high electrical potential on an insulated electrode in the mist issuing from the inlet of the capillary tube, a plasma discharge can be struck. The spray of droplets passes through the plasma and electrons are stripped from neutral molecules such that the final yield of ions is greatly increased. Ionized molecules collide with neutral molecules, leading to hydrogen transfer and the formation of protonated molecular ions (as with chemical ionization). This process has led to the method known as atmospheric-pressure chemical ionization (APCI). Both solvent and solute (sample) ions are generated, but, as before, mostly only the heavier sample ions are extracted into the analyzer of the mass spectrometer. The addition of a plasma device turns thermospray into a much more sensitive inlet/ion source known as "plasmaspray." This name is somewhat misleading in that the plasma does not cause the spray, which is still generated thermally.

Another way of improving ion yield is to include a repeller electrode (Figure 11.1). This electrode slows lighter ions more than heavier ones, which catch up and collide, causing enhanced chemical ionization.

Nature of the Ions Produced

Most of the ions produced by either thermospray or plasmaspray (with or without the repeller electrode) tend to be very similar to those formed by straightforward chemical ionization with lots of protonated or cationated positive ions or negative ions lacking a hydrogen (see Chapter 1). This is because, in the first part of the inlet, the ions continually collide with neutral molecules in the early part of their transit. During these collisions, the ions lose excess internal energy,

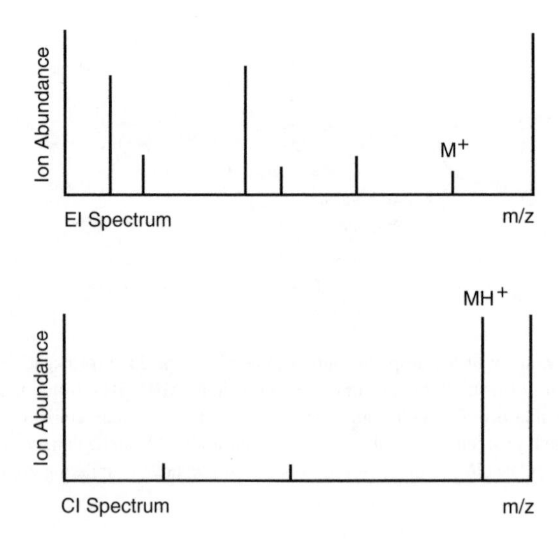

Figure 11.3.
Comparison of EI and CI mass spectra illustrating the greater degree of fragmentation in the former and the greater abundance of quasi-molecular ions in the latter.

interchange protons, and lose hydrogen atoms. These are the predominant processes in chemical ionization. Plasmaspray mass spectra show some fragment ions, and use of the repeller electrode leads to more.

The ions passing into the mass spectrometer analyzer from thermospray or plasmaspray have little excess of internal energy and therefore not enough energy to fragment. Most of the ions are of the form $[M + X]^+$ or $[M - H]^-$, in which X can be hydrogen or some other element such as sodium or potassium. While these quasi-molecular ions are excellent for giving accurate molecular mass information — which may be all that is required — they give little or no information concerning the actual molecular structure of the substance being investigated, which is provided by fragment ions. Compare the typical EI and CI mass spectra illustrated in Figure 11.3.

This is entirely analogous to the problem with simple chemical ionization, and the solution to it is similar. To give the quasi-molecular ions the extra energy needed for them to fragment, they can be passed through a collision gas and the resulting spectra analyzed for metastable ions or by MS/MS methods (see Chapters 20 through 23).

Uses of Plasmaspray and Electrospray

Although simple solutions *can* be examined by these techniques, for a single substance dissolved in a solvent, straightforward evaporation of the solvent outside the mass spectrometer with separate insertion of the sample is usually sufficient. For mixtures, the picture is quite different. Unless the mixture is to be examined by MS/MS methods, it will be necessary to separate it into its individual components. This separation is most often done by gas or liquid chromatography (GC or LC).

In the latter, small quantities of emerging mixture components dissolved in elution solvent would be laborious to deal with if each component were to be first isolated by evaporation of solvent before its introduction into the mass spectrometer. In such circumstances, the direct introduction, removal of solvent, and ionization provided by plasmaspray or electrospray is a boon and puts LC/MS on a par with GC/MS for mixture analysis. Furthermore, GC is normally concerned with volatile, low-molecular-weight compounds and is of little or no use for the many polar, water-soluble, high-molecular-mass substances such as the peptides, proteins, carbohydrates, nucleotides, and similar substances found in biological systems. In contrast, LC/MS with plasmaspray is used frequently in biochemical analysis for research and medicine.

General Comments

Thermospray alone is not a very sensitive technique and is mostly of use only for polar compounds. It is now rarely used on its own because its plasma-discharge-assisted variant, plasmaspray, gives far better results and ion yields. Plasmaspray complements electrospray. Furthermore, thermospray alone needs a solution containing electrolytes (ionic substances) if ions are to be formed. Since these electrolytes prevent reversed-phase liquid chromatography from being used and often lead to blocking of the capillary inlet, there are added cogent reasons for moving to plasmaspray or electrospray.

Conclusion

By rapidly vaporizing a solution by heat, a spray is produced from which the solvent can be removed, leaving sample ions that pass straight into the analyzer region of a mass spectrometer. Plasmaspray is very similar, but ion yield is vastly improved through use of a corona discharge.

Chapter 12

Particle-Beam Interface

Background

The combined technique of gas chromatography and mass spectrometry (GC/MS), whereby the gaseous effluent from a gas chromatograph can be passed into a mass spectrometer, has proved to be an enormously successful analytical method for volatile substances. Similar development for high-performance liquid chromatography and MS (HPLC/MS) or liquid chromatography and MS (LC/MS) as an analytical method for less-volatile or polar materials was severely hindered by the need to remove the liquid solvent from the HPLC eluant. During transition from a liquid eluant at atmospheric pressure to the vaporized eluant needed for injection into the mass spectrometer, so much solvent vapor is formed that the spectrometer vacuum system is overwhelmed. A number of attempts have been made to separate solvent from solute without losing much of the solute (substrate). The particle-beam interface (LINC™) is one such successful method.

The Nebulizer

As a first stage, the stream of liquid from an HPLC eluant is passed through a narrow tube toward the LINC interface. Near the end of the tube, the liquid stream is injected with helium gas so that it leaves the end of the tube as a high-velocity spray of small drops of liquid mixed with helium. From there, the mixture enters an evacuation chamber (Figure 12.1). The formation of spray (nebulizing) is very similar to that occurring in the action of aerosol spray cans (see Chapter 19).

The Evacuation Chamber

The small drops from the nebulizer have a large surface-area-to-volume ratio, and solvent begins to evaporate rapidly from their surfaces; generally, solute or substrate is much less volatile than solvent and starts to concentrate as the drops get smaller (Figure 12.2). The chamber is heated to prevent the evaporating solvent from condensing on its inner walls. The spray of drops finally leaves the evacuation chamber as a fast-moving, slightly spreading beam — the beginning of the particle beam.

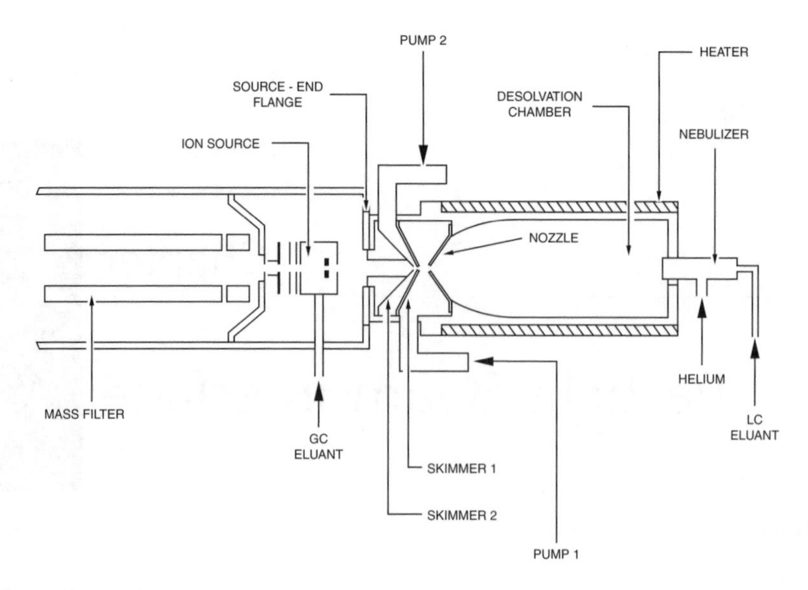

Figure 12.1
A typical arrangement for producing a particle beam from a stream of liquid, showing: (1) the nebulizer, (2) the desolvation chamber, (3) the wall heater, (4) the exit nozzle, (5, 6) skimmers 1, 2, (7) the end of the ion source, (8) the ion source, and (9) the mass analyzer. An optional GC inlet into the ion source is shown.

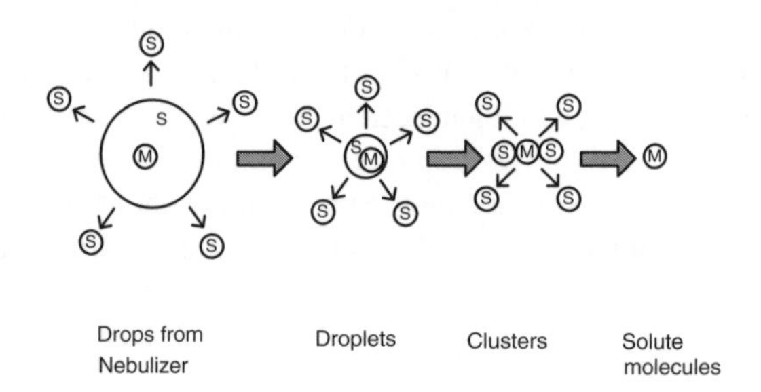

Figure 12.2
The passage of drops of solvent (S) containing a solute (M) through the evacuation chamber, the exit nozzle, skimmers 1 and 2, and into the ion chamber. Molecules of solvent evaporate throughout this passage, causing the drops to get smaller until only solute molecules remain.

The First Skimmer

The flow of droplets is directed through a small orifice (Skimmer 1; Figure 12.1) and across a small region that is kept under vacuum by rotary pumps. In this region, approximately 90% of solvent and injected helium is removed from the incipient particle beam. Because the rate of diffusion of a substance is inversely proportional to its molecular mass, the lighter helium and solvent molecules diffuse away from the beam and are pumped away. The heavier solute molecules diffuse more slowly and pass through the first skimmer before they have time to leave the beam; the solute is accompanied by residual solvent and helium.

The Second Skimmer

The beam from the first skimmer is directed toward a second one (Figure 12.1), again across an evacuated region where almost all of the residual solvent and helium are pumped away to leave a beam now consisting mostly of heavier solute molecules, together with small clusters of solvent and solute molecules. Gas pressure after this region is then reduced from atmospheric to about 10^{-4} mbar.

The Ion Source

The particle beam — after linear passage from the evacuation chamber nozzle, through the first and second skimmers, and into the end of the ion source — finally passes through a heated grid immediately before ionization. The heated grid has the effect of breaking up most of the residual small clusters, so residual solvent evaporates and a beam of solute molecules enters the ionization chamber.

Ionization

The beam entering the ion chamber is suitable for both electron (EI) and chemical (CI) ionization, and either mode can be used (Figure 12.3). Mass analysis follows in the usual way, typically using quadruple or magnetic-sector instruments.

Efficiency

The efficiency of separation of solvent from solute varies with their nature and the rate of flow of liquid from the HPLC into the interface. Volatile solvents like hexane can be evaporated quickly and tend not to form large clusters, and therefore rates of flow of about 1 ml/min can be accepted from the HPLC apparatus. For less-volatile solvents like water, evaporation is slower, clusters are less easily broken down, and maximum flow rates are about 0.1–0.5 ml/min. Because separation of solvent from solute depends on relative volatilities and rates of diffusion, the greater the molecular mass difference between them, the better is the efficiency of separation. Generally, HPLC is used for substances that are nonvolatile or are thermally labile, as they would otherwise be analyzed by the practically simpler GC method; the nonvolatile substances usually have molecular masses considerably larger than those of commonly used HPLC solvents, so separation is good.

Similarity to Other Interfaces

The nebulization and evaporation processes used for the particle-beam interface have closely similar parallels with atmospheric-pressure ionization (API), thermospray (TS), plasmaspray (PS), and electrospray (ES) combined inlet/ionization systems (see Chapters 8, 9, and 11). In all of these systems, a stream of liquid, usually but not necessarily from an HPLC column, is first nebulized

$$M + e^- \longrightarrow M^{\bullet+} + 2e^- \quad \text{(a)}$$

$$M + RH^+ \longrightarrow MH^+ + R \quad \text{(b)}$$

Figure 12.3
The ionization of solute molecules (M) by (a) EI and (b) CI. A typical reagent gas is shown as RH^+.

and the solvent is then selectively removed from the solute. However, LINC™ aims to provide a particle beam that is ready for ionization (by EI or CI), while the other methods provide a particle beam that is already ionized. Thus, the particle beam (LINC™) interface is strictly an inlet system while API, TS, PS, and ES provide both inlet and ionization combined. The major difference between these last and LINC™ lies in the additional measures taken to ensure ionization in the inlet so that no separate ion source is needed. From a practical viewpoint, techniques such as electrospray tend to give only quasi-molecular and no fragment ions, so an additional means must be provided to cause fragmentation. The LINC™ method, with its separate ion source, provides molecular, quasi-molecular, and fragment ions by using a conventional EI or CI source.

Conclusion

The particle-beam interface (LINC) works by separating unwanted solvent molecules from wanted solute molecules in a liquid stream that has been broken down into droplets. Differential evaporation of solvent leaves a beam of solute molecules that is directed into an ion source.

Chapter 13

Dynamic Fast-Atom Bombardment and Liquid-Phase Secondary Ion Mass Spectrometry (FAB/LSIMS) Interface

Introduction

The basic principles of fast-atom bombardment (FAB) and liquid-phase secondary ion mass spectrometry (LSIMS) are discussed only briefly here because a fuller description appears in Chapter 4. This chapter focuses on the use of FAB/LSIMS as part of an interface between a liquid chromatograph (LC) and a mass spectrometer (MS), although some theory is presented.

Bombardment of a liquid surface by a beam of fast atoms (or fast ions) causes continuous desorption of ions that are characteristic of the liquid. Where the liquid is a solution of a sample substance dissolved in a solvent of low volatility (often referred to as a matrix), both positive and negative ions characteristic of the solvent and the sample itself leave the surface. The choice of whether to examine the positive or the negative ions is effected simply by the sign of an electrical potential applied to an extraction plate held above the surface being bombarded. Usually, few fragment ions are observed, and a sample of mass M in a solvent of mass S will give mostly $[M + H]^+$ (or $[M - H]^-$) and $[S + H]^+$ (or $[S - H]^-$) ions. Therefore, the technique is particularly good for measurement of relative molecular mass.

The FAB source operates near room temperature, and ions of the substance of interest are lifted out from the matrix by a momentum-transfer process that deposits little excess of vibrational and rotational energy in the resulting quasi-molecular ion. Thus, a further advantage of FAB/LSIMS over many other methods of ionization lies in its gentle or mild treatment of thermally labile substances such as peptides, proteins, nucleosides, sugars, and so on, which can be ionized without degrading their structures.

Liquid chromatography is a separation method that is often applied to nonvolatile, thermally labile materials such as peptides, and, if their mass spectra are required after the separation step, then a mild method of ionization is needed. Since FAB/LSIMS is mild and works with a liquid matrix, it is not surprising that attempts were made to utilize this ionization source as both an inlet

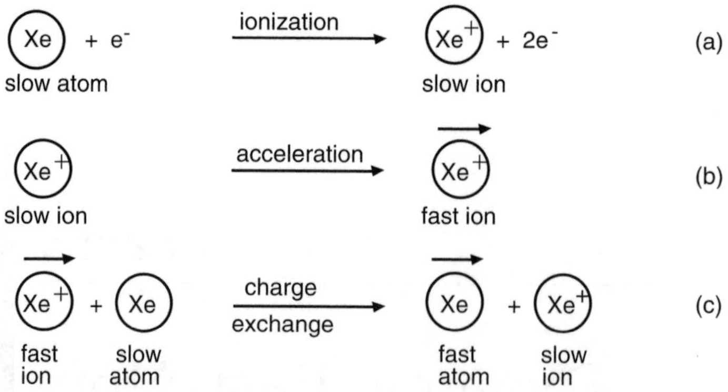

$Xe + e^-$ (slow atom) —ionization→ $Xe^+ + 2e^-$ (slow ion) (a)

Xe^+ (slow ion) —acceleration→ Xe^+ (fast ion) (b)

Xe^+ (fast ion) + Xe (slow atom) —charge exchange→ Xe (fast atom) + Xe^+ (slow ion) (c)

Figure 13.1
Xenon atoms are ionized to Xe^+ using electrons. These ions are relatively slow and move in all directions. (b) Xe^+ ions are accelerated through a large electric potential so that they attain a high speed in one direction. (c) Charge exchange between fast Xe^+ ions and slow Xe atoms gives a beam of fast Xe atoms and slow ions. The latter are removed by electric deflector plates, leaving just a beam of fast atoms.

and an ion source capable of linking a liquid chromatograph directly to a mass spectrometer. Dynamic FAB is the adaptation of FAB/LSIMS to form an interface between a liquid chromatograph and a mass spectrometer.

Atom or Ion Beams

A gun is used to direct a beam of fast atoms (often Xe) or fast ions (often Cs^+) onto a small metal target area where the solution of interest is placed. Production of an atom beam is described in Figure 13.1.

In dynamic FAB, this solution is the eluant flowing from an LC column; i.e., the target area is covered by a flowing liquid (dynamic) rather than a static one, as is usually the case where FAB is used to examine single substances. The fast atoms or ions from the gun carry considerable momentum, and when they crash into the surface of the liquid some of this momentum is transferred to molecules in the liquid, which splash back out, rather like the result of throwing a stone into a pond (Figure 13.2). This is a very simplistic view of a complex process that also turns the ejected particles into ions (see Chapter 4 for more information on FAB/LSIMS ionization).

Properties of the Solvent (Matrix Material)

Liquids examined by FAB are introduced into the mass spectrometer on the end of a probe inserted through a vacuum lock in such a way that the liquid lies in the target area of the fast atom or ion beam. There is a high vacuum in this region, and there would be little point in attempting to examine a solution of a sample in one of the commoner volatile solvents such as water or dichloromethane because it would evaporate extremely quickly, probably as a burst of vapor when introduced into the vacuum. Therefore it is necessary to use a high-boiling solvent as the matrix material, such as one of those listed in Table 13.1.

The solvents used for liquid chromatography are the commoner ones such as water, acetonitrile, and methanol. For the reasons just stated, it is not possible to put them straight into the ion source without problems arising. On the other hand, the very viscous solvents that qualify as matrix material are of no use in liquid chromatography. Before the low-boiling-point eluant from the LC column is introduced into the ion source, it must be admixed with a high-boiling-point matrix

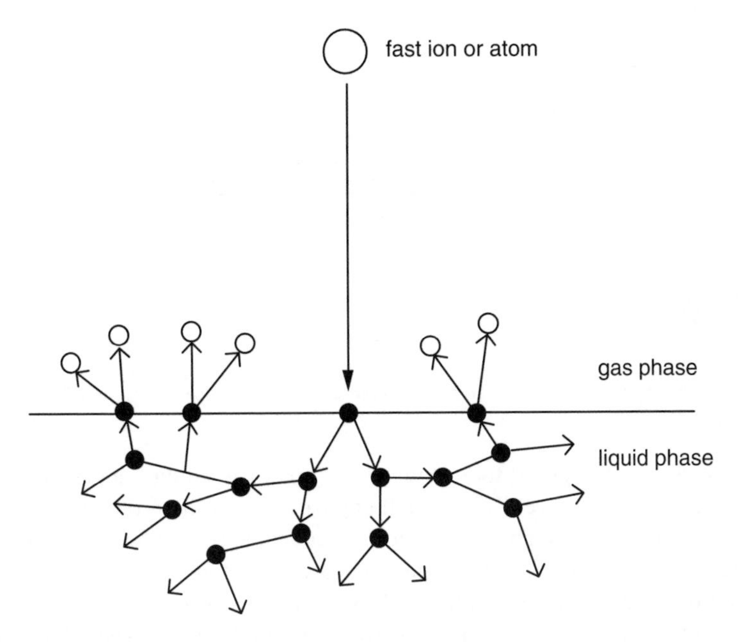

Figure 13.2
A typical cascade process. A fast atom or ion collides with surface molecules, sharing its momentum and causing the struck molecules to move faster. The resulting fast-moving particles then strike others, setting up a cascade of collisions until all the initial momentum has been dissipated. The dots (●) indicate collision points. Ions or atoms (○) leave the surface.

TABLE 13.1
Some Commonly Used Solvents for FAB or LSIMS

Solvent	Protonated molecular ions (m/z)
Glycerol	93
Thioglycerol	109
3-NOBA [a]	154
NOP [b]	252
Triethanolamine	150
Diethanolamine	106
Polyethylene glycol (mixtures)	- - - - [c]

[a] 3-Nitrobenzyl alcohol.
[b] n-Octyl-3-nitrophenyl ether.
[c] Wide mass range depending on which glycol is used.

material. Generally, addition of some 10% of matrix material (e.g., glycerol; see Table 13.1) is sufficient. It can be added with the solvents entering the LC column or admixed with the eluant leaving the column. The latter mode is better for LC separations.

Having considered the various parts of a dynamic-FAB system (atom gun, ionization, and matrix), it is now necessary to see how these are put together in a working inlet/ion source interface.

Dynamic-FAB/LSIMS Interface

One of the earliest models is illustrated in Figure 13.3, which clearly shows the principles used in later improvements. The LC effluent was pumped along a length of silica capillary tubing inside

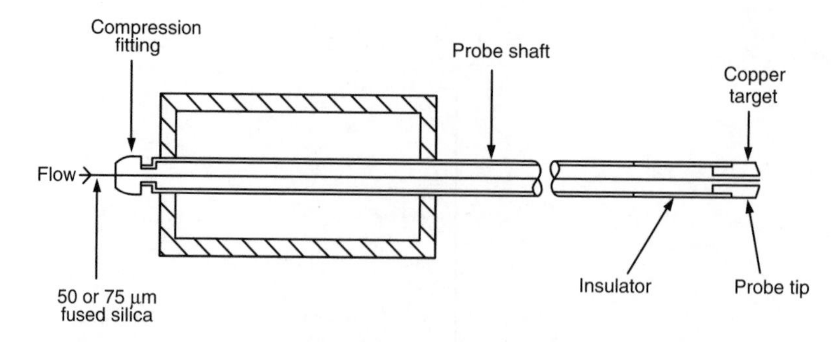

Figure 13.3
A dynamic-FAB probe having a simple copper target. The narrow fused-silica tube passes through the shaft, its end lying flush with the target surface.

a protective metal sheath (the probe shaft; Figure 13.3) that passed through a vacuum lock. At the end of the capillary, the effluent reaches a small copper target at the probe tip and flows out from there. If flow rates are not too high (5 ml/min), so just enough liquid reaches the tip to balance that evaporating in the vacuum, then the system is reasonably stable, and fresh solution is continuously presented to the atom or ion beam. With only eluting solvent flowing, the resulting mass spectrum consists of ions arising only from the solvent and added matrix. These ions are of relatively small mass, and by operating the mass spectral scan only above about m/z 200, few extraneous ions are recorded. When a component from a mixture that has been separated on the LC column reaches the target area, it too is ionized and affords a mass spectrum. Thus, as components elute from the column and reach the target area, the recorded ion current rises from its background level and then falls again as the component passes beyond the target area. The ion current traces out a roughly triangular peak with time. A chart plotting the ion current (y-axis) vs. time (x-axis) is then a total ion current (TIC) chromatogram, an example of which is shown in Figure 13.4. The area

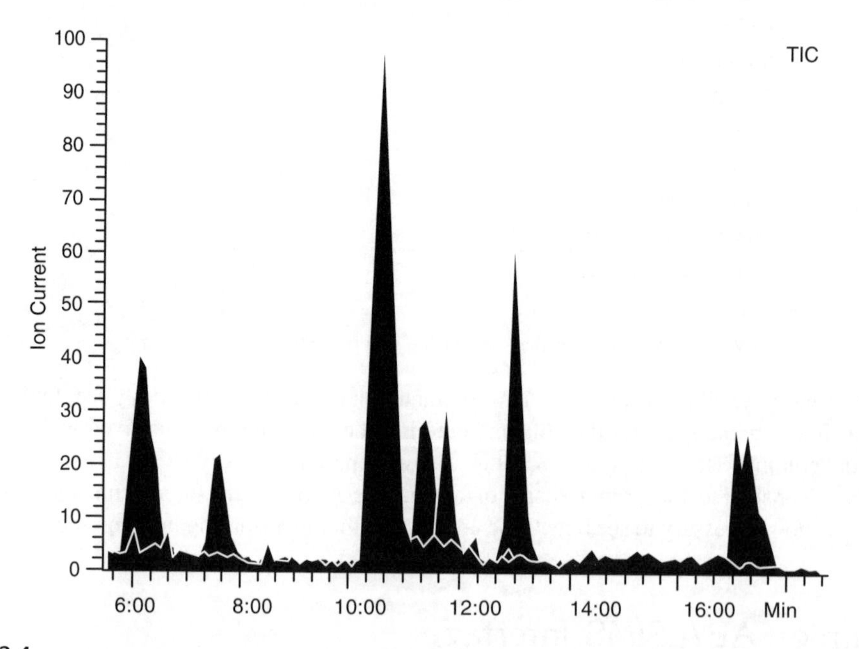

Figure 13.4
A typical TIC chromatogram from an analysis of peptides resulting from enzymatic digest of myoglobin. The peaks represent individual peptides eluting from an LC column and being mass measured by a spectrometer coupled to it through a dynamic-FAB inlet/ion source.

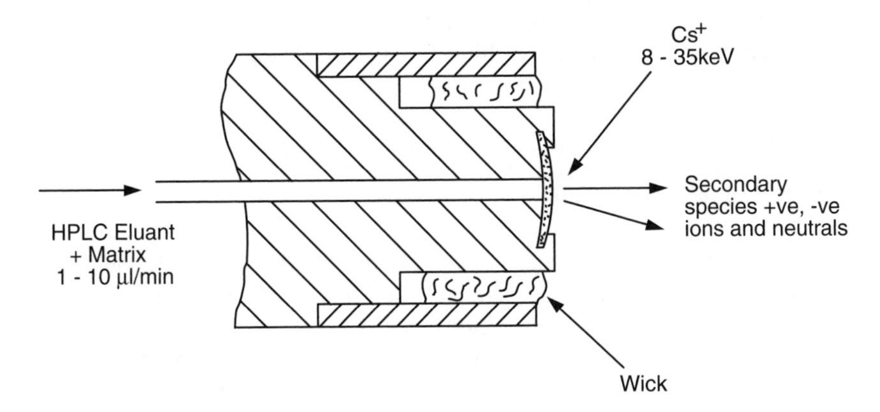

Figure 13.5
A dynamic-FAB probe tip incorporating a screen and wick assembly at the target surface.

under a peak is a measure of the amount of eluting component, and the elapsed time from the start of the chromatogram to the maximum of the peak is its retention time.

More recent versions of this type of probe include some refinements, such as the provision of a wick to aid evaporation of the solvent and matrix from the probe tip (Figure 13.5). Such improvements have allowed greater flow rates to be used, and rates of 1 to 10 ml/min are possible. For these sorts of low flow rates, minibore LC columns must be employed.

As mentioned above, matrix material must be added to the normal eluting solvent used in the LC system before the solvent stream enters the FAB inlet. The simplest way of adding the solvent is to premix about 10% of the matrix material with the solvent before it enters the LC column, by dissolving about 10% of matrix in the elution solvent. Although very convenient, this preaddition often leads to undesirable changes in the effectiveness of the column in separating components of a mixture. For example, introduction of the commonly used glycerol as matrix material can lead to serious tailing (distortion of the peak shape) of the mixture components on the LC column. The best compromise is to add the matrix material to the solvent as it elutes from the column and before it enters the inlet of the dynamic-FAB probe. Figure 13.6 presents one simple way in which this postcolumn addition can be done through

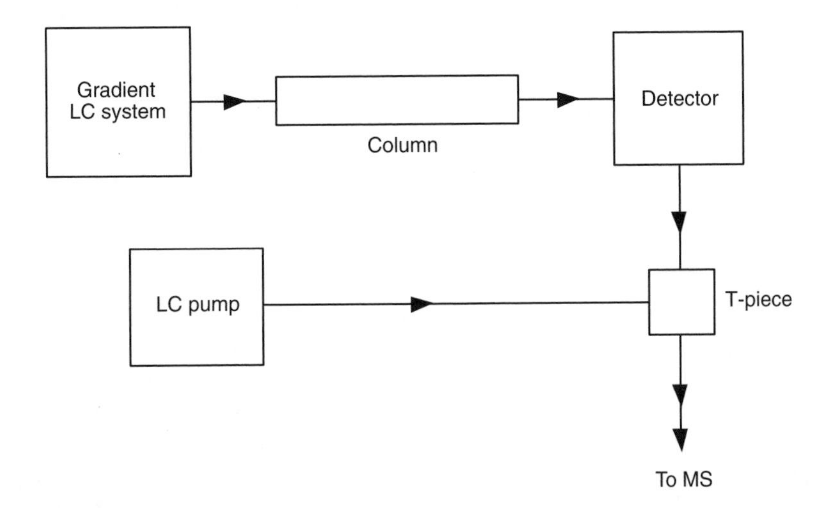

Figure 13.6
HPLC system incorporating a postcolumn matrix-addition facility. Note the pump used for postcolumn addition must be essentially pulse free.

use of a second pump. The total setup may take about 30 min to stabilize before it can be used for analysis and is checked by watching for the ion source pressure reading to become constant.

Types of Ions Produced

Mostly, positive-ion FAB yields protonated quasi-molecular ions $[M + H]^+$, and the negative-ion mode yields $[M - H]^-$. In the presence of metal salts (e.g., KCl) that are sometimes added to improve efficiency in the LC column, ions of the type $[M + X]^+$ are common, where X is the metal. Another type of ion that is observed is the so-called cluster, a complex of several molecules with one proton, $[M_n + H]^+$ with n = 1, 2, 3, … , etc. Few fragment ions are produced.

In static FAB/LSIMS, small abundances of ions appear at almost every m/z value and are colloquially referred to as grass. These ions represent general degradation of the liquid bombarded by the fast-atom or -ion beam. If the ion corresponding to the molecular mass of the sample under investigation has only low abundance, it may be difficult to differentiate it from the grass. This effect is particularly serious at high mass. In dynamic FAB, these degradation products do not have time to accumulate because, as they are formed, they are swept away by the flowing liquid, so their ions are correspondingly much less abundant. Dynamic-FAB spectra are characterized by having far less grass compared with the static mode, and therefore sample ions are much easier to observe. Therefore the dynamic form of FAB is more sensitive than the static one, even though the basic ionization processes are the same.

Conclusion

By passing a continuous flow of solvent (admixed with a matrix material) from an LC column to a target area on the end of a probe tip and then bombarding the target with fast atoms or ions, secondary positive or negative ions are ejected from the surface of the liquid. These ions are then extracted into the analyzer of a mass spectrometer for measurement of a mass spectrum. As mixture components emerge from the LC column, their mass spectra are obtained.

Chapter 14

Plasma Torches

Introduction

This chapter should be read in conjunction with Chapter 6, "Coronas, Plasmas, and Arcs." A plasma is defined as a gaseous phase containing neutral molecules, ions, and electrons. The numbers of ions and electrons are usually almost equal. In a plasma torch, the plasma is normally formed in a monatomic gas such as argon flowing between two concentric quartz tubes (Figure 14.1).

In a plasma, collisions between atoms, positive ions, and electrons thermolize their kinetic energies; the distribution of kinetic energies corresponds to what would be present in a hot gas thermally heated to an equivalent temperature. Direct heating of a gas is not used to form the plasma. Instead, a high-frequency electromagnetic field is applied through the load coil. This rapidly oscillating electromagnetic field interacts inductively with the charged species; electrons and ions try to follow the field and are speeded up, gaining kinetic energy. In the rapidly oscillating field, random collisions of electrons and ions with neutral species redistributes this extra kinetic energy, and the whole ensemble becomes hotter. If the rapidly oscillating electromagnetic field is maintained, the ions and electrons continue to follow a chaotic motion as they gain more speed and undergo more collisions, which continue to redistribute the kinetic energies. Eventually, the kinetic energies become so high that the plasma reaches temperatures of 8,000–10,000°C. At these high temperatures, the plasma behaves like a flame issuing from the ends of the concentric quartz tubes, hence the derivation of the name *plasma torch*. As in a normal flame, atoms and ions are excited electronically in the collision processes and emit light. In the plasma torches used for mass spectrometry, the argon that is used emits pale-blue-to-lilac light as excited atoms relax and as a proportion of the electrons and ions recombine.

If a sample solution is introduced into the center of the plasma, the constituent molecules are bombarded by the energetic atoms, ions, electrons, and even photons from the plasma itself. Under these vigorous conditions, sample molecules are both ionized and fragmented repeatedly until only their constituent elemental atoms or ions survive. The ions are drawn off into a mass analyzer for measurement of abundances and m/z values. Plasma torches provide a powerful method for introducing and ionizing a wide range of sample types into a mass spectrometer (inductively coupled plasma mass spectrometry, ICP/MS).

Because light emitted from inductively coupled plasma torches is characteristic of the elements present, the torches were originally introduced for instruments that optically measured the frequencies and intensities of the emitted light and used them, rather than ions, to estimate the amounts and types of elements present (inductively coupled plasma atomic emission spectroscopy,

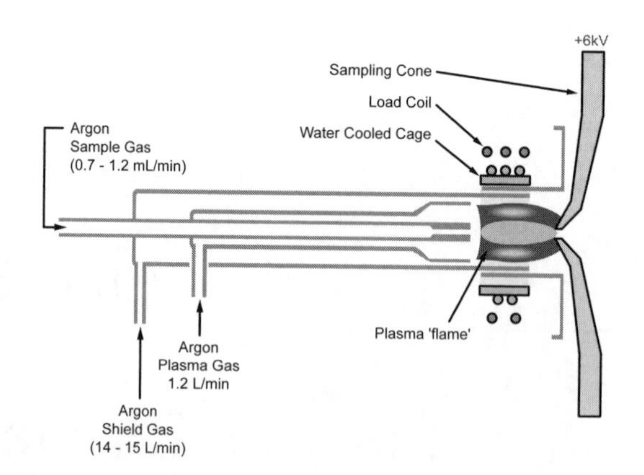

Figure 14.1
Argon gas flows at a rate of about 1 to 2 l/min along the second of three concentric quartz tubes and is ignited to form a plasma by introducing a few sparks from a piezoelectric device. The plasma is maintained and heated by a high-frequency electromagnetic field passing through a load coil that is wound around the outside of the torch. Note the annular space between the load coil and the outermost quartz tube; a water-cooled cage (discussed later) can be placed in this location. Because the very hot plasma could melt the outermost quartz tube if it impinged on it, a second flow (14 to 15 l/min) of fast-moving coolant argon gas is used to shield the walls of the tube from the plasma. Finally, there is a third flow (0.7 to 1.2 l/min) of argon through the central quartz tube; this gas first passes through the sample, which is carried into the flame. The plasma is cooler at its center and hottest at its outside edge, where it is exposed to the high-frequency field. The end of the plasma flame is shown impinging onto the orifice of the sampling cone, which is part of a thick nickel-plated copper disc used to dissipate heat quickly. Electrons, ions, and neutrals pass through the sampler cone orifice and into the interface region before mass analysis. The positive ions are accelerated into the interface region by the large positive potential of about 6000 V while electrons are pulled out and neutralized at the sampling cone.

ICP/AES). The mass spectrometric approach has introduced a wider ranging and more sensitive system for estimating element types and abundances in a huge range of sample types.

Construction of the Plasma Torch

Figure 14.1 illustrates a typical arrangement for construction of a plasma torch. Essentially, it consists of three concentric quartz tubes. Argon gas flows through all three tubes. The middle tube is used to start the plasma, with the shield gas flowing to prevent the plasma from impinging onto the walls of the outer tube. The innermost tube is used to introduce the sample into the center of the plasma flame. The end of the torch is surrounded with a few turns of copper coil that carry a high-frequency (27 or 40 MHz) electromagnetic field, which couples inductively with charged species in the plasma.

When the plasma has formed at the exits to the concentric tubes, it assumes a shallow, rounded cone shape with a hollow center, into which the sample is introduced. The back of the plasma is prevented from touching and melting the inner concentric tubes by adjusting the flow of argon to a sufficiently high rate so that the plasma is not fully developed until it is almost into the coil region (Figure 14.1). Some torches also include a concentric water-cooled cage lying outside the outermost quartz tube but inside the high-frequency copper coil. This cage cools the flames and improves performance in some applications. (The use of a cage is discussed later.)

The end or front of the plasma flame impinges onto a metal plate (the cone or sampler or sampling cone), which has a small hole in its center (Figure 14.2). The region on the other side of the cone from the flame is under vacuum, so the ions and neutrals passing from the atmospheric-pressure hot flame into a vacuum space are accelerated to supersonic speeds and cooled as rapid expansion occurs. A supersonic jet of gas passes toward a second metal plate (the skimmer) containing a hole smaller than the one in the sampler, where ions pass into the mass analyzer. The sampler and skimmer form an interface between the plasma flame and the mass analyzer. A light

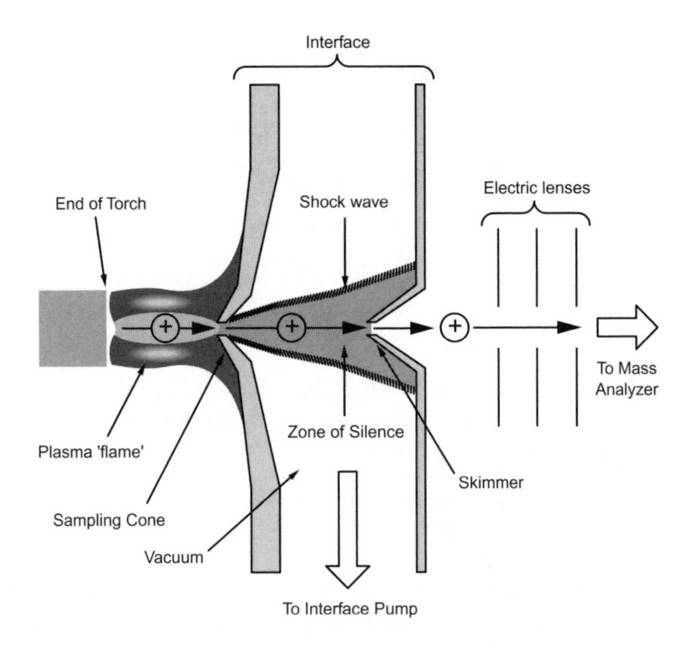

Figure 14.2
Hot ions and atoms formed in the plasma pass through the sampling cone orifice and into a vacuum of about 10^{-5} torr produced between it and the skimmer by an interface pump. Electrons are captured by the sampler, which is held at +6000 V. As the gases expand rapidly into the evacuated space, they produce a shock wave in the form of a hollow cone. The required ions are in the center of this cone (the zone of silence), forming a supersonic jet traveling at about 1000 m/sec. The beam of ions passes through a second orifice (the skimmer) into the mass analyzer. A system of lenses after the skimmer collimates the ion beam before it enters the analyzer. The passage of the positive ions is indicated (----). There may be a direct line-of-sight from the plasma flame to the ion collector, and the emitted light would strike the ion collector, producing a spurious background current. To prevent this background current production, a small metal plate (a light stop) is inserted into the line-of-sight, but then the ion beam must be deflected around it before going on to the analyzer. Alternatively, the ion beam can be deflected away from the line-of-sight via a hexapole and into the analyzer.

stop must be used to prevent photons from the plasma flame reaching the ion collector, which would produce a spurious high background signal.

Plasma Flame

The argon gas, which flows through the concentric quartz tubes (shown in Figure 14.1) and through the high-frequency field, does not become a plasma until a few electrons have been introduced near the flame end of the concentric tubes. The following sequence of events occurs within a few milliseconds. A hot spark, usually produced piezoelectrically, contains electrons that are carried by the flowing argon gas into an oscillating high-frequency electromagnetic field, where they are accelerated rapidly back and forth by its changing magnetic and electric components. The oscillating electrons collide with neutral argon atoms, and their motions become chaotic. Nevertheless, the electrons continue to be accelerated until they gain enough energy to cause ionization of some argon atoms. At this crucial stage, more electrons and ions are produced in a cascade process (Figure 14.3), so within a few milliseconds a high concentration of ions and electrons is produced in the flowing argon gas. The result is a plasma that glows with light emitted from excited atoms and ions and from recombination of electrons with ions (see Chapter 6). This glow gives the argon plasma its characteristic pale-blue-to-lilac coloration. There are approximately equal numbers of positive ions and electrons in a plasma, so there is not much of a space charge. The number of ions and electrons reaches a number density of about 10^{15} to 10^{16} per milliliter.

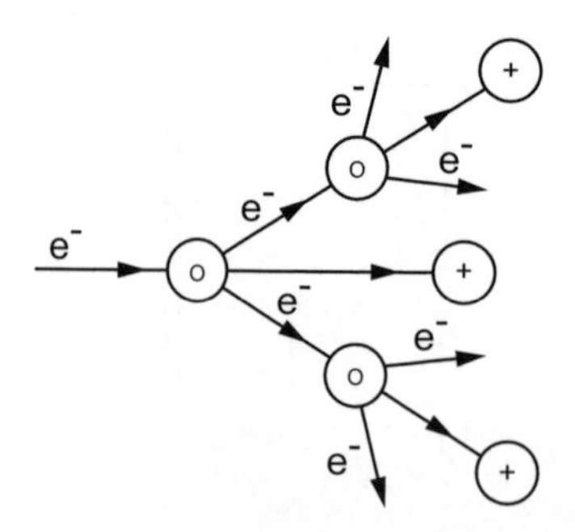

Figure 14.3
In a cascade process, one incident electron (e^-) collides with a neutral atom ((◎)) to produce a second electron and an ion ((⊕)). Now there are two electrons and one ion. These two electrons collide with another neutral atom to produce four electrons and three ions. This process continues rapidly and — after about 20 successive sets of collisions — there are millions of electrons and ions. (The mean free path between collisions is very small at atmospheric pressures.) A typical atmospheric-pressure plasma will contain 10^{16} each of electrons and ions per milliliter. Some ions and electrons are lost by recombination to reform neutral atoms, with emission of light.

Because the flowing argon gas passes through the high-frequency electromagnetic field placed at the ends of the concentric tubes, the plasma actually appears as a flame near the end of the torch, residing inside the outermost concentric tube. By arranging suitable argon gas flows through the three tubes, the very hot plasma flame is prevented from actually contacting the three tubes, which would otherwise melt. The plasma is self-sustaining in that there is no further need to seed it with electrons, which are in sufficient supply to ensure that the process continues until either the gas flow or the high-frequency field is switched off. Because the flame is formed mostly from the gas flowing through the second of the three concentric tubes, there is a small hollow space at its center due to the presence of the innermost tube, from which a cooling flow of sample entrained in argon gas enters and mixes with the flame (Figure 14.2). The outer skin of the plasma is very hot, since this is where most of the high-frequency heating takes place. To shield the outermost concentric quartz tube (Figure 14.1) from this very hot region of the flame, a third flow of cold argon cools the outermost tube and sweeps the plasma flame away. The uses and effects of the three gas flows are illustrated further in Figure 14.4.

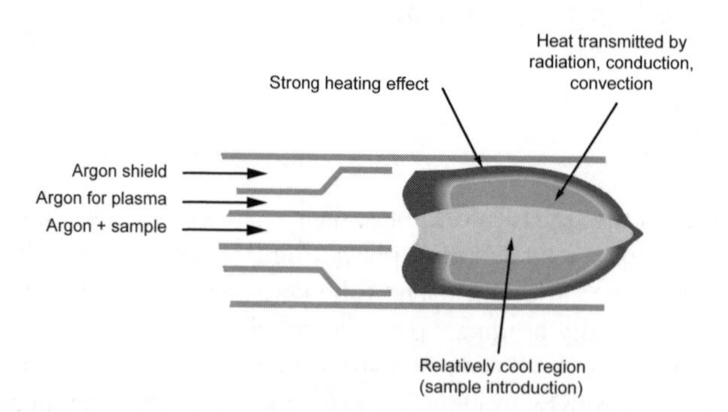

Figure 14.4
In the plasma flame, much of the heating effect of the high-frequency electromagnetic field occurs in the outer skin of the flame, in common with all high-frequency electromagnetic heating. The hot electrons, ions, and neutrals mingle with cooler materials toward the center of the flame, and energy is transferred by conduction, convection, and radiation. The inner regions of the flame are somewhat cooler than the outer, but they are still very hot. At the center of the flame is a flow of cold argon gas carrying the sample, and it is the coolest region. At the outer periphery of the flame, coolant argon gas is used to prevent the very hot plasma from impinging on and melting the outer quartz tube. The rapidly flowing argon gas carries the really hot parts of the plasma beyond the ends of the concentric tubes so the latter do not become too hot. The three gas flows are shown on the diagram, together with an indication of the hotness of the flame in different regions. From hottest to coolest, the temperature ranges between 7000–8000 and 4000–5000 K, but these ranges depend a great deal on the actual conditions of argon gas flow and power input from the high-frequency field.

Temperature of the Plasma

If a gas is contained in a vessel and then heated thermally, the constituent atoms of the gas gain thermal energy upon striking the hot walls of the containment vessel. The heat energy transferred to molecules increases their thermal motions (rotation, vibration, translation). Under equilibrium conditions, the added thermal energy is distributed among the three modes, but for monatomic gases, all of this increase in energy appears as kinetic (translational) energy because there are no bonds to vibrate or rotate in atoms. Thus the process of thermally heating a gas is characterized by an increase in the kinetic energy of its constituent atoms or molecules. Atomic or molecular mass is fixed, so the increased kinetic energy (= $mv^2/2$) appears as increased velocity of motion; viz., as the gas gets hotter, its constituent atoms or molecules move faster. As the atoms move faster and collide with other atoms, thermal energy is shared at each collision, with hot atoms or molecules passing on some of their kinetic energy to cooler atoms or molecules.

The total kinetic energy added through heating becomes distributed throughout the bulk of the gas. At atmospheric pressure, the mean free path between collisions of atoms or molecules is very short, so the extra energy is rapidly distributed through simple collision mechanisms. However, at atmospheric pressures, bulk movement of gas (convection) also leads to mixing. Finally, some of the heating effect arises directly through absorption of radiant energy from hotter gases. Thus, at any given temperature attained by a gas, its atoms or molecules have a distribution of kinetic energies, which is characterized by the Boltzmann equation if the gas has attained equilibrium conditions (Figure 14.5). To obtain a temperature of 8000 to 10,000 K by externally heating a gas, the walls of the containment vessel would have to be heated to an even higher temperature, but there are no materials that can withstand such temperatures.

In a plasma, the constituent atoms, ions, and electrons are made to move faster by an electromagnetic field and not by application of heat externally or through combustion processes. Nevertheless, the result is the same as if the plasma had been heated externally; the constituent atoms, ions, and electrons are made to move faster and faster, eventually reaching a distribution of kinetic energies that would be characteristic of the Boltzmann equation applied to a gas that had been

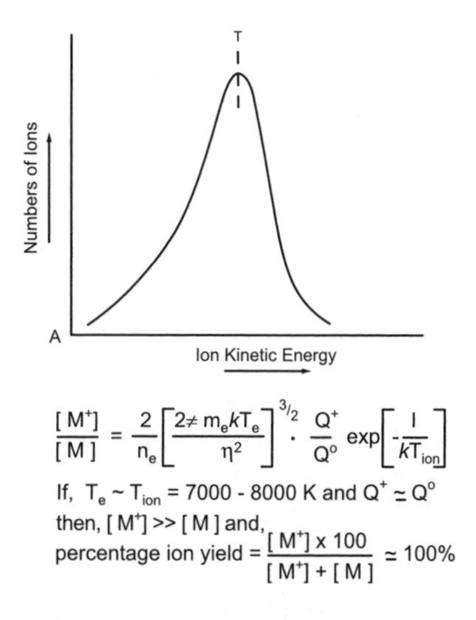

Figure 14.5
At equilibrium, at some temperature T, the atoms or molecules of a gas will have a range of kinetic energies characteristic of the temperature. The diagram illustrates this range for one temperature. At one end of the curve, there is a fairly sharp cutoff in numbers of atoms or molecules having high kinetic energies. At the other end, there is a more gradual drop in numbers of molecules having lower energies. The position of the top of the peak is a characteristic of the temperature and moves from left to right as temperature increases. [T_e = electron temperature, T_{ion} = ion temperature, m_e = mass of an electron at rest, k = Boltzmann constant, h = Planck's constant, Q^+, $\approx Q^0$ = partition functions (approximately equal), I = ionization energy for the element considered.]

heated externally. In other words, the electromagnetically heated gas attains a near-equilibrium condition that is equivalent to some temperature T that might have been achieved thermally. The plasma can be described as having a temperature T defined by the speed of movement of its constituent atoms, ions, and electrons. For an atmospheric-pressure plasma struck in argon, the kinetic energies of the ions, atoms, and electrons are equivalent to a gas that has been heated thermally to 6000 to 8000 K, depending on the conditions used. A plasma flame can be characterized by its temperature, although there are other details that also need to be considered.

The first limitation arises because of the three species present (atoms, ions, and electrons). Electrons have very small mass compared with atoms and ions, and they can be accelerated more easily by the electromagnetic field than can the ions. Thus the electrons move very much faster than the ions and can be said to have a higher temperature (kinetic energy) than the ions. In turn, the movement of ions is affected by the electromagnetic field, while the movement of atoms is not. The ions move somewhat faster than the atoms, so their temperature is somewhat higher than that of the atoms. The atoms gain in velocity through collisions with hot ions and electrons. Therefore, unlike a gas heated externally and having mostly only atoms or molecules present, a plasma contains three species, each with its own temperature, and $T_e > T_i > T_a$, where T_e is the mean temperature of electrons, T_i that of ions, and T_a that of atoms. Often T_i and T_a are very similar and are simply described as a gas temperature, T_g. T_e normally lies in the range 6000–8000 K and T_g in the range 5000–7000 K. Due to convective, conductive, and radiative effects, these temperatures tend to be equalized, and it is often sufficiently accurate to describe the plasma flame as having a single temperature.

The second limitation arises from the method of heating. For a contained gas heated externally, rapid collisions and convection currents mean that the temperature throughout the gas rapidly becomes uniform (equilibrated). A plasma heated by inductive coupling to a high-frequency electromagnetic field is not heated uniformly. For high-frequency heating of any material, a skin effect is typically found, with most of the direct heating occurring near the surface of the plasma (Figure 14.4). However, this heat soon becomes distributed to the remainder of the flowing gas through diffusion, convection, conduction, and radiation. Because the gas is flowing, there is insufficient time for the skin heating to attain full equilibration conditions, so the plasma has a temperature profile from hot on the outside to cooler on the inside. The central regions of the plasma, into which sample is introduced, are somewhat cooler than the outer reaches, but still hot.

A calculation of ion yields is shown in Figure 14.5. The equation in Figure 14.5 illustrates how the number of singly charged positive ions $[M^+]$ can be calculated in relation to the number of neutral particles [M]. For example, with potassium at 8000 K, the ratio of ions to neutral atoms in a plasma is about 2000:1. Thus, at high temperatures, $[M^+] >> [M]$, and the yield of ions reaches close to 100% for all elements. Therefore high sensitivities for detection of ions depends partly on having a high plasma flame temperature. However, at the higher temperatures, new molecular ions such as ArO^+ begin to form, and to suppress them it may be necessary to operate the flame at a lower temperature.

The degree of ionization increases with temperature, and at 6000–8000 K, where ionization efficiencies are 90 to 95%, nearly all atoms exist as ions in the plasma. Operation of the plasma torch under conditions that produce a cooler flame (cold plasma) has important advantages for some applications, which are discussed after the next section.

Processes Occurring in the Plasma after Introduction of a Sample

If a sample substance is introduced into a plasma, its constituent molecules experience a number of radiative, convective, and collisional processes that break the molecules into their constituent atoms, which appear mostly in ionized form.

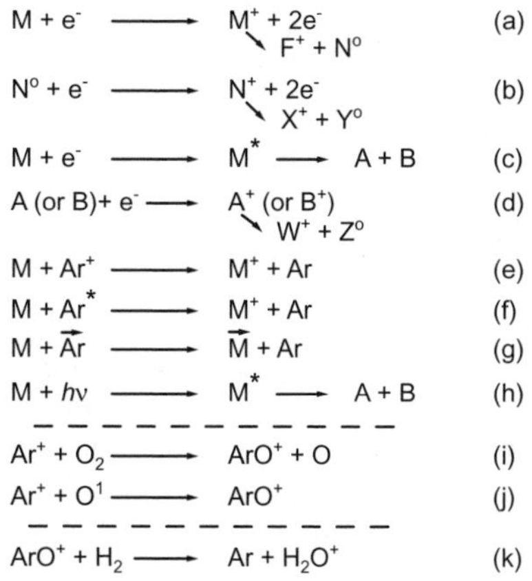

$M + e^- \longrightarrow M^+ + 2e^-$ (a)
$\searrow F^+ + N^o$

$N^o + e^- \longrightarrow N^+ + 2e^-$ (b)
$\searrow X^+ + Y^o$

$M + e^- \longrightarrow M^* \longrightarrow A + B$ (c)

$A\ (\text{or } B) + e^- \longrightarrow A^+\ (\text{or } B^+)$ (d)
$\searrow W^+ + Z^o$

$M + Ar^+ \longrightarrow M^+ + Ar$ (e)

$M + Ar^* \longrightarrow M^+ + Ar$ (f)

$M + \vec{Ar} \longrightarrow \vec{M} + Ar$ (g)

$M + h\nu \longrightarrow M^* \longrightarrow A + B$ (h)

- - - - - - - - - - - - -

$Ar^+ + O_2 \longrightarrow ArO^+ + O$ (i)

$Ar^+ + O^1 \longrightarrow ArO^+$ (j)

- - - - - - - - - - - - -

$ArO^+ + H_2 \longrightarrow Ar + H_2O^+$ (k)

Figure 14.6
The various reactions illustrate some of the major routes to fragmentation of molecules inside a plasma: (a) Ionization and fragmentation following collision of a molecule M with an electron to produce molecular ions M^+, fragment ions F^+, and neutrals N^0. (b) Ionization and fragmentation of neutrals N^0 from (a) collision with other electrons to give ions N^+, X^+ and neutrals Y^0. (c) Vibrational excitation of molecules M by electrons to give M^*, followed by bond breaking to give neutral fragments A, B. (d) Ionization of neutrals A, B from the latter process. (e) Charge exchange of an argon ion and a molecule M. (f) Ionization by collision of an excited argon atom with a molecule M. (g) A fast argon atom colliding with a molecule M and passing on some of its kinetic energy (designated by the arrow). (h) Absorption of radiation $h\nu$ leading to bond breaking in the molecule M after excitation to M^*. (i, j) Two ways in which ArO^+ can be formed by collisional processes. (k) Collision of ArO^+ ions with H_2 to form Ar^+ and H_2O.

Any sample to be introduced into the center of the plasma flame is often first nebulized (broken down into small droplets) by using argon gas as a spraying medium. The argon gas and the spray of droplets flow down a central tube of the plasma torch (Figures 14.1 and 14.4) and into the center of the plasma flame. Given the high temperatures in the plasma, the droplets rapidly lose low-boiling solvent molecules, and both solvent and solute molecules diffuse into the plasma, where a number of processes occur leading to molecular fragmentation.

Fast-moving, high-temperature electrons in the plasma collide with sample molecules and cause ionization (Figure 14.6); the energy transferred in this process also causes some fragmentation. More collisions of the products from this first step with electrons cause more extensive fragmentation. Additionally, collisions of fragments or intact molecules with fast-moving hot ions and atoms from the plasma lead to the transfer of thermal (rotational and vibrational) energy, which also gives rise to thermal cracking of sample molecules. The cracked fragments are themselves subjected to collision with electrons, ions, and atoms, and the fragments are broken down and ionized even more. The numbers of collisions and the total energy transferred during the time the sample molecules are in the plasma is so high that, within about 2 to 3 msec, every sample molecule is broken into its constituent atoms, and most of these have been ionized. Essentially, it can be said that at the high temperatures within the plasma, sample molecules are rapidly torn apart into ions of their constituent elements. Conditions inside the plasma are so energetically fierce that all kinds of sample molecules are literally shredded into their constituent elements.

If the elements appear as excited atoms, emission of light occurs, which can be used to examine the elemental composition of a sample by using a spectrophotometer to sample the light emitted (ICP/AES), by which it identifies elements from their characteristic atomic emission lines and the amounts of the elements from the intensities of the emission lines. However, a very large proportion of the elements appear as ions, which are best examined by mass spectrometry (ICP/MS). The ions are mostly singly charged, but, depending on ionization energies and the temperature of the flame, there are also doubly charged ions.

Note that the plasma emits light of high intensity inside the plasma itself and over a wide spectrum of wavelengths; some of this light is absorbed by the sample molecules or their fragments and is rapidly converted into internal vibrational and rotational energy. This extra energy alone can lead to fragmentation of the sample molecules or further breakdown of the fragments already present.

In addition to the fragmentation reactions, there are also some interactions of various species that can synthesize unexpected and unusual molecular ions. For example, the argon gas of the plasma can combine with oxygen to give stable ArO^+ ions or with hydrogen to give ArH^+. These unexpected ions have strong ionic bonds and are formed at the high temperatures within the plasma. Their masses coincide with (are isobaric with) isotopes of other elements, which complicates the task of analysis. For example, ArO^+ (mass = 56) is isobaric with the main isotope of iron (mass = 56), and ArH^+ (mass = 41) is isobaric with an isotope of potassium. One way of reducing the numbers of these unwanted so-called molecular ions is to run the plasma under somewhat cooler conditions using a "cold" plasma. This term is relative because the plasma still reaches temperatures of several thousand degrees.

Cold Plasma Conditions and the Plasma Cage

Although the plasma has been described as producing a very low background of extraneous ions, the interfering molecular ions discussed above are produced in quantities sufficient to lead to difficulties in accurate measurement of isotopic ratios for certain elements. The plasma gas itself produces Ar^+ (m/z = 40, which interferes with calcium), ArH^+ (m/z = 41, interferes with calcium), and Ar_2^+ (m/z = 80, interferes with selenium). Other typical interfering ions arise from the solvents and give O^+, NO^+, ArO^+, O_2^+, and so on. Under cold plasma conditions, the abundance of these interfering isobaric ions can be reduced by several orders of magnitude.

There are several methods of producing cold plasmas. For example, the power input from the coil can be reduced and the sample injector gas flow increased. Reducing the power reduces the heating effect but still produces energetic electrons and ions. Increasing the gas flow to the inner regions of the plasma via the sample inlet reduces the temperature by mixing more cold gas with the plasma gas. Similarly, the level of the interfering molecular ions can be reduced simply by introducing another gas into the system, as with the inclusion of some nitrogen into the argon gas flow.

One way of producing a cold plasma is to place a cage or shield near the outermost tube of the plasma gun, inside the high-frequency coil (Figure 14.1). The shielding greatly reduces penetration of the electromagnetic field into the interior of the plasma. The heating at the outermost skin continues, but electromagnetic heating within the body of the plasma from the high-frequency field is reduced, although this region still receives heat from mixing (conduction, convection, and radiation) and is still quite hot at 4500 to 5000 K.

Although cold plasmas have benefits in removing interfering ions such as ArO^+, they are not necessary for other applications where interferences are not a problem. Thus, in laboratories where a range of isotopes needs to be examined, the plasma has to be changed between hot and cold conditions, whereas it is much simpler if the plasma can be run under a single set of conditions. For this reason, some workers use warm plasmas, which operate between the hot and cold conditions.

The cold plasmas tend to be unstable, are sometimes difficult to maintain, and provide ion yields that are less than those of the hot plasmas. To obviate the difficulties of the interfering isobaric molecular ions from hot plasmas, it has been found highly beneficial to include a collision cell (hexapole; see Chapter 22) before the mass analyzer itself. This collision cell contains a low pressure of hydrogen gas. Ion/molecule collisions between the hydrogen and, for example, ArO^+

lead to the removal of the offending ion through formation of neutral Ar atoms and H_2O^+ (Figure 14.6k). This principle can be extended into the flame itself. For example, C^+ ions formed from carbonaceous materials in the flame interact with Se atoms to give Se^+ ions and C atoms, and this indirect process enhances the yield of Se^+ ions formed in the flame directly.

The Interface

Ions produced in the plasma must be transferred to a mass analyzer. The flame is very hot and at atmospheric pressure, but the mass analyzer is at room temperature and under vacuum. To effect transfer of ions from the plasma to the analyzer, the interface must be as efficient as possible if ion yields from the plasma are to be maintained in the analyzer.

A typical interface is shown in Figure 14.2. The orifice in the sampler must be large enough to sample the center of the plasma flame while disturbing the plasma as little as possible. A diameter of about 1 mm is used normally, which is well in excess of the Debye length (Λ_D) but not so large as to let too much plasma gas or even air into the interface region, as this would make it difficult to maintain a suitable vacuum behind the sampler. The Debye length is a measure of the width of the effective electric field of an ion and is given approximately by Equation 14.1, in which T_e is the electron temperature and N_e is the number density of electrons (per milliliter).

$$\Lambda_D = 6.9(T_e/N_e)^{1/2} \tag{14.1}$$

For a plasma temperature of 8000 K and $N_e = 10^{14}$/ml, Λ_D is about 0.0006 mm, which is very much smaller than the 1-mm sampler orifice, so ions can pass through easily. Hot gases from the plasma impinge on the edges of the sampler orifice so deposits build up and then reduce its diameter with time. The surrounds of the sampler orifice suffer also from corrosive effects due to bombardment by hot species from the plasma flame. These problems necessitate replacement of the sampler from time to time.

As the gas leaves the other side of the sampler orifice, it experiences a vacuum of about 10^{-5} torr, and the expanding jet of gas cools very rapidly and reaches supersonic speeds. There is a resulting conical shock wave, inside which is the zone of silence (Figure 14.2). Air and other gases diffuse from outside and into the zone of silence, and the amount eventually reaches a level at which the shock wave heats it along a frontal zone called the *Mach disc*. Therefore, to sample the fast-moving ions issuing from the sampler, the entrance to the next orifice (the skimmer) needs to be sited within the zone of silence and before the Mach disc, which means it is normally placed about 5 to 6 mm from the sampler. The skimmer orifice needs to be small enough that it prevents too much exterior neutral gas getting through it and into the analyzer region and yet large enough that it exceeds the mean free path of the particles passing through it so as not to interrupt the flow. For these reasons, the skimmer orifice is about 0.8 mm in diameter.

After the skimmer, the ions must be prepared for mass analysis, and electronic lenses in front of the analyzer are used to adjust ion velocities and flight paths. The skimmer can be considered to be the end of the interface region stretching from the end of the plasma flame. Some sort of light stop must be used to prevent emitted light from the plasma reaching the ion collector in the mass analyzer (Figure 14.2).

Conclusion

A plasma of electrons, ions, and neutrals produced in gas flowing through concentric tubes is maintained and heated to 5000 to 8000 K by inductive coupling to a high (radio) frequency electromagnetic field. Sample substances introduced into the hot plasma are torn apart into their constituent atoms, most of which are ionized. The ions are measured for abundance and m/z values by a suitable mass analyzer, often a quadrupole or time-of-flight instrument.

Chapter 15

Sample Inlets for Plasma Torches, Part A: Gases

Introduction

To examine a sample by inductively coupled plasma mass spectrometry (ICP/MS) or inductively coupled plasma atomic-emission spectroscopy (ICP/AES), it must be transported into the flame of a plasma torch. Once in the flame, sample molecules are literally ripped apart to form ions of their constituent elements. These fragmentation and ionization processes are described in Chapters 6 and 14. To introduce samples into the center of the (plasma) flame, they must be transported there as gases, as finely dispersed droplets of a solution, or as fine particulate matter. The various methods of sample introduction are described here in three parts — A, B, and C; Chapters 15, 16, and 17, respectively — to cover gases, solutions (liquids), and solids. Some types of sample inlets are multipurpose and can be used with gases and liquids or with liquids and solids, but others have been designed specifically for only one kind of analysis. However, the principles governing the operation of inlet systems fall into a small number of categories. This chapter discusses specifically substances that are normally gases or very volatile liquids at ambient temperatures.

Problems of Sample Introduction

The two major difficulties facing the analyst/mass spectrometrist concern firstly how to get the whole of the sample into the plasma flame efficiently and secondly how to do so without destabilizing or extinguishing the flame. Although plasma flames operate at temperatures of 6000 to 8000 K, the mass of gas in the flame is very small, and its thermal capacity is correspondingly small (Figure 15.1).

Therefore, if a large quantity of sample is introduced into the flame over a short period of time, the flame temperature will fall, thus interfering with the basic processes leading to the formation and operation of the plasma. Consequently introduction of samples into a plasma flame needs to be controlled, and there is a need for special sample-introduction techniques to deal with different kinds of samples. The major problem with introducing material other than argon into the plasma flame is that the additives can interfere with the process of electron formation, a basic factor in keeping the flame self-sustaining. If electrons are removed from the plasma by

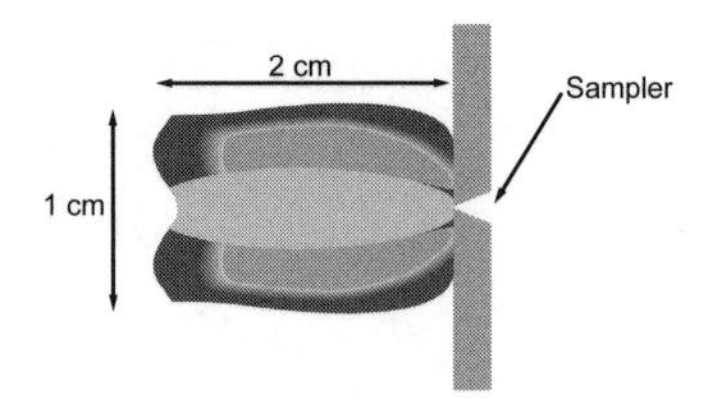

Density of argon gas (300 K)	=	1.78×10^{-3} g/mL
Specific heat of argon gas	=	0.124 cal/g/K
Flame temperature	=	5300 K
Flame dimensions (approx)	=	$\pi (0.5)^2 \times 2 = 1.6$ mL
Volume of 1.6 mL argon at 300K	= ≅	1.6 x 300/5300 0.1 mL
Heat content of flame	= =	$0.1 \times 1.78 \times 10^{-3} \times 0.124 \times 5300$ 0.1 cal
Power output of flame (2 ms)	=	$\frac{0.1}{2 \times 10^{-3}}$ = 50 watts
Power input to flame	≃	1 - 2 kW

Figure 15.1
A plasma flame commonly has a diameter of about 1 cm and a length of about 2–3 cm. If this flame is regarded as approximately cylindrical, the volume of the flame at about 5300 K is 1.6 ml and at 300 K is 0.1 ml. With a specific heat for argon of 0.124 cal/g/K and a density of 1.78×10^{-3} g/ml (at 300 K), the heat content of the flame is 0.1 cal. However, since gas flow through the hot flame occurs in a period of about 2 msec, the power output of the flame is about 50 W. This output should be compared with a power input from the high-frequency electromagnetic field of about 1 kW. The seeming inconsistency between the high temperature and the low heat content arises because of the low number density of hot particles. (The concentration of electrons and other particles in the hot flame is approximately 10^{-8} M.)

secondary processes faster than they can be replaced by the primary generation process, then the plasma process ceases and the flame will go out. Fluctuations in flame temperature and performance lead to significant variations in sample ion yield, often over short periods of time, and these fluctuations affect accurate measurement of isotope ratios. Thus sample preparation and manipulation are important and, for any one type of inlet system, judicious choice of inlet conditions and sample preparation by the operator of the instrument can avoid the worst aspects of the problems just described.

Introduction of Gases

Fundamentally, introduction of a gaseous sample is the easiest option for ICP/MS because all of the sample can be passed efficiently along the inlet tube and into the center of the flame. Unfortunately, gases are mainly confined to low-molecular-mass compounds, and many of the samples that need to be examined cannot be vaporized easily. Nevertheless, there are some key analyses that are carried out in this fashion; the major one is the generation of volatile hydrides. Other methods for volatiles are discussed below. An important method of analysis uses lasers to vaporize nonvolatile samples such as bone or ceramics. With a laser, ablated (vaporized) sample material is swept into the plasma flame before it can condense out again. Similarly, electrically heated filaments or ovens are also used to volatilize solids, the vapor of which is then swept by argon makeup gas into the plasma torch. However, for convenience, the methods of introducing solid samples are discussed fully in Part C (Chapter 17).

Arsenic	AsH_3
Antimony	SbH_3
Bismuth	BiH_3
Germanium	GeH_4
Lead	PbH_2
Mercury	Hg [a]
Phosphorus	PH_3 [b]
Selenium	SeH_2
Sulphur	SH_2
Telurium	TeH_2
Tin	SnH_2

a. Mercury does not form a hydride but any mercury compounds present are reduced to the element itself, which being a volatile liquid is carried along into the plasma flame.
b. This hydride is formed from phosphates and similar anions of phosphorus only on heating the phosphate to 500 °C with $NaBH_4$ in the dry state.

Figure 15.2
A number of elements form volatile hydrides, as shown in the table. Some elements form very unstable hydrides, and these have too transient an existence to exist long enough for analysis. Many elements do not form stable hydrides or do not form them at all. Some elements, such as sodium or calcium, form stable but very nonvolatile solid hydrides. The volatile hydrides listed in the table are gaseous and sufficiently stable to allow analysis, particularly as the hydrides are swept into the plasma flame within a few seconds of being produced. In the flame, the hydrides are decomposed into ions of their constituent elements.

Analysis of Elemental Hydrides (MH_n)

The elements listed in the table of Figure 15.2 are of importance as environmental contaminants, and their analysis in soils, water, seawater, foodstuffs and for forensic purposes is performed routinely. For these reasons, methods have been sought to analyze samples of these elements quickly and easily without significant prepreparation. One way to unlock these elements from their compounds or salts, in which form they are usually found, is to reduce them to their volatile hydrides through the use of acid and sodium tetrahydroborate (sodium borohydride), as shown in Equation 15.1 for sodium arsenite.

$$NaAsO_2 + NaBH_4 + 2HCl + H_2O = AsH_3 + H_2 + H_3BO_3 + 2NaCl \tag{15.1}$$

The volatile hydride (arsine in Equation 15.1) is swept by a stream of argon gas into the inlet of the plasma torch. The plasma flame decomposes the hydride to give elemental ions. For example, arsine gives arsenic ions at m/z 75. The other elements listed in Figure 15.2 also yield volatile hydrides, except for mercury salts which are reduced to the element (Hg), which is volatile. In the plasma flame, the arsine of Equation 15.1 is transformed into As^+ ions. The other elements of Figure 15.2 are converted similarly into their elemental ions.

A major advantage of this hydride approach lies in the separation of the remaining elements of the analyte solution from the element to be determined. Because the volatile hydrides are swept out of the analyte solution, the latter can be simply diverted to waste and not sent through the plasma flame itself. Consequently potential interference from sample-preparation constituents and by-products is reduced to very low levels. For example, a major interference for arsenic analysis arises from ions $ArCl^+$ having m/z 75,77, which have the same integral m/z value as that of As^+ ions themselves. Thus, any chlorides in the analyte solution (for example, from sea water) could produce serious interference in the accurate analysis of arsenic. The option of diverting the used analyte solution away from the plasma flame facilitates accurate, sensitive analysis of isotope concentrations. Inlet systems for generation of volatile hydrides can operate continuously or batchwise.

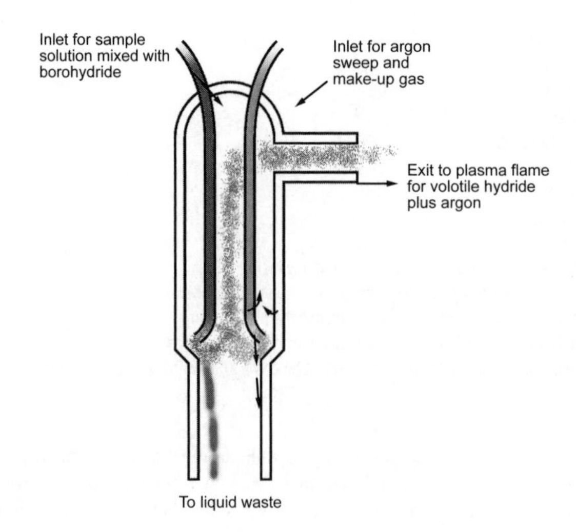

Figure 15.3
A schematic illustration of a typical inlet apparatus for separating volatile hydrides from the analyte solution, in which they are generated upon reduction with sodium tetrahydroborate. When the mixed analyte solution containing volatile hydrides enters the main part of the gas/liquid separator, the volatiles are released and mix with argon sweep and makeup gas, with which they are transported to the center of the plasma. The unwanted analyte solution drains from the end of the gas/liquid separator. The actual construction details of these gas/liquid separators can vary considerably, but all serve the same purpose. In some of them, there can be an intermediate stage for removal of air and hydrogen from the hydrides before the latter are sent to the plasma.

Unfortunately, the borohydride method of forming volatile hydrides also produces hydrogen gas as a by-product (Equation 15.1) which can interfere with the performance of the plasma flame. In sufficiently large concentrations, hydrogen can even cause the flame to go out because it has a high ionization potential, which interferes with the primary cascade production of electrons necessary to maintain the plasma (see Chapters 6 and 14). Other gases produced as by-products of the formation of volatile hydrides are CO_2 and H_2O, and these can also lead to instability in the plasma if present in too large a concentration. Devices for separating unused analyte or hydrogen have been produced for both batchwise and continuous use. One such separator is illustrated in Figure 15.3.

Hydrogen and entrained air can be removed from the gas stream entering the plasma by freezing out the volatile hydrides in a trap while diverting the hydrogen out of the system; the sample hydrides are released by warming the condensate and sweeping them to the flame with argon as they volatilize. Alternatively, hydrogen can be removed by passing the newly produced hydride gases through a tubular membrane separator, which allows easy diffusion of the small hydrogen molecules through its walls but not those of the much larger hydrides, which pass along the tube and into the plasma flame.

Other Volatile Materials Produced Chemically

Chemical ingenuity in using the properties of the elements and their compounds has allowed analyses to be carried out by processes analogous to the generation of hydrides. Osmium tetroxide is very volatile and can be formed easily by oxidation of osmium compounds. Some metals form volatile acetylacetonates (acac), such as iron, zinc, cobalt, chromium, and manganese (Figure 15.4). Iodides can be oxidized easily to iodine (another volatile element in itself), and carbonates or bicarbonates can be examined as CO_2 after reaction with acid.

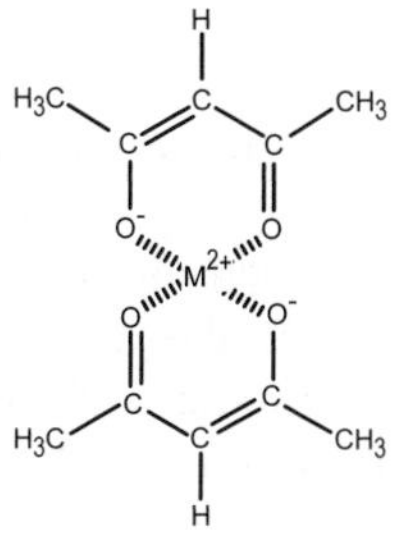

Figure 15.4
The chemical structure of a typical divalent metal acetylacetonate, for which the abbreviation would be $M(acac)_2$. These compounds are internally bonded ionically and complexed to oxygen at the same time. Thus, their intramolecular forces are very strong (they are stable), but their intermolecular forces are weak (they are volatile).

Introduction of Gases Directly

The direct insertion of gases is probably the easiest method of introduction. A gas sample can be introduced through the inlet into the center of the plasma flame, usually after first mixing the sample with argon to dilute it and carry it forward. (The argon is then known as a makeup gas.) For example, semiconductor-grade gases such as silane (SiH_4) can be analyzed by ICP/MS by mixing the gas with argon and admitting it into the plasma flame in small defined amounts (microliters) through use of a gas sampling loop. Where organic compounds are used, it is often beneficial to add low concentrations of oxygen or air to help burn them. Nitrogen or xenon can be used to reduce interferences caused by the formation of carbides. An example of the latter is an interference in chromium determination at m/z 52 caused by formation in the plasma of the carbide, ArC^+ (also at m/z 52). In one commercial instrument, this sort of interference is handled more efficiently with a hexapole collision chamber located after extraction of the ions from the plasma but before passage of the ions into the mass analyzer proper for measurement of m/z values. This collision chamber, containing low pressure hydrogen as the collision gas, causes the decomposition of the carbide species through ion/molecule reactions, as shown in Equation 15.2.

$$\begin{array}{ccc} ArC^+ + H_2 \rightarrow & Ar + & CH_2^+ \\ m/z\ 52 & & m/z\ 14 \end{array} \qquad (15.2)$$

Introduction of Gases and Vapors by Coupling to a Gas Chromatograph (GC/ICP/MS)

Ostensibly, a gas chromatograph (GC) apparatus should be ideal as an inlet for a plasma torch because the effluents from the chromatographic column are already in the gas phase and can be passed straight into the plasma flame. However, most analyses carried out by GC involve carbon compounds, with oxygen, nitrogen, and halogens as commonly occurring constituents. The gas flow from the capillary GC column is normally increased by mixing in more argon gas with the effluent before it reaches the flame. Introduction of a sample into a GC column as a solution means there is usually a large solvent peak at the start of any gas chromatogram. Because the eluting solvent contains a large amount of material, it must be diverted from the flame; otherwise, it would make the plasma very unstable or even extinguish it. Therefore, all GC connections to an ICP/MS instrument contain a diverter to switch the gas flow from the flame during the few seconds required for the solvent to elute from the end of the GC column.

Gas chromatography is mostly used for the analysis of carbon-containing compounds, which may or may not contain oxygen, nitrogen, sulfur, or halogens as the principal coelements. If ICP/MS is simply used only as a detector for GC effluents, without any other purpose, it becomes a very expensive coupled device compared with GC using a simple flame ionization detector. Therefore, to couple GC with an ICP/MS instrument, there must be factors so attractive as to override the cost disadvantages. Element ratio (isotope) analyses and multiple identification of elements are major reasons for using GC/ICP/MS. When mixtures of volatile compounds must be separated and examined for their element compositions and ratios, GC/ICP/MS becomes an important on-line technique. Such conjoint favorable economic circumstances are not too common, and, probably mostly for this reason, GC/ICP/MS is not as widely used as might be imagined, given its attractions. However, many applications of its use or potential use have been described in research papers, and it continues to be developed. For example, organo lead, tin, and mercury compounds have been separated and analyzed on-line by GC/ICP/MS.

Other Vapor Introduction Systems

Other vapor introduction systems are discussed in Parts B and C (Chapters 16 and 17) because, although liquids and solids are ultimately introduced to the plasma flame as vapors, these samples are usually prepared differently from naturally gaseous ones. For example, electrothermal (oven) or laser heating of solids and liquids to form vapors is used extensively to get the samples into the plasma flame. At one extreme with very volatile liquids, no heating is necessary, but, at the other extreme, very high temperatures are needed to vaporize a sample. For convenience, the electrothermal and laser devices are discussed in Part C (Chapter 17) rather than here.

Conclusion

Gases and vapors of volatile liquids can be introduced directly into a plasma flame for elemental analysis or for isotope ratio measurements. Some elements can be examined by first converting them chemically into volatile forms, as with the formation of hydrides of arsenic and tellurium. It is important that not too much analyte pass into the flame, as the extra material introduced into the plasma can cause it to become unstable or even to go out altogether, thereby compromising accuracy or continuity of measurement.

Chapter 16

Sample Inlets for Plasma Torches, Part B: Liquid Inlets

Introduction

To examine a sample by inductively coupled plasma mass spectrometry (ICP/MS) or inductively coupled plasma atomic-emission spectroscopy (ICP/AES) the sample must be transported into the flame of a plasma torch. Once in the flame, sample molecules are literally ripped apart to form ions of their constituent elements. These fragmentation and ionization processes are described in Chapters 6 and 14. To introduce samples into the center of the (plasma) flame, they must be transported there as gases, as finely dispersed droplets of a solution, or as fine particulate matter. The various methods of sample introduction are described here in three parts — A, B, and C; Chapters 15, 16, and 17 — to cover gases, solutions (liquids), and solids. Some types of sample inlets are multipurpose and can be used with gases and liquids or with liquids and solids, but others have been designed specifically for only one kind of analysis. However, the principles governing the operation of inlet systems fall into a small number of categories. This chapter discusses specifically substances that are normally liquids at ambient temperatures. This sort of inlet is the commonest in analytical work.

Problems of Sample Introduction

The two major difficulties facing the analyst/mass spectrometrist concern firstly how to get the whole of the sample into the plasma flame efficiently and secondly how to do so without destabilizing or extinguishing the flame. Although plasma flames operate at temperatures of 6000 to 8000 K, the mass of gas in the flame is very small, and its thermal capacity is correspondingly small (Figure 16.1).

Therefore, if a large quantity of sample is introduced into the flame over a short period of time, the flame temperature will fall, thus interfering with the basic processes leading to the formation and operation of the plasma. Consequently introduction of samples into a plasma flame needs to be controlled, and there is a need for special sample-introduction techniques to deal with different kinds of samples. The major problem with introducing material other than argon into the plasma flame is that such additives can interfere with the process of electron formation, a basic factor in keeping the flame self-sustaining. If electrons are removed from the plasma by

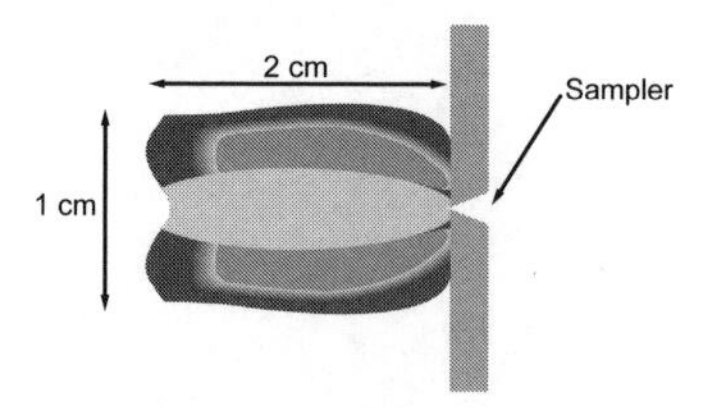

Density of argon gas (300 K)	=	1.78×10^{-3} g/mL
Specific heat of argon gas	=	0.124 cal/g/K
Flame temperature	=	5300 K
Flame dimensions (approx)	=	$\pi (0.5)^2 \times 2 = 1.6$ mL
Volume of 1.6 mL argon at 300K	=	$1.6 \times 300/5300$
	≅	0.1 mL
Heat content of flame	=	$0.1 \times 1.78 \times 10^{-3} \times 0.124 \times 5300$
	=	0.1 cal
Power output of flame (2 ms)	=	$\frac{0.1}{2 \times 10^{-3}} = 50$ watts
Power input to flame	≅	1 - 2 kW

Figure 16.1
A plasma flame commonly has a diameter of about 1 cm and a length of about 2–3 cm. If this flame is regarded as being approximately cylindrical, the volume of the flame at about 5300 K is 1.6 ml and at 300 K is 0.1 ml. With a specific heat for argon of 0.124 cal/g/K and a density of 1.78×10^{-3} g/ml (at 300 K), the heat content of the flame is 0.1 cal. However, since gas flow through the hot flame occurs in a period of about 2 msec, the power output of the flame is about 50 W. This output should be compared with a power input from the high-frequency electromagnetic field of about 1 kW. The seeming inconsistency between the high temperature and the low heat content arises because of the low number density of hot particles. (The concentration of electrons and other particles in the hot flame is approximately 10^{-8} M.)

secondary processes faster than they can be replaced by the primary generation process, then the plasma process ceases and the flame goes out. Fluctuations in flame temperature and performance lead to significant variations in sample ion yield, often over short periods of time, and these fluctuations affect accurate measurement of isotope ratios. Thus sample preparation and manipulation are important, and for any one type of inlet system, judicious choice of inlet conditions and sample preparation by the operator of the instrument can avoid the worst aspects of the problems just described.

Liquid Inlets

Suitable inlets commonly used for liquids or solutions can be separated into three major classes, two of which are discussed in Parts A and C (Chapters 15 and 17). The most common method of introducing the solutions uses the nebulizer/desolvation inlet discussed here. For greater detail on types and operation of nebulizers, refer to Chapter 19. Note that, for all samples that have been previously dissolved in a liquid (dissolution of sample in acid, alkali, or solvent), it is important that high-purity liquids be used if cross-contamination of sample is to be avoided. Once the liquid has been vaporized prior to introduction of residual sample into the plasma flame, any nonvolatile impurities in the liquid will have been mixed with the sample itself, and these impurities will appear in the results of analysis. The problem can be partially circumvented by use of blanks, viz., the separate examination of levels of residues left by solvents in the absence of any sample.

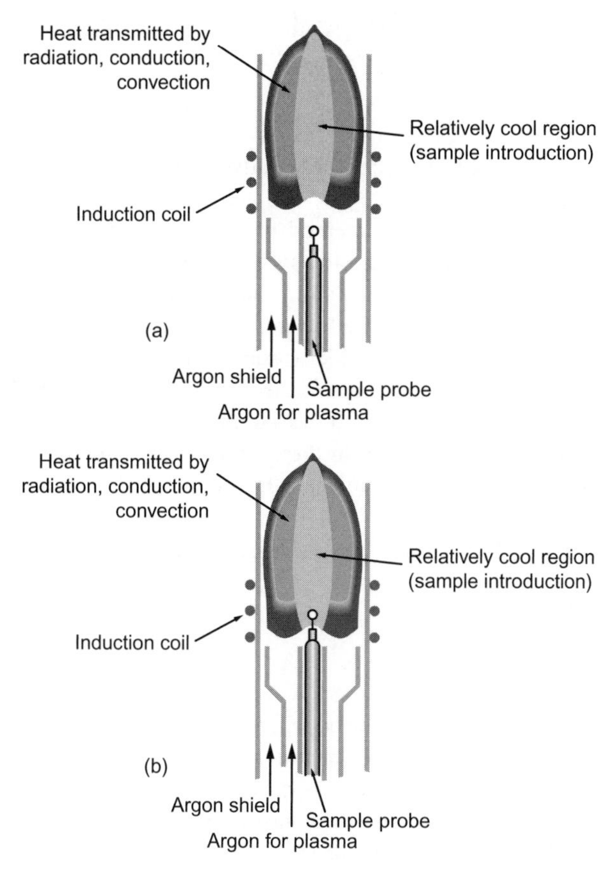

Figure 16.2
Diagram illustrating the method of direct insertion of a sample into a plasma flame. Initially, the sample is held on a wire or in a small graphite or metal boat and then placed just below the flame until conditions have stabilized. The probe is then moved up so the sample holder enters the flame, where ablation occurs and the sample is simultaneously fragmented and ionized. The plasma flame is shown in a vertical configuration (instead of the usual horizontal) because this position eliminates the possibility of the sample's dropping out of the cup used for sample insertion. If there is no problem of containing or holding the sample, then the flame can be in its more normal horizontal alignment.

Direct Insertion Methods (Direct Solids Insertion, DSI)

DSI is discussed in Part C (Chapter 17), since the approach usually requires an initial evaporation of solvent from a solution by moderate heating in a gas stream so as to leave the solute (the analytical sample). The resulting residual sample is then heated strongly to vaporize it. Typically, a solution is placed onto a heat-resistant wire or onto a graphite probe, and then the solvent is allowed to evaporate or is encouraged to do so by application of heat, directly or indirectly. The residual solid on its metal or graphite support is placed just below the plasma flame, which is allowed to stabilize for a short time. The probe and sample are then driven into the high-temperature flame, which causes vaporization, fragmentation, and ionization (Figure 16.2). Because the heat capacity of the flame is relatively small, the sample holder and sample should have as low a thermal mass as possible so as not to interfere with the operation of the flame. With the direct-insertion method, samples appear transiently in the flame; therefore, if a wide range of elements is to be examined, the mass spectrometer should be one that can span a wide m/z range in the short space of time the sample takes to pass through the flame (quadrupole, time-of-flight). Further details of the DSI technique are discussed in Part C (Chapter 17).

Electrothermal Heating Methods (Electrothermal Vaporization, ETV)

This topic is described in Part C (Chapter 17) because application of heat to remove the solvent from a solution results in a residual solid (or other relatively nonvolatile material), which must then be heated strongly to effect vaporization to form an aerosol. The latter is swept into the plasma flame with a flow of argon gas. The ETV method involves placement of a sample solution on a wire, into a boat or cup, or in an oven, where the solvent is evaporated by application of moderate heat. The resulting solvent vapors are routed from the plasma flame, or the evaporation is done slowly to avoid extinguishing the flame. After evaporation of the solvent, the flame is allowed to stabilize if necessary, and then the sample holder (probe) is heated strongly by electrical means to vaporize the residual solid sample as an aerosol, the droplets of which are led into the flame by argon gas flowing through a transfer line. In one variation of this method, the sample on its sample holder (a wire) is placed close to the flame in the sample inlet tube, and the wire is then heated electrically to drive off first the solvent and then the sample. (This apparatus is similar to that shown in Figure 16.2.)

Further details of the ETV technique are described in Part C (Chapter 17).

Nebulizer Methods

By far the most widely used method of introducing a liquid sample into a plasma flame is by splitting the liquid into a stream of droplets (nebulization), which is led along a transfer line and then into the center of the plasma flame. During transfer, the droplets evaporate and become very small, often consisting of only residual analyte. Once in the flame, the small drops of sample are fragmented and ionized. The process can be used to generate a transient or a continuous signal.

The nebulization concept has been known for many years and is commonly used in hair and paint spays and similar devices. Greater control is needed to introduce a sample to an ICP instrument. For example, if the highest sensitivities of detection are to be maintained, most of the sample solution should enter the flame and not be lost beforehand. The range of droplet sizes should be as small as possible, preferably on the order of a few micrometers in diameter. Large droplets contain a lot of solvent that, if evaporated inside the plasma itself, leads to instability in the flame, with concomitant variations in instrument sensitivity. Sometimes the flame can even be snuffed out by the amount of solvent present because of interference with the basic mechanism of flame propagation. For these reasons, nebulizers for use in ICP mass spectrometry usually combine a means of desolvating the initial spray of droplets so that they shrink to a smaller, more uniform size or sometimes even into small particles of solid matter (particulates).

Nebulizers can be divided into several main types. The pneumatic forms work on the principle of breaking up a stream of liquid into droplets by mechanical means; the liquid stream is forced through a fine nozzle and breaks up into droplets. There may be a concentric stream of gas to aid the formation of small droplets. The liquid stream can be directed from a fine nozzle at a solid target so that, on impact, the narrow diameter stream of liquid is broken into many tiny droplets. There are variants on this approach, described in the chapter devoted to nebulizers (Chapter 19).

Thermospray nebulizers operate on the principle of converting a thin stream of liquid into a rapidly expanding vapor through application of heat. As the liquid stream to be nebulized reaches the end of a capillary tube, it meets a short region of tubing that has been preheated to such a temperature that the solvent begins to boil rapidly. The resultant expanding gas stream mixes with unvaporized liquid to give a fine spray of droplets from the end of the capillary tube. These sorts of nebulizers are used to introduce solution eluants from liquid chromatography instruments, particularly those using the very narrow nanobore columns. (Thermospray is described in Chapter 11.) Electrospray forms another mode of nebulization suitable for liquid chromatography and is discussed in detail in Chapter 8.

Droplets can be produced ultrasonically. Application of a rapidly oscillating electric potential (200 to 1000 kHz) to certain types of inorganic crystal causes a face of the crystal to oscillate at a similar rate (piezoelectric effect). For use in piezoelectric devices, the moving face of the crystal is usually protected by fastening a thin metal plate to it. Since erosive and corrosive effects are frequently common with such devices, chemically and physically robust metals such as titanium are used to provide this facing to the piezoelectric crystal. Any thin stream of liquid directed at a steep angle onto the rapidly oscillating surface of a piezoelectric crystal is subjected to a standing acoustic wave, which breaks the stream into droplets. The rate at which droplets are produced is much greater than for pneumatic nebulizers, and a desolvation chamber is necessary to avoid overly large amounts of solvent and sample entering the flame and causing instability.

Once an aerosol has been produced, it is usually in the form of vapor (gas) from the solvent plus droplets of the original solution and sometimes also particulate (solid) matter. The droplets usually cover a wide range of sizes, which are not static but vary as the droplets proceed toward the plasma flame. As the aerosol is swept toward the flame by a flow of argon gas, small droplets grow bigger by collision with others, but, overall, droplets become smaller as the solvent evaporates. Thus, the initial aerosol produced at the nebulizer will have changed its size distribution by the time it meets the plasma flame. Most of these changes will have occurred within about 100 to 200 msec. The changes are frequently assisted by incorporation of a desolvation chamber between the nebulizer and the flame.

Desolvation Chambers

The solutions introduced into an ICP/MS instrument are commonly water-based (acids, alkalies, salts, or sea water) but can also be organic based, as with effluents from liquid chromatography columns. The two types of solvent cause different problems for a plasma flame, but high concentrations of either need to be avoided. A plasma flame has low thermal capacity and depends critically on a continuous formation of electrons and ions. If the flame is diluted by large amounts of vapor or by the presence of certain elements, its temperature is reduced, with a subsequent reduction in ionization efficiency. At best, a variable efficiency of ionization causes problems with the measurement of accurate abundances of ions, and, at worst, the flame may be extinguished altogether. Apart from these problems, inside the flame, water can give rise to interferences due to oxide formation within droplets before evaporation is complete. This effect is particularly marked for the elements vanadium, molybdenum, lanthanum, cerium, thorium, and uranium. For organic solvents, flame efficiency can even be increased by chlorinated solvents which give better efficiencies than do hydrocarbons. An excess of organic solvent can lead to a buildup of carbon deposits on the sampling cone situated at the tip of the plasma flame (see Chapter 14, "Plasma Torches").

These factors make it necessary to reduce the amount of solvent vapor entering the flame to as low a level as possible and to make any droplets or particulates entering the flame as small and of as uniform a droplet size as possible. Desolvation chambers are designed to optimize these factors so as to maintain a near-constant efficiency of ionization and to flatten out fluctuations in droplet size from the nebulizer. Droplets of less than 10 μm in diameter are preferred. For flow rates of less than about 10 μl/min issuing from micro- or nanobore liquid chromatography columns, a desolvation chamber is unlikely to be needed.

The simplest desolvation chambers consist simply of a tube heated to about 150°C through which the spray of droplets passes. During passage through this heated region, solvent evaporates rapidly from the droplets and forms vapor. The mixed vapor and residual small droplets or particulates of sample matter are swept by argon through a second cooled tube, which allows vapor to

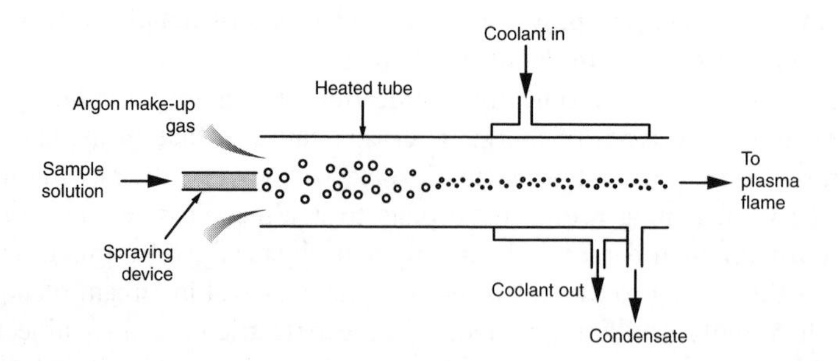

Figure 16.3
In a typical desolvation chamber, the initial sample solution is nebulized by some form of spraying device. The resultant aerosol — a mix of gas, vapor, and droplets having a wide range of sizes — is swept through a heated tube by a flow of argon gas. The tube is typically held at about 150°C. Here, much of the solvent is vaporized from the droplets, which are greatly reduced in size to become small multimolecular aggregates (very small droplets). This mixture of small gaseous solvent molecules and the larger analyte particles or aggregates passes into a cool region, where the solvent molecules, diffusing more rapidly from the stream of gas and droplets, condense onto the walls of the tube and are run off as liquid waste. The remaining small droplets and particulates (together with traces of solvent) pass on into the center of the plasma flame, where fragmentation and ionization of sample occurs. In other devices, the desolvation chamber consists of a membrane through which the small solvent molecules can diffuse but the larger droplets and particulates cannot; the materials retained on the membrane are passed on to the flame.

condense on its walls and to be run off to waste (Figure 16.3). This second tube can be maintained at a temperature of about 0 to –10°C. Other, more elaborate systems that subject the sample to alternate heating and cooling treatments are used to remove almost all of the solvent.

A second form of desolvation chamber relies on diffusion of small vapor molecules through pores in a Teflon® membrane in preference to the much larger droplets (molecular agglomerations), which are held back. These devices have proved popular with thermospray and ultrasonic nebulizers, both of which produce large quantities of solvent and droplets in a short space of time. Bundles of heated hollow polyimide or Naflon® fibers have been introduced as short, high-surface-area membranes for efficient desolvation.

Conclusion

Solutions can be examined by ICP/MS by (a) removing the solvent (direct and electrothermal methods) and then vaporizing residual sample solute or (b) nebulizing the sample solution into a spray of droplets that is swept into the plasma flame after passing through a desolvation chamber, where excess solvent is removed. The direct and electrothermal methods are not as convenient as the nebulization inlets for multiple samples, but the former are generally much more efficient in transferring samples into the flame for analysis.

Chapter 17

Sample Inlets for Plasma Torches, Part C: Solid Inlets

Introduction

To examine a sample by inductively coupled plasma mass spectrometry (ICP/MS) or inductively coupled plasma atomic-emission spectroscopy (ICP/AES), it must be transported into the flame of a plasma torch. Once in the flame, sample molecules are literally ripped apart to form ions of their constituent elements. These fragmentation and ionization processes are described in Chapters 6 and 14. To introduce samples into the center of the plasma flame, they must be transported there as gases or finely dispersed droplets of a solution or as fine particulate matter (aerosol). The various methods of sample introduction are described here in three parts — A, B, and C; Chapters 15, 16, and 17 — to cover gases, solutions (liquids), and solids. Some types of sample inlets are multipurpose and can be used with gases and liquids or with liquids and solids, but others have been designed specifically for only one kind of analysis. However, the principles governing the operation of inlet systems fall into a small number of categories. This chapter deals specifically with substances that are normally solids at ambient temperatures.

Problems of Sample Introduction

The two major difficulties facing the analyst/mass spectrometrist concern firstly how to get the whole of the sample into the plasma flame efficiently and secondly how to do so without destabilizing or extinguishing the flame. Although plasma flames operate at temperatures of 6000 to 8000 K, the mass of gas in the flame is very small, and its thermal capacity is correspondingly small (Figure 17.1).

Therefore, if a large quantity of sample is introduced into the flame over a short period of time, the flame temperature will fall, thus interfering with the basic ionization processes leading to the formation and operation of the plasma. Consequently, introduction of samples into a plasma flame needs to be controlled, and there is a need for special sample-introduction techniques to deal with different kinds of samples. The major problem with introducing material other than argon into the plasma flame is that such additives can interfere with the process of electron formation, a basic factor in keeping the flame self-sustaining. If electrons are removed from the plasma by secondary processes faster than they can be replaced by the primary generation process, then the plasma process ceases and the flame goes out. Fluctuations in flame temperature and performance lead to significant varia-

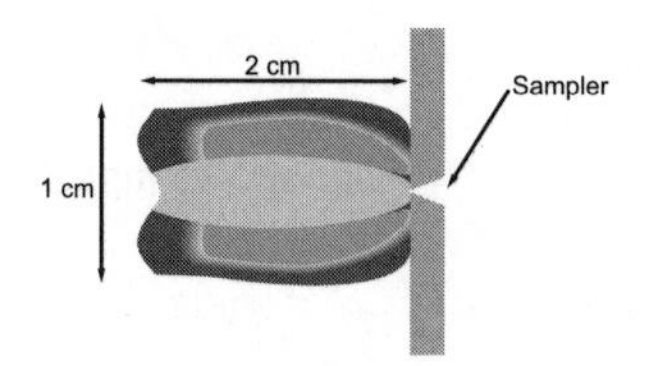

Density of argon gas (300 K)	=	1.78×10^{-3} g/mL
Specific heat of argon gas	=	0.124 cal/g/K
Flame temperature	=	5300 K
Flame dimensions (approx)	=	$\pi (0.5)^2 \times 2 = 1.6$ mL
Volume of 1.6 mL argon at 300K	= ≅	$1.6 \times 300/5300$ 0.1 mL
Heat content of flame	= =	$0.1 \times 1.78 \times 10^{-3} \times 0.124 \times 5300$ 0.1 cal
Power output of flame (2 ms)	=	$\frac{0.1}{2 \times 10^{-3}} = \underline{50}$ watts
Power input to flame	≃	1 - 2 kW

Figure 17.1
A plasma flame commonly has a diameter of about 1 cm and a length of about 2–3 cm. If this flame is regarded as being approximately cylindrical, the volume of the flame at about 5300 K is 1.6 ml and at 300 K is 0.1 ml. With a specific heat for argon of 0.124 cal/g/K and a density of 1.78×10^{-3} g/ml (at 300 K), the heat content of the flame is 0.1 cal. However, since gas flow through the hot flame occurs in a period of about 2 msec, the power output of the flame is about 50 W. This output should be compared with a power input from the high-frequency electromagnetic field of about 1 kW. The seeming inconsistency between the high temperature and the low heat content arises because of the low number density of hot particles. (The concentration of electrons and other particles in the hot flame is approximately 10^{-8} M.)

tions in sample ion yield, often over short periods of time, and these fluctuations affect accurate measurement of isotope ratios. Thus sample preparation and manipulation are important and, for any one type of inlet system, judicious choice of inlet conditions and sample preparation by the operator of the instrument can avoid the worst aspects of the problems just described.

Introduction of Solids

In some cases, it may be convenient to dissolve a solid and present it for analysis as a solution that can be nebulized and sprayed as an aerosol (mixed droplets and vapor) into the plasma flame. This aspect of analysis is partly covered in Part B (Chapter 16), which describes the introduction of solutions. There are vaporization techniques for solutions of solids other than nebulization, but since these require prior evaporation of the solvent, they are covered here. There are also many solid samples that need to be analyzed directly, and this chapter describes commonly used methods to do so.

Basically, there is only one method for dealing with solids, and that is to vaporize them in some way. Because solids vary from highly volatile (e.g., iodine) to highly nonvolatile (e.g., ceramic materials), it is not surprising that different methods have been devised for vaporizing solid samples. Although desirable, it is often the case that a solid cannot simply be put into a vaporization chamber. For example, if a solid has been dissolved first in acid, it is necessary to remove excess acid and/or solvent from the resulting liquid sample by selective heating so the more volatile components are vaporized first and any solid residue is vaporized later.

The various heating methods produce a vapor that is a mixture of gas, very small droplets, and small particles of solid matter (particulates). Before droplets or particulates can coalesce, the whole vapor is swept into the plasma flame for analysis. Clearly, the closer the heating source is to the sample flame, the less are the losses due to deposition on surrounding walls in the instrument or on lead-ins to the flame. However, this does not prevent carriage of vapors over quite long distances (up to 20 m is possible for some inlets). The more important methods used for introduction of solids are based on lasers, arcs (and sparks), and conventional electrical heating. In some instances, the sample can be heated directly by the plasma flame.

Laser Devices (Laser Ablation, LA)

Laser desorption to produce ions for mass spectrometric analysis is discussed in Chapter 2. As heating devices, lasers are convenient when much energy is needed in a very small space. A typical laser power is 10^{10} W/cm^2. When applied to a solid, the power of a typical laser beam — a few tens of micrometers in diameter — can lead to very strong localized heating that is sufficient to vaporize the solid (ablation). Some of the factors controlling heating with lasers and laser ablation are covered in Figure 17.2.

Typical data for a laser running as a pulsed beam (Q-mode) could be:

Power output = 10^9 watts.cm^{-2} = 10^9 J.s^{-1} cm^{-2}
Pulse length = 15 ns = 15×10^{-9} s
Laser beam radius = 0.5 m = 0.5×10^{-3} cm

Area of sample exposed = $\pi r^2 = 3.14 \times (0.5 \times 10^{-3})^2 = 0.8 \times 10^{-6}$ cm^2

Typical data for an iron sample could be:

Density of iron = 8 g.cm^{-3}
Specific heat of iron = 0.1 cal.g^{-1}

1 J = 4.2 cal

Let ΔT = the rise in temperature (K), when an iron sample is heated by 1 laser pulse and a pit of 4×10^{-4} cm is produced.

Then,

the volume of iron ablated = $0.8 \times 4 \times 10^{-10}$ cm^3 = 3.2×10^{-10} cm^3
the mass of iron ablated = 2.6×10^{-9} g
heat required for ablation = $2.6 \times 10^{-10} \times \Delta T$ cal

But,

heat input by laser in one pulse = $10^9 \times 15 \times 10^{-9} \times 4.2$ cal.cm^{-2} = 63 cal.cm^{-2}
heat input over the ablated area = $63 \times 0.8 \times 10^{-6}$ cal = 5×10^{-5} cal

If,

heat input from laser = heat used for ablation,
5×10^{-5} = $2.6 \times 10^{-10} \times \Delta T$,

And,

$\Delta T = 2 \times 10^5$ K

At 2% efficiency, estimated ΔT = 4000 K

Figure 17.2
These data are typical of lasers and the sorts of samples examined. The actual numbers are not crucial, but they show how the stated energy in a laser can be interpreted as resultant heating in a solid sample. The resulting calculated temperature reached by the sample is certainly too large because of several factors, such as conductivity in the sample, much less than 100% efficiency in converting absorbed photon energy into kinetic energy of ablation, and much less than 100% efficiency in the actual numbers of photons absorbed by the sample from the beam. If the overall efficiency is 1–2%, the ablation temperature becomes about 4000 K.

Suffice it to say at this stage that the surfaces of most solids subjected to such laser heating will be heated rapidly to very high temperatures and will vaporize as a mix of gas, molten droplets, and small particulate matter. For ICP/MS, it is then only necessary to sweep the ablated aerosol into the plasma flame using a flow of argon gas; this is the basis of an ablation cell. It is usual to include a TV monitor and small camera to view the sample and to help direct the laser beam to where it is needed on the surface of the sample.

With a typical ablated particle size of about 1-μm diameter, the efficiency of transport of the ablated material is normally about 50%; most of the lost material is deposited on contact with cold surfaces or by gravitational deposition. From a practical viewpoint, this deposition may require frequent cleaning of the ablation cell, transfer lines, and plasma torch.

There are different types of laser, which can be categorized according to the wavelength of the emitted radiation and by whether or not the lasers are used in a pulsed or continuous mode. Major characteristics of some commonly used lasers are given in Chapter 18, which should be consulted for further details. For laser ablation, short-pulsed (Q-mode) or continuous (free-running) mode can be important for achieving a desired result; however, in a practical sense, ultimate sensitivity of detection is not strikingly different as pulsed lasers generally give approximately ten-times-lower sensitivity. Pulsed lasers tend to give less total ablated material, but that material contains a greater proportion of gas to particulate matter. A proper comparison of the effects of pulsed and free-running lasers should take into account the total energy absorbed by the sample in unit time.

The degree of focusing of the laser beam is important. A tightly focused laser beam delivers its energy to a very small area of sample so the density of energy deposition is very high. This method leads to the formation of a pit in the sample solid where material has been ablated. Successive pulses deepen the pit. This mode of operation is used to produce a depth profile, viz., a profile of the composition of a sample throughout its thickness. When the beam is defocused, the area of sample irradiated is greater than in the focused mode, and the density of energy deposition is much lower. The resulting pit is very much shallower and covers a larger area. Therefore, the two kinds of beam, focused or defocused, have different analytical consequences. In the focused mode, the thickness of a sample is examined as the laser works its way down an increasingly deeper pit. The defocused mode is mostly used to survey variations in sample composition across its surface (surface profile). The two methods of operation (depth and surface profiling) are complementary, and each is useful in its own right.

A further important factor controlling the use of lasers for ablation purposes concerns the wavelength of the laser light and the sort of material irradiated. For ablation to occur, the sample should have one or more absorption bands overlapping the laser wavelength. The absorbed photons are converted rapidly into vibrational and rotational energy in the sample, and, in turn, some of this internal energy is converted into kinetic energy, leading to ejection of material from the surface (ablation). If the sample does not absorb light at the laser wavelength, most of the laser beam will be reflected (or will pass through a transparent sample) without causing any or much heating. If there is a substantial overlap of laser wavelength and absorption wavelength for the sample, much of the laser beam will be absorbed, with a concomitant rapid (about 10^{-7} to 10^{-8} sec) increase in temperature and production of ablated material. At suitable infrared wavelengths, most substances have absorption bands, so the efficiency of energy absorption can be quite high. However, at these longer wavelengths the energy of each photon is much less than at ultraviolet wavelengths. Therefore, with any one type of laser, the efficiency and amount of ablation can vary considerably from sample to sample, but the variation tends to be less at infrared wavelengths.

In considering the use of a laser for ablative sampling in ICP/MS, major criteria that must be considered include laser power, pulsed or continuous laser operation, pulse repetition rate, focused or defocused laser modes, depth or surface profiling, and the absorption characteristics of the sample range. To obtain the greatest ablation yields, all of these factors should be optimized. Of these, the most difficult is the focusing; therefore, ablation cells normally have some sort of microscope to observe the surface of the sample continuously and to select the areas to be examined. With modern technology, it is more convenient and safer to use a charge-coupled television (CCTV) camera to view the surface and a TV monitor to display the effects of the laser beam on the sample.

Electrical Discharge Ablation

Under the right conditions, an electrical potential placed on two electrodes (anode and cathode) separated by a short distance (usually 1 to 5 mm) in a gas at normal pressures will lead to an electrical discharge as the insulating properties of the gas break down and a current flows between the electrodes. The discharge can be intermittent (sparking) or continuous (arcing). These processes are discussed in greater detail in Chapter 6. Generally, sparking occurs when the insulating properties of the gas between the electrodes is beginning to break down but the discharge cannot maintain itself for long periods. Conditions leading to this behavior are low current flow associated with high voltages. Arcing occurs when the discharge becomes self-sustaining (more electrons are formed than are discharged at the anode) and electrically runs with a high current flow at low voltages. Sparks last for only a few micro- or milliseconds, but arcs can last for several minutes or more. In arcs or sparks, electrons flow to the anode, and positive ions bombard the cathode to maintain overall electrical neutrality.

Because a spark or an arc has a narrow diameter but contains a large number of ions and electrons, the heating effect caused by the current flow of a spark or arc leads to the electrodes' becoming very hot over a small area. The anode is usually shaped to a sharp tip to promote the discharge (compare the effect of a lightning conductor) and is cooled to prevent its temperature from becoming too high; nothing is then ablated at this point. The heating effect of ion bombardment at the cathode leads to ablation of electrode material as particulate matter, molten droplets, and vapor (an aerosol). Therefore, if a sample is included in the cathode material, it too becomes vaporized. The point of contact of the discharge at the cathode tends to wander over its surface, and therefore any small heterogeneities in the sample are smoothed during an analysis. As with the laser source discussed in the previous section, the ablated material is swept into a plasma torch for analysis of the elements present and their isotope ratios (ICP/MS). The argon gas, which is used to sweep the aerosol into the torch, prevents air getting to the sample being heated and therefore prevents oxidation or burning of the sample.

One problem with the spark or arc sources lies in sample preparation for nonconducting materials. If there is sufficient sample and it is conducting, then it can be machined into a cathode or simply placed on the cathode surface (in electrical contact). Thus, this method of ablation is very useful for examining metallurgical samples, which are normally conducting. Nonconducting samples must be thoroughly mixed with a conducting substance such as powdered copper or graphite, and then pressed into a disc before being placed on the cathode. The added conducting material must be of ultrahigh purity to prevent the introduction of impurities into the analysis. A typical generalized spark ablation source is shown in Figure 17.3.

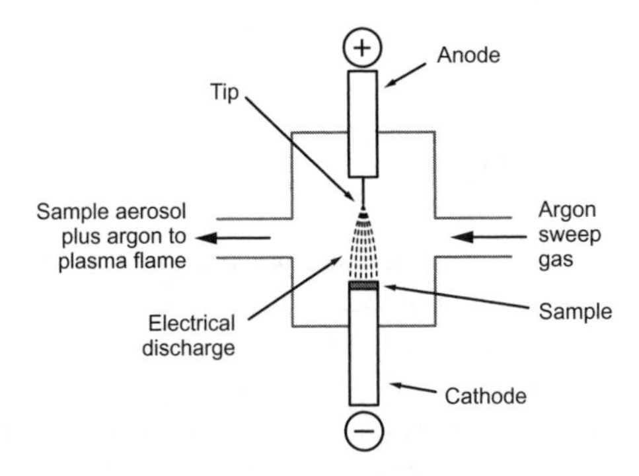

Figure 17.3
A schematic diagram showing the general construction of an arc or spark source. Actual construction details depend partly on whether samples need to be analyzed automatically. The sample material can be placed on the cathode or can even compose the whole of the cathode. If graphite is used, the sample needs to be pressed into the shape of a cathode after admixture with the carbon.

In operation, a spark source is normally first flushed with argon to remove loose particulate matter from any previous analysis. The argon flow is then reduced, and the cathode is preheated or conditioned with a short burn time (about 20 sec). The argon flow is then reduced once more, and the source is run for sufficient time to build a signal from the sample. The spark is then stopped, and the process is repeated as many times as necessary to obtain a consistent series of analyses. The arc source operates continuously, and sample signal can be taken over long periods of time.

Calibration of an arc or spark source is linear over three orders of magnitude, and detection limits are good, often within the region of a few micrograms per gram for elements such as vanadium, aluminum, silicon, and phosphorus. Furthermore, the nature of the matrix material composing the bulk of the sample appears to have little effect on the accuracy of measurement.

Electrical Heating (Electrothermal Vaporization, ETV)

Electrothermal heating is basically the use of an electric current to heat any suitable material, which can be in the form of a wire or filament (direct heating), cup, or shape of a small oven. Whatever the kind of sample support, the sample to be examined is heated on the wire, in the cup, or in the oven to a temperature high enough to effect its evaporation (usually between about 100 and 2000°C). Volatile samples are dealt with easily by using a quartz holder, which is then heated externally. More generally, the sample is placed directly onto a filament prior to heating or put into a boat, cup, or crucible, which is then heated. Sometimes a sample solution is placed in the boat, or it is sprayed onto a thin graphite rod or into a graphite tube. After absorption of the sample solution, solvent is evaporated before residual solid sample is heated to vaporization, and the resulting aerosol is swept into the plasma flame for analysis. Typically, about 5–10 mg of sample are needed. Clearly, if the sample is already a solution (analyte dissolved in a solvent), the ETV device can be used as a liquid inlet (Chapter 16) to examine the solution if the more volatile solvent is first evaporated at lower temperatures and then the residual analyte is vaporized at higher temperatures. To achieve variable heating, the electrical supply for ETV analysis is (normally) continuously variable.

Limits of detection for an ETV source are in the picogram to nanogram range and are often better than for direct solution introduction via a nebulizer. As with direct-insertion methods (DSI, see below), graphite supports for solid samples often lead to carbide formation with elements that resist volatilization. In these cases, a thermal reagent is commonly added to convert the sample elements into a more volatile form. For example, chlorine- or fluorine-containing compounds, such as halothanes, can be added to the argon gas (about 2 to 5% by volume), or sodium chloride can be added directly to the sample preparation so the solids react upon heating. Interferences from oxide formation is greatly reduced because no solvent is passed to the plasma flame and air (oxygen) is kept out of the system.

Direct Sample Insertion (DSI)

In principle, DSI is the simplest method for sample introduction into a plasma torch since the sample is placed into the base of the flame, which then heats, evaporates, and ionizes the sample, all in one small region. Inherent sensitivity is high because the sample components are already in the flame. A diagrammatic representation of a DSI assembly is shown in Figure 17.4.

In practice, direct insertion of samples requires a somewhat more elaborate arrangement than might be supposed. The sample must be placed on an electrode before insertion into the plasma flame. However, this sample support material is not an electrode in the usual meaning of the term since no electrical current flows through it. Heating of the electrode is done by the plasma flame. The electrode or probe should have small thermal mass so it heats rapidly, and it must be stable at the high temperatures reached in the plasma flame. For these reasons, the sort of materials used

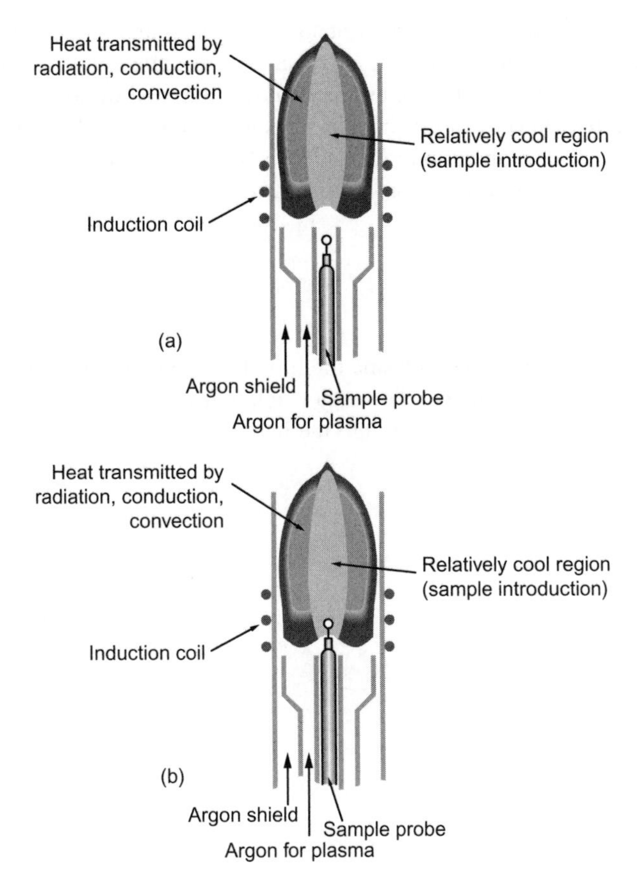

Figure 17.4
Diagram illustrating the method of direct insertion of a sample into the plasma flame. (a) Initially, the sample is held on a wire or in a small graphite or metal boat and then placed just below the flame until conditions have stabilized. (b) The probe is then raised so the sample holder enters the flame, where ablation occurs, and the sample is simultaneously fragmented and ionized. The plasma flame is shown in a vertical configuration (instead of the usual horizontal) because this position eliminates the possibility of the sample's dropping out of the cup used for sample insertion. If there is no problem of containing or holding the sample, then the flame can be in its more usual horizontal alignment.

for thermal ionization sources in isotope analysis (see Chapter 7) can be used. Graphite is frequently the material of choice for the probe material, both because of its good thermal properties and because it can be machined easily. However, graphite (carbon) forms carbides with many elements at high temperatures, and these can lead to serious interferences in some instances. For example, the analysis of chromium at m/z 52 is made difficult by the formation of ArC^+, which is also at the same m/z value. Other materials have been used, such as tantalum, tungsten, or molybdenum. The probe can be in the form of a cup in which to place the sample, or it can be simply a thin wire loop on which a sample solution has been evaporated, leaving behind a solid residue. Probe temperatures of 2000 to 3000°C have been measured.

These direct-insertion devices are often incorporated within an autosampling device that not only loads sample consecutively but also places the sample carefully into the flame. Usually, the sample on its electrode is first placed just below the load coil of the plasma torch, where it remains for a short time to allow conditions in the plasma to restabilize. The sample is then moved into the base of the flame. Either this last movement can be made quickly so sample evaporation occurs rapidly, or it can be made slowly to allow differential evaporation of components of a sample over a longer period of time. The positioning of the sample in the flame, its rate of introduction, and the length of time in the flame are all important criteria for obtaining reproducible results.

Generally, sample sensitivities of several nanograms per gram can be attained, but precision may not be as good as with other introduction techniques. Volatile elements such as cadmium or zinc on a probe of small thermal mass are evaporated over a period of about one second, giving a sharp transient signal, but slower-evaporating elements give wider signals, and sensitivity may not be as high. This variation in evaporation rate can even be used to achieve excellent sensitivity, as in the determination of volatile lead in nonvolatile nickel.

Conclusion

Solid samples can be analyzed using a plasma torch by first ablating the solid to form an aerosol, which is swept into the plasma flame. The major ablation devices are lasers, arcs and sparks, electrothermal heating, and direct insertion into the flame.

Chapter 18

Lasers and Other Light Sources

Introduction

Before discussing light sources generally, it may be useful to consider some basic characteristics of light.

Visible light comprises a very small section of the electromagnetic energy spectrum, which ranges approximately from cosmic rays to radio waves (Figure 18.1). The wavelengths of electromagnetic energies are related to the velocity of light (c) through the formula, $c = \nu\lambda$, in which λ is the wavelength and ν is the frequency of the radiation. The highest energies are associated with the highest frequencies and the lowest energies with lowest frequencies (Figure 18.1). Wavelengths follow the inverse order.

The smallest unit (packet) of electromagnetic energy (a photon) is related to frequency by the formula, $E = h\nu$, in which E is the energy and h is Planck's constant. Alternatively, the relation can be written, $E = hc/\lambda$. Frequency (ν) is a number with units of cycles per second (cps, the number of times a wavefront passes a given point in unit time, sec^{-1}) and is given the name Hertz (Hz). Planck's constant is a fundamental number, measured in J·sec or erg·sec.

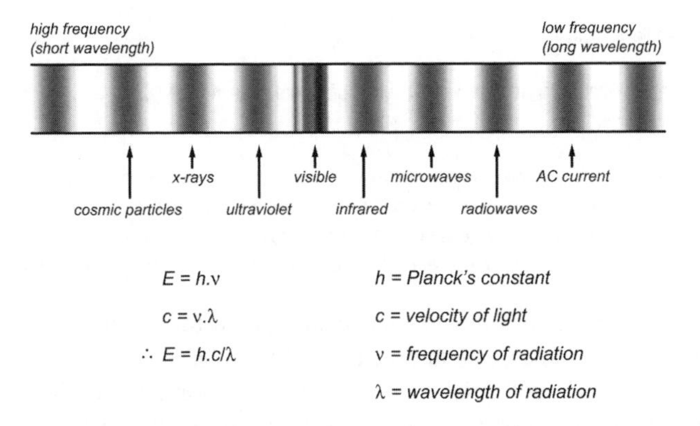

Figure 18.1
The electromagnetic spectrum ranges from high-energy cosmic rays (high frequency, ν) to low-energy alternating current (low frequency, ν). Different parts of the spectrum need different detection systems. The human eye detects light (blue to red), and the human body can detect heat (infrared), but humans can detect no other sections of the electromagnetic spectrum without the aid of instruments. Frequency and wavelength are inversely related, so high frequencies imply short wavelengths and vice versa. The above equations relate energy, frequency, wavelength, and Planck's constant.

Units of light

The units in which light is measured are summarized briefly here.

The energy emitted by a light source and the energy falling on an object are two of the most important measures of the amount of light. Since the energy of a photon varies according to its wavelength (E = h.c/l), measurement of the numbers of photons emitted or received gives a measure of energy emitted or received. The standard for light sources is emission at 555 nm. At this wavelength, the energy of one photon is 3.6 x 10^{-19} J.

The standard unit of light measurement (the light flux) is the lumen, which is the amount of energy (power, watts) emitted or received (Joules per second). At 555 nm, 1 lumen = 0.00147 watts = 0.00147 J.s^{-1}. This is sometimes called a "lightwatt". Since the energy for one photon at this wavelength is 3.6 x 10^{-19} J, then the number of photons represented by 1 lumen is approximately 4 x 10^{15} per second, radiated or received. Thus, the luminous flux (lumens, lm) gives the power radiating from an object or the power received by an object.

The number of lumens indicates the total amount of power but gives no indication of the "density" of that power. This last measure is given by the luminous intensity:

Luminous intensity = lumens/steradians (candelas)

The intensity is the flux of light passing through a solid angle of 1 steradian (1 steradian = the angle subtended at the surface of a sphere, for which the area illuminated is the square of the radius). Thus, 1 steradian (sr) is 1 cm^2 at the center of a sphere 1 cm in radius or it can be 1 m^2 at the center of a sphere of radius 1 m, and so on. The luminous intensity is a measure of the power density being radiated and the unit is the candela (cd). For any one power setting (lumens), the greater the solid angle, the less the luminous intensity. A laser beam has an extremely small angle of divergence (10^{-8} sr) and therefore its luminous intensity is very high even though the actual power may be low.

Luminance is the luminous intensity divided by the area of emission of light (lumens/steradian/m^2). This is the power density emitted per unit area.

The luminance relates to the luminous intensity radiated from an entire surface of a light source:

Luminance = lumens/seradians/m^2 = luminous intensity/m^2

If the amount of light is measured over an area of receiving surface, the energy falling on the surface is measured in lumens per unit area ("lux" or "phot"). Thus, the number of lux = lumens/m^2 and this measures the power received per square meter of surface (energy per second/per unit area) and phot = lumens/cm^2 and measures the light power received per square centimeter of surface.

An international "candle" = 1 lumen.

Some examples of approximate luminances of various light sources are given in Table 18.1.

Figure 18.2
The common units of light intensity or power density of light emitted or received are as shown above. Care should be taken in comparing luminances. For example, Table 18.1 reveals that a tungsten filament lamp has about a tenth of the luminance of the sun, but the area of the sun's emitting surface is massively greater than that of a filament lamp, and therefore the luminous intensity of the sun is massively greater than the luminous intensity of the filament lamp.

Emission of light from various devices has been known for millions of years (sun, lightning, fireflies). Fires, oil lamps, candles, gas lamps, etc. all use chemical reactions (combustion in air) to make hot (active) atoms or molecules, which emit light upon cooling. More recently, electrical production of light has become common, as in tungsten lamps, fluorescent lights, and arc lamps. The tungsten filament lamp uses electrical resistive heating to make a wire glow white hot, but fluorescent and arc lights use forms of electrical discharge, similar in principle to the natural phenomenon of lightning (see Chapter 6). Photodiodes convert electrical energy directly into light without the need for heating or discharge. In the period 1950 to 1960, a new form of light emission was developed, for which the acronym LASER was used (light amplification by stimulated emission of radiation).

TABLE 18.1.
Some Approximate Luminances of Various Light Sources

Light source	Luminance (candelas/cm^2)
Sun (noon)	160,000
Tungsten filament lamp	12,000
Fluorescent lamp	0.82
Mercury discharge lamp	970
Metal halogen bulb	810
Photoflash light	20,000

TABLE 18.2.
Some Typical Lasers and Their Power Outputs

Lasing substances [a]	Physical state	Laser wavelength (nm) [b]	Pulse length or continuous wave	Typical maximum power output (watts) [c]
Cr/alumina (ruby)	solid	694	nanoseconds	100 MW
Nd/glass	solid	1,060	picoseconds	100 TW
Ga/As	solid	840	continuous	10 mW
Rh6G (dye)	liquid	600	femptoseconds	10 kW
He/Ne	gas	633	continuous	1 mW
CO_2	gas	10,600	continuous	200 W
Ar/F	gas	193	nanoseconds	10 MW

[a] These designations are popularly used to describe the basis of the laser but are not accurate descriptions of the chemical states.

[b] As the wavelength moves into the infrared region, it is more common to change units from nanometers to micrometers (microns). For example, 10,600 nm would be written as 10.6 μm.

[c] The maximum or peak power depends critically on the pulse length. An energy output of 1 J in one second is the power of 1 W, but if the 1 J is emitted in a picosecond, the power rises to $1/10^{-12}$, which is 10^{12} W = 1 TW.

The amount of light emitted by a source is measured by its luminance or by its luminous intensity, which are defined in Figure 18.2. Intrinsic light emission relates to the amount of light emitted per unit area (luminance). Table 18.1 lists approximate luminances for some common light sources.

Almost all of the oldest light sources give light that covers a range of wavelengths and is not coherent; the light waves are not propagated in phase with each other. The development of the laser has provided light sources that emit sharply monochromatic, coherent, and intense radiation ranging from the ultraviolet (UV) to the infrared (IR). Apart from their use in research, lasers have found important applications in a large range of everyday devices, from CD players to metal plate cutters and welders. This chapter cannot cover this huge range of applications. Instead, it concentrates on principles and descriptions of the most commonly used, commercially available lasers. Table 18.2 indicates some of the power outputs obtainable with various types of laser.

Until about the 1990s, visible light played little intrinsic part in the development of mainstream mass spectrometry for analysis, but, more recently, lasers have become very important as ionization and ablation sources, particularly for polar organic substances (matrix-assisted laser desorption ionization, MALDI) and intractable solids (isotope analysis), respectively.

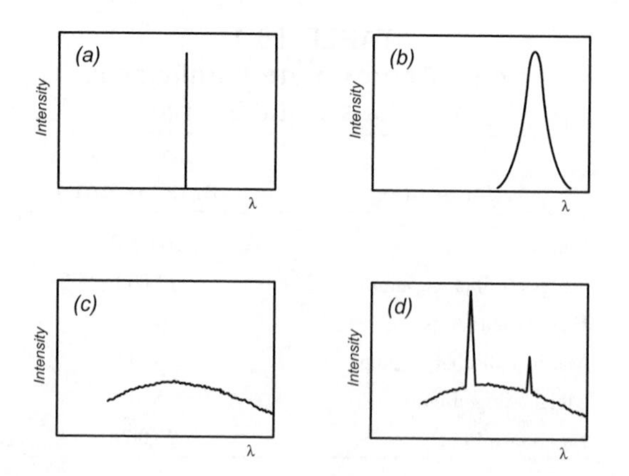

Figure 18.3
The diagrams depict some typical light outputs in relation to wavelength for various sources. (a) The sharp line (single wavelength) output from a laser or an atomic emission line. (b) The broader but still sharp wavelengths obtained by using interference filters, monochromators, and diffraction gratings or from molecular emission bands. (c) Broadband light output covering a wide range of wavelengths from such sources as the tungsten filament lamp. (d) A mixed output of a typical discharge lamp, having some intense narrow bands from atomic emission superimposed on broadband radiation from ion/electron recombination.

Some Characteristics of Light as a Waveform

Wavelength

Light sources can emit photons over a wide range of wavelengths (e.g., electric light bulbs) or at a single wavelength (e.g., lasers). Therefore, a light source can emit a single wavelength, multiple wavelengths, or broadband radiation (Figure 18.3). Various devices (filters, interference devices, diffraction gratings, and monochromators) are available for selecting particular wavelengths from broadband radiation to obtain selectively more monochromatic light.

Intensity

The intensity of a light beam refers to the number of photons it contains passing through a unit area (flux). The intensity also refers to the amplitude of a multiphoton waveform, which varies with the number of photons and whether or not the light is coherent or incoherent. The power emitted by a light source or received by an object (light watts) relates to the energy of each photon (Figure 18.1), the number of photons, and the time for which light is emitted or received (Figure 18.2).

Coherence

Consider two trains of light waves (Figure 18.4). If two photon waves are coherent, the waves are in phase; therefore the wave intensity (amplitude) is doubled at all points. If the waves are not coherent, the two waveforms are out of step (out of phase), and the amplitude does not equal the maximum attainable by coherent waves. It is even possible for light waves to overlap in such a way that there is no light whatsoever; the waves cancel each other (they are totally out of step), and darkness results (interference). A coherent laser beam can be contrasted with an incoherent incandescent light beam. In the former, all of the maxima in the waveform occur together in a tightly bundled stream (a light beam), i.e., the power density is high; for incandescent light, all the waveforms overlap randomly, and the power density is much lower (Figure 18.4).

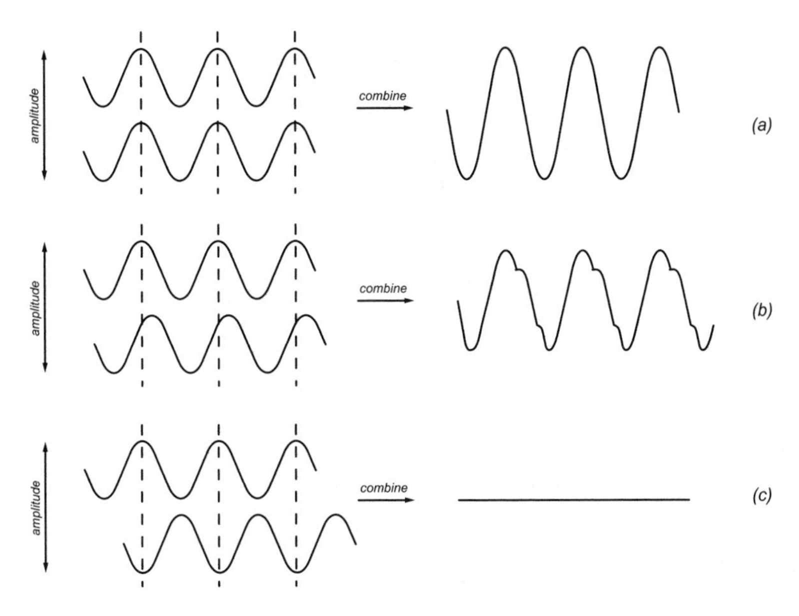

Figure 18.4
In (a), two photon waves combine to give a new waveform, which has the same appearance and frequency as the initial separate waves. The photons are said to be coherent, and the amplitude of the waves (light intensity) is simply doubled. In (b), the two photon waves are shown out of step in time (incoherent). Addition of the two waveforms does not lead to a doubling of amplitude, and the new waveform is more complex, composed of a doubled overlapping frequency. In (c), the two waveforms are completely out of step (out of phase) and completely cancel each other, producing darkness rather than light (an interference phenomenon).

Directionality and Divergence

A light source can be relatively large, emit light in all directions, and have a low luminous intensity (Figure 18.2), typical of an incandescent electric light bulb, for which mirrors, reflectors, and lenses must be used to effect any focusing or collimating needed to provide a beam. In contrast, a laser source emits light as a very narrow beam having very little divergence. For the same power output, an incandescent lamp emits far fewer photons per unit volume in any one direction than does a laser. Unless focused, light from an incandescent lamp is spread more or less uniformly over all space. In contrast, a laser source emits a light beam that is tightly focused in one direction.

Conversion of Energy into Light

Given suitable mechanisms, all forms of energy can be converted into light as long as the initial energy is equal to or greater than the energy of photons of light required. For example, chemical reactions can emit light, usually when an intermediate or product of reaction is formed in an energetically excited state and needs to lose energy to return to its ground state. Mostly, this loss of energy is vibrational and rotational and occurs through collision with other molecules or the walls of the containment vessel. In other instances, the excess of energy can be lost by emission of photons of visible light (chemiluminescence). This phenomenon can be observed naturally in the light emitted by glow-worms or fireflies. Electrical energy can be converted into light either directly by heating a wire (filament lamp) or through its discharge through a gas (corona, plasma, arc, lightning, photoflash). The shockwave of an electron traveling near the velocity of light in a particle accelerator or upon nuclear disintegration emits light (Czerenkov radiation). Atomic particles from the sun strike the earth's upper atmosphere, exciting nitrogen and oxygen to emit the beautiful lights of the aurorae.

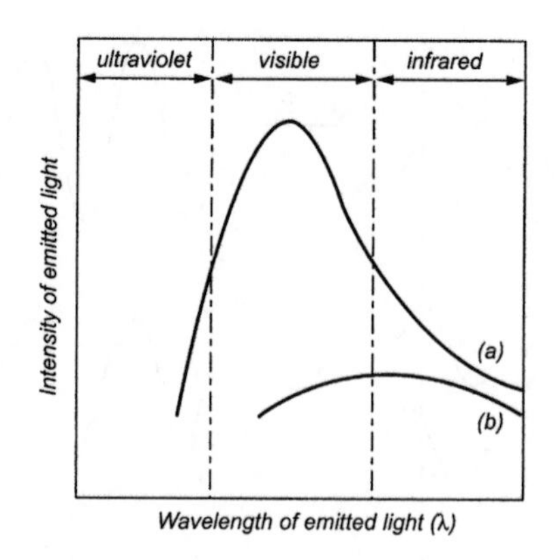

Figure 18.5
Curve (a) represents the intensity (number of photons) of light emitted from a hot incandescent object at about 5000°C plotted against wavelength. The emission covers all visible wavelengths but not evenly and appears white. Curve (b) represents the radiation from a hot body at about 2000°C (typical of an incandescent filament lamp). More light is emitted toward the red end of the visible spectrum, and the light appears yellowish (less blue light). Compared with the amount of heat energy emitted, the proportion of light energy is small.

Nonlaser Light Sources

Incandescent Lamps

As mentioned briefly above, when an electric current is passed through a resistance wire, it becomes hot and emits photons of heat and light. The higher the temperature of the wire, the greater is the proportion of light to heat, but this form of lighting is inefficient in that much of the energy input is wasted as unwanted heat output. The emitted radiation is quantized, but, because the quanta are separated by very small energy gaps, there is in effect a continuum of light stretching from the infrared to the ultraviolet region. The very high temperatures needed to cause thermal emission of ultraviolet light are not generally available, but such emission is observed during welding, for example. Short-wavelength radiation emitted at high temperatures can be found in some stars that are so hot as to emit a significant fraction of light toward the ultraviolet end of the spectrum and therefore appear blue rather than the common white, orange, or red colors of stars like the sun. With everyday incandescent lamps, the emitted light tends to be yellowish because there is an excess of light emitted toward the red end of the spectrum (Figure 18.5).

As described already, light from incandescent sources is emitted in all directions, is multichromatic (covers many wavelengths), and is incoherent. It is possible to select a single wavelength or even a narrow band of wavelengths by using interference filters or diffraction gratings, but only at the expense of rejecting most of the unwanted light. Focusing of selected light wavelengths is also inefficient because of the need to use lenses and mirrors. Thus, when specific wavelengths are required, incandescent lamps are very inefficient, and, unless they are extremely powerful, the output at any selected wavelength is weak. This light is in marked contrast to laser light, which is produced as a tightly collimated, narrow beam of intense monochromatic radiation.

Arc Lamps

The basic principles of coronas, plasmas, and arcs are discussed in Chapter 6. Essentially, light emitted from such sources arises from gases at low, medium, or high pressures, which are excited by the passage of electrons. Gas atoms are electronically excited (gain energy) or are ionized through collision with electrons. Light is emitted when the excited atoms return to their ground state by ejection of photons or when ions recombine with electrons and return to the ground state. The light is characteristic of the gas used for the discharge and consists of a mixture of narrow and broadband emissions. The narrow bands arise from atomic emission lines and the wide bands from recombination of electrons and ions (Figure 18.3a). Corona, plasma, and arc sources are produced under a range of pressure conditions and current flows. Their emissions stretch from the ultraviolet to the infrared region of the electromagnetic spectrum.

Arc lamps are more efficient than incandescent light sources and radiate from a small region of space, so the light is much easier to focus or collimate than is the light from a heated filament. For that reason — and the possibility of obtaining an intense light beam — arc lamps are used for many purposes, as in photolithography, photocuring of polymers, or theater lighting, but they cannot match the laser for the amount of energy deposited into a small space in a short time.

Laser Light Sources

Absorption and Emission of Light

To understand the production of laser light, it is necessary to consider the interaction of light with matter. Quanta of light (photons) of wavelength λ have energy E given by Equation 18.1, in which h is Planck's constant (6.63×10^{-34} J·sec) and c is the velocity of light (3×10^{8} m·sec–^{1}).

$$E = h\nu = hc/\lambda \qquad (18.1)$$

Thus, green light from near the middle of the electromagnetic spectrum (Figure 18.1) has a wavelength of about 500 nm. From Equation 18.1, this wavelength corresponds to a frequency $\nu = 0.6 \times 10^{15}$ Hz and an energy per photon $E = 3.98 \times 10^{-19}$ J. If an atom or molecule has a ground state (A^0) and a suitable excited state (A^*) separated by an energy gap of 3.98×10^{-19} J; then a photon of green light can be absorbed (Figure 18.6). The molecule is then in an excited state. Such atoms or molecules exist in the excited state for lifetimes which vary with the structures involved but are generally of the order 10^{-13} to 10^{-8} sec. After this time, the excited molecule returns to the ground state (A^0) by a variety of mechanisms. Sometimes this energy loss occurs simply by emission of a photon of the same or similar energy (wavelength) to that which excited it in the first place (Figure 18.6). Apart from this normal emission, ejection of a photon can be induced (stimulated) if the atom in its excited state is irradiated by a second photon.

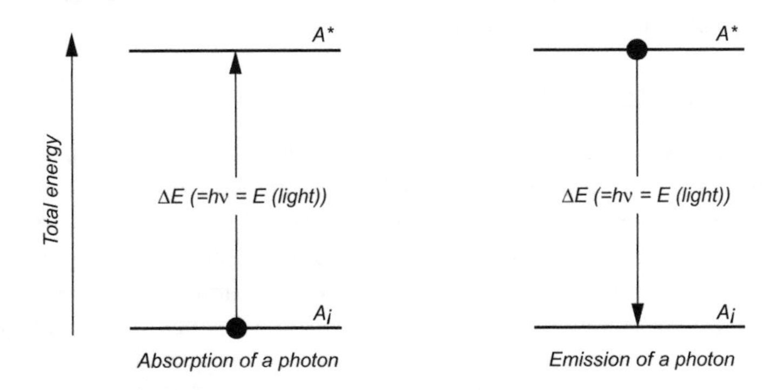

Figure 18.6
If an atom or molecule (A) in its ground state (A^0) has a more excited state (A^*, higher energy), then the energy separation is ΔE. If a photon of light has energy E(light), which equals ΔE, then the photon can be absorbed by A. On absorption, the atom moves from energy state (A^0) to the higher energy state (A^*), and $\Delta E = E(\text{light}) = h\nu = hc/\lambda$. Thus, if λ is the wavelength of light that corresponds to ΔE, the light can be absorbed. The state (A^*) can emit a photon of energy $E = h\nu$ to return to the ground state (A^0).

A major difference between normal emission and stimulated emission lies in the number of photons involved. For normal emission, a ground-state atom or molecule first absorbs a photon and moves into an excited state and, some time later, emits a photon to return to the ground state (one photon absorbed, one photon emitted). For stimulated emission, a ground-state molecule absorbs a photon and again enters an excited state. Before it can decay to the ground state, it interacts with another photon leading to two photons emitted together. The two emerging photons travel in the same direction, are in phase (coherent), and have identical wavelengths. This stimulated emission can lead to a cascade effect in that one initial photon becomes two, and then these become four, then eight, and so on as they stimulate emission from other atoms or molecules in the excited state (A*). This cascade of emerging photons, all traveling in the same direction, all in phase, and all of the same wavelength, is the basis of laser light emission, whereby one photon event is amplified. Before this stimulated emission can effectively produce a laser beam, there are other requirements, which are considered next.

Populations of Ground and Excited States

The Boltzmann equation (Equation 18.2) shows that, under equilibrium conditions, the ratio of the number (n) of ground-state molecules (A^0) to those in an excited state (A*) depends on the energy gap E between the states, the Boltzmann constant k (1.38×10^{-23} J·K^{-1}), and the absolute temperature T (K).

$$n(A^*)/n(A^0) = \exp(-E/kT) \qquad (18.2)$$

For the energy of excitation discussed above (3.98×10^{-19} J) and a temperature of 20°C (293 K), the ratio of excited states to ground states is $n(A^*)/n(A^0) = 10^{-36.7}$. In other words, there is only one excited-state atom or molecule for every 5×10^{36} molecules in the ground state! Thus, the chances of observing a natural or stimulated emission are vanishingly small because there are almost no atoms or molecules in the excited state. For every photon entering such an assembly of molecules, there are billions of chances that it will be absorbed and only one that it will induce emission. Any photons so emitted would meet only ground-state molecules and not another excited-state atom or molecule required to start a cascade. Thus, an incipient cascade would stop before it could develop.

If the temperature were raised, more molecules would attain the excited state, but even at 50,000°C there would be only one excited-state atom for every two ground-state atoms, and stimulated emission would not produce a large cascade effect. To reach the excess of stimulated emissions needed to build a large cascade (lasing), the population of excited-state molecules must exceed that of the ground state, preferably at normal ambient temperatures. This situation of an excess of excited-state over ground-state molecules is called a population inversion in order to contrast it with normal ground-state conditions.

Maintaining a Population Inversion

If the excited state has a very short lifetime, there is little chance of building up a population excess in the excited state because absorption is followed almost immediately by emission. This situation is typical for a simple two-level system of ground and excited states, as shown in Figure 18.6. To build up a population of excited-state molecules, it is necessary that the excited state have a sufficiently long lifetime before it returns to the ground state naturally. The easiest way of achieving that lifetime is to decay the initial excited state (A*) to another excited state of lower energy (A′*), which has a relatively long lifetime. This occurs frequently by spin inversion of an electron, as shown below, and is the basis of a three-level system (two upper and one ground state; Figure 18.7).

If all spins (±1/2) in an atom or molecule are paired (equal numbers of spin +1/2 and −1/2), the total spin must be zero, and that state is described as a singlet (total spin, S = 0 and the state is described by the term 2S + 1 = 1). When a singlet ground-state atom or molecule absorbs a photon, a valence electron of spin 1/2 moves to a higher energy level but maintains the same

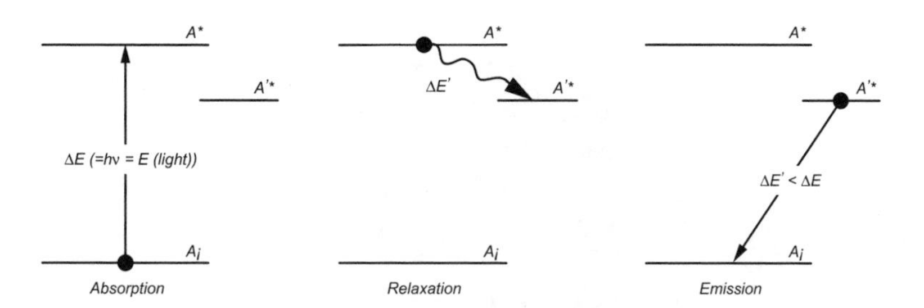

Figure 18.7
In a typical three-level system, initial absorption of light changes a ground-state atom (A^0) to an excited state (A^*), as discussed in Figure 18.5. However, in the three-level system, the excited-state atom or molecule can lose energy by collision and electronic-state crossing to reach a more stable excited state (A'^*). Emission from this state gives photons of lower energy than the ones that led to the initial absorption ($\Delta E = h\nu$, $\Delta E'' = \Delta E - \Delta E'$).

spin of 1/2, and therefore the total spin remains zero; this excited state is also a singlet. Therefore, absorption of a photon is said to give a singlet-to-singlet transition. Emission of a photon to return to the ground state is allowed, giving a lifetime for the excited state of only a few picoseconds to nanoseconds. However, if the electron in the higher energy state can invert its spin, the system becomes a triplet (S now equals 1 because all spins are paired except for the two in the same spin state, and therefore $2S + 1 = 3$). Return to the ground state is now not easy. (It becomes a forbidden transition.) In these circumstances, the lifetime of the excited state increases markedly to microseconds or milliseconds, as the system has to find some other way of returning to the ground state. If sufficiently long, this increased lifetime allows a high population of atoms or molecules in the excited state to build rapidly. The population inversion needed for lasing can be achieved in various ways (Figure 18.7).

Lasing (Emission of Laser Light)

Once a large inverted population of excited-state atoms or molecules has been created, stimulated emission will occur readily. If one excited-state molecule emits a photon naturally, there will be a very high probability that this photon will interact with another excited-state molecule before it has traveled very far. This first interaction leads to stimulated emission of two identical photons. Before proceeding more than a few Ångstroms, each of these two photons interacts with other excited-state molecules to give a total of four photons, and so on; a cascade begins. Since the velocity of light is 3×10^8 m·sec^{-1} then, within a few nanoseconds, a cascade of photons is produced in which millions of photons are emitted from the overpopulated excited state and make up an emerging laser beam (Figure 18.8).

Excited-State Levels

It was shown above that the normal two-level system (ground to excited state) will not produce lasing but that a three-level system (ground to excited state to second excited state) can enable lasing. Some laser systems utilize four- or even five-level systems, but all need at least one of the excited-state energy levels to have a relatively long lifetime to build up an inverted population.

Pulsing and Continuous Wave

The timing of the emission is clearly dependent on the system in use. For example, if pumping is relatively slow and stimulated emission is fast, then the emergent beam of laser light will appear as a short pulse (subsequent lasing must await sufficient population inversion). This behavior is

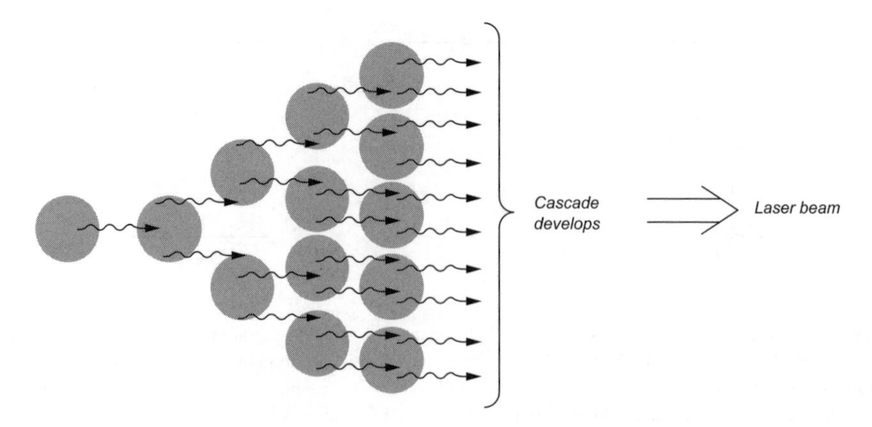

Figure 18.8
Interaction of an excited-state atom (A′*) with a photon stimulates the emission of another photon so that two coherent photons leave the interaction site. Each of these two photons interacts with two other excited-state molecules and stimulates emission of two more photons, giving four photons in all. A cascade builds, amplifying the first event. Within a few nanoseconds, a laser beam develops. Note that the cascade is unusual in that all of the photons travel coherently in the same direction; consequently, very small divergence from parallelism is found in laser beams.

characteristic of certain types of pulsed lasers. Alternatively, if emission and population inversion (pumping) are more or less in time with each other, then the laser light appears more or less continuously, which is characteristic of continuous-wave lasers.

Pumping (Amplification)

Once laser emission has been generated, the excited state is depleted and must be repopulated before lasing can again take place. Excited-state molecules can be generated by a variety of methods, but they need to be replenished repeatedly to maintain the population inversion. This repeated replenishment is commonly known as pumping. Thus, where a flash of light is used to obtain a population inversion, the flash must be repeated as long as the lasing action is needed. In effect, the flashes of light pump photons into one or more relatively long-lived excited states. Methods of pumping the excited state include electric discharges in gases, injection of electric current in semiconductors, flash lamps for solid-state lasers, chemical reactions, and even the light from other lasers.

The Laser Cavity

Once a population inversion has been built up, any naturally emitted photon can initiate the lasing action. To improve the overall effect, the laser generator is normally enclosed within a resonance cavity with mirrors at each end (Figure 18.8). The purpose of the mirrors is to cause the lasing action to travel up and down the cavity (less loss of light to the walls of the cavity), increasing the cascade in one direction so that most of the inverted population is stimulated to emit in a very short space of time. However, this action can prevent the excited-state population from building up to a large excess, resulting in a low intensity of laser light output. There are ways of improving the population of excited-state molecules, two of which are Q-switching and mode locking.

Q-Switching

As shown in Figure 18.9, any one emitted photon inside a laser resonance cavity can begin a cascade of photons. If the cascade reflects back and forth between the mirrors of the resonance cavity, most of the inverted population will be stimulated to emit laser photons. If one of the mirrors is temporarily made nonreflective, the initial cascade can no longer bounce back and forth inside the cavity, and the numbers of stimulated emissions will be quite small (there is little laser light), providing time for the excited-state population to build. If, after a short time, the nonreflective mirror is made reflective again, there will be a rapid formation of a large cascade of photons back and forth inside the cavity, causing all of the overpopulated excited states to emit coherent photons. A giant pulse of laser light is emitted from the cavity.

One of the ways of making a mirror temporarily nonreflective is to place in front of it a solution of a dye that has suitable absorption bands. As long as the dye absorbs photons, none can get through to the mirror to be reflected. However, at some stage, as the number of photons in the cavity builds up, the dye molecules are all raised to their excited state in a short time and the dye loses its capacity to absorb any more light. It is said to be bleached. At this stage, the laser photons can reach the mirror, which is reflective. The laser photons can now bounce back and forth very quickly between the two mirrors of the cavity, stimulating emission from the large inverted population that has had time to build while the dye was absorbing radiation. There is then a large pulse of intense coherent laser light lasting a few nanoseconds. After this event, most of the dye molecules return to the ground state and the mirror once more becomes nonreflective until just before the next giant pulse is emitted as the process repeats. This sort of Q-switching is said to be *passive* because it happens without any external intervention.

Active Q-switching occurs when laser light access to one of the mirrors in the cavity is controlled by an electro-optical cell, which works on the principle of affecting the passage of polarized light (see below; Kerr or Pockels cell). These last devices are able to turn on and off the transmission of light by using electric stimuli to alter the optical characteristics of the medium comprising the cell. In effect the cell acts as a very high-speed shutter controlled by a change in voltage. Application of a suitable electric potential prevents the passage of light and its removal allows it (or vice versa). Thus, by placing such a cell in front of one of the mirrors, laser light can be prevented from reaching the mirror until a large inverted population has built up inside the laser cavity. The cell is then switched to allow passage of photons, which pass up and down between the two mirrors and produce a giant pulse of laser light.

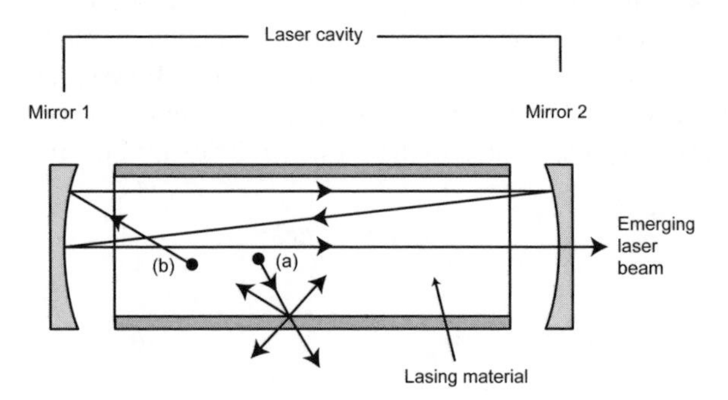

Figure 18.9
Once an inverted population has been built up, any one excited-state atom or molecule can start a cascade of stimulated emissions. In (a), one such excited-state atom has emitted a photon in the direction of the walls of the laser cavity. Although a cascade begins, when it reaches the walls, much of the light is scattered or travels through them, and no laser beam emerges from the cavity. In (b), the excited atom emits a photon toward an end mirror (1); the developing cascade is focused onto the second mirror (2) and then back to the first and so on. In a very short time, a laser beam emerges from one of the mirrors, which is manufactured to be partly transparent instead of being totally reflecting.

Mode Locking

Mode locking is similar to the passive Q-switching except that the bleaching is effected by the differing modes of light. For any passage of light up and down the laser cavity, a standing wave must be built up — if d is the distance between the mirrors, then the frequencies of the standing waves are given by Equation 18.3, in which n is the refractive index of the laser material and i is an integer (1, 2, 3, …).

$$\nu = ix(c/2d{\cdot}n) \tag{18.3}$$

If d·n matches ν, then standing waves are generated. The less good the match, the fewer standing waves are formed. If, as with passive Q-switching, there is an absorbing dye in front of one of the mirrors, the dye will absorb light as before. However, since frequencies ν that do not match d·n very well are fewer than those that do match, this small number of frequencies is absorbed totally by the dye, leaving an excess of matching frequencies. Eventually, the dye is bleached by a cascading effect of the selected modes (mode-locking; better matched frequencies). The time needed for a photon to make one trip up and down the cavity occurs as the dye becomes completely bleached. The resulting emergent laser pulse is very short and is highly monochromatic since the low population of unmatched frequencies has been largely eliminated. The pulse also lasts only a very short time before the inverted population is lost and the dye returns to its unbleached state. The time taken for a photon to move up and down the cavity is the "pulse repetition frequency," not to be confused with the frequency of the emerging laser light.

Brewster Angle

When a beam of nonpolarized light meets a surface separating two optically clear materials of differing refractive index (n_1,n_2) such as air and glass, part of the light is refracted (transmitted) and part is reflected at the surface. The refracted part of the beam is polarized in a plane at right angles to the surface (parallel to the plane of incidence; Figure 18.10), and the reflected part of the beam is polarized in the plane of the surface (perpendicular to the parallel component; Figure 18.8). As the angle of incidence of the incoming beam is decreased (Figure 18.10), less of the parallel component is reflected until, at the Brewster angle (θ_B), all of the parallel component is refracted (transmitted), and the reflected component is almost completely polarized in a direction parallel to the surface. At the Brewster angle, there is maximum transmission of light (a minimum of reflection), viz., the transmission is most efficient. Two relationships are characteristic of the Brewster angle. The first is that the refracted and reflected parts of the incident beam are at right angles to each other (Equation 18.4), and, second, the Brewster angle is related to the refractive index by Equation 18.5 (Figure 18.10a).

$$\theta_B + \theta_r = \pi/2 \tag{18.4}$$

$$\tan \theta_B = n_2/n_1 \tag{18.5}$$

For an air/glass interface, $\tan \theta_B = n$, the refractive index of glass. In a gas laser, the light must be reflected back and forth between mirrors and through the gas container hundreds of times. Each time the beam passes through the cavity, it must pass through transparent windows at the ends of the gas container (Figure 18.10b), and it is clearly important that this transmission be as efficient as possible.

By placing these windows in a plane set at the Brewster angle with respect to the light beam, maximum transmission is assured (minimum reflection). The above considerations regarding polarization of refracted and reflected beams must be modified for nontransparent media such as metals, for which there is little polarization of a reflected beam.

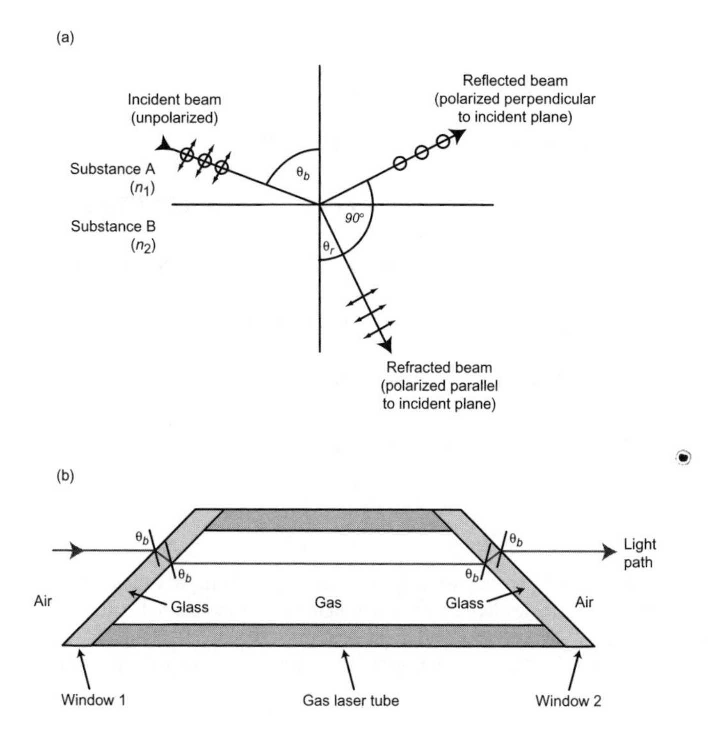

Figure 18.10
(a) A light beam incident at the surface separating two transparent substances (A,B) is shown to be partly refracted and partly reflected. When the incident beam is at the Brewster angle (θB), there is maximum transmission of the part of the beam polarized in a plane parallel to the plane of incidence (at right angles to the surface). The reflected part of the beam is polarized perpendicular to the plane of incidence (parallel to the surface). (b) For two windows at the ends of a gas laser tube, substance A is air and B is glass. The glass end plates are set at the Brewster angle for an air/glass interface so a transmitted beam is passed through with maximum efficiency.

Methods of Pumping to Obtain Excited States

Only a few of the many methods that have been used (those for which commercial lasers are obtainable) are described here.

Gas-State Lasers

Molecular (Chemical) Reaction. The principle of emission can be illustrated by means of an example. Suppose a mixture of two gases, xenon and chlorine, is contained within a laser cavity that also has two metal electrodes. Under normal conditions, no compounds are formed between the two gases (for example, ground-state XeCl is not stable and cannot be isolated). If a strong electric field is applied across the two electrodes, some electrons are produced, and they interact with the xenon so as to ionize it, producing more electrons (for more detail, see Chapter 6, "Coronas, Plasmas, and Arcs"). Some of the electrons attach themselves to chlorine (Figure 18.11). The net result is that, in a very short time, a population of positive xenon ions and negative chlorine ions forms. These ions combine to give XeCl in an electronically excited state (an excited-state complex known as an exciplex, XeCl*), in which the XeCl* is stable for a short time, providing a population inversion of excited-state XeCl molecules. On stimulated emission, this electronically excited state returns to ground-state XeCl by emission of light and, because it is not stable, the resulting ground-state molecule dissociates into Xe and Cl (Figure 18.11).

By varying the types of gases inside the cavity, the wavelength of the laser emission can be varied (Table 18.3). These gas lasers are useful because the emitted light lies mostly in the ultraviolet

$$Xe + e^- \longrightarrow Xe^{+\bullet} + 2e^-$$

$$Cl_2 + e^- \longrightarrow Cl^{-\bullet} + Cl^\bullet$$

$$Xe^{+\bullet} + Cl^{-\bullet} \longrightarrow XeCl^* \quad \text{(excited state)}$$

$$Xe^{+\bullet} + Cl_2 \longrightarrow XeCl^* + Cl^\bullet \text{ (excited state)}$$

$$XeCl^* \longrightarrow Xe + Cl^\bullet + hr$$

$$Cl^\bullet + Cl^\bullet \longrightarrow Cl_2$$

$$\text{or } Cl^\bullet + e^- \longrightarrow Cl^-$$

Figure 18.11
The reaction path shows how Xe and Cl_2 react with electrons initially to form Xe cations. These react with Cl_2 or Cl– to give electronically excited-state molecules XeCl, which emit light to return to ground-state XeCl. The latter are not stable and immediately dissociate to give xenon and chlorine. In such gas lasers, translational motion of the excited-state XeCl gives rise to some Doppler shifting in the laser light, so the emission line is not as sharp as it is in solid-state lasers.

TABLE 18.3.
Wavelength of Laser Light Emitted from Some Chemical Lasers

Chemical reactants	Excited state species	Laser wavelength (nm)
Argon, fluorine	ArF	193
Krypton, fluorine	KrF	248
Xenon, fluorine	XeF	351
Krypton, chlorine	KrCl	222
Xenon, chlorine	XeCl	308

region of the electromagnetic spectrum. Because of the translational motions of the molecules of gas, the emitted laser beam is not as highly collimated or quite as monochromatic as many other types of laser light. In these lasers, excited-state overpopulation is supplied continuously, and the laser light is also emitted continuously, so it is termed a continuous wave (cw) laser.

Molecular Interaction. The examples of gas lasers described above involve the formation of chemical compounds in their excited states, produced by reaction between positive and negative ions. However, molecules can also interact in a formally nonbonding sense to give complexes of very short lifetimes, as when atoms or molecules collide with each other. If these "sticky" collisions take place with one of the molecules in an electronically excited state and the other in its ground state, then an excited-state complex (an exciplex) is formed, in which energy can be transferred from the excited-state molecule to the ground-state molecule. The process is illustrated in Figure 18.12.

This type of energy transfer is common and is used to promote some substances into excited states that are not easy to obtain by other means. For example, normal oxygen molecules exist in an electronic triplet state, which is relatively unreactive (which is just as well, or you would not be reading this book!). However, oxygen can be activated into its singlet state, which it is extremely reactive, causing the oxygen chemically to attack all kinds of substances. Direct irradiation of oxygen

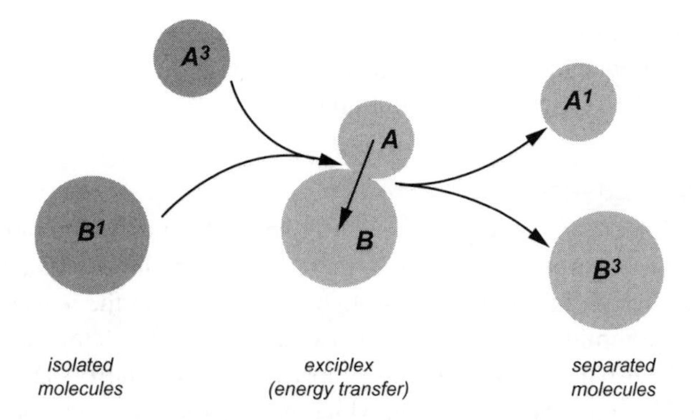

Figure 18.12
If a triplet-state molecule (A^3) meets a singlet-state molecule (B^1), a short-lived complex can be formed (an exciplex). In the latter, the molecules exchange energy, A^3 returning to its singlet state (A^1) and B^1 raised to its triplet state (B^3). If the new triplet state is relatively long-lived, it can serve to produce the population inversion needed for lasing, as in the He/Ne laser.

with light is very inefficient for effecting this triplet/singlet conversion, which is formally disallowed because an electron must invert its spin in the process. However, there are substances (sensitizers) that can be excited efficiently into triplet states. If one of these excited sensitizer molecules collides with a ground-state oxygen molecule, then energy transfer occurs in the resulting exciplex whereby the oxygen is translated into its triplet state with high efficiency while the sensitizer returns to its ground state. Gas lasers can work on a similar principle, in particular the He/Ne laser (Figure 18.13).

If helium is part of a discharge gas, it is transformed into metastable ions of energy 20.4 eV. If, before it can return to the ground state, the excited atom collides with a neon atom, energy transfer occurs in such a way that the neon atom is excited into a 20.66-eV electronic state and the helium atom returns to the ground state. The small energy difference of 0.26 eV is taken up as a kinetic energy change in the exciplex. The excited state of the neon is relatively long-lived, and an inverted population of this excited state builds. Stimulated emission causes the production of

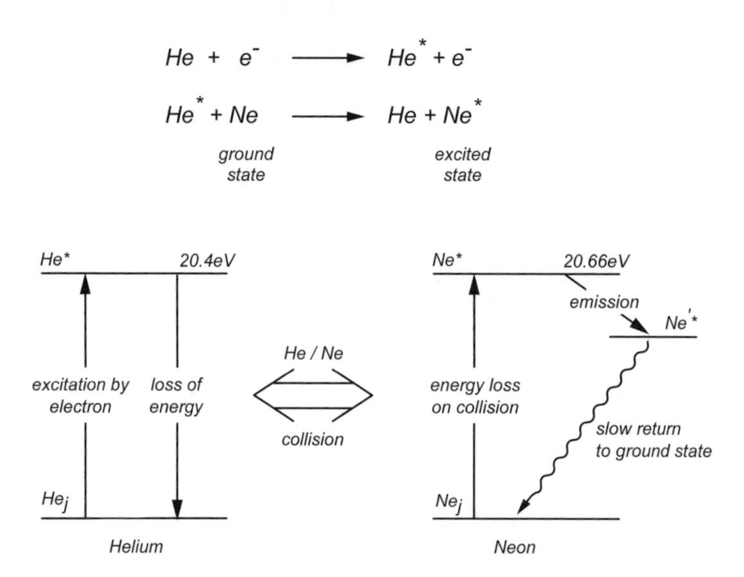

Figure 18.13
In a mixture of helium and neon, excited-state helium atoms are formed upon collision with electrons (electric discharge effect); the energy of this state is 20.4 eV above the ground state. When the He* atoms collide with neon, energy exchange takes place to give excited-state neon atoms (Ne*) and ground-state He. This excited state of neon is 20.66 eV above the ground state and is relatively long-lived, so an inverted population of Ne* builds until stimulated laser emission at 632.8 nm moves the Ne* atoms to a lower energy state (Ne′*), which quickly returns to the ground state (Ne).

laser light at a wavelength of 632.8 nm (Figure 18.13). The steady formation of helium atoms in excited states and the frequency of their collision with neon atoms produce a steady supply of inverted population so the lasing action is continuous (cw).

Liquid-State (Dye) Lasers

Dye lasers use delayed fluorescence to produce a laser beam. If suitable dye molecules are irradiated with light at their visible absorption wavelength, there is an interval of time until light is emitted as fluorescence. The wavelength of the fluorescent light is greater than the wavelength of the incident light (Figure 18.14). If the emitted light is passed between mirrors, then stimulated emission can occur to give laser light in the usual way. Rather than simply having two mirrors at the ends of the cavity, one mirror is replaced by a diffraction grating, which not only acts as a mirror but also can be used to select a narrow range of wavelengths from the light falling on it (Figure 18.15). Therefore, after the first fluorescent photons (which cover a range of wavelengths) have been selected at the diffraction grating, the remaining reflected beam of photons covers only a very narrow range of wavelengths and passes through the dye solution, inducing stimulated emission of more photons of the same wavelength. Thus, the stimulated emission (emergent laser beam) covers a narrow range of wavelengths. To prevent complete bleaching of the dye during this process, the dye solution is arranged to flow through the cavity and then circulate back to a holding vessel before passing through the cavity again. To obtain high powers from the laser, the light used for pumping the lasing levels to overpopulation often comes from an argon laser.

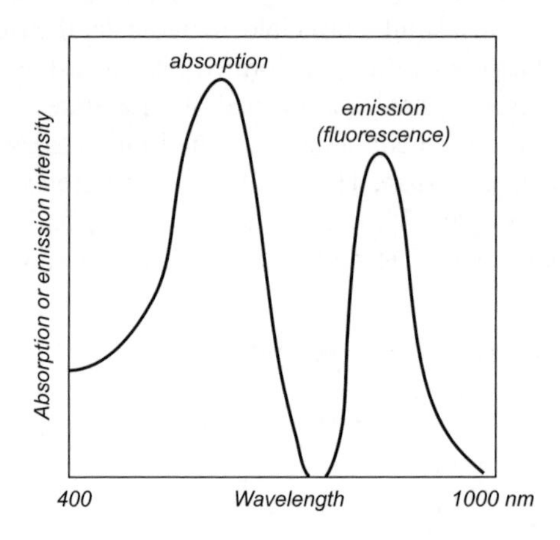

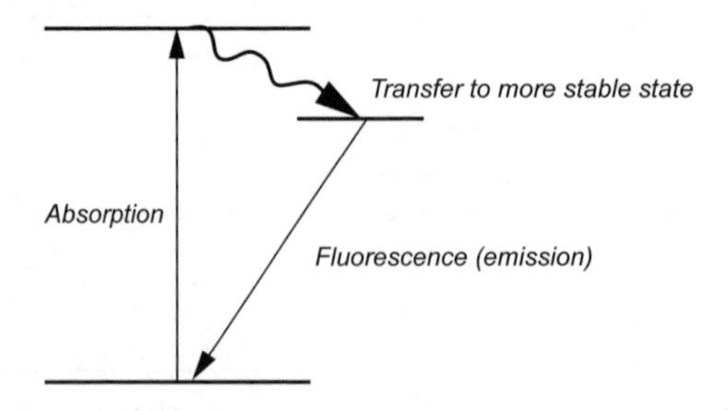

Figure 18.14
A dye molecule has one or more absorption bands in the visible region of the electromagnetic spectrum (approximately 350–700 nm). After absorbing photons, the electronically excited molecules transfer to a more stable (triplet) state, which eventually emits photons (fluoresces) at a longer wavelength (composing three-level system.) The delay allows an inverted population to build up. Sometimes there are more than three levels. For example, the europium complex (Figure 18.15) has a four-level system.

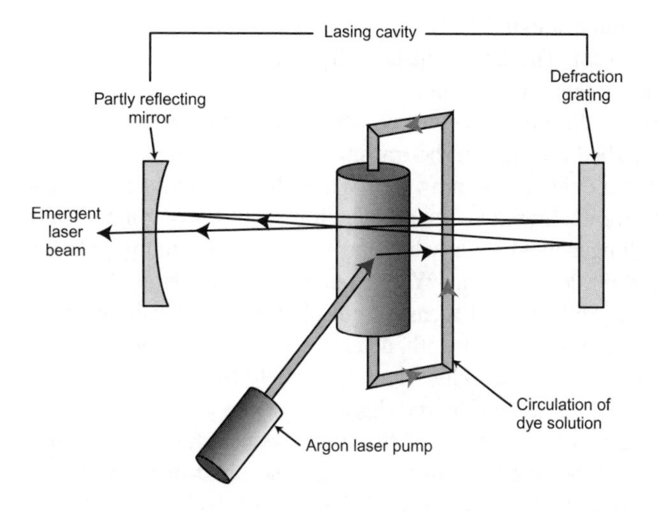

Figure 18.15
An argon laser is used to pump dye molecules into electronically excited states, which fluoresce. The emitted light travels to a diffraction grating, where the desired wavelength is selected. These selected photons pass back into the laser cavity, causing a cascade of stimulated emission, which is mode-locked at the selected wavelength. Because the diffraction grating can be turned to reflect other selected wavelengths, the dye laser can be used to tune the emergent laser beam to whatever wavelength is required. (The range is set by the fluorescence emission spectrum of the dye.)

The sorts of "dyes" that are used are materials that absorb visible light and then emit visible fluorescent light. They are mostly rigid organic or metallo/organic materials having extended π-systems, such as those shown in Figure 18.16. They are characterized by having a high quantum efficiency for converting incident light into fluorescent light. Rhodamine G is one of the most efficient of such fluorescent molecules.

Dye lasers are very useful because their output can be tuned over a range of wavelengths.

Solid-State Lasers

Ruby Laser. Ruby (essentially alumina) owes its well-known color to the presence of very small proportions of chromium ions (Cr^{3+}) distributed through it. Ruby lasers do not use natural rubies because of the imperfections they contain. Instead, synthetic single crystals of chromium

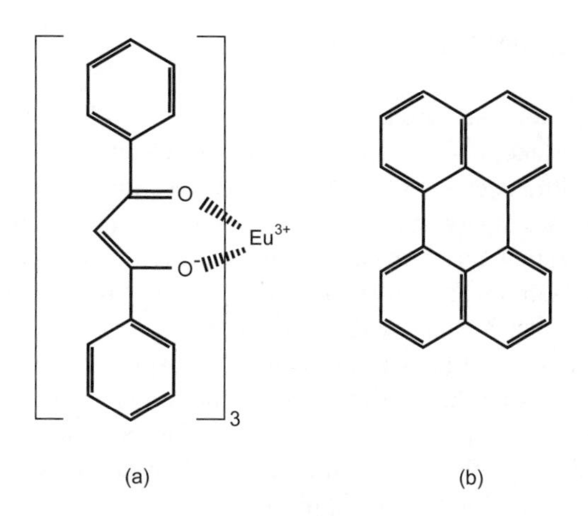

Figure 18.16
Two typical dye molecules. The europium complex (a) transfers absorbed light to excited-state levels of the complexed Eu^{3+}, from which lasing occurs. The perylene molecule (b) converts incident radiation into a triplet state, which decays slowly and so allows lasing to occur.

ions (0.05%) in alumina are used. These lasers emit light at 694 nm, in the red end of the electromagnetic spectrum. The emergent laser light is produced in pulses of a few milliseconds, and peak power can be tens of kilowatts.

Chromium ions produce the red color of ruby because they absorb blue/green light from white light, leaving unabsorbed red light to be transmitted. For a ruby laser, white light is produced by a flash lamp situated alongside the ruby rod. When the chromium ions absorb the light, they are promoted to an electronically excited state, which rapidly loses some vibrational energy to cross into a more stable, longer-lived state, which provides the inverted population required for lasing. The ruby laser is a three-level system. When spontaneous emission occurs, some of these emitted photons stimulate the production of a cascade of photons, appearing as red laser light.

The ruby rod is cut to a precise length, determined by the wavelength of the laser light. Mirrors are at each end of the rod, one of which is only partially reflective so as to allow the laser beam to emerge. As lasing begins within the rod, light reflects up and down the rod, and the intensity of the resulting cascade builds up. Some of the light escapes during the buildup of the cascade of photons (amplification), but the buildup to a pulse is so rapid that, within a few milliseconds, the main part of the pulse emerges as the laser beam. Figure 18.9 illustrates the principle. The flash lamp continues to pump up the excited state, ready for the next pulse. Thus, the emergent beam of laser light consists of a series of short pulses of intense visible radiation, with the time between pulses controlled partly by the rate at which the pump light flashes.

If the flash lamp is pulsed very rapidly, the emergent beam appears at a rate governed by the lifetime of the inverted population. The resulting laser beam becomes almost continuous because the pulses follow each other so rapidly. However, such a solid-state laser should not be pulsed too rapidly because, if it is, the rod heats to an unacceptable extent, causing distortion and even fracture. Generally, solid-state lasers are not used in continuous mode because of this heating aspect. Liquid or gas lasers do not suffer from this problem.

Neodymium and YAG Lasers. The principle of neodymium and YAG lasers is very similar to that of the ruby laser. Neodymium ions (Nd^{3+}) are used in place of Cr^{3+} and are often distributed in glass rather than in alumina. The light from the neodymium laser has a wavelength of 1060 nm (1.06 μm); it emits in the infrared region of the electromagnetic spectrum. Yttrium (Y) ions in alumina (A) compose a form of the naturally occurring garnet (G), hence the name, YAG laser. Like the ruby laser, the Nd and YAG lasers operate from three- and four-level excited-state processes.

Lasers in Mass Spectrometry

Until recently, lasers were not much used in mass spectrometry. It has been known for many years that light can ionize substances if the light energy (wavelength) is sufficient. Generally, this wavelength is in the far-ultraviolet end of the UV/visible spectrum, in which region all substances absorb the radiation. The mass analyzers in mass spectrometers operate under vacuum, so any light sources to be used for ionization must also operate under similar vacuum conditions if the irradiation is to reach and ionize the sample molecules to be analyzed. For this reason, ion sources based on ionization of sample molecules by light were mostly research curiosities. When lasers first became commercially available, the useful laser sources emitted mostly visible light of an energy that is insufficient for ionization. As the laser beams became more intense (more photons per unit area per unit time), multiphoton absorption events could be achieved easily. Whereas one photon can be absorbed by a molecule but with insufficient energy for ionization, two photons absorbed within a short space of time can cause ionization (Figure 18.17).

At this stage, lasers began to be used to excite molecules and even ions in flight in the mass spectrometer. This last absorption of light by molecular ions is enough to cause them to fragment. Multiphoton absorption and the ease of shining laser light through suitable windows into a mass

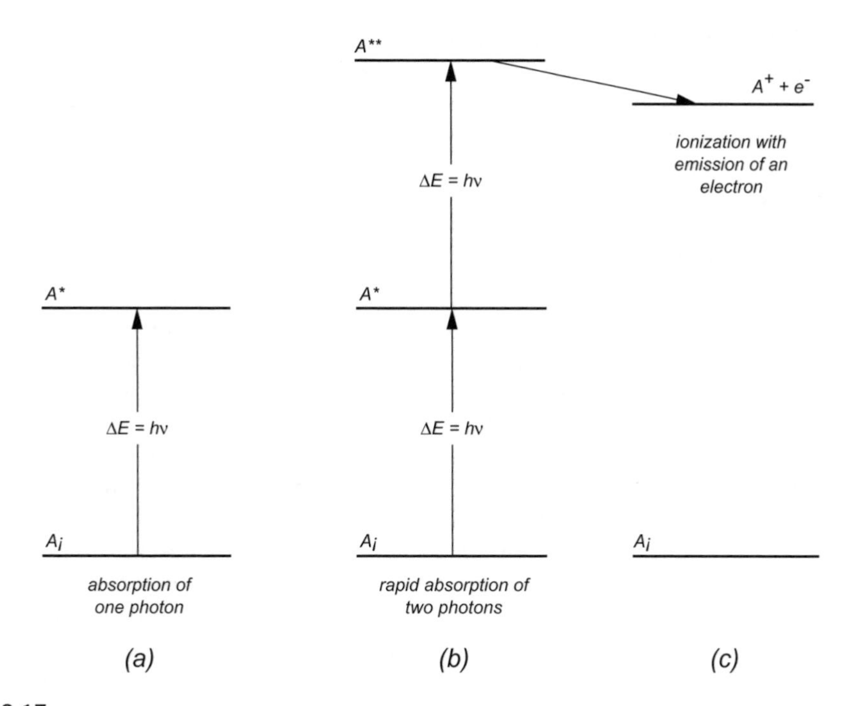

Figure 18.17
(a) Absorption of one photon raises an atom or molecule from the ground state (A) to an excited state (A*), with energy gain ΔE. (b) Rapid successive absorption of two photons raises the atom or molecule to a doubly excited-state level (A**), with energy 2ΔE. This state is energetically above the normal ionization energy (A^+) for A. Therefore, an electron is ejected as the cation forms; the energy of the emitted electron is the difference in energy between A** and A^+.

spectrometer led to the first commercial ionization sources. These ionization sources were generally not too efficient in the numbers of ions formed per unit mass of substance irradiated, but the process has been developed considerably and is particularly well known in MALDI (see below) and in the formation of aerosols from solids in plasma-torch isotope analysis.

For irradiation of solids, the intensity of a laser beam means that a great deal of energy is absorbed by the substance irradiated in a very short space of time. Multiphoton absorption in a few pico- or nanoseconds leads to elevating many molecules in the sample to highly excited states. As the latter begin to equilibrate, much of this added energy is channeled into rotational and vibrational modes so the sample molecules behave as if they had been subjected to very high temperatures for a brief interval of time. There is rapid expansion at the irradiation site, and volatilization occurs (ablation). This expansion is so rapid that a shock wave travels through the sample, and underlying layers of the sample substance are temporarily subjected to very high temperatures and pressures. Usually, before the sample molecules can relax back to their ground states, a second laser pulse arrives, and the process is repeated. This pulsing is so rapid that the vaporizing sample is subjected to the next laser pulse before all of the vapor (a plasma of neutrals, ions, and electrons) has dissipated, and the vapor molecules are further excited by absorbing more photons. Many of the excited states reached by multiple photon absorption lead to ejection of an electron (ionization) so ablation produces some ions, which can be extracted by suitable electric fields and mass analyzed.

Another consequence of the rapidity with which photons are absorbed lies in the amount of fragmentation observed. The rapid dissipation of energy into vibrational modes might be expected to lead to bond cleavage and fragmentation of the sample molecules. Some does occur, but before there is time for much bond-breaking to occur, there is conversion of the vibrational energy into kinetic energy as the sample molecules vibrate against each other. The molecules vaporize due to the kinetic energy, and much of the excess vibrational energy is dissipated this way. In these circumstances, fragmentation is limited. Ionization itself is still not highly likely, and few ions are formed. Many of the ions are protonated molecules rather than singly ionized species. A big advance was made when a matrix was used to improve the numbers of ions formed on laser irradiation. This technique became known as MALDI (matrix-assisted laser desorption ionization; see below). Ablation is important when used with a secondary ionization mode. Intractable solids such as ceramics can be ablated, and the resulting vapor passed into, for example, a plasma torch for complete ionization (see Chapter 14). This mode is extremely useful in the examination of atomic isotope patterns and ratios in samples that are otherwise difficult to analyze. Direct laser desorption ionization is used to examine the composition of surfaces and for depth profiling through specimens (see Chapter 2).

For MALDI, the matrix is a substance of low molecular mass that is mixed intimately with the sample to be analyzed. The matrix is preferably a good source of protons and is usually acidic, but other Lewis acids (X^+), as with Ag^+, can be used to promote formation of $[M + X]^+$ ions. When the matrix/sample mix is irradiated by laser light of a suitable wavelength, the matrix absorbs most of the energy and ablates or vaporizes. Sample molecules (M) accompany this expanding cloud of matrix, and, if the basicities are right, transfer of protons between matrix and sample occurs, leading to the formation of $[M + H]^+$ ions. Thus, MALDI mass spectra are characterized by protonated molecular ions and few fragment ions. The technique is so useful with even thermally sensitive molecules that it has come into widespread use for proteins and similar polar molecules. It might be noted that nonlinear molecules have 3N – 6 vibration modes, in which N is the number of atoms. A molecule with 200 atoms has 594 vibration modes. If each of these is raised to the next excited vibrational level, each vibration requires little energy, but a total of 594 vibrators contain enough total energy for bond breaking. For such bond breaking to occur, all or most of the excess of vibrational energy has to arrive in one bond at the same instance of time (about 10^{-13} sec). The chances of this occurrence in a multiatomic molecule are very low, and, together with radiative loss of excess of vibrational energy, this effect is important in preserving the structural integrity of large, normally thermosensitive molecules. This phenomenon applies to isolated molecules and ions, as in ablated plasmas and vapors. For assemblies of molecules, as in solids, melts, or dense gases, there is continual interaction between the molecules (collision) with energy transfer, and there are constraints on vibration and rotation because of the close packing. This, and the longer time scales before vaporization can occur (a few milliseconds to infinity), means there is ample time for excess of vibrational energy to accumulate into one or more bonds so as to cause bond cleavage. This is what would happen if a protein sample were simply heated slowly in a pot to try to effect volatilization before mass spectral analysis.

The advent of lasers and MALDI into mass spectrometry has had a major effect, especially in the analysis of large, polar biochemicals. Whereas electron ionization gives many fragment ions, which carry structural information, direct laser ionization and MALDI give mostly protonated molecular ions, so MS/MS with collision-induced dissociation becomes necessary to stimulate the fragmentation needed to obtain the same structural information. Of course, electron ionization methods are not useful for vaporizing and ionizing large biomolecules if they need to be first vaporized thermally, since this leads to their decomposition and therefore loss of molecular mass and structural data.

Conclusion

Modern commercial lasers can produce intense beams of monochromatic, coherent radiation. The whole of the UV/visible/IR spectral range is accessible by suitable choice of laser. In mass spectrometry, this light can be used to cause ablation, direct ionization, and indirect ionization (MALDI). Ablation (often together with a secondary ionization mode) and MALDI are particularly important for examining complex, intractable solids and large polar biomolecules, respectively.

Chapter 19

Nebulizers

Background

Samples to be examined by inductively coupled plasma and mass spectrometry (ICP/MS) are commonly in the form of a solution that is transported into the plasma flame. The thermal mass of the flame is small, and ingress of excessive quantities of extraneous matter, such as solvent, would cool the flame and might even extinguish it. Even cooling the flame reduces its ionization efficiency, with concomitant effects on the accuracy and detection limits of the ICP/MS method. Consequently, it is necessary to remove as much solvent as possible which can be done by evaporation off-line or done on-line by spraying the solution as an aerosol into the plasma flame.

The Nature of an Aerosol

If a liquid is vaporized rapidly or its vapor is cooled rapidly, it can form an aerosol consisting of a mixture of purely gaseous components, small droplets, and, sometimes, small particles of solid matter (particulates). Aerosols are dynamic systems, with evaporation from some droplets that become smaller and coalescence of other droplets that grow bigger. Given time and suitable temperatures, the components of aerosols may condense to form a liquid or evaporate to form vapor. In the form of a mist of fine droplets, this condensation is not fast, even with strong cooling, and an aerosol produced in a gas stream can be swept for quite long distances without serious losses of components of the aerosol. For example, clouds formed in the atmosphere by rapid cooling of warm moist air consist of small droplets of water, which scatter light and hence give rise to the typical opacity of clouds; a similar effect near ground level at low temperatures gives rise to fog or mist. A good example of man-made aerosols are the clouds of water vapor formed above the cooling towers of power stations. Such clouds as well as natural ones are swept along by the wind. Given the right cooling conditions, the small water droplets can coalesce to form larger drops, which eventually may be large enough to fall as rain. Under warmer conditions, the small droplets evaporate completely and the clouds disperse.

An aerosol produced instrumentally has similar properties, except that the aerosol is usually produced from solutions and not from pure liquids. For solutions of analytes, the droplets consist of solute and solvent, from which the latter can evaporate to give smaller droplets of increasingly concentrated solution (Figure 19.1). If the solvent evaporates entirely from a droplet, the desolvated dry solute appears as small solid particles, often simply called particulate matter.

For a droplet of initial radius r_0, its radius r_t at time t may be estimated from equation (1), the various terms of which are listed below.

$$r_0^3 - r_t^3 = 980 \times D.\sigma.\Delta p\ (M/d{\cdot}R{\cdot}T)^2 \times t$$

D = diffusivity of the solvent vapour from the sample solution ($cm^2{\cdot}s^{-1}$)
σ = surface tension of sample solution ($erg{\cdot}cm^{-2}$)
p = vapour pressure of sample solution (mm Hg x 1·359 = $g{\cdot}cm^{-2}$)
M = relative molecular mass of the liquid (solvent; $g{\cdot}mol^{-1}$)
d = density of sample solution ($g{\cdot}cm^{-3}$)
T = absolute temperature of droplet (K)
R = gas constant = 8.31×10^7 ($erg{\cdot}K^{-1}{\cdot}mol^{-1}$)

The factor of 980 changes g.cm units into ergs.

With typical values for water at 20°C, the time taken for a droplet to shrink to 10% of its original radius may be calculated as an example of the use of equation (1).
With D = 0.24, σ = 73, p = 15.5 mm Hg, M = 18, d = 1, T = 293 K then, if $t_t = 0.1 \times t_0$.

$$t_0^3 - t_t^3 = 0.999 \times t_0^3 = 0.224 \times 10\text{-}12 \times t$$

$$\therefore t = \underline{4.5}\ s$$

At 40°C, assuming the other terms do not change too much, p = 55 mm Hg and,

$$t = \underline{1.5}\ s$$

Let the argon sweep flow of the aerosol through a tube of 1 cm radius be 1 L/min.
In 1.5 s, the argon will flow through a distance (l) given by the formula:

$$l = \frac{\text{flow rate x time}}{\text{cross-sectional area}} = \underline{7.9\ cm}$$

Figure 19.1.
The calculation shows how rapidly a droplet changes in diameter with time as it flows toward the plasma flame. At 40°C, a droplet loses 90% of its size within about 1.5 sec, in which time the sweep gas has flowed only about 8 cm along the tube leading to the plasma flame. Typical desolvation chambers operate at 150°C and, at these temperatures, similar changes in diameter will be complete within a few milliseconds. The droplets of sample solution lose almost all of their solvent (dry out) to give only residual sample (solute) particulate matter before reaching the plasma flame.

Aerosols can be produced as a spray of droplets by various means. A good example of a nebulizer is the common household hair spray, which produces fine droplets of a solution of hair lacquer by using a gas to blow the lacquer solution through a fine nozzle so that it emerges as a spray of small droplets. In use, the droplets strike the hair and settle, and the solvent evaporates to leave behind the nonvolatile lacquer. For mass spectrometry, a spray of a solution of analyte can be produced similarly or by a wide variety of other methods, many of which are discussed here. Chapters 8 ("Electrospray Ionization") and 11 ("Thermospray and Plasmaspray Interfaces") also contain details of droplet evaporation and formation of ions that are relevant to the discussion in this chapter. Aerosols are also produced by laser ablation; for more information on this topic, see Chapters 17 and 18.

The term *nebulizer* is used generally as a description for any spraying device, such as the hair spray mentioned above. It is normally applied to any means of forming an aerosol spray in which a volume of liquid is broken into a mist of vapor and small droplets and possibly even solid matter. There is a variety of nebulizer designs for transporting a solution of analyte in droplet form to a plasma torch in ICP/MS and to the inlet/ionization sources used in electrospray and mass spectrometry (ES/MS) and atmospheric-pressure chemical ionization and mass spectrometry (APCI/MS).

General Principles of Aerosol Formation

For use in ICP/MS, an aerosol of analyte solution is produced in a nebulizer by mechanically breaking the solution into a spray of droplets and solvent vapor. This spray is swept along to the plasma flame by a flow of argon gas. The droplets have a range of diameters, depending on the type of nebulizer used. Frequently, before entering the torch, the aerosol first passes through a spray

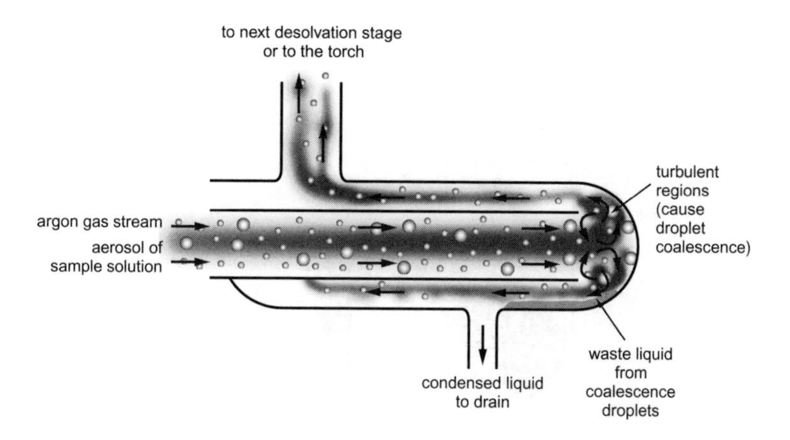

Figure 19.2.
The aerosol, consisting of solvent vapor and droplets having a range of diameters immediately after formation, can be passed through a spray chamber (which may or may not be present in any one apparatus). In the spray chamber, a combination of gravity and turbulent flow induces the larger droplets to hit the walls of the chamber, where the deposited droplets coalesce into a liquid waste stream that is drained away. The smaller droplets of the aerosol do not fall out of the gas stream and are swept along to the next stage, either the plasma flame or a desolvation chamber placed before the flame.

chamber designed to remove many of the larger droplets through their collision with the walls of the spray chamber under the influence of turbulence and gravity (Figure 19.2).

During passage to the flame, all of the droplets lose solvent by evaporation desolvation, becoming smaller as they do so (Figure 19.1). Droplets coming from the spray chamber or directly from the nebulizer become smaller through evaporation of solvent as they flow in the carrier gas stream. To assist this process, the gas stream is passed through a heating chamber held at about 150°C, which leads to much more rapid desolvation of the droplets. It is preferable to remove most or all of the solvent vapor before it reaches the plasma flame. To do this, a cooling chamber following the hot desolvation chamber induces the solvent to condense. The solvent condensate flows to waste, and residual dry particulate matter is swept into the plasma flame by a flow of argon gas. The flame promotes atomization and ionization to produce ions of the elements present in the sample (see Chapter 14). The efficiencies with which solutions are transported from the inlet through the nebulizer and along to the flame vary widely with the design of the nebulizer. Efficiency is typically within the range of 10 to 20%, although a few nebulizers are capable of much higher transport efficiencies.

The solution to be nebulized can be a one-off sample, pumped or drawn into the nebulizer at a rate varying from a few microliters per minute to several milliliters per minute. Alternatively, the supply of solution can be continuous, as when the nebulizer is placed on the end of a liquid chromatographic column.

Many designs of nebulizer are commonly used in ICP/MS, but their construction and mode of operation can be collated into a small number of groups: pneumatic, ultrasonic, thermospray, APCI, and electrospray. These different types are discussed in the following sections, which are followed by further sections on spray and desolvation chambers.

Pneumatic Nebulizers (PN)

Principles of Operation

If a gas flows over the surface of a liquid, certain effects ensue. Only the relative velocity of the liquid surface and gas is important in giving rise to nebulization. Thus, some pneumatic nebulizers

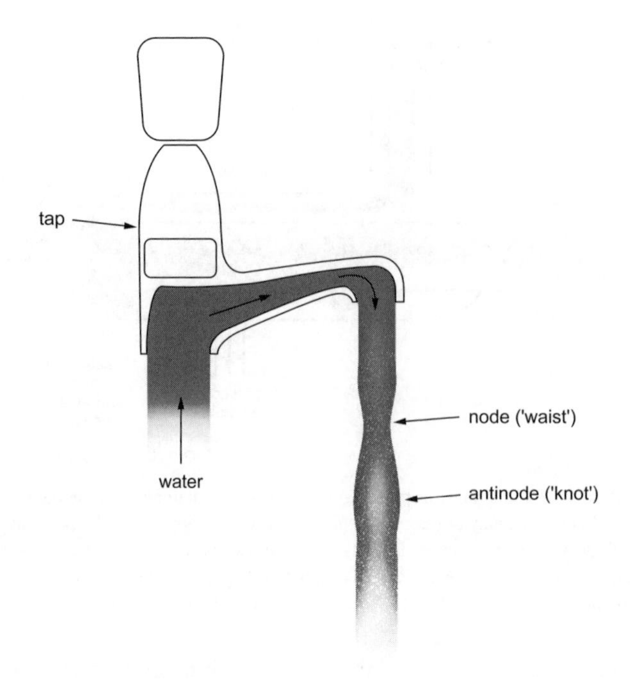

Figure 19.3.
Water flowing from a tap begins as a steady, parallel-sided stream, but upon flowing through the air, this stream begins to develop waists and knots as the flow becomes uneven. Together with associated turbulent flow, this unevenness leads to the stream's breaking up completely, first into sections and then into large drops. This effect is best seen with a long, narrow waterfall. The water initially flows in an unbroken stream, but by the time it has accelerated and reached the end of its fall, it has broken apart. Where it hits the ground, a spray of smaller droplets is produced, exactly similar to the fine spray produced by an impactor bead placed in the path of an aerosol generated by a nebulizer.

work on the principle of using a high-velocity gas stream (usually argon) passing over a liquid surface or through a liquid to produce an aerosol, while others produce a fast-moving stream of liquid that is passed into an almost stationary gas.

The early stages of pneumatic aerosol formation are familiar in everyday life. As an airplane flies, the flow of air over its wings causes a reduction in pressure just above the wing. The same principle is evident in sailing, in which boats are able to sail against the wind by making use of the reduction in pressure caused by a gas flow over a surface except that, for the boat, the wing (sail) is vertical instead of horizontal. Water flowing from a tap in a steady stream is found to lose its even flow and to form waists and knots (Figure 19.3) as turbulence begins.

In pneumatic nebulizers, the relative velocity of gas and liquid first induces a reduction in pressure above the surface of the liquid (see the calculation in Figure 19.4). The reduction in pressure is sufficient to cause liquids to flow out of capillary tubes, in accord with Poiseuille's formula (Figure 19.5). As the relative velocity of a liquid and a gas increases — particularly if the mass of liquid is small — this partial vacuum and rapid flow cause the surface of the liquid to be broken into droplets. An aerosol is formed.

The size of the droplets formed in an aerosol has been examined for a range of conditions important in ICP/MS and can be predicted from an experimentally determined empirical formula (Figure 19.6). Of the two terms in the formula, the first is most important, except at very low relative flow rates. At low relative velocity of liquid and gas, simple droplet formation is observed, but as the relative velocity increases, the stream of liquid begins to flutter and to break apart into long thinner streamlets, which then break into droplets. At even higher relative velocity, the liquid surface is stripped off, and the thin films so-formed are broken down into

Reduced pressure caused by a flowing stream

When a gas or liquid flows over a surface, the pressure at the surface is reduced according to the formula shown in equation (1), in which *d* is the density and *v* is the linear flow velocity of the moving stream.

$$\Delta p = d.v^2/2 \qquad (1)$$

If *d* is measured in $g{\cdot}cm^{-3}$ and *v* in $cm{\cdot}s^{-1}$ then Δp is in $dyn{\cdot}cm^{-2}$. This result can be converted aproximately into atmospheres on multiplication by 10^{-6}.

Consider a flow of argon of 0.5 L/min through an annular space of 4×10^{-4} cm^2 between two concentric capillary tubes at normal ambient temperatures. If the density of argon is taken to be 1.2×10^{-3} $g{\cdot}cm^{-3}$ then,

Flow of argon = 0.5 L/min = 500 cm^3/min = 8.7 $cm^3{\cdot}s^{-1}$

Linear flow = $8.7 / (4 \times 10^{-4})$ $cm{\cdot}s^{-1}$ = 2.2×10^4 $cm{\cdot}s^{-1}$

$\Delta p = d{\cdot}v^2/2$ = $1.2 \times 10^{-3} \times 4.8 \times 10^8 / 2 = 2.9 \times 10^5$ $dyn{\cdot}cm^{-2}$

= 0.3 atmospheres

Figure 19.4.
The drop in pressure when a stream of gas or liquid flows over a surface can be estimated from the given approximate formula if viscosity effects are ignored. The example calculation reveals that, with the sorts of gas flows common in a concentric-tube nebulizer, the liquid (the sample solution) at the end of the innermost tube is subjected to a partial vacuum of about 0.3 atm. This vacuum causes the liquid to lift out of the capillary, where it meets the flowing gas stream and is broken into an aerosol. For cross-flow nebulizers, the vacuum created depends critically on the alignment of the gas and liquid flows but, as a maximum, it can be estimated from the given formula.

Poiseuille's formula

This predicts the rate of flow of liquid through a smooth tube under the effect of a pressure difference between the ends of the tube (equation 1).

$$F = (\pi \times r^4 \times \Delta p) / (8 \times \eta \times l) \qquad (1)$$

In the equation, *F* is the flow of liquid ($cm^3{\cdot}s^{-1}$) of viscosity η(poise or $dyn{\cdot}s{\cdot}cm^{-2}$) under a pressure drop (Δp $dyn{\cdot}cm^{-2}$) through a tube (inner radius, *r* cm) of length *l* (cm).

From the example of Figure 19.1, a flow of argon can cause a pressure drop of 0.3 atmosphere (3×10^5 $dyn{\cdot}cm^{-2}$). Let the viscosity of the sample solution be the same as that of water 0.01 poise), the radius be 0.01 cm and the length of capillary be 10 cm.

$$F = (\pi \times 0.01^{-4} \times 3 \times 10^5) / (8 \times 0.01 \times 1) = 1.2 \times 10^{-2} \ cm^3{\cdot}s^{-1} \ (0.7 \text{ mL/min})$$

Under these conditions, the flow of sample liquid is predicted to be shown and this is similar to the flows observed with concentric nebulizers.

Figure 19.5.
Using Poiseuille's formula, the calculation shows that for concentric-tube nebulizers, with dimensions similar to those in use for ICP/MS, the reduced pressure arising from the relative linear velocity of gas and liquid causes the sample solution to be pulled from the end of the inner capillary tube. It can be estimated that the rate at which a sample passes through the inner capillary will be about 0.7 ml/min. For cross-flow nebulizers, the flows are similar once the gas and liquid stream intersection has been optimized.

droplets. Thus, droplet formation occurs over a wide range of relative velocities of liquid and gas, so aerosols (droplets plus vapor) can be formed under a wide range of conditions. For ICP/MS purposes, it is preferable to have aerosols with very small droplets covering a narrow distribution of diameters. The various pneumatic nebulizers are constructed on the basis of these principles.

$$D = \frac{585}{V}\sqrt{\frac{s}{d}} + 597\left(\frac{g}{\sqrt{d.s}}\right)^{0.45} x(1000Q)^{1.5}$$

D = mean droplet diameter (μm)
s = surface tension (dyne· cm^{-1})
d = density of liquid (g· cm^{-3})
g = viscosity of liquid (poises)
V = difference in linear velocities of gas and liquid flows (m/s)
Q = ratio of volume flows of liquid / gas

For water solutions, the following approximate values can be used:

$s = 73; d = 1; g = 0.01; Q = 10^{-3}$ (1μL·min^{-1} / 1μL·min^{-1})

For $V = 50$ m/s, $D \simeq 120$ μm

Note that the formula is valid only for $30 < s < 73$; $0.01 < g < 0.3$; $0.8 < d < 1.2$; $5 < V < 50$

Figure 19.6.
This formula for estimating droplet size was determined experimentally. Of the various terms, the first is the most important for small values of V. As V becomes small, the second term gains in importance. Unless the density or viscosity of the sample solution changes markedly from the values for water, mean droplet size can be estimated approximately by using the corresponding values for water, as shown.

Concentric Tubular Nebulizers

Figure 19.7 shows a typical construction of a concentric-tube nebulizer. The sample (analyte) solution is placed in the innermost of two concentric capillary tubes and a flow of argon is forced down the annular space between the two tubes. As it emerges, the fast-flowing gas stream causes a partial vacuum at the end of the inner tube (Figure 19.4), and the sample solution lifts out (Figure 19.5). Where the emerging solution meets the fast-flowing gas, it is broken into an aerosol (Figure 19.7), which is swept along with the gas and eventually reaches the plasma flame. Uptake of sample solution is commonly a few milliliters per minute.

The dimensions of concentric-tube nebulizers have been reduced to give microconcentric nebulizers (MCN), which can also be made from acid-resistant material. Sample uptake with these microbore sprayers is only about 50 μl/min, yet they provide such good sample-transfer efficiencies that they have a performance comparable with other pneumatic nebulizers, which consume about 1 ml/min of sample. Careful alignment of the ends of the concentric capillary tubes (the nozzle)

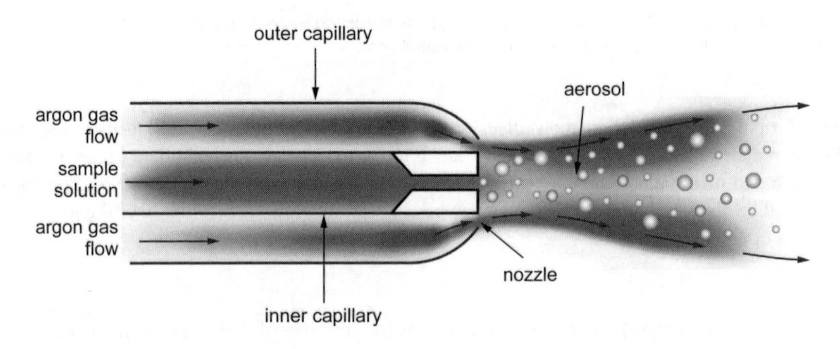

Figure 19.7.
In a concentric-tube nebulizer, the sample solution is drawn through the inner capillary by the vacuum created when the argon gas stream flows over the end (nozzle) at high linear velocity. As the solution is drawn out, the edges of the liquid forming a film over the end of the inner capillary are blown away as a spray of droplets and solvent vapor. This aerosol may pass through spray and desolvation chambers before reaching the plasma flame.

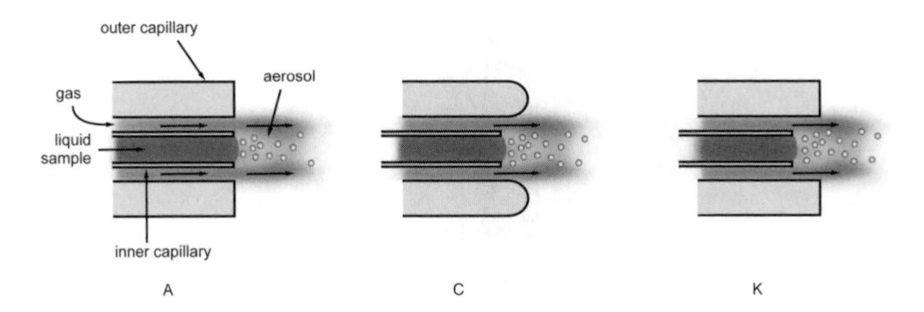

Figure 19.8.
Three common types of nozzle are shown diagrammatically. Types A and K are similar, with sharp cutoffs on the ends of the outer and inner capillaries to maximize shear forces on the liquid issuing from the end of the inner tube. In types K and C, the inner capillary does not extend to the end of the outer tube, and there is a greater production of aerosol per unit time. These concentric-tube nebulizers operate at argon gas flows of about 1 l/min.

and structural stability is essential. Commercially, three main types of arrangement are available: A, C, or K (Figure 19.8).

The actual gas flow needed to produce a fine spray depends mainly on the nozzle geometry and the diameters of the two concentric tubes. Generally, gas flows of about 1 l/min support a sample solution flow of about 0.5–4 ml/min. A special design of a high-efficiency nebulizer (HEN) uses much narrower capillary tubes that are about 100 μm in diameter. These nebulizers or narrower tubes afford sample flows of about 10 μl/min and give a narrow distribution of droplet sizes of mean diameter close to 10 μm. This small droplet size means that transfer efficiency of the sample into the plasma flame is high. To attain the required gas flow for nebulization of analyte solution at a rate of about 1 l/min, a much higher argon gas operating pressure is needed. One problem with the very narrow bore tubes is that they can block easily due to small pieces of particulate matter in the liquid solution sticking inside them. This problem can be avoided by using some form of prefilter.

In an alternative arrangement, the relative velocity of liquid and gas is achieved differently. The sample is pumped at high pressure (6000 psi) through a fine capillary tube (30-μm diameter), so it emerges as a very fast-moving thin stream which meets a flow of argon gas. The latter is arranged to flow relatively slowly to the plasma flame. Where the fast-moving liquid meets the slow-moving gas, aerosol formation occurs as described above, and this aerosol is transported to the plasma flame.

In these and other pneumatic nebulizers, the first formed aerosol is sometimes broken down into even smaller droplets by simply allowing it to impinge onto a solid target. The mechanical effect of small droplets' being flung against a solid object leads to further fragmentation of the droplets (Figure 19.9). The target is usually a glass bead because this shape encourages suitable gas flow around it. The target is known as an impactor bead. One problem with the very fine capillaries used in these and other such devices is the possibility of blockages by particulate matter

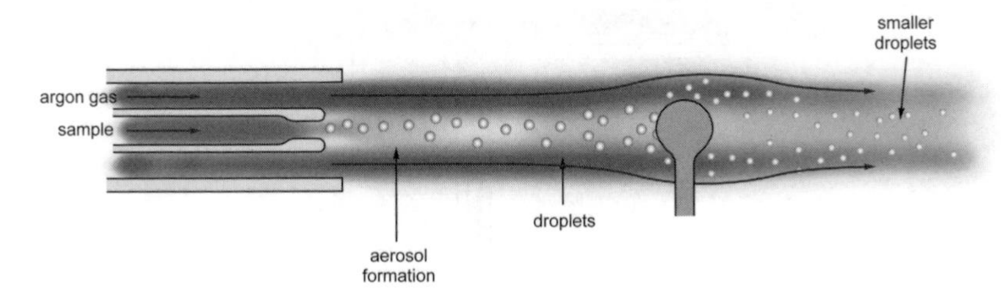

Figure 19.9.
The fast-flowing narrow liquid stream has a high relative linear velocity with respect to the slower flow of the argon gas stream. This leads to breaking up the liquid stream into fast-moving droplets, which strike the impactor bead and form much smaller droplets.

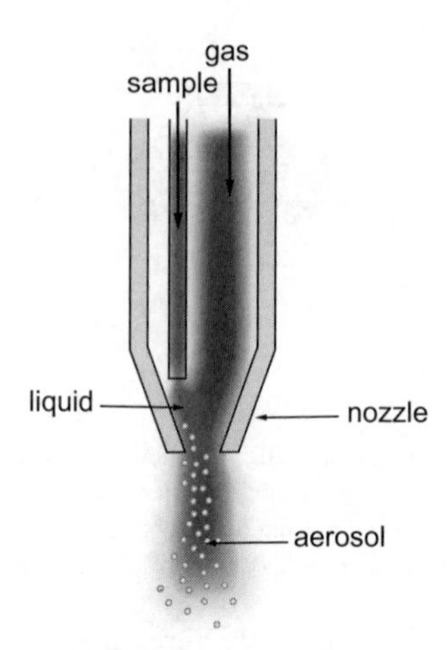

Figure 19.10.
The liquid sample flows into the nozzle and coats the inside walls. The sample stream arrives at the orifice (the nozzle outlet is about 0.01 cm diameter), where it meets the argon stream and is nebulized.

suspended in the sample solutions. With some solution types, the ends of the capillaries can become blocked by deposition of solids from the evaporating solution.

In an even simpler arrangement of the concentric-tube concept, the liquid and fast-flowing argon gas streams are simply mixed before they issue from the nebulizer. For this to work well, the solution needs to be pumped into the nozzle area. A higher gas pressure is needed to achieve a good nebulization. This device (Figure 19.10) gives a very finely dispersed spray and provides a good sample-transfer rate.

Cross-Flow Nebulizers

Capillary Tubes

The flows of gas and liquid need not be concentric for aerosol formation and, indeed, the two flows could meet at any angle. In the cross-flow nebulizers, the flows of gas and sample solution are approximately at right angles to each other. In the simplest arrangement (Figure 19.11), a vertical capillary tube carries the sample solution. A stream of gas from a second capillary is blown across this vertical tube and creates a partial vacuum, so some sample solution lifts above the top of the capillary. There, the fast-flowing gas stream breaks down the thin film of sample

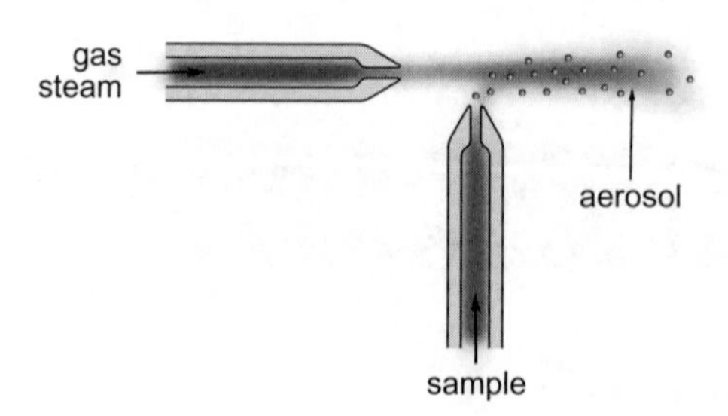

Figure 19.11.
In the cross-flow arrangement, the argon gas flows at high linear velocity across the face of an orthogonal capillary tube containing sample solution. The partial vacuum causes liquid to lift above the end of the capillary. Here, it meets the argon and is nebulized.

solution to form an aerosol, which is swept to the plasma flame. Again, an impactor bead can be used to enhance breakdown into fine droplets. The cross-flow arrangement is less prone to blocking from buildup of solids from the evaporating analyte. For best performance, the capillaries need to be placed very carefully into an optimum relationship to each other and, for this reason, may need to be adjustable. Such an arrangement may lead to the capillaries' oscillating in the gas stream; to obviate this effect, one design (MAK) uses thick-walled, small-diameter, rigid capillaries and a high argon gas pressure of about 200 psi to achieve an argon gas flow of about 0.5 l/min. Like other cross-flow devices, this last is particularly free from clogging and gives good long-term stability.

Liquid Films

This form of cross-flow nebulizer produces excellent aerosol sprays. In general, with these devices, a thin film of the sample solution attempts to cover a small orifice through which argon gas flows at right angles to the putative film. The thin film of liquid around the rim of the orifice is blown apart to form an aerosol (Figure 19.12). The sample solution can either flow under gravity onto the orifice or be pumped there. This aerosol apparatus is frequently called a Babington nebulizer. The cross-flow design again provides freedom from clogging, and it can even be used with slurries and not just solutions. The efficiency in aerosol formation with these devices arises from the thin line

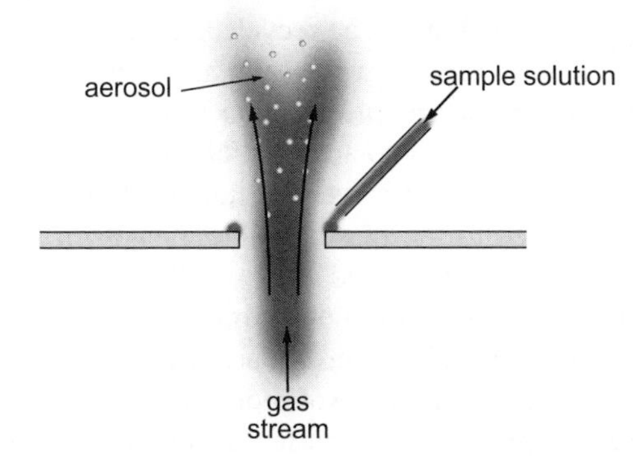

Figure 19.12.
In this cross-flow arrangement, a thin film of sample solution is obtained as it flows around the edge of a small opening, through which there is a fast linear flow of argon. The liquid film is rapidly nebulized along the rim of the orifice.

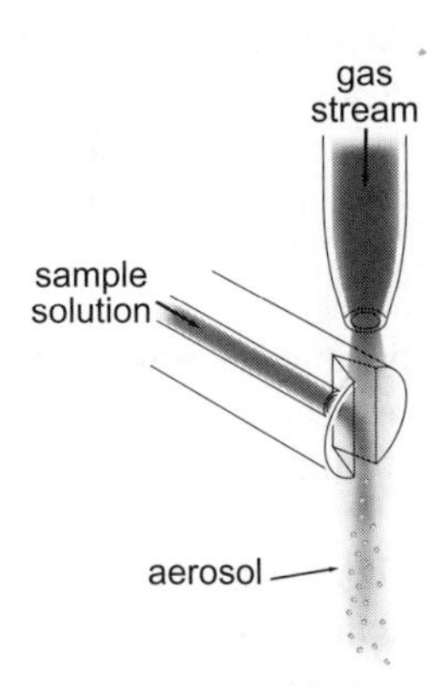

Figure 19.13.
A sample solution is drawn or pumped into a V-shaped groove cut into the end of a capillary tube. The crossed gas and liquid streams form an aerosol. An impactor bead can be used to provide an even smaller droplet size.

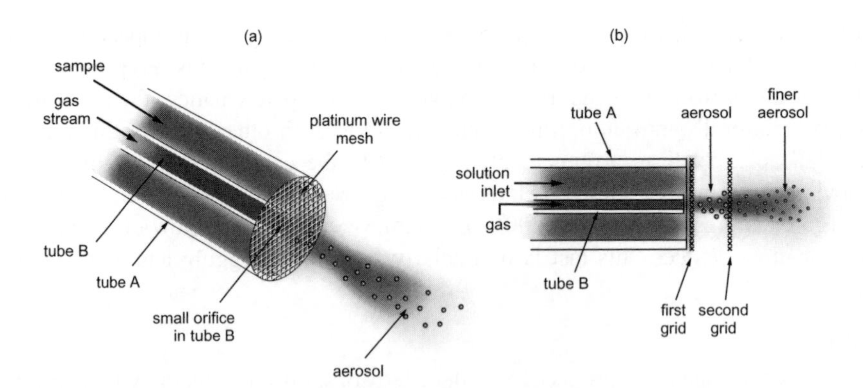

Figure 19.14.
(a) The sample solution flows to the end of a tube A, where it enters a groove cut into the end of the tube. The liquid runs as a thin film down the grid wires of a fine platinum mesh. Argon flows through the inner tube B and issues from a small orifice placed just in front of the mesh. The high linear velocity of the argon passing through the mesh and over the thin film of liquid forms an aerosol. (b) To aid in further reducing droplet size, a second grid is placed after the first and is used as an impactor, similar to the use of impactor beads.

of liquid surrounding the orifice, which allows the fast-flowing gas stream to break it up into a fine spray. With this simplest of arrangements, much of the sample solution does not reach the orifice and flows away to a drain, from which it can be recycled. This procedure is complicated, however, and other ways have been found to provide a thin film and a gross flow of gas.

In the so-called V-groove device, sample solution flows as a thin stream along a V-shaped groove cut into a metal block. A gas flow is arranged almost coincident with and in the direction of the groove. The fast-flowing gas passing over the liquid stream produces a fine aerosol. Much more of the solution is transferred to the plasma flame than is the case for a simple Babington nebulizer. The V-groove can also be incorporated into a capillary cross-flow nebulizer (Figure 19.13), or, in yet another arrangement, the liquid film flows from a V-groove and onto a platinum grid (fine wire mesh), where the liquid presents a high surface area as it covers the wires. Fast-flowing argon gas flows through the grid, producing an aerosol. For even better droplet dispersion and smaller size, there is normally a second platinum grid placed behind the first. This second grid acts in the same way as an impactor bead, breaking small drops into even smaller ones (Figure 19.14).

Another variant (the cone spray) allows the sample solution to flow down the sides of an inverted cone and through a hole at the bottom of which flows a fast stream of argon gas. As the liquid film meets the gas, it is ripped apart into a finely dispersed aerosol (Figure 19.15).

Finally, in yet another variant, the sample liquid stream and the gas flow are brought together at a shaped nozzle into which the liquid flows (parallel-path nebulizer). Again, the intersection of liquid film and gas flow leads to the formation of an aerosol. Obstruction of the sample flow by formation of deposits is not a problem, and the devices are easily constructed from plastics, making them robust and cheap.

Frit Nebulizers

The aim of breaking up a thin film of liquid into an aerosol by a cross flow of gas has been developed with frits, which are essentially a means of supporting a film of liquid on a porous surface. As the liquid flows onto one surface of the frit (frequently made from glass), argon gas is forced through from the undersurface (Figure 19.16). Where the gas meets the liquid film, the latter is dispersed into an aerosol and is carried as usual toward the plasma flame. There have been several designs of frit nebulizers, but all work in a similar fashion. Mean droplet diameters are approximately 100 nm, and over 90% of the liquid sample can be transported to the flame.

There are problems in use of the frit nebulizer. Memory effects tend to be severe, and each sample needs to be followed by several wash-outs with clean solvent before the pores of the frit become free of residual sample. Biological samples frequently contain detergent-like materials, and

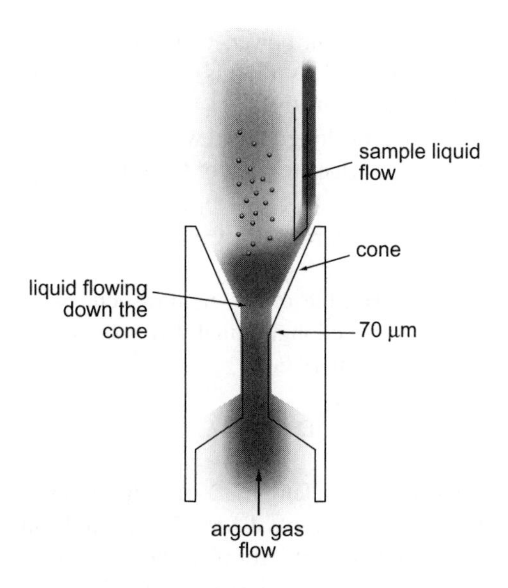

Figure 19.15.
This arrangement provides a thin film of liquid sample solution flowing down to a narrow orifice (0.007-cm diameter) through which argon flows at high linear velocity (volume flow is about 0.5–1 l/min). A fine aerosol is produced. This particular nebulizer is efficient for solutions having a high concentration of analyte constituents.

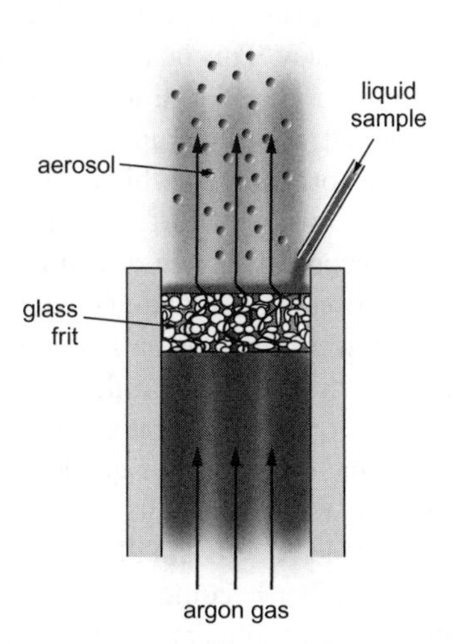

Figure 19.16.
The sample solution flows onto a piece of fritted glass through which argon gas flows. The flow of argon is broken down into narrow parallel streams of high linear velocity, which meet the thin film of liquid percolating into the pores of the frit. At the interfaces, an aerosol is formed and is blown from the top of the frit.

these give rise to frothing on the frit when gas is blown through them. However, in suitable instances, the frits perform well, especially with small volumes of sample, which can be discrete or continuous.

Ultrasonic Nebulizers (USN)

Principles of Operation

If ultrasound of a frequency in the approximate range 0.2–10 MHz is passed through a liquid, then longitudinal rarefaction and compression waves break up the liquid surface. If the amplitude of the waves is sufficient, viz., if there is sufficient power, then the surface of the liquid is disrupted to form a stream of droplets (an aerosol). This idea is the basis of the ultrasonic nebulizer. The ultrasound is usually produced piezoelectrically.

For any ultrasound of frequency ν, the longitudinal wavelength λ at the surface can be calculated from Equation 19.1, in which σ is the surface tension of the liquid and ρ is the density of the liquid.

$$\lambda = (8\pi\sigma/\rho\nu^2)^{1/3} \tag{19.1}$$

For example, at a frequency of 1 MHz, the effect of ultrasound in water of surface tension 73 dyn·cm and density 1 $g{\cdot}cm^{-3}$ is to produce longitudinal waves of about 12 μm. The resulting mean droplet diameter (D) is given by Equation 19.2.

$$D = 0.34\lambda \tag{19.2}$$

For a longitudinal disturbance of wavelength 12 μm, the droplets have a mean diameter of about 3–4 μm. These very fine droplets are ideal for ICP/MS and can be swept into the plasma flame by a flow of argon gas. Unlike pneumatic forms of nebulizer in which the relative velocities of the liquid and gas are most important in determining droplet size, the flow of gas in the ultrasonic nebulizer plays no part in the formation of the aerosol and serves merely as the droplet carrier.

Piezoelectric Transducer Nebulizers

Application of an AC voltage of high frequency to a piezoelectric crystal causes faces of the crystal to move back and forth at the same frequency. This is a piezotransducer. If one of the piezocrystal faces is immersed in a liquid, the oscillatory motion transmits longitudinal ultrasound waves into the liquid. At the surface, the longitudinal waves disrupt it as compression and rarefaction waves arrive there. The breakup of the surface forms drops which spray away from the surface. The rate of formation of drops is similar to the frequency of surface disruption. Thus, a 100-kHz ultrasound wave produces about 10^5 droplets per second. This rate of formation of droplets is some thousands of times greater than their rate of formation in pneumatic nebulizers. Therefore, with normal argon flows, much more material is carried toward the plasma flame per unit time than is the case with pneumatic devices. Even so, there may be a need for a desolvation chamber to remove as much solvent as possible before the droplets or particulates reach the flame, if the performance of the latter is not to be affected. The ultrasonic devices have greater need still for desolvation because of their higher rate of droplet formation and higher rate of transfer of sample solution. However, at the highest frequencies (1 MHz), droplet size is small, and natural desolvation by evaporation is so rapid that a special desolvation chamber becomes unnecessary.

The transfer efficiencies for ultrasonic nebulizers (USN) are about 20% at a sample uptake of about 1 ml/min. Almost 100% transfer efficiency can be attained at lower sample uptakes of about 5–20 μl/min. With ultrasonic nebulizers, carrier gas flows to the plasma flame can be lower than for pneumatic nebulizers because they transfer sample at a much higher rate. Furthermore, reduction in the carrier-gas flow means that the sample remains in the mass measurement system for a longer period of time which provides much better detection limits.

Ultrasonic nebulizers are almost free of clogging from solute, have better detection limits, and have become popular despite their high cost relative to the pneumatic forms. A typical construction of an ultrasonic nebulizer is shown in Figure 19.17.

To accommodate smaller liquid flows of about 10 μl/min, micro-ultrasonic nebulizers have been designed. Although basically similar in operation to standard ultrasonic nebulizers, in these micro varieties, the end of a very-small-diameter capillary, through which is pumped the sample solution, is in contact with the surface of the transducer. This arrangement produces a thin stream of solution that runs down and across the center of the face of the transducer. The stream of sample

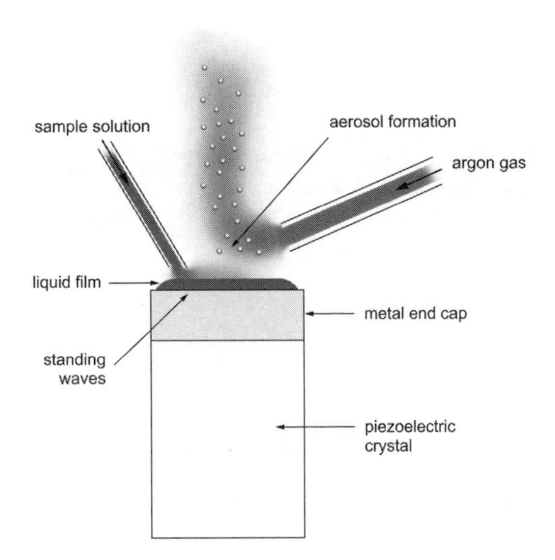

Figure 19.17.
Graphical representation of an ultrasonic nebulizer showing the sample solution spreading as a thin film across the face of a piezoelectric transducer. Standing longitudinal waves produce a spray of very small droplets, which are transported to the plasma flame by a flow of argon gas. The argon plays no part in manufacturing the aerosol and is used merely as a carrier gas flowing at 0.5–1 l/min.

solution is broken down into small-diameter droplets by standing waves in the analyte solution, which is transferred to the plasma flame with very high efficiencies.

Oscillating Capillary

In appearance, this device resembles the concentric-tube capillary pneumatic nebulizers, but it operates on a different principle. As with the pneumatic type described above, the sample solution passes through the inner capillary (diameter about 50 μm), and argon gas is blown through the annular space between the inner and outer capillaries (diameter about 250 μm). Unlike the pneumatic concentric-tube nebulizers, the inner capillary tube is not held rigidly but is allowed to vibrate as gas flows past the end of it. This transverse or fluttering motion can be likened to the fluttering of a flag in a strong wind. The solution stream leaving the end of the inner capillary is subjected to this transverse vibration and begins to break up into segments. At the same time as the transverse wave is being produced mechanically, an acoustic wave is generated in the liquid (again, this can be compared to the noise made by a fluttering flag). The acoustic wave travels longitudinally along the liquid stream while the latter is undergoing transverse mechanical vibration. At a frequency of about 1 kHz, this longitudinal wave has a wavelength of about 15 μm and leads to the stream being broken down into smaller droplets (Equation 19.2). This type of nebulizer delivers good performance over a range of sample solution flows (1 μl to 1 ml/min) and with a range of solvents. Accordingly, the oscillating-capillary nebulizer provides a convenient interface between a liquid chromatographic apparatus and a plasma flame.

Thermospray Nebulizers (TN)

Principles of Operation

In one sense, the thermospray nebulizer could be considered a pneumatic device, in which a fast-flowing argon gas stream is replaced by a very rapidly vaporizing flow of solvent from the sample solution. A typical arrangement of a thermospray device is shown in Figure 19.18.

The sample solution is pumped along a narrow capillary tube, the end of which becomes the nozzle of the nebulizer. On the outside of the capillary near its nozzle end, an electrical heater rapidly

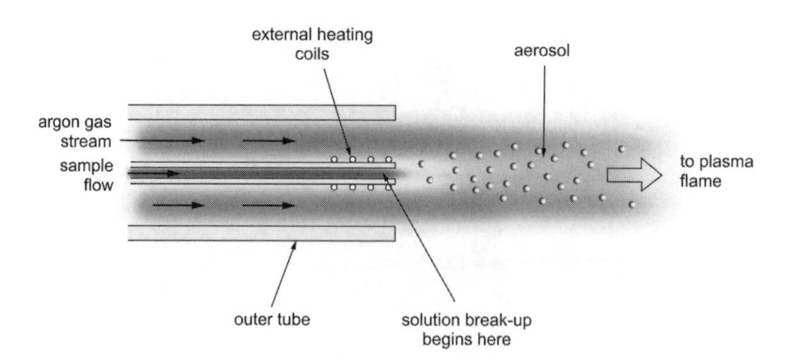

Figure 19.18.
The sample solution is pumped (e.g., from the end of a liquid chromatographic column) through a capillary tube, near the end of which it is heated strongly. Over a short length of tube, some of the solvent is vaporized and expands rapidly. The remaining liquid and the expanding vapor mix and spray out the end of the tube as an aerosol. A flow of argon carries the aerosol into the plasma flame.

warms the thin stream of liquid inside the capillary so the solvent vaporizes very quickly. The rapidly expanding vapor mixes with unvaporized solution and blows it out the end of the capillary as an aerosol. An argon carrier gas is used to ferry the aerosol to the plasma flame; the gas generally plays little or no role in formation of the droplet spray. As the droplets are already warm, further evaporation of solvent is rapid. More discussion of this process can be found in Chapter 11.

The thermospray device produces a wide dispersion of droplet sizes and transfers much of sample solution in unit time to the plasma flame. Therefore, it is essential to remove as great a proportion of the bigger droplets and solvent as possible to avoid compromising the flame performance. Consequently, the thermospray device usually requires both spray and desolvation chambers, especially for analyte solutions in organic solvents.

Thermospray nebulizers are somewhat expensive but can be used on-line to a liquid chromatographic column. About 10% of sample solution is transferred to the plasma flame. The overall performance of the thermospray device compares well with pneumatic and ultrasonic sprays. When used with microbore liquid chromatographic columns, which produce only about 100 μl/min of eluant, the need for spray and desolvation chambers is reduced, and detection sensitivities similar to those of the ultrasonic devices can be attained; both are some 20 times better than the sensitivities routinely found in pneumatic nebulizers.

Electrospray Nebulization (EN)

For a discussion of droplet and ion formation in electrospray mass spectrometry, please see Chapter 8.

Electrospray nebulizers were used for the formation of ionic aerosols before they were used as general inlets in organic chemical applications, particularly in conjunction with liquid chromatography. Two effects may operate: one electrical, the other pneumatic (Figure 19.19). The sample solution flows or is pumped along a capillary tube, the end of which is held at a high positive or negative electrical potential. Because of the electrical charge, the surface of the solution at the outlet of the capillary also becomes charged and is repelled by the existing electric field of the same sign. If the capillary tube is narrow enough, the liquid inside is forced out of the end of the capillary, and the surface of the liquid is rounded with a high radius of curvature. This point of liquid leads to repelling a steady stream of charged droplets into a desolvation chamber. If the charged capillary tube is also surrounded by an uncharged concentric capillary, argon or other gas can be blown through the annular space between the capillaries, which can be used to aid droplet formation, as with concentric capillary pneumatic nebulizers.

The electric field and gas flow work together to provide a finely dispersed spray of charged droplets. An electric potential of about 3–5 kV is used for capillaries of about 50–100-μm diameter.

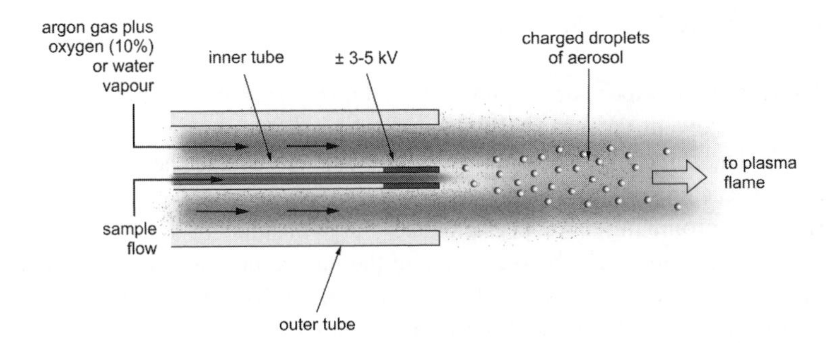

Figure 19.19.
At the high potential on the end of the inner capillary tube, the analyte solution is forced out the end of the tube and sprays into the argon stream as a charged aerosol. Droplet formation can be assisted by arranging the gas flow through a concentric outer tube (uncharged), similar to a concentric-tube nebulizer. Oxygen or water vapor are added to the argon stream to remove electrical charges from the droplets.

Even without a rapid gas flow, the droplets produced are very small, at about 1-μm diameter or less, and they are produced at a rate of about 10^8 per second, similar to the rate of production of droplets in an ultrasonic device.

For use in ICP/MS, the charged droplets produced by the electrospray nebulizer pose a problem. The droplets do not coalesce because they carry the same sign of charge; viz., they are either all positively or all negatively charged, depending on the sign of the applied electric potential. This same charge leads to the droplets' repelling each other and spreading the aerosol spray; many droplets will be attracted to the opposite electrode or to ground potential, causing them to migrate to the walls of the nebulizer or desolvation chamber. Beyond the loss of material, the charged species also provide an electrically conducting path in an otherwise nonconducting gaseous medium. As described in Chapter 6, "Coronas, Plasmas, and Arcs," this situation is a recipe for an electrical discharge. Such discharges can interfere with the sensitive detection electronics of the instrument to produce a spike in the measurement of m/z values. It is essential to remove the charges from the droplets before they can be used in ICP/MS, which can be done by adding air or water vapor to the argon carrier gas. Removal of charge from the droplets is achieved by their reaction with oxygen or water (Figure 19.20). Although this scheme reduces the problem of electrical discharges, it does not remove the problem entirely.

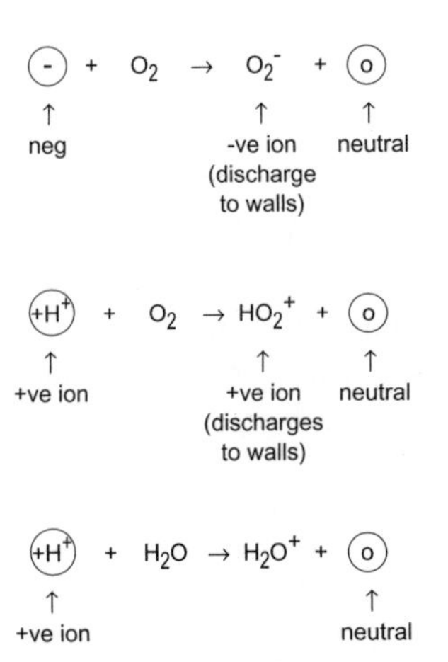

Figure 19.20.
During charge neutralization, oxygen or water molecules accept electrons or protons or other charged atoms from the droplets. Having lost their electrical charges and become neutral, the droplets remain in the main gas stream and are carried into the plasma flame. The small charged species formed by these exchange reactions migrate rapidly from the main gas stream and are discharged at the walls of the containment vessel.

Spray and Desolvation Chambers

Spray and desolvation chambers are adjuncts to nebulizer systems. The aerosol produced by a nebulizer consists of solvent vapor and droplets of solution, with sometimes even small pieces of particulate matter. The solvent diffuses into the argon carrier gas and the droplets are swept along by the gas. Before reaching the plasma flame, it is necessary to remove as much solvent vapor as possible to avoid compromising flame performance. Large droplets in the aerosol do not have time to desolvate by natural evaporation before reaching the flame, and these large amounts of material entering the flame can cause serious instabilities. Therefore, as well as nebulizing a solution, it may be necessary to modify the aerosol before it reaches the plasma flame. Some or even all of this modification can be avoided by good design of the nebulizer, but in the cases when such modification is needed it normally takes one or both of two forms.

The first form of aerosol modifier is a spray chamber. It is designed to produce turbulent flow in the argon carrier gas and to give time for the larger droplets to coalesce by collision. The result of coalescence, gravity, and turbulence is to deposit the larger droplets onto the walls of the spray chamber, from where the deposited liquid drains away. Since this liquid is all analyte solution, clearly some sample is wasted. Thus when sensitivity of analysis is an issue, it may be necessary to recycle this drained-off liquid back through the nebulizer.

Having removed the larger droplets, it may remain only to encourage natural evaporation of solvent from the remaining small droplets by use of a desolvation chamber. In this chamber, the droplets are heated to temperatures up to about 150°C, often through use of infrared heaters. The extra heat causes rapid desolvation of the droplets, which frequently dry out completely to leave the analyte as small particles that are swept by the argon flow into the flame.

Having assisted desolvation in this way, the carrier gas then carries solvent vapor produced in the initial nebulization with more produced in the desolvation chamber. The relatively large amounts of solvent may be too much for the plasma flame, causing instability in its performance and, sometimes, putting out the flame completely. Therefore, the desolvation chamber usually contains a second section placed after the heating section. In this second part of the desolvation chamber, the carrier gas and entrained vapor are strongly cooled to temperatures of about 0 to –10°C. Much of the vapor condenses out onto the walls of the cooled section and is allowed to drain away. Since this drainage consists only of solvent and not analyte solution, it is normally directed to waste.

Introduction of sample solution via a nebulizer may need both spray and a desolvation chamber, but a well-designed, efficient nebulizer needs neither.

Conclusion

Nebulizers are used to introduce analyte solutions as an aerosol spray into a mass spectrometer. For use with plasma torches, it is necessary to produce a fine spray and to remove as much solvent as possible before the aerosol reaches the flame of the torch. Various designs of nebulizer are available, but most work on the principle of interacting gas and liquid streams or the use of ultrasonic devices to cause droplet formation. For nebulization applications in thermospray, APCI, and electrospray, see Chapters 8 and 11.

Chapter 20

Hybrid Orthogonal Time-of-Flight (oa-TOF) Instruments

Introduction

With time-of-flight (TOF) analyzers, the time taken for ions to travel the length of an evacuated tube is used to deduce the m/z values of the ions and thus obtain a spectrum. The ion beam can start from an ion source and be directed straight into the analyzer, as shown in Figure 20.1a for a simple matrix-assisted laser desorption ionization (MALDI) TOF apparatus. This is the in-line or stand-alone TOF instrument and, as shown in the more extended description (Chapter 26, "Time-of-Flight Ion Optics"), it can be somewhat more complicated to enable tandem mass spectrometry (MS/MS) measurements. Alternatively, an ion beam produced by some other analyzer can be deflected into a TOF analyzer placed at right angles (orthogonal) to the beam (Chapter 27, "Orthogonal Time-of-Flight Ion Optics"), as shown in Figure 20.1b. This arrangement constitutes the basis of a hybrid TOF instrument.

For either the in-line or hybrid analyzers, the ions injected into the TOF section must all begin their flight down the TOF tube at the same instant if arrival times of ions at a detector are to be used to measure m/z values (see Chapter 26, "TOF Ion Optics"). For the hybrid TOF instruments, the ion detector is usually a microchannel plate ion counter (see Chapter 30, "Comparison of Multipoint Collectors (Detectors) of Ions: Arrays and Microchannel Plates").

This chapter provides brief descriptions of analyzer layouts for three hybrid instruments. More extensive treatments of sector/TOF (AutoSpec-TOF), liquid chromatography/TOF (LCT or LC/TOF with Z-spray), and quadrupole/TOF (Q/TOF), are provided in Chapters 23, 22, and 21, respectively.

Q/TOF

The Q in Q/TOF stands for quadrupole (see Chapter 25, "Quadrupole Ion Optics"). A Q/TOF instrument — normally used with an electrospray ion inlet — measures mass spectra directly to obtain molecular or quasi-molecular mass information, or it can be switched rapidly to MS/MS mode to examine structural features of ions. The analyzer layout is presented in Figure 20.2.

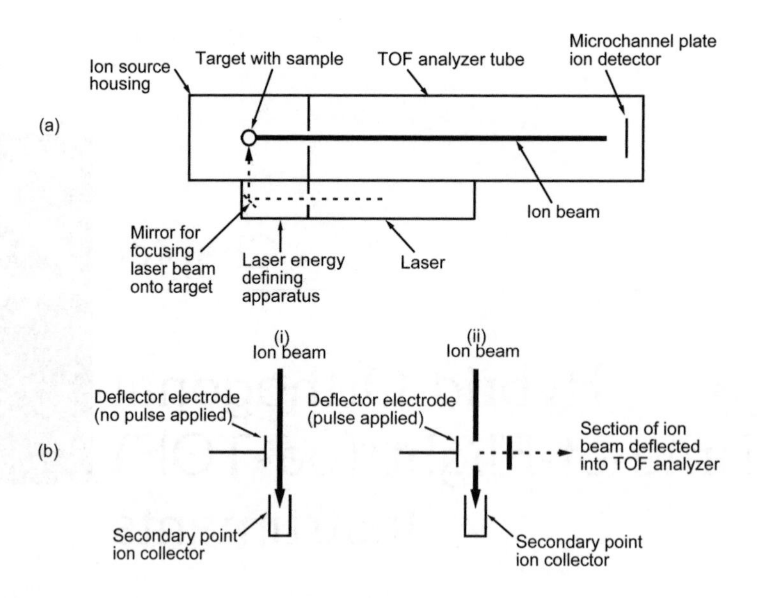

Figure 20.1
(a) The outline of a simple in-line TOF mass spectrometer is shown. Ions produced by a 4-nsec pulse of laser light onto the sample produces ions, which are accelerated through about 25 kV and directed into the flight tube. The times taken for the ions to reach a microchannel plate collector and the number arriving are recorded, and these times are then converted into m/z values and abundances to obtain a mass spectrum. After several microseconds, the laser can be fired again to obtain another spectrum. The process can be continued as long as there is a sufficient supply of sample. (b) A continuous ion beam produced in one source can be detected by a point ion collector placed in line with the beam; this detector is usually a secondary collector. There is no electric field applied to the deflector electrode in (i). However, in (ii), an electric field of 1–5 kV has been applied to the electrode for about 10 nsec. This detaches a short length of the main ion beam, which is accelerated almost at right angles to its original direction and proceeds into a TOF analyzer. This last analyzer produces a full spectrum, scanned in a few milliseconds. After the pulse, the main ion beam continues as it did before the pulse was applied. Pulses can be applied at a rate of several kilohertz.

LC/TOF

A liquid chromatograph (LC) is combined with a TOF instrument through a Z-SPRAY™ ion source. Two hexapoles are used to focus the ion beam before it is examined by a TOF analyzer, as described in Figure 20.3.

AutoSpec-TOF

An AutoSpec-TOF mass spectrometer has a magnetic sector and an electron multiplier ion detector for carrying out one type of mass spectrometry plus a TOF analyzer with a microchannel plate multipoint ion collector for another type of mass spectrometry. Either analyzer can be used separately, or the two can be run in tandem (Figure 20.4).

Conclusion

A TOF analyzer can be used alone or in conjunction with other analyzers to function as a hybrid mass spectrometer. The hybrids provide advantages not attainable, or difficult to attain by the

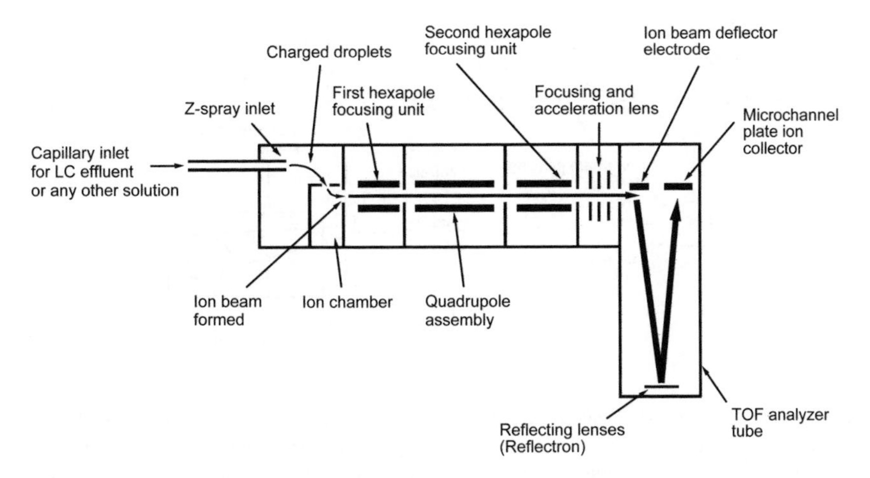

Figure 20.2
A solution containing the analyte of interest is sprayed from the end of a capillary by application of a high electric potential. The resulting charged droplets are stripped of solvent, and ions formed from analyte molecules are directed electrically into the mass spectrometer (Z-spray). The ion beam passes into a quadrupole analyzer, which can be operated in a narrow band-pass mode so as to transmit ions of defined m/z values or in its wide band-pass mode, in which all ions are transmitted regardless of m/z value. There is a further focusing hexapole, after which the ion beam is focused and accelerated by an electric lens before being passed into the TOF analyzer in front of a deflector electrode. A high electric potential is applied to this electrode in pulses so that, at each pulse, a section of the ion beam is deflected and accelerated into the TOF analyzer. After reflection by the reflectron, the ions are detected at a microchannel plate multipoint collector. The reflectron is used mainly to increase the time intervals at which successive m/z values are detected at the collector and to improve focusing. The quadrupole is operated in the narrow band-pass mode for MS/MS and in its wide band-pass mode for obtaining a full spectrum by the TOF analyzer.

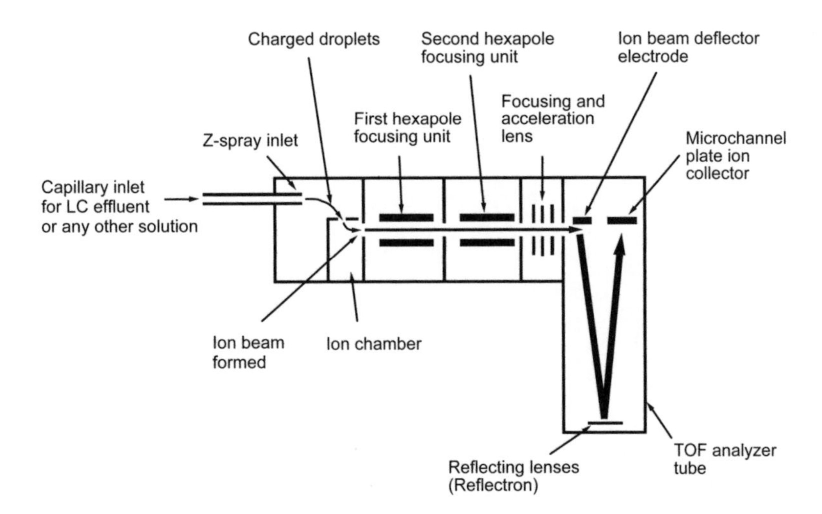

Figure 20.3
A solution containing the analyte of interest is sprayed from the end of a capillary by application of a high electric potential. The resulting charged droplets are stripped of solvent, and ions formed from analyte molecules travel into the mass spectrometer (Z-spray). The hexapoles do not separate ions according to m/z value but contain them into a beam. Finally, the slow-moving ions are focused and accelerated through a potential of about 40 V before passing in front of a deflector plate (electrode). A large electric potential of several kilovolts is applied to this electrode in pulses so that, at each pulse, a section of the ion beam is deflected and accelerated into the TOF analyzer. After reflection by the reflectron, the ions are detected at a microchannel plate multipoint collector.

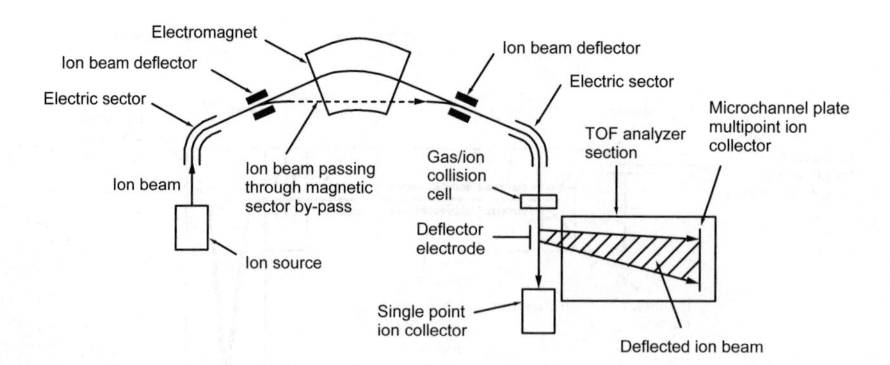

Figure 20.4
The main features of the AutoSpec-TOF® hybrid layout. In one mode of operation, the instrument is used with the deflector electrode switched off and with the main ion beam passing through the magnet sector for the separation of ions into their individual m/z values; a full mass spectrum is obtained by scanning the magnetic sector. In a second mode, the magnetic field is switched off, and the beam deflectors transmit the whole ion beam through the magnet bypass; the mass spectrum is obtained by application of a pulsed electric field to the deflector electrode, which allows m/z values of ions to be measured in the TOF analyzer. In yet another mode, the ions of one m/z value can be selected from the main beam by the magnetic sector. A gas is placed in the collision cell and causes the ions to fragment. This fragmentation (MS/MS) spectrum is measured in the TOF analyzer. Other modes of operation are possible.

analyzers used separately. When used in the hybrid mode, the TOF analyzer is usually placed at right angles (orthogonal) to a main ion beam emanating from another analyzer.

Chapter 21

Hybrid Magnetic-Sector Time-of-Flight (Sector/TOF) Instruments

Introduction

The hybrid mass spectrometer formed by using a magnetic/electric-sector instrument coupled with a time-of-flight (TOF) instrument is termed here the AutoSpec-TOF®, after the terminology of its manufacturer. However, the principles used could be any form of hybrid where a magnetic-sector mass spectrometer precedes a TOF mass spectrometer. The principles underlying the uses of magnetic/electric sectors and TOF instruments are extensively described in Chapters 24 and 27, "Ion Optics of Magnetic/Electric-Sector Mass Spectrometers" and "Orthogonal Time-of-Flight Ion Optics," respectively).

In the magnetic-sector/TOF hybrid, ions produced in an ion source pass through the magnetic sector first and then might enter the TOF section, depending on how the hybrid is operated. The hybrid can be used as two separate instruments or as two instruments in conjunction with each other.

In general terms, the main function of the magnetic/electric-sector section of the hybrid is to be able to resolve m/z values differing by only a few parts per million. Such accuracy allows highly accurate measurement of m/z values and therefore affords excellent elemental compositions of ions; if these are molecular ions, the resulting compositions are in fact molecular formulae, which is the usual MS mode. Apart from accurate mass measurement, full mass spectra can also be obtained. The high-resolution separation of ions also allows ions having only small mass differences to be carefully selected for MS/MS studies.

Also in general terms, the TOF part of the hybrid is used mostly for MS/MS studies in which ions produced in the magnetic sector are collided with neutral gas molecules to induce decomposition (see Chapter 23). In this mode the instrument produces more highly resolved product ion spectra than can be attained in simple magnetic-sector instruments.

A further important property of the two instruments concerns the nature of any ion sources used with them. Magnetic-sector instruments work best with a continuous ion beam produced with an electron ionization or chemical ionization source. Sources that produce pulses of ions, such as with laser desorption or radioactive (Californium) sources, are not compatible with the need for a continuous beam. However, these pulsed sources are ideal for the TOF analyzer because, in such a system, ions of all m/z values must begin their flight to the ion detector at the same instant in order that their flight times can be measured accurately. Therefore, a magnetic-sector/TOF hybrid can be used with a wide variety of ion sources.

In the following discussion, the separate use of the magnetic sector and the TOF sector are examined briefly, followed by a discussion of the hybrid uses.

AutoSpec-TOF® Ion Optics

The arrangement of magnetic and electric fields is shown in Figure 21.1.

Magnetic/Electric-Sector

For this instrument there are actually two electric sectors associated with the magnetic sector. If, for the moment, the TOF section is ignored, the remainder of Figure 21.1 reveals a magnetic/electric-sector instrument (the AutoSpec®). Ions are accelerated from an ion source through an electric field of several thousand volts. This continuous ion beam is then deflected toward the magnetic sector by the first electric sector. No ion separation occurs in the electric-sector stage, but the beam is focused for kinetic energy. In the following magnetic analyzer (sector), the ions are deflected according to their m/z values. By changing the magnetic field strength, ions of individual m/z values are selected and pass into a the second electric sector, where they are again focused and deflected. Again, no separation by m/z values occurs at this stage because all the separation into m/z values occurred in the magnetic sector. The ions then pass through an empty gas-collision cell and into a single-point ion collector (an electron multiplier). Thus, this section can be operated as a magnetic-sector instrument having an electric sector (a high-resolution mass spectrometer). The magnetic field can be changed continuously so ions of successive m/z values are selected one after the other and pass into the ion detector. This mode gives a full mass spectrum.

By including a mass measurement standard substance, the mass spectrometer can be operated at high resolution to give accurate masses of the ions. These accurate masses give the elemental compositions of the ions. If they are molecular ions, the elemental composition is the molecular formula. The mass spectrometer is then being operated as a high-resolution magnetic/electric-sector instrument. It is in a high-resolution MS mode.

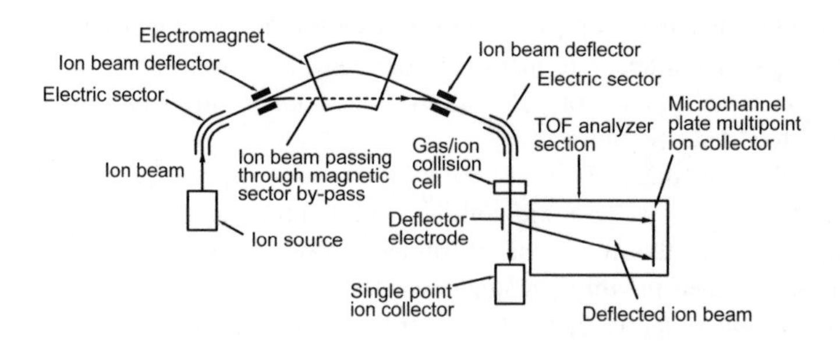

Figure 21.1
The main features of an AutoSpec®-TOF hybrid layout. In one MS mode of operation, the instrument is used with the ion beam deflector electrodes switched off and the main ion beam passes through the magnet sector so as to separate the ions into their individual m/z values; a full mass spectrum is obtained by changing the magnetic field strength. In a second MS mode, the magnetic field is switched off, and the beam deflectors transmit the whole ion beam through the magnet bypass; the mass spectrum is obtained by application of a pulsed electric field to the deflector electrode, which causes the ionic m/z values to be measured in the TOF analyzer. In an MS/MS mode, the ions of one m/z value can be selected from the main beam by the magnetic sector. A gas is placed in the collision cell and causes the selected ions to fragment. This fragmentation spectrum is measured in the TOF analyzer. Other modes of operation are possible.

Alternatively, ions of any one selected m/z value can be chosen by holding the magnetic field steady at the correct strength required to pass only the desired ions; any other ions are lost to the walls of the instrument. The selected ions pass through the gas cell and are detected in the single-point ion collector. If there is a pressure of a neutral gas such as argon or helium in the gas cell, then ion-molecule collisions occur, with decomposition of some of the selected incident ions. This is the MS/MS mode. However, without the orthogonal TOF section, since there is no further separation by m/z value, the new ions produced in the gas cell would not be separated into individual m/z values before they reached the detector. Before the MS/MS mode can be used, the instrument must be operated in its hybrid state, as discussed below.

Time-of-Flight

The TOF analyzer can be used in its stand-alone mode just as the magnetic sector could be used. For this usage, the magnetic sector is unnecessary because all of the ions are required for the TOF analyzer and not just selected ones. Accordingly, the AutoSpec uses two beam deflectors to bypass the magnetic sector (Figure 21.1). Now, ions produced in the source are accelerated again through several thousand volts and focused as before for kinetic energy in the first electric sector. The beam deflector passes the ion beam outside the magnetic field and into a second beam deflector, which redirects the beam into the second electric sector. Thus, the beam deflectors act like a bypass for the ion beam, and no separation of the ions into individual m/z values can occur. All of the ions head toward the single-point collector.

As far as the orthogonal TOF section is concerned, there is simply an ion beam passing the end of the analyzer section. (Note that the vacuum is continuous from the ion source right through the magnetic sector and into the TOF sector.) The TOF analyzer requires that, for measurement of m/z values, all ions must start their journey down the flight tube at the same instant. Therefore, there is a deflector or pusher electrode placed alongside the beam of ions passing the end of the TOF analyzer (Figure 21.1). A very short electric-field pulse of several thousand volts is applied to this electrode, causing a small section of the beam to be deflected and accelerated into the TOF section (Figure 21.2).

This small section of the beam contains a representative selection of the ions in the original ion beam, but now the ions have all been started off down the TOF flight tube at the same instant (the

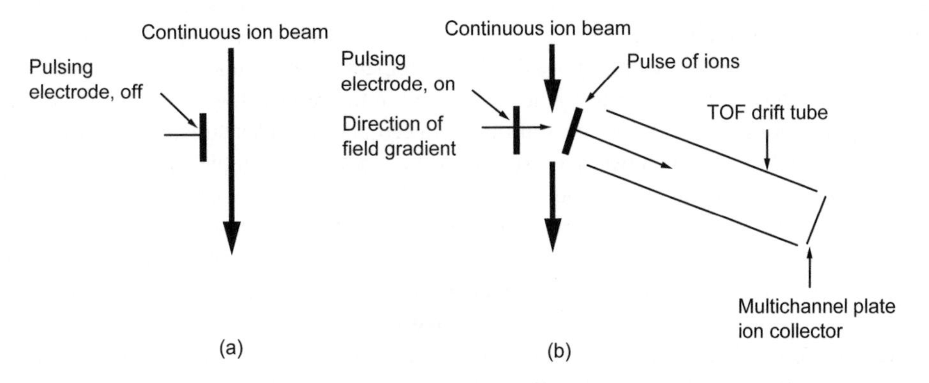

Figure 21.2
(a) The pulsing electrode is switched off, and a continuous ion beam passes by it. (b) The electrode has been pulsed for a few nanoseconds with a field gradient at right angles to the main beam which has caused a short section of the ion beam to detach and to travel in the direction shown. The detached segment of the main beam enters the flight tube of a TOF instrument. The m/z values are determined from the times the ions take to reach the microchannel plate ion collector after initiation of the pulse.

pulsed electric field was like a starter's gun in a race). Ions travel down the flight tube at different speeds (depending on the square roots of their m/z values) and reach a type of microchannel multipoint ion collector (time-to-digital converter, TDC) at different times. The flight times give the m/z values. By applying a pulse to the deflector (pusher) electrode, a full mass spectrum of the ions in the beam can be obtained. This process constitutes operation of the TOF section in an MS-only mode. The resolution in the TOF mode is quite high but is not as good as for the magnetic-sector MS mode.

Because the resolution in the MS mode is better with the magnetic sector, which is the preferred mode of operation for obtaining a routine mass spectrum or when accurate mass measurements need to be made. However, this is not always the case. The ion beam emerging from the ion source may be unsteady, depending on the source, or it may be pulsed as with laser desorption ionization. Obtaining a full mass spectrum from the magnetic sector in these conditions is at best difficult and at worst virtually impossible. The TOF section has an operational speed of just a few microseconds for a complete mass spectrum. Thus, for a fluctuating or pulsed ion source, the TOF analyzer can capture a full mass spectrum in as short a time as the fluctuations or pulses. Further, the TOF sector can be operated at a speed of several kHz, and it is possible to accumulate several thousand spectra in one second. This is an ideal arrangement for smoothing out fluctuations in ion yield or for capturing mass spectra from pulsed ion sources, and the TOF section alone can be used for MS mode operation in some circumstances.

Like the magnetic sector, the TOF section by itself is not capable of MS/MS operation, but allied with the sector, the two make an excellent MS/MS instrument.

Operation of the Combined Magnetic and TOF Sectors as a Hybrid Instrument

The hybrid mode of operation is used for MS/MS studies. For mass spectrometry, molecular ions are valuable because their m/z values give molecular mass information and, if measured accurately, they give molecular formulae. However, fragmentation of the molecular ions is useful because it gives information about the structure of the sample put into the mass spectrometer. Electron ionization gives both molecular and fragment ions, so it provides two batches of information. In contrast, many ionization sources — such as those based on chemical, electrospray, fast-atom bombardment, or laser ionization — give mostly or entirely molecular or quasi-molecular ions, and there is no fragmentation. Consequently, structural information is lost. To offset this drawback to the so-called soft ionization methods, the molecular ions are made to decompose (fragment) by passing them through a cell operating at a gas pressure somewhat above that for the remainder of the mass spectrometer. Collisions between ions and gas molecules (often helium or argon) in the cell cause some of the molecular ions to form fragment ions; viz., a mass spectrum is produced and provides structural information. This is the basis of MS/MS, in which two mass spectrometers are used, as in the AutoSpec®-TOF hybrid.

In operation, the magnetic section of the hybrid is used to select ions of a particular m/z value. For example, if a mixture of two substances gives two molecular ions, M_1 and M_2, the magnetic sector is used to select one or the other. The selected ions collide with gas in the collision cell (Figure 21.1), and some of them decompose to yield fragment ions, say F_1, F_2, and F_3. Thus, a stream of ions M_1 (some of which have not been decomposed) plus F_1, F_2, and F_3 leave the collision cell (Figure 21.3). If this beam went straight to the single-point ion collector, there would be no separation into the individual m/z values, and it would not be possible to measure their m/z values. However, by pulsing the pusher electrode placed just after the collision cell, a section of the beam is sent orthogonally down the TOF analyzer tube, which does separate them according to m/z value, which is related to the length of time they take to reach the multipoint microchannel plate collector (Figure 21.1). Therefore, molecular ions and fragment ions are obtained in this MS/MS mode.

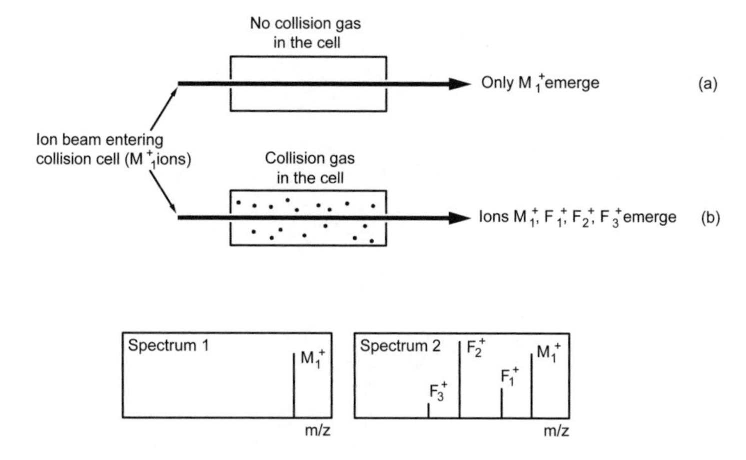

Figure 21.3
The diagram represents a beam of M_1 ions from the magnetic sector entering a gas-collision cell. (a) There is no collision gas in the cell, so only M_1 ions emerge and the TOF analyzer would give spectrum 1. (b) There is collision gas in the cell. Now ion-molecule collisions lead to some fragment ions (F_1, F_2, F_3) being formed, and they, along with undecomposed M_1 ions, leave the collision cell to give spectrum 2, as measured by the TOF analyzer. The hybrid has given an MS/MS result.

A big advantage of the AutoSpec®-TOF hybrid lies in its ability to examine ions at high resolution. In the above example, suppose that the two molecular ions M_1 and M_2 had the same integer m/z values. At low resolution, the two ions could not be distinguished. However, their accurate masses might differ by, for example, 0.0005 mass units. Now, at high resolution, the two sets of ions can be separated in the magnetic sector, and it again becomes possible to distinguish M_1 and M_2.

Other Advantages of the Hybrid

The hybrid can be used with EI, CI, FI, FD, LSIMS, APCI, ES, and MALDI ionization/inlet systems. The nature of the hybrid leads to high sensitivity in both MS and MS/MS modes, and there is rapid switching between the two. The combination is particularly useful for biochemical and environmental analyses because of its high sensitivity and the ease of obtaining MS/MS structural information from very small amounts of material. The structural information can be controlled by operating the gas cell at high or low collision energies.

Conclusion

The AutoSpec®-TOF hybrid mass spectrometer combines the advantages of a magnetic/electric-sector instrument with those of time-of-flight to give a versatile instrument capable of MS or MS/MS at high or low resolution.

Chapter 22

Hybrid Hexapole Time-of-Flight (Hexapole/TOF) Instruments

Introduction

The term *hybrid hexapole/TOF* describes a type of hybrid mass spectrometer system in which a hexapole assembly in conjunction with a time-of-flight analyzer (TOF) is used to obtain mass spectra of ions emerging from an atmospheric-pressure inlet/ion source. The latter source is frequently used for dealing with solutions eluting from a liquid chromatographic (LC) column. Thus, as a substrate emerges from an LC column, it frequently passes through some sort of detector (UV/visible, refractive index, etc.) to reveal when substrates are emerging. Although these detectors reveal when a substrate is eluting, they are not generally useful for giving the sort of detailed structural information available from a mass spectrum. To obtain such a spectrum, first the solvent must be stripped from the substrate, which itself needs to be ionized. These two operations of desolvation and ionization are conveniently effected concurrently by using an atmospheric-pressure inlet/ionization source such as ESI or APCI (see Chapters 8 and 9, "Electrospray Ionization" and "Atmospheric-Pressure Chemical Ionization," respectively). Once the solvent has been stripped away and the substrate (solute) has been ionized, the ions are examined for m/z values by a TOF analyzer placed orthogonally to the main ion beam (see Chapter 26, "Time-of-Flight Ion Optics"). The solution to be examined does not need to be from an LC column and equally well could be simply a static solution in a container. Other aspects of this hybrid are given in Chapter 20, "Hybrid Orthogonal Time-of-Flight Instruments."

Inlet Systems

The LC/TOF instrument was designed specifically for use with the effluent flowing from LC columns, but it can be used also with static solutions. The initial problem with either of these inlets revolves around how to remove the solvent without affecting the substrate (solute) dissolved in it. Without this step, upon ionization, the large excess of ionized solvent molecules would make it difficult if not impossible to observe ions due only to the substrate. Combined inlet/ionization systems are ideal for this purpose. For example, dynamic fast-atom bombardment (FAB), plasmaspray, thermospray, atmospheric-pressure chemical ionization (APCI), and electrospray (ES)

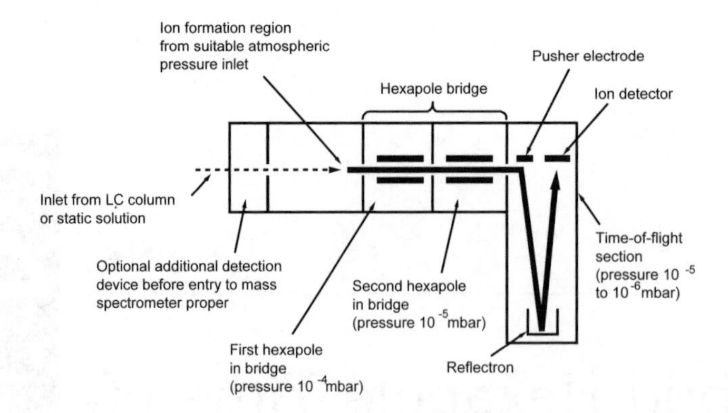

Figure 22.1
A diagrammatic representation of a hexapole/TOF instrument showing the consecutive arrangement of solution inlet at atmospheric pressure, the ion source (in which the pressure begins to be reduced), the hexapole bridge (in which the pressure is further reduced), and finally the m/z measurement in the TOF section (in which the vacuum is the highest). The path of the ion beam is shown as a bold black line. After reaching the pusher electrode, sections of the beam are accelerated into the TOF unit, where m/z separation takes place. After passing through the reflectron, the ions are detected at a microchannel plate collector.

have all been used, but only the last two are now common. The ES inlet has evolved into the highly efficient Z-spray (see Chapter 10, "Z-Spray Combined Inlet/Ionization Source"). After passing through such inlets, most of the solvent is removed.

Residual solvent and ions from the substrate are passed into the beginning of the hexapole/TOF instrument proper. At this point, background air pressure and the remaining few solvent molecules must be removed while keeping the substrate ion beam intact and on line to pass into the TOF mass analyzer. This connection between the inlet and the analyzer is composed of a hexapole bridge (Figure 22.1). In this mode of operation with atmospheric-pressure inlet systems, other measuring devices can still be used at the end of an LC column. Thus, the common UV/visible, refractive index, radioactive, and light-scattering devices can be used without affecting the performance of the mass spectrometer, thereby giving more analytical information (Figure 22.1).

Hexapole Bridge

A quadrupole mass analyzer can be used in two modes: narrow and wide band-pass modes. In the first, two opposed poles have an applied radio-frequency (RF) voltage applied, with the other two opposed poles having a constant DC voltage. Depending on the voltages and the frequency of the RF field, ions of selected m/z values pass right through the four poles while others cannot. The device is used as a mass analyzer. In the wide band-pass or RF-only mode, all four poles have only an applied RF electric field, and all ions are transmitted so the quadrupolar assembly acts as an ion guide, containing and directing an ion beam along the length of the poles. A hexapole assembly of rods (poles) is built similarly to the quadrupole, but now there are six rods evenly spaced around a central axis (Figure 22.2).

The hexapole cannot act as a mass filter by applying a DC field and is used only in its all-RF mode, in which it allows all ions in a beam to pass through, whatever their m/z values. In doing so, the ion beam is constrained, so it leaves the hexapole as a narrow beam. This constraint is important because the ion beam from the inlet system tends to spread due to mutual ion repulsion and collision with residual air and solvent molecules. By injecting this divergent beam into a hexapole unit, it can be refocused. At the same time, vacuum pumps reduce the background pressure to about 10^{-4} mbar (Figure 22.1). The pressure needed in the TOF analyzer is about 10^{-5}

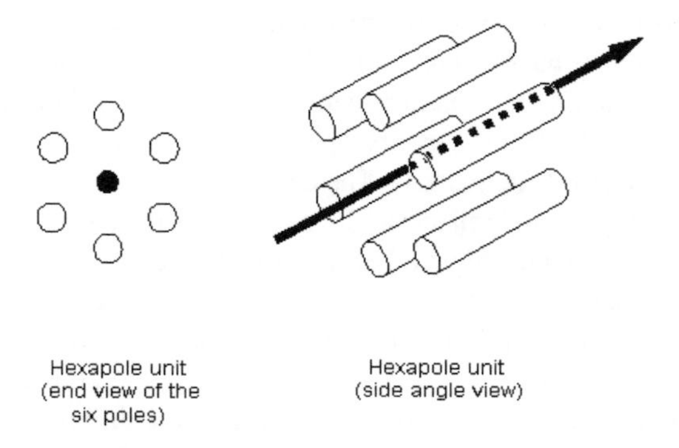

Figure 22.2
A diagrammatic representation of a hexapole unit. The left-hand figure shows the end view of the unit, the ends of the six poles being represented as open circles and the ion beam by the black circle. The right-hand figure is an angled side view of the unit, showing the six parallel poles (or rods) with the ion beam passing through the middle of them. If it is the first hexapole, then the emergent ion beam passes through a narrow orifice into the second hexapole unit. If it is the second hexapole unit, then the ion beam passes through a narrow orifice into an acceleration (electric) lens before passing in front of the pusher electrode of the TOF tube. Note that the small lens system is not shown in Figure 22.1 for the sake of clarity.

to 10^{-6} mbar, and, usually, a second hexapole unit is provided to maintain the ion beam on track and collimated. The passage from the first hexapole into the second is through a very-small-diameter orifice, which allows the passage of the ion beam but restricts entry of gas molecules. Thus, a second vacuum pump deals with residual gas in the second hexapole chamber, bringing down the pressure to about 10^{-5} mbar.

After this second hexapole stage, the ion beam is focused, accelerated slightly, and passed into the TOF analyzer (Figure 22.1). It can be seen that the two hexapoles act as a means of channeling the ions from the inlet (near-atmospheric pressure) to the TOF analyzer (under high vacuum) but without introducing any selection by m/z value. It is for this reason that the hexapoles are said to bridge the inlet and analyzer sections of the mass spectrometer.

Time-of-Flight Analyzer

If ions of different m/z values are instantaneously accelerated by an electric field, they attain velocities determined by their m/z values and the accelerating field to which they have been subjected. After flying along an evacuated tube, the ions arrive at a detector in proportion to the velocities they have, the faster ions arriving before slower ones. Because an ion velocity depends upon its m/z value, the ions of smaller m/z value arrive first and those of larger value arrive last. By measuring the flight time of the ions along a TOF tube (analyzer), the m/z values can be deduced, and, therefore, a mass spectrum can be measured. Among the advantages of TOF measurement of mass spectra is the very short time interval needed to measure a full mass spectrum. Typically, a few microseconds is all that is needed to scan from 1 to 2000 mass units. On the human timescale, this speed of scanning appears to be instantaneous. For sake of illustration, a TOF analyzer could be likened to a camera taking snapshots of the m/z values of an assembly (beam) of ions; the faster the repetition rate at which the camera shutter is clicked, the greater is the number of mass spectra that can be taken in a very short time. For TOF analyzers, it is not uncommon to measure several thousand mass spectra in one second! All such spectra can be added to each other digitally, a process that improves the signal-to-noise ratio in the final accumulated total.

Clearly, the longer the flight tube, the longer it will take for ions to traverse it. For two ions of velocities v_1 and v_2, their flight times will be t_1 and t_2 for a flight path of length d. If the flight path is doubled in length, the flight times become $2t_1$ and $2t_2$, and the difference in arrival times for the two ions at a detector is doubled; viz., it changes from $(t_1 - t_2)$ to $2 \times (t_1 - t_2)$. As differences in flight arrival times are typically measured in nanoseconds for a flight tube of 50 cm, then doubling this length doubles the differences and makes their measurement that much easier. In effect, the resolution of the instrument is greatly improved by making the ions pass twice along a TOF flight tube instead of only once before reaching the detector (Figure 22.1). The device that reflects the ions is called a reflectron.

There is another advantage of the reflectron. Ions of one particular m/z value should all have the same velocity when traveling along the flight tube. However, because of various small inhomogeneities in the electric optics and other applied fields, there will be a small spread of velocities. Figure 22.3 shows typical spreads in velocities (and therefore in kinetic energies) for two sets of ions centered at m/z A and m/z B. Without the reflectron, these ions would reach the detector over small intervals of time rather than tightly bunched at two specific times. If the time difference between arrivals of the ions at m/z A and the following ones at m/z B is only small, then there will be some overlap between the slowest ions of the first m/z value and the fastest ions of the second m/z value (Figure 22.3a). In effect, this means that the ions of different m/z value are inadequately separated; the resolution of the analyzer is impaired. In the reflectron, faster ions travel further into it compared with any slower ones before being reversed and sent back out. Thus, the slower ions get a chance to catch up with the faster ones, so when both slower ones and faster ones reach the detector, they do so at about the same time. This effect appears to reduce the velocity spread in the ions of any one particular m/z value and significantly improves the resolution of the TOF instrument (Figure 22.3b).

Operation of the Hybrid

The effluent from an LC column or from a static solution supply is passed into an atmospheric-pressure inlet/ion source (Figure 22.1), where ions are produced. The ions are transmitted rapidly

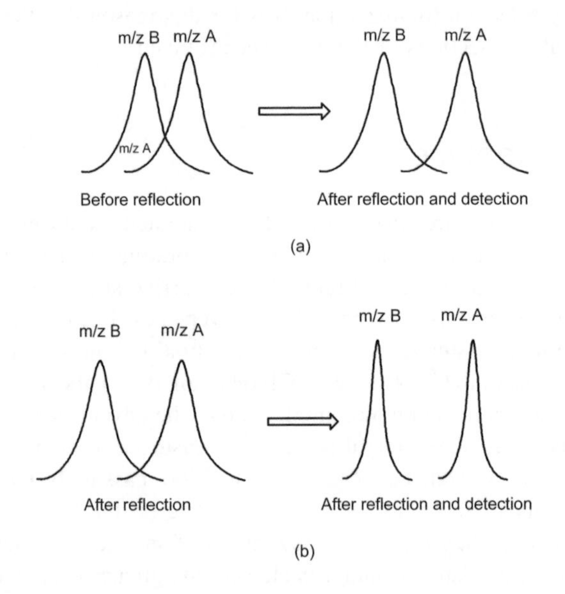

Figure 22.3
(a) Two peaks are shown, representing kinetic energy distributions for two sets of ions of m/z values A and B. The two peaks are shown as overlapping extensively. If there were no improvement, the resolution of the instrument would be quite poor. After doubling the flight path by use of the reflectron, the separation of the tops of the peaks is doubled, which improves their resolution. (b) A second effect of the reflectron is shown. Now, the kinetic energy spread appears to be much narrower, and the resolution of the peaks at m/z A and B is improved again. The combination of doubling the length of the flight path and decreasing the separation of ions of each m/z value leads to greatly improved instrumental resolution of m/z values.

to the TOF analyzer, which is set to record a mass spectrum at a rate of once every 30 to 40 μsec. Since most of the background of air and solvent molecules has been removed, when there is no solute present, there is only a very small amount of background noise. When a solute appears, mass spectra are obtained.

Some Advantages of the Hybrid

Ions formed in an electrospray or similar ion source are said to be thermolized, which is to say that their distribution of internal energies is close to that expected for their normal room-temperature ground state. Such ions have little or no excess of internal energy and exhibit no tendency to fragment. This characteristic is an enormous advantage for obtaining molecular mass information from the stable molecular ions, although there is a lack of structural information.

By being able to obtain an unequivocal relative molecular mass, or even a molecular formula derived from that mass, the hybrid mass spectrometer becomes a powerful tool for investigating single substances or mixtures of substances. With an APCI inlet, fragmentation can be induced to obtain structural information (see Chapter 9).

Other instrumental advantages include its high sensitivity and a linear mass scale to m/z 10,000 at full sensitivity. The linearity of the mass scale means that it is necessary to calibrate the spectrometer using a single or sometimes two known mass standards. Some calibration is necessary because the start of the mass scale is subject to some instrumental zero offset. The digitized accumulation of spectra provides a better signal-to-noise ratio than can be obtained from one spectrum alone.

Conclusion

In one instrument, ions produced from an atmospheric-pressure ion source can be measured. If these are molecular ions, their relative molecular mass is obtained and often their elemental compositions. Fragment ions can be produced by suitable operation of an APCI inlet to obtain a full mass spectrum for each eluting substrate. The system can be used with the effluent from an LC column or with a solution from a static solution supply. When used with an LC column, any detectors generally used with the LC instrument itself can still be included, as with a UV/visible diode array detector sited in front of the mass spectrometer inlet.

Chapter 23

Hybrid Quadrupole Time-of-Flight (Q/TOF) Instruments

Introduction

The term *Q/TOF* is used to describe a type of hybrid mass spectrometer system in which a quadrupole analyzer (Q) is used in conjunction with a time-of-flight analyzer (TOF). The use of two analyzers together (hybridized) provides distinct advantages that cannot be achieved by either analyzer individually. In the Q/TOF, the quadrupole is used in one of two modes to select the ions to be examined, and the TOF analyzer measures the actual mass spectrum. Hexapole assemblies are also used to help collimate the ion beams. The hybrid orthogonal Q/TOF instrument is illustrated in Figure 23.1.

A brief description of this hybrid system appears in Chapter 20. For further information on the quadrupole or TOF instruments, see Chapters 25 and 26, "Quadrupole Ion Optics" and "Time-of-Flight Ion Optics."

The Separate Quadrupole and Time-of-Flight Analyzers

A Quadrupole Assembly in Narrow Band-Pass Mode

A quadrupole analyzer can be operated in two main configurations of electric fields: RF and DC. If an RF potential is applied to the poles (rods) and a DC potential is applied to the two opposed poles, then by adjusting the voltages and frequencies of the applied fields, ions of any selected m/z value injected at a small velocity into the front end of the quadrupole can pass right through the interpole space and out the other end. Ions of m/z values not selected do not pass through but strike the poles (rods) and are lost. By continuously adjusting the voltages on the poles, ions of successive m/z values can be allowed to pass through the assembly in turn to obtain a mass spectrum of all of the ions injected into the front end (e.g., from an ion source). This configuration of electric fields is often termed the narrow band-pass mode and effectively acts as an electronic gate (sometimes

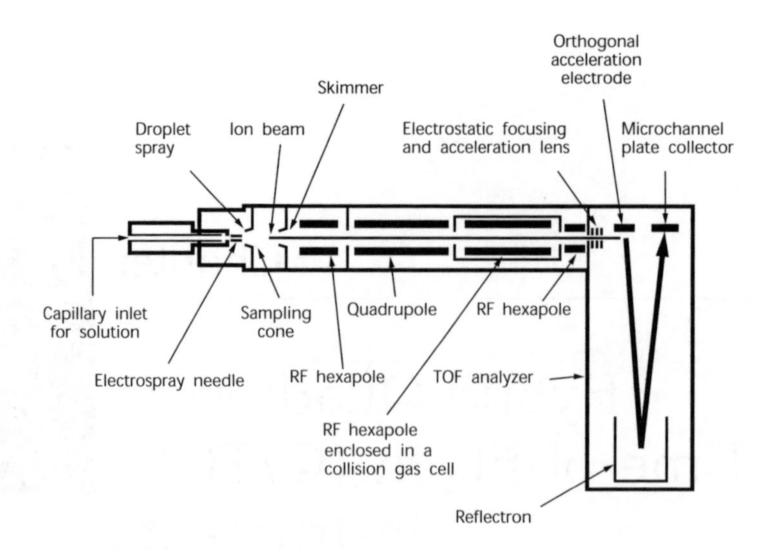

Figure 23.1
Schematic diagram of an orthogonal Q/TOF instrument. In this example, an ion beam is produced by electrospray ionization. The solution can be an effluent from a liquid chromatography column or simply a solution of an analyte. The sampling cone and the skimmer help to separate analyte ions from solvent. The RF hexapoles cannot separate ions according to m/z values and are instead used to help confine the ions into a narrow beam. The quadrupole can be made to operate in two modes. In one (wide band-pass mode), all of the ion beam passes through. In the other (narrow band-pass mode), only ions selected according to m/z value are allowed through. In narrow band-pass mode, the gas pressure in the middle hexapole is increased so that ions selected in the quadrupole are caused to fragment following collisions with gas molecules. In both modes, the TOF analyzer is used to produce the final mass spectrum.

erroneously called a mass filter). By opening or closing the gate, ions of selected m/z values can be allowed to traverse from one end of the quadrupole assembly to the other.

A Quadrupole in Wide Band-Pass Mode

Alternatively, the quadrupole can be operated with only RF potentials applied to all four poles (rods). In this wide band-pass mode, no selection of ions occurs because the electric gate is, in effect, held open. Now, all ions produced in an ion source and injected into the front end of the quadrupole assembly traverse it and emerge from the other end. No ions are lost by striking the poles. Instead, the rod assembly acts to contain the ion beam, so it does not spread (the beam is collimated). Thus, all ions emerging from the quadrupole represent those produced in the ion source. The wide- and narrow band-pass modes can be switched very quickly, so at one instant the quadrupole may be letting through selected ions of m/z 100, m/z 200, m/z 300 or so on. After rapidly switching to the wide band-pass mode, all of the ions of m/z 100, 200, and 300 are allowed through the quadrupole together.

Hexapoles

A hexapole assembly of rods (poles) is built similarly to the quadrupole, but now there are three sets of opposed rods evenly spaced around a central axis. The hexapole cannot act as a mass filter by applying a DC field and is used only in its all-RF mode. It is therefore a wide band-pass filter and is used to collimate an ion beam. (Like-charged particles repel each other, and an electrically charged beam will tend to spread apart because of mutual repulsion of ions unless steps are taken to reduce the effect.)

Time-of-Flight Analyzer

If ions of different m/z values are instantaneously accelerated by an electric field, they attain velocities determined by their m/z values and the accelerating field to which they have been subjected. After flying along an evacuated tube, the ions arrive at a detector in proportion to the velocities they have, the faster ions arriving before slower ones. Because an ion velocity (v) depends upon the inverse of the square root of its m/z value, the ions of smaller m/z value arrive first and those of larger value arrive last. Thus, by measuring the flight time of the ions along a TOF tube (analyzer), the m/z values can be deduced, and therefore a mass spectrum can be measured. Among the advantages of TOF measurement of mass spectra is the very short interval of time needed to measure a full mass spectrum. Typically, a few microseconds is all that is needed to scan from 1 to 2000 mass units.

For the sake of illustration, a TOF analyzer could be likened to a camera taking snapshots of the m/z values of an assembly (beam) of ions; the faster the repetition rate at which the camera shutter is clicked, the greater is the number of mass spectra that can be taken in a very short time. For TOF analyzers, it is not uncommon to measure several thousand mass spectra in one second! All such spectra can be added to each other digitally, a process that improves the signal-to-noise ratio in the final accumulated total.

Clearly, the longer the flight tube, the longer it will take for ions to traverse it. For two ions of velocities v_1 and v_2, their flight times will be t_1 and t_2 for a flight path of length d. If the flight path is doubled in length, the flight times become $2t_1$ and $2t_2$, and the difference in arrival times for the two ions at a detector changes from $(t_1 - t_2)$ to $2 \times (t_1 - t_2)$; it is doubled. Because differences in flight arrival times are typically measured in nanoseconds for a flight tube of 50 cm doubling this length doubles the differences and makes their measurement that much easier. In effect, the resolution of the instrument is greatly improved by passing the ions twice along a TOF flight tube instead of only once before reaching the detector (Figure 23.2a). The device that reflects the ions is called a reflectron (Figure 23.1).

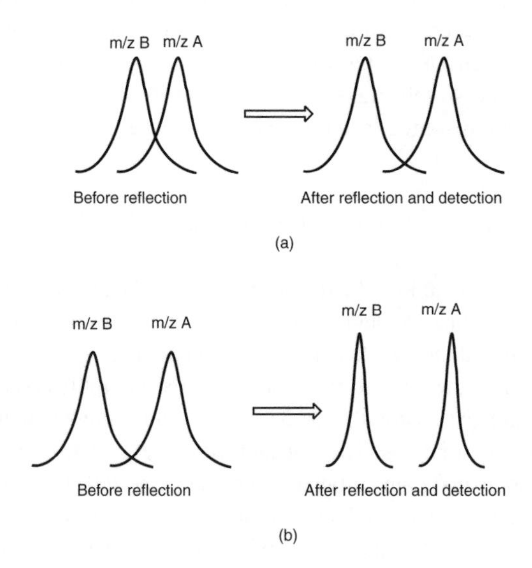

Figure 23.2
Two peaks are shown, representing kinetic energy distributions for two sets of ions of m/z values A and B. The two peaks are shown as overlapping extensively. If there were no improvement, the resolution of the instrument would be quite poor. After doubling the flight path by use of the reflectron, the separation of the tops of the peaks is doubled, which improves their resolution. (b) A second effect of the reflectron is shown. Now, the kinetic energy or velocity spread is much narrower, and the resolution of the peaks at m/z A and B is improved again. The combination of doubling the length of the flight path and decreasing the velocity spread of ions of each m/z value leads to greatly improved instrumental resolution of m/z values.

There is another advantage of the reflectron. Ions of one particular m/z value should all have the same velocity when traveling along the flight tube. However, because of various small inhomogeneities in the electric optics and other applied fields, there will be a small spread of velocities. Figure 23.2 shows typical spreads in velocities (and therefore in kinetic energies) for two sets of ions centered at m/z A and m/z B. Without the reflectron, these ions would reach the detector over small intervals of time rather than tightly bunched at two specific times. If the time difference between arrivals of the ions at m/z A and the following ones at m/z B is only small, then there will be some overlap between the slowest ions of the first m/z value and the fastest ions of the second m/z value (Figure 23.2a). In effect, this means that the ions of different m/z value are inadequately separated; the resolution of the analyzer is impaired. In the reflectron, faster ions travel further into it compared with any slower ones before being reversed and sent back out. Thus, the slower ions get a chance to catch up with the faster ones, so when both slower ones and faster reach the detector, they do so at about the same time. This effect appears to reduce the velocity spread in the ions of any one particular m/z value and significantly improves the resolution of the TOF instrument (Figure 23.2b).

Operation of the Hybrid

With the Quadrupole in Wide Band-Pass Mode

All ions from an ion source are injected into the quadrupole via a hexapole unit (Figure 23.1). In the ion source, the ions produced have thermal energies approximately corresponding to a room-temperature ground state. Upon passing through the skimmer, the ions are accelerated to supersonic speeds by the gas expansion that occurs. The first hexapole has a high enough gas pressure to slow the ions to thermal energies once more. Next, the quadrupole allows all of the ions to pass through in its wide band-pass mode. The ion beam is then collimated by two further hexapoles before passing through an electric lens, which defines the beam and accelerates the ions to a kinetic energy of about 40 V. The ions reach the pusher electrode, and a full mass spectrum is obtained in the TOF analyzer. Thus, in this configuration, ions are simply transmitted from an ion source, through the quadrupole, and into the TOF analyzer. The ion source is shown as a solution electrospray inlet in Figure 23.1.

With the Quadrupole in Narrow Band-Pass Mode

Ions from an ion source are injected into the quadrupole via the hexapole unit as before. Now, however, ion selection is made. By adjusting DC and RF voltages on the poles, ions of a particular m/z value are chosen (Figure 23.3). The middle hexapole (Figure 23.1) becomes a gas collision cell by increasing the background gas pressure within this assembly, which is why the assembly is shown encased in a container having holes just large enough to transmit the ion beam but small enough to maintain a small gas pressure against the effect of the vacuum pumps. The selected ions emerging from the quadrupole collide with gas molecules in this hexapole and gain sufficient internal energy to fragment. Thus, selected precursor ions are caused to dissociate to give product (fragment) ions. The product ions and any unchanged precursor ions continue to the TOF analyzer, in which a mass spectrum is obtained. This sequence illustrates how the Q/TOF hybrid can be operated in MS/MS precursor-ion mode.

Some Advantages of the Hybrid

Figure 23.1 shows ions' being produced from an electrospray source. Generally, these ions are said to be thermolized, which is to say that their distribution of internal energies is close to their normal

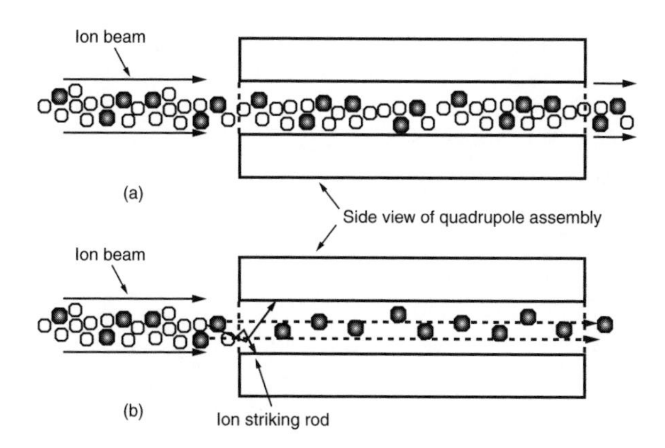

Figure 23.3
(a) A schematic drawing of ions in a beam entering a quadrupole assembly. The shaded circles (●) represent ions of one particular m/z value that are to be selected. The open circles (○) represent all ions of other m/z values that will not be selected. Diagram (a) represents the wide band-pass mode, and all ions (shaded and unshaded) are transmitted. (b) In the narrow band-pass mode, the shaded ions have been selected and pass through the quadrupole assembly while other ions follow trajectories that lead them to striking the quadrupole rods and being lost. For the sake of clarity, only two such incidents are shown.

room-temperature ground state. Such ions have little or no excess of internal energy and exhibit no tendency to fragment which is an enormous advantage for obtaining molecular mass information from the stable molecular ions. However, a molecular mass gives relatively little information about how the constituent atoms are associated in the ion or in the original molecule from which it came. There is a lack of structural information. Fragmentation of the molecular ions yields product, which are characteristic of the structure of the original (precursor) ion and therefore of the original molecule. With the ability to obtain an unequivocal relative molecular mass — or even a molecular formula derived from that mass — and then obtain structural information, the hybrid mass spectrometer becomes a powerful tool for investigating single substances or mixtures of substances.

Other instrumental advantages include its high sensitivity and a linear mass scale to 10,000 at full sensitivity. The linearity of the mass scale means that it is necessary to calibrate the spectrometer using a single or sometimes two known mass standards. Some calibration is necessary because the start of the mass scale is subject to some instrumental zero offset. The digitized accumulation of spectra provides a better signal-to-noise ratio than can be obtained from one spectrum alone. In MS/MS mode, the good peak shapes obtained for product ions, with a resolution of about 500 at half-peak height, afford excellent mass spectra.

Conclusion

A single instrument — a hybrid of a quadrupole and a TOF analyzer — can measure a full mass spectrum of ions produced in an ion source. If these are molecular ions, their relative molecular mass is obtained. Alternatively, precursor ions can be selected for MS/MS to give a fragment-ion spectrum characteristic of the precursor ions chosen, which gives structural information about the original molecule.

Chapter 24

Ion Optics of Magnetic/Electric-Sector Mass Spectrometers

Introduction

In the ion source, substances are converted into positive or negative ions having masses (m_1, m_2, … , m_n) and a number (z) of electric charges. From a mass spectrometric viewpoint, the ratio of mass to charge (m_1/z, m_2/z, … , m_n/z) is important. Generally, z = 1, in which case, $m_1/z = m_1$, $m_2/z = m_2$, … , $m_n/z = m_n$, so the mass spectrometer measures masses of ions. To do this, a stream of ions (the ion beam) is injected into the mass analyzer region, which is a series of electric and magnetic fields known as the ion optics. In this region, the ion beam is focused, corrected for aberrations in shape, and the individual m/z ratios measured. The ion beam finally arrives at a collector that measures the number (abundance) of ions at each m/z value. The width and shape of the ion beam is controlled by a series of slits (object or source, collector, alpha, etc.) situated between the ion source and the collector. A chart of m/z values and their respective abundances makes up the mass spectrum.

Mass Analysis of Ions

Magnetic Sector

When moving charged species (ions) experience a magnetic field, they are deflected. The direction of the deflection can be described by Fleming's left-hand rule (Figure 24.1). The magnitude of the deflection is governed by the momentum of the ion and is described by Equations 24.1 and 24.2. Firstly, the kinetic energy of the ion is equal to the energy gained through acceleration from the ion source (Equation 24.1).

$$zV = 1/2mv^2 \quad (24.1)$$

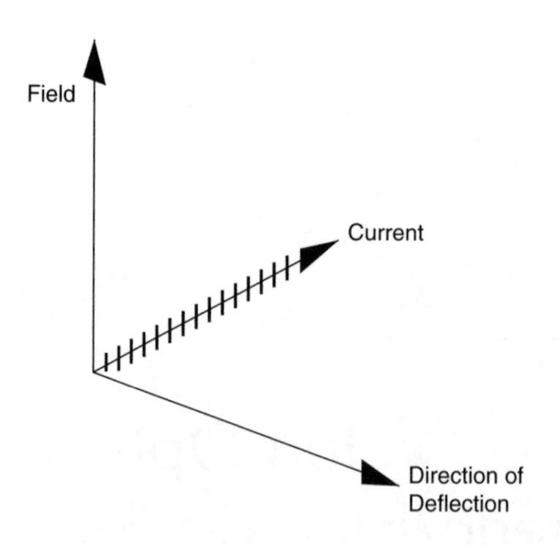

Figure 24.1
Fleming's left-hand rule.

Secondly, the centrifugal force on the ion as its path is deflected by a magnetic field is equal to the force exerted by the field on a moving charge (Equation 24.2).

$$mv/R = zB \tag{24.2}$$

From Equations 24.1 and 24.2, the velocity of the ion can be eliminated to give Equation 24.3,

$$m/z = B^2r^2/2v \tag{24.3}$$

where:

r = radius of arc of ions deflected in the magnetic field
V = accelerating potential applied to ions leaving the ion source
B = magnetic field strength
z = number of charges on an ion
m = mass of any one ion
v = velocity of an ion after acceleration through the electric potential (V).

If only ions with a single charge (z = 1) are considered, then with a constant magnetic field strength and constant accelerating voltage, the radius of arc depends on mass and, from Equation 24.3, Equation 24.4 is obtained.

$$r = \sqrt{\frac{2mV}{zB^2}} \tag{24.4}$$

Thus, it is possible to separate ions of different mass (Figure 24.2), because ions arriving at position 1 (greater deflection) have lower mass than those arriving at position 5 (lesser deflection).

In the modern scanning mass spectrometer, it is more convenient that ions arrive at a single point for monitoring (collection), so r (or r^2) is kept constant. Therefore, B or V must be varied to bring all ions to the same focus; viz., one of the relationships in Equation 24.5 must apply:

$$m \propto B^2 \text{ (V constant)}$$

$$m \propto 1/V \text{ (B constant)}$$

$$m \propto B^2/V \tag{24.5}$$

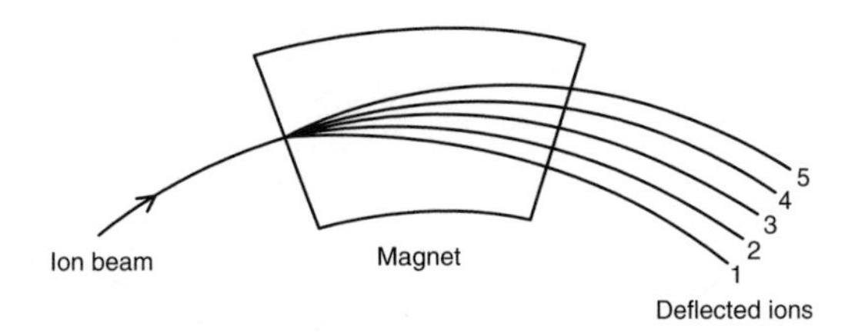

Figure 24.2
Deflection in a magnetic field of an ion beam consisting of increasing mass-to-charge ratios, m_1/z, … , m_5/z, and split into different trajectories (1–5), respectively.

From these relationships in Equation 24.5, it can be seen that, if either the magnetic field (B) or the voltage (V) or both B and V are scanned, the whole range of masses of the ions can be brought into focus sequentially at a given point, the collector. Generally, a scanning magnetic-sector mass spectrometer carries out mass analysis by keeping V constant and varying the magnetic field (B).

A further property of the magnetic field is that a diverging ion beam entering that field leaves with the beam converging. Thus, the magnet is said to be directional (or angular) focusing (Figure 24.3).

So far, it has been assumed that all ions leaving the source have exactly the same kinetic energy, which is not really the case. In electron ionization (EI), the spread in kinetic energy can be as much as 1 V, and with fast-atom bombardment (FAB) the spread can be as much as 4 V. This spread results in a blurred image at the collector because the magnet has no energy focusing, and ions of different kinetic energies are brought to slightly different foci. Thus a single magnetic sector has only directional (or angular) focusing and therefore is said to be single focusing.

Electrostatic Analyzer (Electric Sector)

An electrostatic analyzer (ESA) is an energy-focusing device (Figure 24.4). As shown in Equations 24.1 and 24.2, the energy gained by ions accelerated from the ion source is zV (= $1/2mv^2$). In the electric sector, the centrifugal force acting on the ions is given by Equation 24.6.

$$zV = 1/2(mv^2) \qquad (24.6)$$

Now if

E = electric potential (voltage) between the inner and outer ESA plates
R = radius of curvature of the ion trajectory

then the relationship in Equation 24.7 is obtained:

$$zE = mv^2/R \qquad (24.7)$$

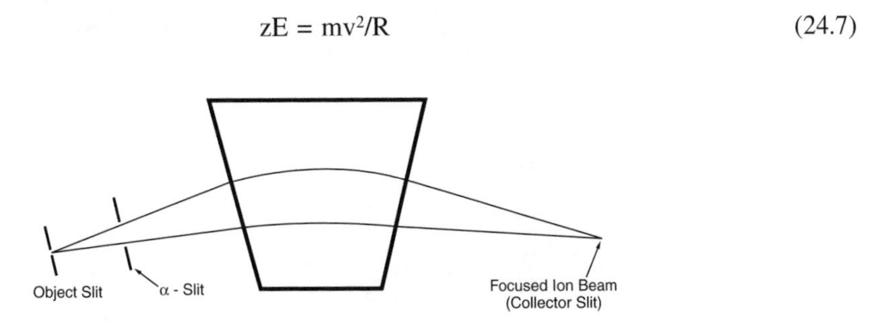

Figure 24.3
Directional (or angular) focusing of a magnet.

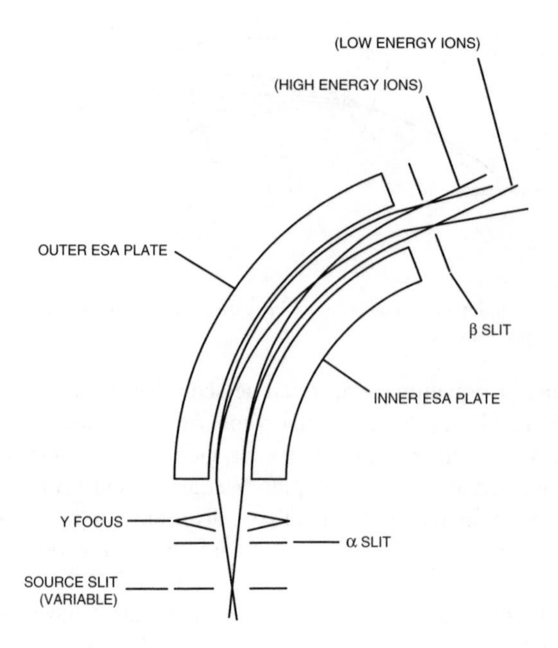

Figure 24.4
Focusing and dispersion properties of an electrostatic analyzer.

No mass or charge appears in this equation, so in the electric sector the ion flight path bends in an arc that depends only on the accelerating voltage (V) and the ESA voltage (E). The ion beam is focused for kinetic energy; viz., all ions having the same kinetic energy are focused but those having different kinetic energies are dispersed.

Magnetic/Electrostatic Analyzer Combination

The ion beam is collimated when a magnetic analyzer is combined with an ESA. The combination can be made both energy and mass focused; viz., the ion beam is focused for energy in the ESA and focused for angular dispersion in the magnetic field (Figure 24.5). The combination is called double focusing because it is both directional (or angular) and energy focusing. The double-focusing mass spectrometer is designed such that ions of different energies (but of the same mass) converge at the collector (Figure 24.5).

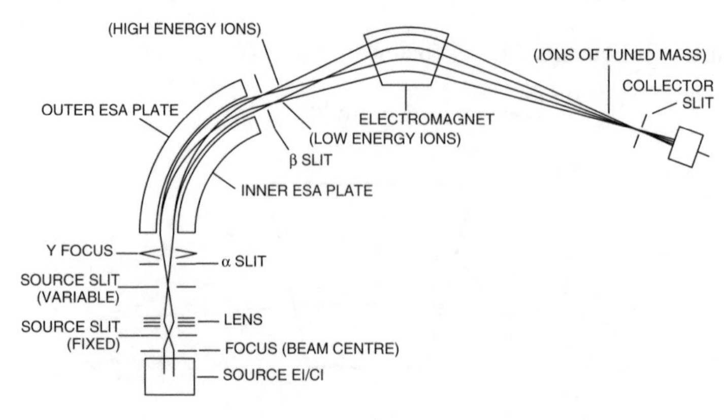

Figure 24.5
Double-focusing ion optics (forward geometry).

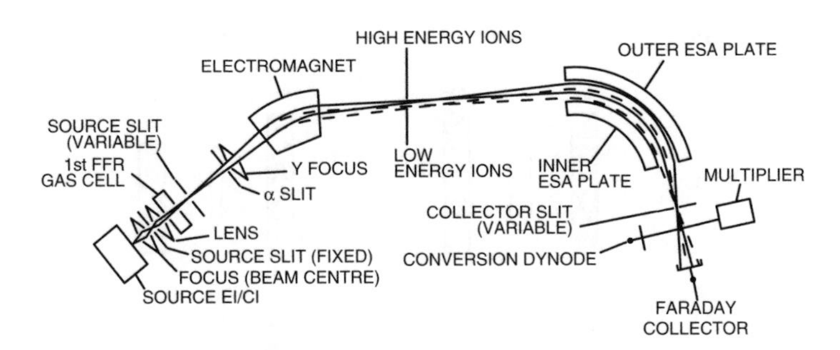

Figure 24.6
Double-focusing ion optics (reverse geometry).

Double-focusing forward geometry ion optics is a combination in which the ESA is placed before the magnet, as shown in Figure 24.5.

Double-focusing reverse geometry ion optics is a combination in which the magnet is placed before the ESA, as shown in Figure 24.6.

The double-focusing combination of electrostatic- and magnetic-sector analyzers allows the inherent energy spread of the beam to be compensated for by design and ensures that there is no spread in the beam at the collector.

Electric Focusing Lenses

Y-Focus, Z-Focus, and Deflection Lenses

It was stated above that the focus of all masses will occur at a single position, the collector slit (Figures 24.5 and 24.6). However, because the actual shape of the field within and around the pole tips of the magnet varies with changing field, especially at higher field strengths, the final focal point of the beam shifts as field strength changes. This focal point shift leads to a change of focus with mass and affects the ability of the instrument to resolve small mass differences. On early mass spectrometers, the problem could be corrected by physically adjusting the position of the magnet for any given mass. On modern instruments, an electric field called the *Y-focus* is used to compensate for these imperfections (Figures 24.5 and 24.6). The aim of this lens is to focus the ions at the same position (the collector slit) throughout the mass range. Thus, using the electric and magnetic sectors with a Y-focus lens ensures all ions are brought to the same focus and allows small differences in mass to be detected; the resolution of the instrument is enhanced.

On VG instruments, these lenses are sited before and after the magnetic sector. The focus and deflection lenses steer the beam so it coincides with the gap in the collector slit. The Z-focus lenses change the divergence of the beam by adjusting voltages on lens plates situated on either side of the beam. The deflection lenses move the whole beam; the Y-deflection lens is used to move the beam from one side to the other, and the Z-deflection lens moves the beam up or down. The two lenses allow the ion beam to be aligned correctly with the collector slit. Voltages on the lens plates are adjusted to effect such movements of the beam (Figure 24.7).

Curvature and Rotation Lenses

Curvature and rotation lenses correct for any imperfections (aberrations) in the cross-sectional shape of the beam before it reaches the collector slit. The curvature lens provides a means of changing any banana-shaped beam cross-section into a rectangular shape (Figure 24.8). The rotation lens rotates the beam such that the sides of the beam become parallel with the long axis of the collector slit (Figure 24.8).

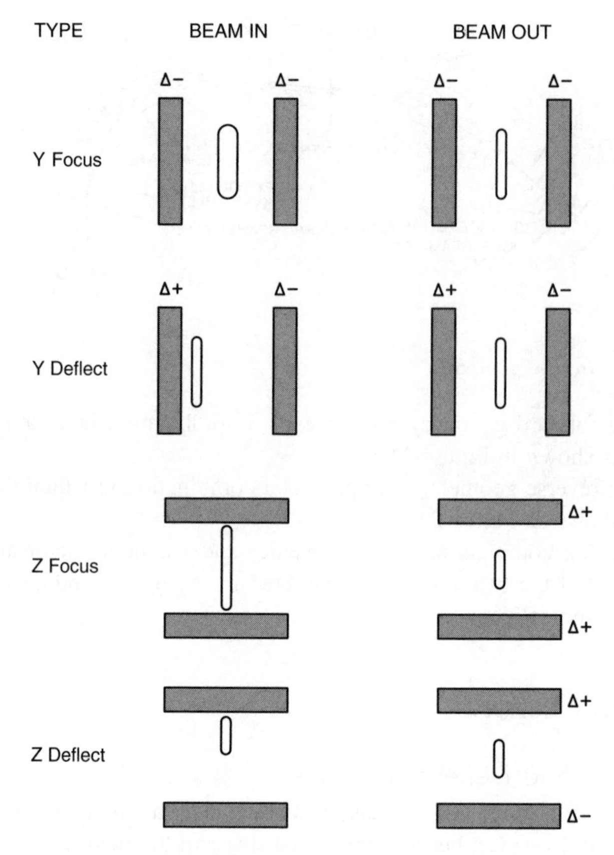

Figure 24.7
Y-focus, Z-focus, and deflect lenses with their effects on the ion beam.

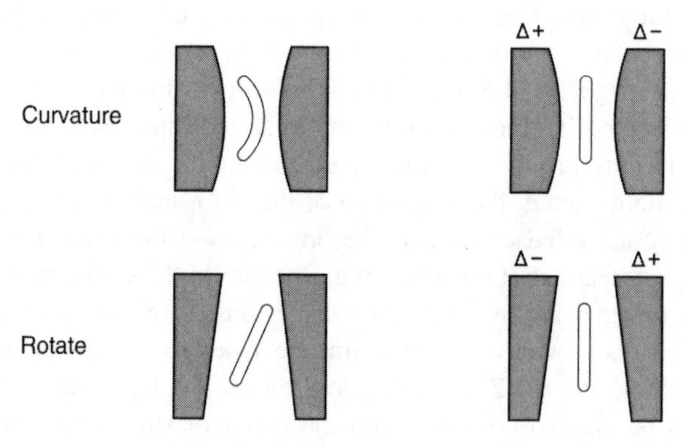

Figure 24.8
Curvature and rotation lenses and their effects on the ion beam.

Metastable Ions

An ion beam mainly comprises normal ions, all having the same kinetic energy gained on acceleration from the ion source, but there are also some ions in the beam with much less than the full kinetic energy; these are called *metastable* ions.

Energy Filter

An energy filter is a system of electrostatic fields that strictly has little to do with the main focusing fields but, rather, provides a means of discriminating between normal and metastable ions. The system is a filter, preventing metastable ions from being detected by the collector, and it consists of a series of parallel lens plates to which is applied a decelerating voltage of 90–99% of the original accelerating potential (V). Normal ions have enough energy to pass through the filter and reach the collector, but metastable ions do not.

Conclusion

Through the use of sequential electric (electrostatic) and magnetic fields (sectors) and various correcting lenses, the ion beam leaving the ion source can be adjusted so that it arrives at the collector in focus and with a rectangular cross-section aligned with the collector slits. For the use of crossed electromagnetic fields, Chapter 25 ("Quadrupole Ion Optics") should be consulted.

Chapter 25

Quadrupole Ion Optics

Background

Charged species such as ions, when passing through magnetic or electric fields (sectors), experience a force that deflects them from their original trajectory. This effect is utilized in magnetic-sector mass spectrometers to separate ions according to mass or, strictly, mass-to-charge ratio (m/z), the deflection being related to m/z and magnetic field strength (see Chapter 24). In a quadrupole instrument, only electric fields are used to separate ions according to mass. The ions are separated as they pass along the central axis of four parallel, equidistant rods (poles) that have DC and alternating (radio frequency, RF) voltages applied to them (Figure 25.1).

Depending on field strengths, it can be arranged that only ions of one selected mass can pass (filter) through the rod assembly while all others are deflected to strike the rods. By changing the strengths and frequencies of electric fields, different masses can be filtered through the system to produce a mass spectrum.

Equations of Motion of Ions

Unlike simple deflections or accelerations of ions in magnetic and electric fields (Chapter 25), the trajectory of an ion in a quadrupolar field is complex, and the equations of motion are less easy to understand. Accordingly, a simplified version of the equations is given here, with a fuller discussion in the Appendix at the end of this chapter.

The four rods of circular cross-section shown in Figure 25.1 should theoretically each have a hyperbolic cross-section (Figure 25.2a), but in practice cylindrical rods (Figure 25.2b) are quite satisfactory if properly spaced apart. One opposed pair of rods has a potential of $+(U + V\cos\omega t)$ applied to it and the other pair has $-(U + V \cos \omega t)$ applied. U is a fixed potential and $V\cos\omega t$ represents a radio frequency (RF) field of amplitude (V) and frequency (ω). As $\cos(\omega t)$ cycles with time (t), the applied voltages on opposed pairs of rods (A, B) change in the manner shown in Figure 25.3. Along the central axis of the quadrupole assembly and along the two planes shown in Figure 25.2b, the resultant electric field is zero.

In the transverse direction of the quadrupoles, an ion will oscillate among the poles in a complex fashion, depending on its mass, the voltages (U, V), and the frequency (ω) of the alternating RF potential. By suitable choices of U, V, and ω, it can be arranged that only ions of one mass will oscillate stably about the central axis; in this case, all other ions will oscillate

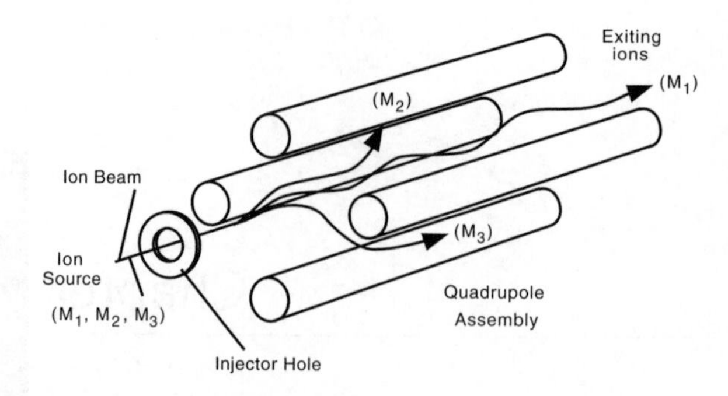

Figure 25.1
For a beam of ions injected into the quadrupole assembly from an ion source, depending on mass, some (M_1) pass along the central axis, but others (M_2, M_3) are deflected to the rods.

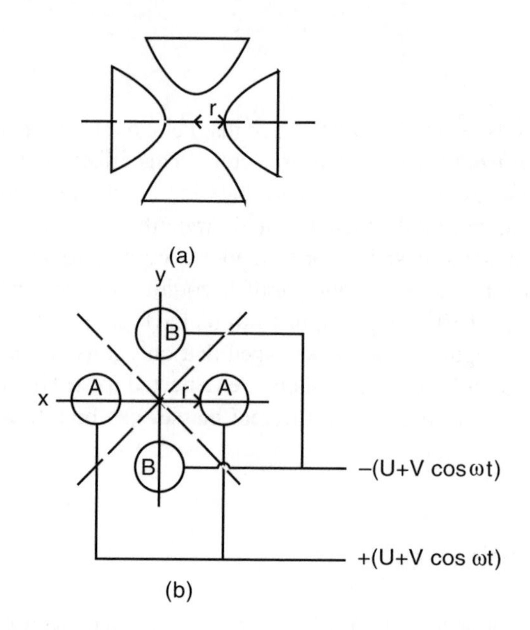

Figure 25.2
End views of the quadrupole assembly (a) showing the theoretically desired cross-section and (b) illustrating the practical system. In (b), a positive potential, +(U + Vcosωt), is applied to two opposed rods (A) and a negative potential, –(U + Vcosωt), to the other two (B). The dotted lines indicate planes of zero electric field. The dimension (r) is typically about 5 mm with rod diameters of 12 mm. The x- and y-axes are indicated, with the z-axis being perpendicular to the plane of the paper.

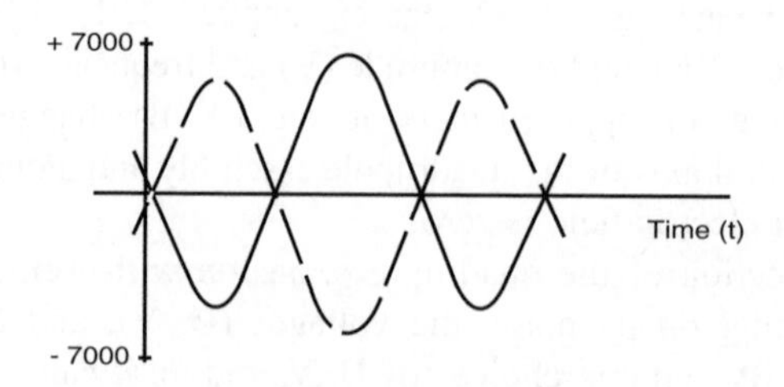

Figure 25.3
Variation of total voltages with time on each pair of rods A (—) and B (- - -); U = 1000 V, V = 6000 V, and RF is ω.

with greater and greater amplitude until eventually they strike one of the poles and are lost (unstable motion). Unlike this transverse situation, in the direction of the central axis of the quadrupole assembly, the applied field is zero and an ion trapped in the quadrupolar field will not pass through the assembly unless it has first been given some momentum (a push) in that direction. To give ions this forward momentum, they are often accelerated through a potential of about 5 V, maintained between the ion source and the quadrupole. Sometimes sufficient momentum arises naturally from space-charge effects from following ions. Usually, this injected ion beam is divergent, and, in order to reduce the spread in the beam and improve mass resolution, it is passed through a small defining hole or aperture placed between the source and the quadrupole (Figure 25.1).

Therefore, ions passing into the quadrupole assembly move along the central axis with a complex oscillatory trajectory; those with the "right" values of m/z have a stable trajectory and can pass right through to the detector, and those with the "wrong" values have unstable trajectories and are lost to the poles. By suitable variation of U, V, and ω, ions of different masses can pass through the quadrupole filter successively and be detected. The numbers (abundances) of ions arriving at the detector for each m/z value give the mass spectrum in the usual way.

In practice, the frequency (ω) is fixed, and typical values are in the range 1–2 MHz. The DC voltage (U) may be 1000 V, and the maximum RF voltage (V) is 6000 V. The pole assemblies range in length from about 50 to 250 mm, depending on application, and pole diameters lie in the range of 6 to 15 mm. Very high mechanical accuracy — to the micron level — in both the rods and their spacing is needed for high performance. For example, with a spacing (r) accurate to 10^{-3} r, the maximum resolution would be about 500 ($0.5r/10^{-3}$ r). Scanning speeds of up to 6000 amu/sec are routine with quadrupoles, and unit mass resolution up to 2000 amu is attainable.

Comparison of Quadrupole and Magnetic Sector Instruments

Major advantages of quadrupoles compared with magnetic instruments include their relatively small size of mass analyzer for comparable resolving power, linear mass scale, fast scanning, simplicity of construction, robustness, and ease of cleaning. The main disadvantage of the quadrupole lies in ultimate attainable resolving power. As the ratio of the voltage (U,V) changes, a limiting resolution (the ability of the spectrometer to separate two adjacent masses) is reached, largely determined by the accuracy with which the various quadrupole parameters are controlled and by the initial motion of the ion as it enters the quadrupole field. A second, less important disadvantage is that the efficiency of transmission of ions through the quadrupole decreases as ion mass increases, leading to a relative loss in sensitivity at higher masses.

Thus, where magnetic sector and quadrupole instruments might be considered for a particular application, ultimate resolving power could be a decisive factor. For example, for use in a gas chromatograph/mass spectrometer combination (GC/MS), because most compounds that are volatile enough to pass through the GC will have relative molecular masses of well under 800, the lower ultimate resolving power of the quadrupole is less important, and its lower cost becomes decidedly advantageous.

There are other characteristics of quadrupoles that make them cheaper for attainment of certain objectives. For example, quadrupoles can easily scan a mass spectrum extremely quickly and are useful for following fast reactions. Moreover, the quadrupole does not operate at the high voltages used for magnetic sector instruments, so coupling to atmospheric-pressure inlet systems becomes that much easier because electrical arcing is much less of a problem.

TABLE 25.1
Some Factors Important in Choosing between Quadrupole and Magnetic-Sector Mass Spectrometers

Factor	Quadrupole	Magnetic
Overall cost	+++	+
Ease of use	+++	+
Fast scanning or peak switching	++	+
Ease of coupling to inlet systems	++	+
Ultimate resolving power (accurate mass determination)	+	+++
Ultimate mass range (over about 4000 amu)	+	+++
Ultimate sensitivity, particularly at high mass	+	+++

Note: The plus (+) signs provide only a qualitative guide and should not be construed as a definitive assessment. The greater the number of plus signs, the more advantageous the system.

The Choice of Quadrupole or Magnetic-Sector Instruments

Modern mass spectrometers are used in a very wide variety of situations, so it is almost impossible to have a simple set of criteria that would determine whether a quadrupole or magnetic sector instrument would be best for any particular application. Nevertheless, some attempt is made here to address major considerations, mostly relating to cost.

The most important deciding factor is undoubtedly the end use of the instrument: Is it to be used for a very wide range of general applications or for a narrow, specialized area of work? Generally, a magnetic-sector instrument is capable of ultimately higher resolution and greater accuracy of mass measurement than is a quadrupole. Other considerations, some of which are shown in the Table 25.1, give an advantage to the quadrupole.

Note that in mass spectrometry/mass spectrometry (MS/MS) applications, quadrupole and magnetic sectors can be used together advantageously. It is also worth noting that the quadrupole can be operated without the DC voltages. In this RF-only mode, no mass separation occurs, and these quadrupoles are used as ion transmission guides, described in Chapter 49.

Conclusion

By injecting ions along the central axis of four parallel, equidistantly rods (poles) to which are applied variable static DC and alternating (RF) electric fields, ions can be separated according to mass. The distance apart of the rods and the magnitudes and frequencies of the applied electric voltages are all important in determining which ion masses traverse the assembly and which instead strike the rods. In the RF-only mode, a quadrupole can be used as an ion transmission guide or as an ion/molecule collision cell.

Appendix — More Details of Equations of Motion

Under the influence of the varying electric fields, $+(U + V\cos\omega t)$ and $-(U + V\cos\omega t)$, the resultant electric potential (F) in the x–y plane (transverse direction) of the quadrupoles is given by Equation 25.1.

$$F = [(x^2 - y^2)/r^2]\,(U + V\cos\omega t) \tag{25.1}$$

Thus, for x = y, F = 0 which gives rise to planes of zero field strength (Figure 25.2b). At all other positions between the poles, the oscillating electric field (F) causes ions to be alternately attracted to and repelled by the pairs of rods (A, B; Figure 25.2).

Note that Equation 25.1 shows that the field (F) has no effect along the direction of the central (z) axis of the quadrupole assembly, so, to make ions move in this direction, they must first be accelerated through a small electric potential (typically 5 V) between the ion source and the assembly. Because of the oscillatory nature of the field (F; Figure 25.3), an ion trajectory as it moves through the quadrupole assembly is also oscillatory.

For the "right" values of the parameters U, V, ω, and m/z, an ion can pass right through the length of the assembly, but with the "wrong" values, the field (F) deflects an ion such that it strikes one of the four rods and is lost (Figure 25.1). Thus, for any particular m/z value, the passage of an ion through the quadrupole depends critically on the DC voltage (U), the RF amplitude (V), the RF frequency (ω), and the distance between the poles (r; Figure 25.2).

Passage through the quadrupole assembly is described as stable motion, while those trajectories that lead ions to strike the poles is called unstable motion. From mathematical solutions to the equations of motion for the ions, based on Equation 25.1, two factors (a and q; Equation 25.2) emerge as being important in defining regions of stable ion trajectory.

$$a = (8zU)/(mr^2\omega^2)$$
$$q = (4zV)/(mr^2\omega^2)$$
$$a/q = 2U/V \qquad (25.2)$$

For small values of a and q, the shaded area in Figure 25.4 indicates an area of stable ion motion; it shows all values for a and q for which ions can be transmitted through the quadrupole assembly.

To gain some idea of the meaning of this shaded area, consider the straight line OA of slope a/q shown in Figure 25.4. The line enters the region of stable motion at P and leaves it at Q. For typical values of U (1000 V), V (6000 V), ω (1.5 MHz), and r (1.0 cm), Equation 25.2 predicts that point P corresponds to an ion of m/z 451 and Q to m/z 392. Therefore, with these parameter values, all ions having m/z between 392 and 451 will be transmitted through the quadrupole.

For a line OB (Figure 25.4) of smaller slope (smaller a/q or smaller U/V), an even greater range of m/z values will be transmitted through the quadrupole assembly. Conversely, for a line OC that passes through the apex (r) of the region of stability, no ions of any m/z value are transmitted.

To ensure that only ions of any one selected m/z are transmitted (maximum resolution), the parameters (U, V, ω) must be chosen such that a/q (or 2U/V) fits a line that passes close to R but

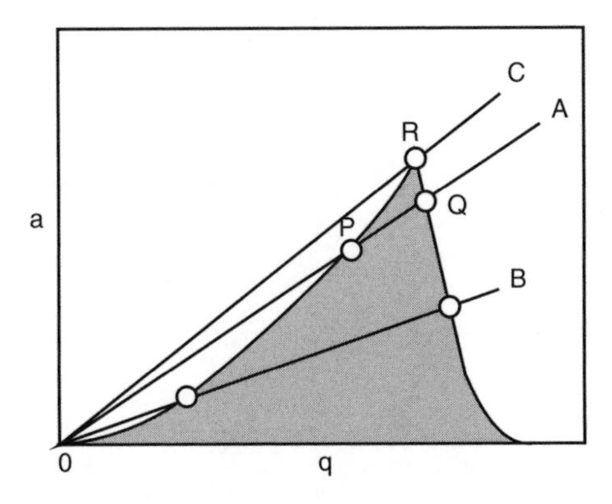

Figure 25.4
Relationship between a and q. The shaded area indicates regions of stable ion motion through the quadrupolar field.

that still lies within the region of stability. Such parameters will give maximum resolution for the instrument. For example, with U (1000 V), V (6000 V), ω (2 MHz), and r (1.0 cm), only ions of m/z 862 would be transmitted. To transmit ions of other m/z, the parameters U, V, ω, and r have to be changed.

For a given assembly, r is fixed; electronically, it is easier to change voltages (U, V) than frequencies (ω). Therefore, to transmit ions of other m/z values, the frequency is kept constant but the voltages (U, V) are varied in such a way that U/V (or a/q) remains constant. By continuously increasing or decreasing U and V while keeping U/V constant, ions of increasing or decreasing m/z successively traverse the quadrupole assembly to give a mass spectrum.

For convenience, different frequencies can be used for different mass ranges; e.g., the RF can be 1.5 MHz for 0–1000 amu and 0.8 MHz for 1000–4000 amu.

Chapter 26

Time-of-Flight (TOF) Ion Optics

Background

The fundamentals of the ion optics for magnetic and electric sector, quadrupole, ion trap, and Fourier-transform ion cyclotron resonance (FTICR) mass spectrometers range from fairly simple to difficult. The basic ion optics of time-of-flight (TOF) instruments are very straightforward. Basically, ions need to be extracted from an ion source in short pulses and then directed down an evacuated straight tube to a detector. The time taken to travel the length of the drift or flight tube depends on the mass of the ion and its charge. For singly charged ions ($z = 1$; $m/z = m$), the time taken to traverse the distance from the source to the detector is proportional to a function of mass: The greater the mass of the ion, the slower it is in arriving at the detector. Thus, there are no electric or magnetic fields to constrain the ions into curved or complicated trajectories. After initial acceleration, the ions pass in a straight line, at constant speed, to the detector. The arrival of the ions at the detector is recorded in the usual way: as a trace of ion abundance against time of arrival, the latter being converted into a mass scale to give the final mass spectrum.

Equations of Motion of Ions

A short pulse of ions is extracted from an ion source (Figure 26.1). It is necessary to use a pulse because when using only time to differentiate among the masses it is important that the ions all leave the ion source at the same instant (like runners in a race upon hearing a starting pistol). The first step is acceleration through an electric field (E volts). With the usual nomenclature (m = mass, z = number of charges on an ion, e = the charge on an electron, v = the final velocity reached on acceleration), the kinetic energy ($mv^2/2$) of the ion is given by Equation 26.1.

$$mv^2/2 = z \cdot e \cdot E \tag{26.1}$$

Equation 26.2 follows by simple rearrangement.

$$v = \sqrt{(2z \cdot e \cdot E)/m} \tag{26.2}$$

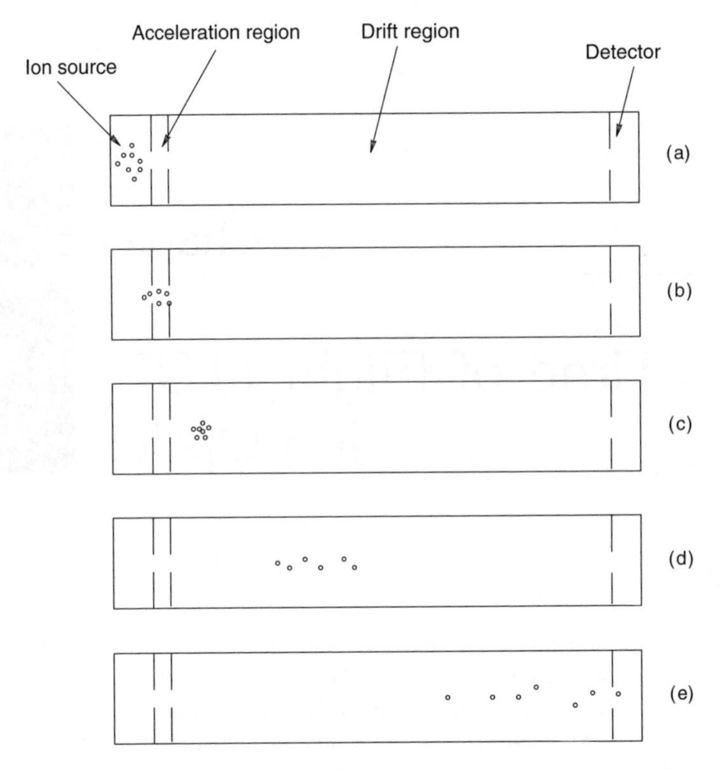

Figure 26.1
The essentials of time-of-flight optics. In (a), a pulse of ions is formed and then accelerated out of the source (b) into the drift region (c). After a short time (d), the ions have separated along the drift region according to m/z value. In (e), the ions with smallest m/z value (fastest moving) begin arriving first at the detector, to be followed by the ions of gradually increasing m/z.

If the distance from the ion source to the detector is d, then the time (t) taken for an ion to traverse the drift tube is given by Equation 26.3.

$$t = d/v = d/\sqrt{(2z \cdot e \cdot E/m)} = d \cdot [\sqrt{(m/z)}]/\sqrt{(2e \cdot E)} \tag{26.3}$$

In Equation 26.3, d is fixed, E is held constant in the instrument, and e is a universal constant. Thus the flight time of an ion t is directly proportional to the square root of m/z (Equation 26.4).

$$t = \sqrt{m/z} \times \text{a constant} \tag{26.4}$$

Equation 26.4 shows that an ion of m/z 100 will take twice as long to reach the detector as an ion of m/z 25: $t_{100}/t_{25} = \sqrt{100}/\sqrt{25} = 10/5 = 2$

Resolution

Generally, the attainable resolving power of a TOF instrument is limited, particularly at higher mass, for two major reasons: one inherent in the technique, the other a practical problem. First, the flight times are proportional to the square root of m/z. The difference in the flight times (t_m and t_{m+1}) for two ions separated by unit mass is given by Equation 26.5.

$$t_m - t_{m+1} = \Delta t = [(\sqrt{m/z}) - (\sqrt{(m+1)/z})] \times \text{a constant} \quad (26.5)$$

As m increases, Δt becomes progressively smaller (compare the difference between the square roots of 1 and 2 (= 0.4) with the difference between 100 and 101 (= 0.05). Thus, the difference in arrival times of ions arriving at the detector become increasingly smaller and more difficult to differentiate as mass increases. This inherent problem is a severe restriction even without the second difficulty, which is that not all ions of any one given m/z value reach the same velocity after acceleration nor are they all formed at exactly the same point in the ion source. Therefore, even for any one m/z value, ions at each m/z reach the detector over an interval of time instead of all at one time. Clearly, where separation of flight times is very short, as with TOF instruments, the spread for individual ion m/z values means there will be overlap in arrival times between ions of closely similar m/z values. This effect (Figure 26.2) decreases available (theoretical) resolution, but it can be ameliorated by modifying the instrument to include a reflectron.

Reflectron

The reflectron consists of a series of ring electrodes, on each of which is placed an electric potential. The first ring has the lowest potential and the last ring the highest to produce an electrostatic field that increases from the front end of the electron to the back.

This field is at the end of the flight path of the ions and has a polarity the same as that of the ions; viz., positive (or negative) ions experience a retarding positive (or negative) potential. In the reflectron, ions come to a stop and are then accelerated in the opposite direction. The ions are reflected by the ion mirror or reflectron. The ion mirror is often at a slight angle (Figure 26.3) to the line of flight of the ions and, when reflected, the ions do not travel back along the same path but along a slightly deflected line. (Some instruments reflect ions back along the path they took to the reflection.)

Consider ions of any given m/z value. The faster ions, having greater kinetic energy, travel further into the electrostatic field before being reflected than do slower ions. As a result, the faster ions spend slightly more time within the reflectron than do the slower ones. Therefore, the faster ions have further to travel than do the slower ions. The faster ions catch up with the slower ones at the ion detector, so they all arrive together. There is a similar effect for all other individual m/z values, and overall resolution is greatly improved. Instead of a typical TOF resolution of about 1000, resolutions of around 10,000 can be achieved. There is a disadvantage in using a reflectron in that the sensitivity of the instrument is decreased through ion losses by collision and dispersion from the main beam. It is a more severe problem for ions of large mass, for which the reflectron is often not used despite the better resolution attainable with its.

Comparison with Other Mass Spectrometers

TOF mass spectrometers are very robust and usable with a wide variety of ion sources and inlet systems. Having only simple electrostatic and no magnetic fields, their construction, maintenance, and calibration are usually straightforward. There is no upper theoretical mass limitation; all ions can be made to proceed from source to detector. In practice, there is a mass limitation in that it becomes increasingly difficult to discriminate between times of arrival at the detector as the m/z value becomes large. This effect, coupled with the spread in arrival times for any one m/z value, means that discrimination between unit masses becomes difficult at about m/z 3000. At m/z 50,000, overlap of 50 mass units is more typical; i.e., mass accuracy is no better than about 50–100 mass

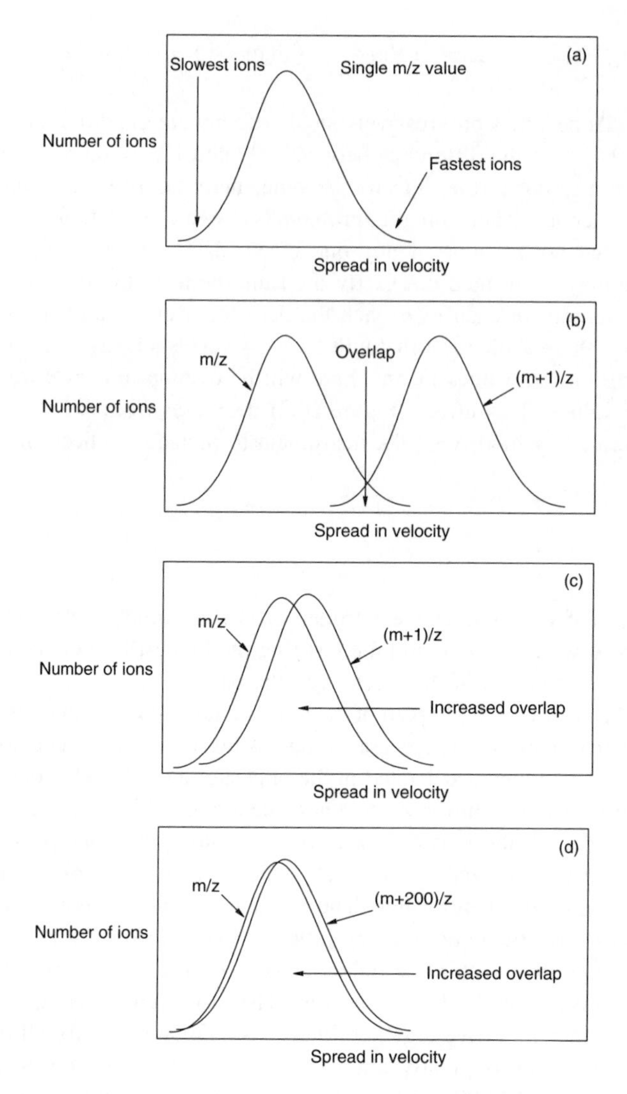

Figure 26.2
(a) Each set of ions at any one m/z value will have a small spread in speed because they are not all formed at exactly the same place in the ion source. (b) Two sets of ions separated by one mass unit (m, m + 1) overlap because the slower ions of smaller mass overlap the faster ions of greater mass. (c) For larger m/z values, this effect leads to almost total disappearance of unit resolution. (d) At still greater m/z values, even mass differences of 200 or more may not be separated.

units (Figure 26.2). Nevertheless, the ability of a TOF instrument to measure routinely and simply masses this large gives it a decided advantage over many other types of mass analyzer.

A major advantage of the TOF mass spectrometer is its fast response time and its applicability to ionization methods that produce ions in pulses. As discussed earlier, because all ions follow the same path, all ions need to leave the ion source at the same time if there is to be no overlap between m/z values at the detector. In turn, if ions are produced continuously as in a typical electron ionization source, then samples of these ions must be utilized in pulses by switching the ion extraction field on and off very quickly (Figure 26.4).

On the other hand, there are some ionization techniques that are very useful, particularly at very high mass, but produce ions only in pulses. For these sources, the ion extraction field can be left on continuously. Two prominent examples are Californium radionuclide and laser desorption ionization. In the former, nuclear disintegration occurs within a very short time frame to give a

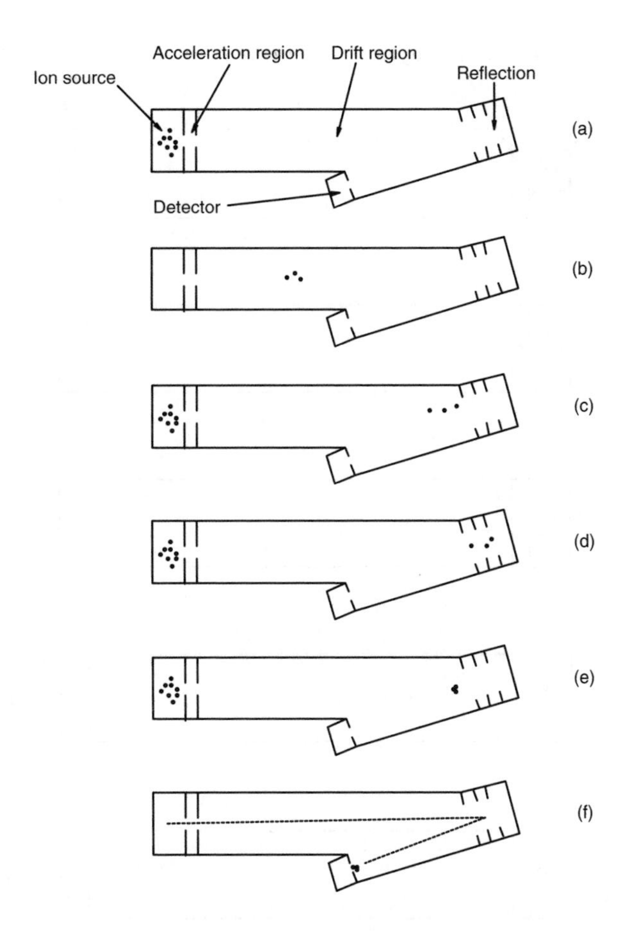

Figure 26.3
In (a), a pulse of ions is formed but, for illustration purposes, all with the same m/z value. In (b), the ions have been accelerated but, because they were not all formed in the same space, they are separated in time and velocity, with some ions having more kinetic energy than others. In (c), the ions approach the ion mirror or reflectron, which they then penetrate to different depths, depending on their kinetic energies (d). The ones with greater kinetic energy penetrate furthest. In (e), the ions leave the reflectron and travel on to the detector (f), which they all reach at the same time. The path taken by the ions is indicated by the dotted line in (f).

short pulse of ions; the same disintegration is used to start the timer (stopwatch) for the race of the ions down the flight tube. Similarly, a laser pulse lasting only a few nanoseconds produces many ions and acts to start the timer also.

Conclusion

By accelerating a pulse of ions into a flight or drift tube, the ions achieve different velocities depending on their individual m/z values; the higher the m/z value, the slower the ion travels down the tube. A detector at the end of the tube records the arrival times of the ions, hence their flight times. The flight times are easily converted into m/z information by a simple formula. The time-of-flight mass spectrometer is particularly useful for ions produced in pulses and where fast response times are needed. It can be used to detect ions up to very large m/z values.

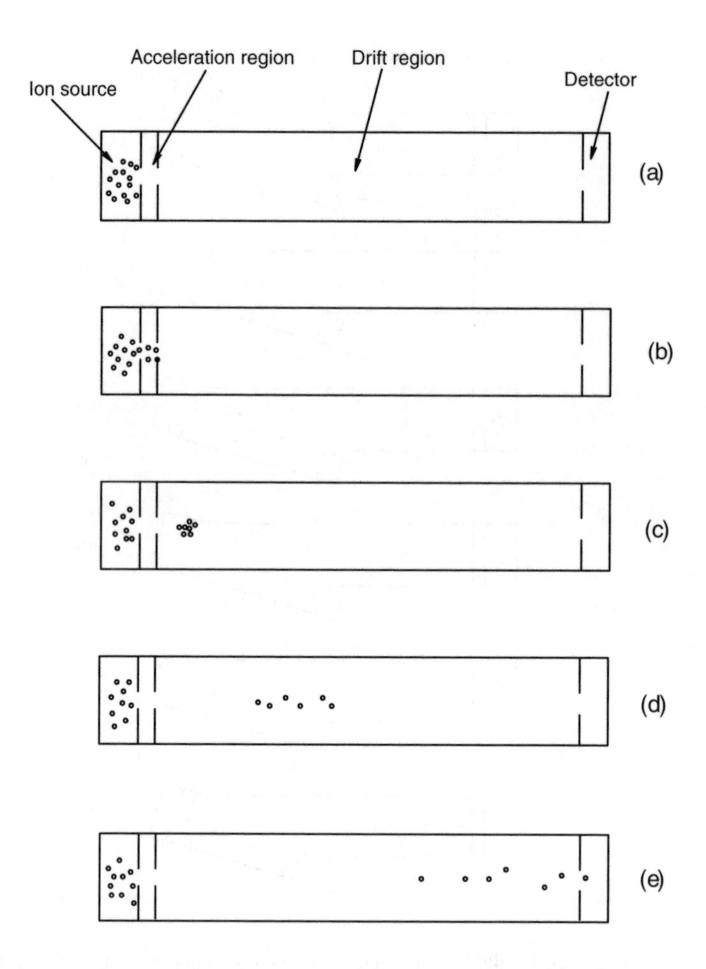

Figure 26.4
In this mode, ions are formed continuously in the ion source (a), but the electrostatic accelerating potential is applied in pulses (b). Thus, a sample of ions is drawn into the drift region (c) with more ions formed in the source. As shown in Figure 26.1, the ions separate according to m/z values (d) and arrive at the detector (e), the ions of largest m/z arriving last.

Chapter 27

Orthogonal Time-of-Flight (oa-TOF) Ion Optics

Introduction

Ions produced in an ion source can be separated into their m/z values by a variety of analyzers. The resultant set of m/z values, along with the numbers (abundances) of ions, forms the mass spectrum. The separation of ions into their individual m/z values has been effected by analyzers utilizing magnetic fields or RF (radio frequency) electric fields. For example, the mass analysis of ions by instruments using a magnetic field is well known, as are instruments having quadrupole RF electric fields (quadrupole, ion trap). Ions can also be dispersed in time, so their m/z values are measured according to their flight times in a time-of-flight (TOF) instrument. These individual pieces of equipment have their own characteristics and are commonly used in mass spectrometry. In addition, combinations of sectors have given rise to hybrid instruments. The earliest of these was the double-focusing mass spectrometer having an electric sector to focus ions according to their energies and then a magnetic sector to separate the individual m/z values. There is now a whole series of hybrid types, each with some advantage over nonhybrids. Ion collectors have seen a similar improvement in performance, and any of the above analyzers may be used with ion detectors based on single-electron multipliers or, in the case of magnetic sectors, on arrays of multipliers, or, in the case of ion cyclotron resonance (ICR), on electric-field frequencies.

Thus, there is a bewildering variety of instruments potentially available. However, except for very highly specialized purposes, most of the possible hybrids are not used in general mass spectrometry, and only a few types are in common use. One of these is the so-called orthogonal time-of-flight (oa-TOF) instrument. This chapter describes the orthogonal arrangement and discusses some of its advantages. Actual operation of hybrid orthogonal TOF instruments is discussed in Chapter 20.

The Physical Basis of oa-TOF

Consider a stream of ions emitted from an ion source as a beam. The ions are produced continuously and are first accelerated through an electric field of V_1 volts. If the number of charges on an ion is z, and e is the charge on the electron, then the energy acquired by an ion after acceleration through a field of V_1 volts is zeV_1. This energy must be equal to the kinetic energy ($mv^2/2$) gained by the ion, where m is the mass of the ion and v is its final velocity. Thus, $v = \sqrt{(2zeV_1/m)}$ and its momentum, $mv = \sqrt{(2zemV_1)}$. Therefore, the beam consists of a range of ions having momenta proportional to the charge, mass, and accelerating voltage. As the beam is produced continuously, there is no separation of ions in time (no temporal separation). This beam is shown as an arrow with velocity $\sqrt{(2zeV_1/m)}$ in Figure 27.1. A magnetic sector, for example, would separate the ions in space because the effect of the magnetic field is to bend the flight path in proportion to mass, charge, and accelerating voltage (see Chapter 24).

Now consider ions emitted from an ion source not as a beam but as a pulse, so all ions are accelerated through the potential V_1, applied as a pulse. Thus all ions start from the ion source at exactly the same time and then pass along a flight tube of length d. The times taken for the ions to reach a collector is given by t = d/v, where $v = \sqrt{(2zeV_1/m)}$, as above. This is the basis of the TOF instrument, in which ions of different m/z values are separated according to the flight times, t.

Let there be an electrode placed so its (pulsed) electric-field gradient (direction) is at right angles to the continuous beam. If the electric potential at this point is V_2 volts, then a pulse of ions will be given additional energy zeV_2, and a velocity $\sqrt{(2zeV_2/m)}$ in a direction at right angles to the main beam. The vectorial resultant velocity of the ions and direction of travel of the pulsed set are shown in Figure 27.1. Effectively, a section of the main ion beam is selected and pulsed away (Figure 27.2). Ions in this pulsed set all start at the same instant and can be timed by a TOF instrument. The flight times give m/z values and the numbers of ions give the abundances; the two are combined to give a mass spectrum.

Pulsed Main Beams of Ions

Although the above has considered only the use of a continuous main ion beam, which is then pulsed, it is not necessary for the initial beam to be continuous; it too can be pulsed. For example, laser desorption uses pulses of laser light to effect ionization, and the main ion beam already

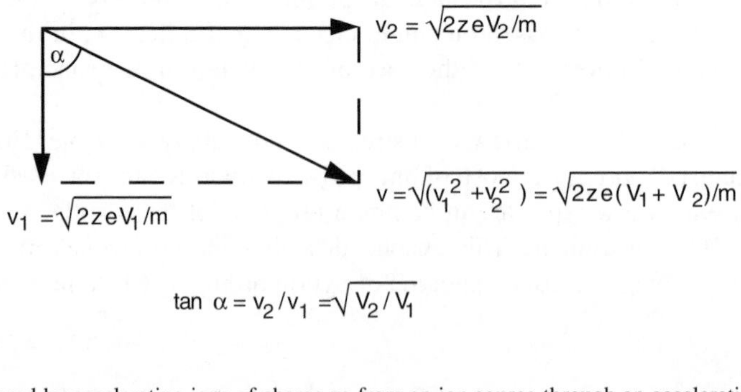

$v_2 = \sqrt{2zeV_2/m}$

α

$v = \sqrt{(v_1^2 + v_2^2)} = \sqrt{2ze(V_1 + V_2)/m}$

$v_1 = \sqrt{2zeV_1/m}$

$\tan\alpha = v_2/v_1 = \sqrt{V_2/V_1}$

Figure 27.1
An ion beam is produced by accelerating ions of charge ze from an ion source through an accelerating voltage of V_1 volts so that they have kinetic energy corresponding to zeV_1 and a velocity $v_1 = \sqrt{(2zeV_1/m)}$. In an orthogonal acceleration chamber, the ions are subjected to a pulsed electric field with accelerating voltage of V_2 volts, which gives them additional kinetic energy zeV_2 with a velocity $v_2 = \sqrt{(2zeV_2/m)}$ at right angles to their original direction. The ion beam has a resultant energy of $ze(V_1 + V_2)$ and a resultant velocity component $v = \sqrt{(2ze(V_2 + V_1)/m)}$ in a direction α degrees from the initial direction, where $\tan\alpha = \sqrt{(V_2/V_1)}$.

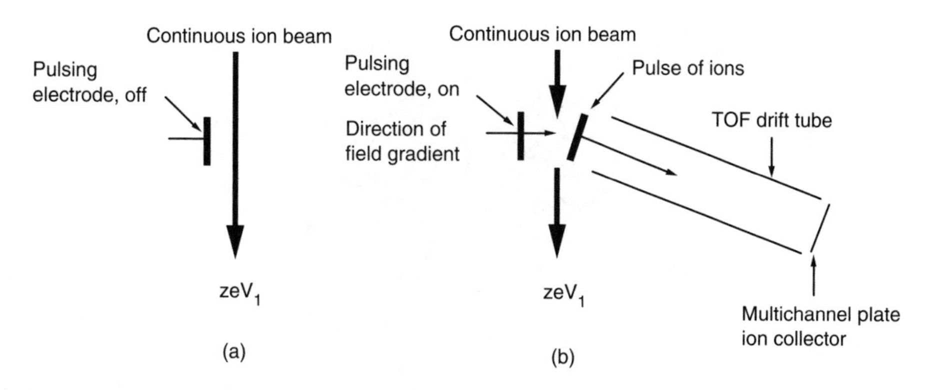

Figure 27.2
(a) The pulsing electrode is switched off, and a continuous ion beam of energy, zeV_1, passes by it. (b) The electrode has been pulsed for a few microseconds, with a field gradient at right angles to the main beam. This has caused a section of the ion beam to travel in the direction shown, the direction being determined by the magnitudes of the voltages V_1 and V_2 (Figure 27.1). The detached segment of the main beam enters the flight tube of a TOF instrument. The m/z values are determined from the times taken for the ions to reach the microchannel plate ion collector after initiation of the pulse.

consists of pulses of ions passing the orthogonal pulsing electrode. As shown below, these ion pulses are of no consequence because they can again be directed into a TOF flight tube just as though they had formed part of a continuous ion beam.

Rate of Application of the Pulsed-Field Gradient

Clearly, the pulsing electrode can be turned on and off at any frequency chosen, but there are some constraints on the frequencies actually used. At the fastest, there is little point in pulsing the electrode at such a rate that one lot of ions has not had time to travel the length of the flight tube before the next lot is on its way. With the physical dimensions of typical flight tubes and the magnitudes of accelerating voltages commonly used in mass spectrometers, an upper limit of about 30 kHz is found. The slowest rate of pulsing the electrode is almost anything! At 30 KHz, the TOF instrument can measure one mass spectrum every 33 μsec or, put another way, in one second the TOF instrument can accumulate and sum 30,000 spectra. Little wonder that the acquisition of a spectrum appears to be instantaneous on the human time scale.

Microchannel Plate Ion Collector

A fuller description of the microchannel plate is presented in Chapter 30. Briefly, ions traveling down the flight tube of a TOF instrument are separated in time. As each m/z collection of ions arrives at the collector, it may be spread over a small area of space (Figure 27.3). Therefore, so as not to lose ions, rather than have a single-point ion collector, the collector is composed of an array of miniature electron multipliers (microchannels), which are all connected to one electrified plate, so, no matter where an ion of any one m/z value hits the front of the array, its arrival is recorded. The microchannel plate collector could be crudely compared to a satellite TV dish receiver in that radio waves of the same frequency but spread over an area are all collected and recorded at the same time; of course, the multichannel plate records the arrival of ions not radio waves.

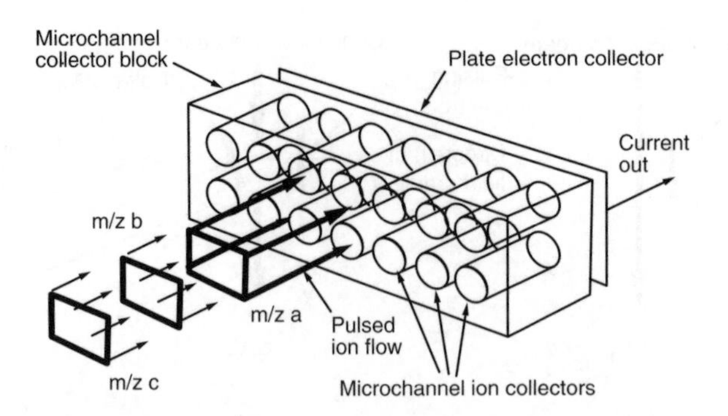

Figure 27.3
Diagram showing a flow of ions of m/z *a, b, c,* etc. traveling in bunches toward the front face of a microchannel array. After each ion strikes the inside of any one microchannel, a cascade of electrons is produced and moves toward the back end of the microchannel, where they are collected on a metal plate. This flow of electrons from the microchannel plate constitutes the current produced by the incoming ions (often called the ion current but actually a flow of electrons). The ions of m/z *a, b, c,* etc. are separated in time and reach the front of the microchannel collector array one set after another. The time at which the resulting electron current flows is proportional to $\sqrt{(m/z)}$.

Resolution by m/z Value

Since the microchannel plate collector records the arrival times of all ions, the resolution depends on the resolution of the TOF instrument and on the response time of the microchannel plate. A microchannel plate with a pore size of 10 μm or less has a very fast response time of less than 2 nsec. The TOF instrument with microchannel plate detector is capable of unit mass resolution beyond m/z 3000.

MS/MS Operation

Figure 27.4 is a diagrammatic representation of a typical MS/MS experiment in which a main ion beam selected for ions (precursor ions) of mass m and having kinetic energy zeV has been directed

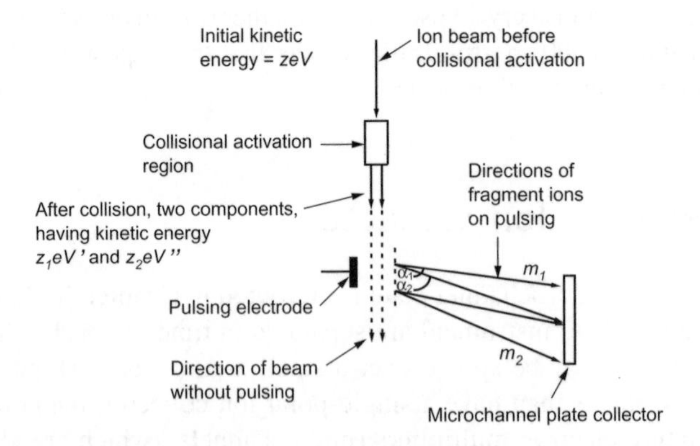

Figure 27.4
Diagram showing a mass-selected main ion beam (precursor ions) of kinetic energy zeV entering a collisional activation region and being fragmented to produce two fragment (product) ions having kinetic energies equal to z_1eV' and z_2eV''. If no electric field is pulsed onto the electrode, the ions continue straight on. If a pulsed electric field is applied, ions of energy z_1eV' will be deflected through an angle α_1, and the ions of energy z_2eV'' will be deflected through an angle α_2 and into a TOF analyzer tube. Both deflected beams are detected at the microchannel plate collector.

into a collision cell so as to cause fragmentation into two new species (products) of mass m_1 and m_2 with charges z_1 and z_2, respectively (z_1 or z_2 can be zero). The kinetic energies of the product ions can be written as z_1eV' and z_2eV'', respectively. Without setting a pulsed electric-field gradient orthogonal to the main beam, these fragment ions continue straight on. Application of a pulsed voltage to the electrode gives the ions a velocity component at right angles to their original direction. The vectorial resultant velocities form angles α_1 and α_2 to the original direction of the beam (Figure 27.1). Although now directed along different paths (Figure 27.4), both beams of fragment ions strike the wide microchannel plate.

The times of arrival at the plate are proportional to the $\sqrt{(m/z)}$ values of the masses involved. A mass spectrum of the ions resulting from collisional activation is produced. It should be recalled that, after a single collision, the momenta and kinetic energies of product ions are different from the momenta or kinetic energies of the precursor ions, but the velocities of the product ions are equal to each other and to that of the initial precursor ions. There is no change in velocities of ions upon fragmentation, only of momentum and kinetic energy. The TOF section can measure this mass spectrum in the normal fashion but, of course, it is a mass spectrum of the product ions resulting from fragmentation of precursor ions in the collision cell.

Advantages of Orthogonal TOF Arrangements

As indicated above, specific orthogonal TOF instruments are covered in greater detail in the section on hybrid instruments (Chapter 20). However, it may be noted that the orthogonal TOF instrument provides significant advantages for MS/MS operation in the examination of trace quantities of materials and as an adjunct to instruments in which the ion sources do not yield a steady ion current but, rather, pulsed sets of ions (laser desorption, radioactive desorption, sputtering). Even for continuous ion sources, vagaries of the ion current are smoothed through the accumulation of, say, 30,000 spectra at 33-μsec time intervals in a space of one second. The summed spectra are printed out as one mass spectrum. There is often a significant gain in signal-to-noise ratio for the orthogonal TOF system.

Conclusion

A snapshot of a beam of ions can be obtained by accelerating sections of the beam in pulses, away from the main stream. The accelerating voltage to do this is applied as electric-field pulses on a pusher electrode. The pulsed field gradient is at right angles (orthogonal) to the direction of the main ion beam. The pulsed ions are analyzed in a TOF tube and collected by a microchannel plate detector. The orthogonal TOF arrangement can be used in connection with a variety of other kinds of mass spectrometer to produce useful hybrid instruments. There are distinct advantages to these hybrids compared with the capabilities of the separate instruments alone.

Chapter 28

Point Ion Collectors (Detectors)

Introduction

Ion detectors can be separated into two classes: those that detect the arrival of all ions sequentially at one point (point ion collector) and those that detect the arrival of all ions simultaneously along a plane (array collector). This chapter discusses point collectors (detectors), while Chapter 29 focuses on array collectors (detectors).

All mass spectrometers analyze ions for their mass-to-charge ratios (m/z values) by separating the individual m/z values and then recording the numbers (abundance) of ions at each m/z value to give a mass spectrum. Quadrupoles allow ions of different m/z values to pass sequentially; e.g., ions at m/z 100, 101, 102 will pass one after the other through the quadrupole assembly so that first m/z 100 is passed, then 101, then 102 (or vice versa), and so on. Therefore, the ion collector (or detector) at the end of the quadrupole assembly needs only to cover one point or focus for a whole spectrum to be scanned over a period of time (Figure 28.1a). This type of point detector records ion arrivals in a time domain, not a spatial one.

A magnetic-sector instrument separates ions according to their m/z values but, unlike the quadrupole, by dispersing them in space (Figure 28.1b). Having dispersed the ions, their arrival can be recorded over a region of space (array detection) or, by increasing (or decreasing) the magnetic field, the ions can be brought sequentially to a focus (point detection). Either point or array detectors are used with magnetic instruments.

Other types of mass spectrometer may use point, array, or both types of collector. The time-of-flight (TOF) instrument uses a special multichannel plate collector; an ion trap can record ion arrivals either sequentially in time or all at once; a Fourier-transform ion cyclotron resonance (FTICR) instrument can record ion arrivals in either time or frequency domains which are inter-convertible (by the Fourier-transform technique).

Types of Point Ion Collector

Three main types point ion collectors are in use for quadrupole, magnetic-sector, and TOF instruments, and they are discussed here. The multichannel plate collector (or time-to-digital converter)

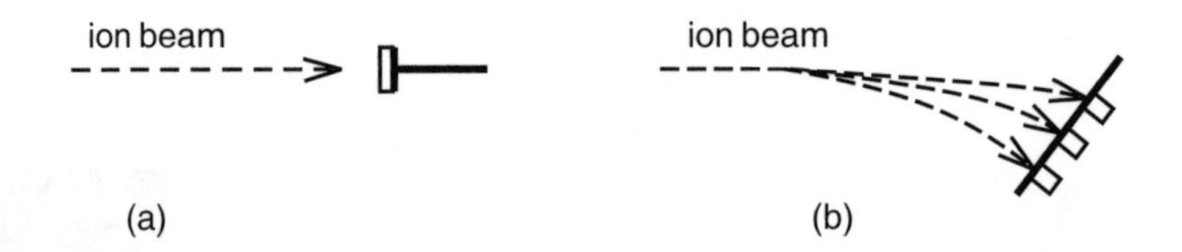

Figure 28.1
(a) Point ion collector. The ion beam is focused at a point, as shown, and ions of different m/z value are directed onto the collector sequentially in the time domain. (b) Array ion collector. The ion beam is dispersed, and all m/z values are recorded simultaneously in a space domain.

commonly used with TOF instruments is discussed in Chapter 31. FTICR requires separate study and is not covered in this book.

Faraday Cup

Ions carry positive or negative electric charges and, upon arrival at an earthed metal plate, they are neutralized either by accepting or donating electrons. The resulting flow of electrons constitutes a tiny electric current that can be amplified and used to drive a recording device. Such a simple plate collector is inefficient because the ions are also often traveling at high velocity. Upon striking a metal surface, fast-moving ions cause a shower of electrons to be emitted (secondary electrons). By using a cup collector rather than a simple plate, the secondary electrons can be collected (Figure 28.2). Thus one positive ion arriving at a (Faraday) cup collector needs one electron for neutralization but causes several electrons to be emitted; this provides a "gearing" or amplification — several electrons for each ion. The Faraday cup detector is simple and robust and is used in when high sensitivity is not required.

Electron Multiplier

As mentioned previously, a particle such as an ion traveling at high speed causes a number of secondary electrons to be ejected when it strikes a metal surface. This principle is utilized in the electron multiplier (Figure 28.3).

Ions are directed onto the first plate (dynode) of an electron multiplier. The ejected electrons are accelerated through an electric potential so they strike a second dynode. Suppose each ion collision causes ten electrons to be ejected, and, at the second dynode each of these electrons causes ten more to be ejected toward a third dynode. In such a situation, arrival of just one ion causes $10 \times 10 = 10^2 = 100$ electrons to be ejected from the second dynode, an amplification of 100. Commercial electron multipliers routinely have 10, 11, or 12 dynodes, so amplifications of 10^6 are

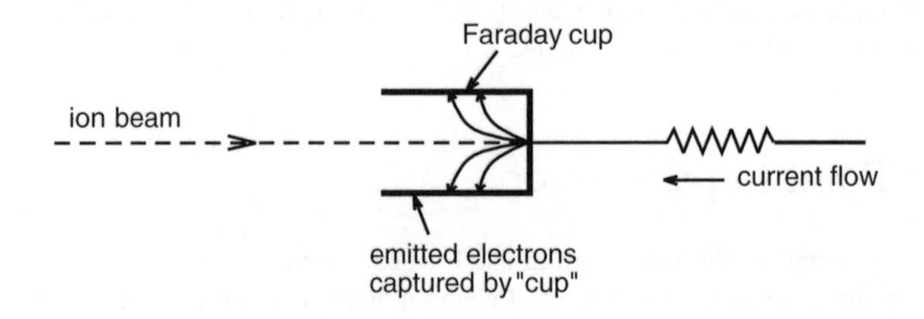

Figure 28.2
Ions traveling at high speed strike the inside of the metal (Faraday) cup and cause secondary electrons to be ejected. This production of electrons constitutes a temporary flow of electric current as the electrons are recaptured.

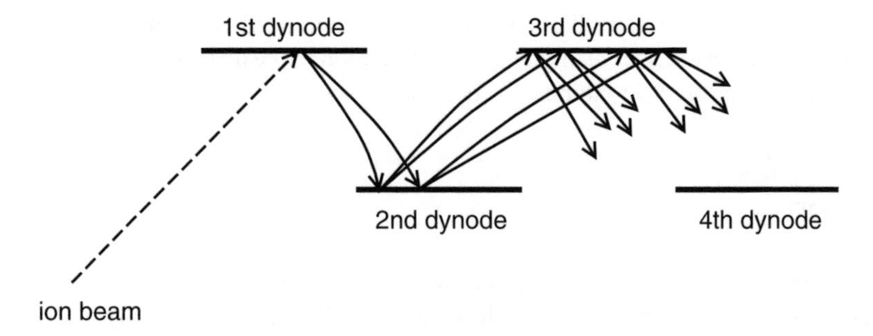

Figure 28.3
Each ion from an incident ion beam causes two electrons to be emitted from the first dynode. These electrons are accelerated to the second dynode, where each causes two more electrons (now four in all) to be ejected. These in turn are accelerated to a third dynode and so on, eventually reaching, say, a tenth dynode, by which time the initial two electrons have become a shower of 2^9 electrons.

readily available by using more dynodes. The final flow of electrons provides an electric current that can be further increased by ordinary electronic amplification. These multipliers provide highly sensitive ion detection systems. They work under a vacuum and can be connected directly to the mass spectrometer. Only electrical potentials on the dynodes are needed to make them work, so they are very robust.

Eventually, multipliers become less sensitive and even fail because of surface contamination caused by the imperfect vacuum in the mass spectrometer and the impact of ions on the surfaces of the dynodes.

Scintillator

A scintillator, sometimes known as the Daly detector, is an ion collector that is especially useful for studies on metastable ions. The principle of operation is illustrated in Figure 28.4. As with the first dynode of an electron multiplier, the arrival of a fast ion causes electrons to be emitted, and they are accelerated toward a second dynode. In this case, the dynode consists of a substance (a scintillator) that emits photons (light). The emitted light is detected by a commercial photon

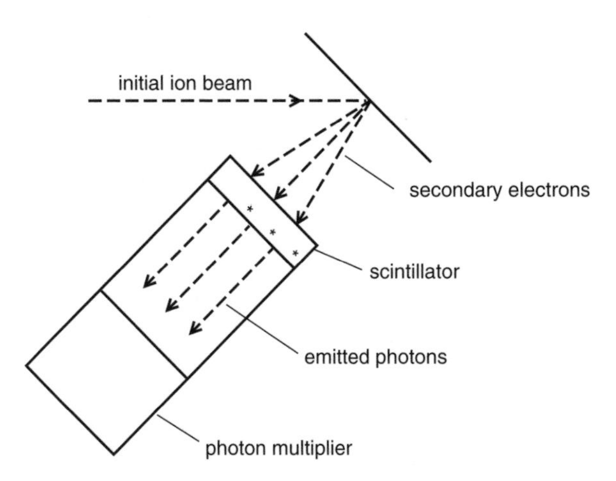

Figure 28.4
An incident ion beam causes secondary electrons to be emitted which are accelerated onto a scintillator (compare this with the operation of a TV screen). The photons that are emitted (like the light from a TV screen) are detected not by eye but with a highly sensitive photon detector (photon multiplier), which converts the photon energy into an electric current.

multiplier placed behind the scintillator screen and is converted into an electric current. Since photon multipliers are very sensitive, high gain amplification of the arrival of a single ion is achieved. These detectors are important in studies on metastable ions.

Conclusion

An ion beam causes secondary electrons to be ejected from a metal surface. These secondaries can be measured as an electric current directly through a Faraday cup or indirectly after amplification, as with an electron multiplier or a scintillation device. These ion collectors are located at a fixed point in a mass spectrometer, and all ions are focused on that point — hence the name, point ion collector. In all cases, the resultant flow of an electric current is used to drive some form of recorder or is passed to an information storage device (data system).

Chapter 29

Array Collectors (Detectors)

Introduction

Ion detectors can be separated into two classes: those that detect the arrival of all ions sequentially at one point (point ion collector) and those that detect the arrival of all ions simultaneously along a plane (array collector). This chapter discusses array collectors (detectors), while Chapter 28 focuses on point ion collectors (detectors).

All mass spectrometers analyze ions for their mass-to-charge ratios (m/z values) and simultaneously for the abundances of ions at any given m/z value. By separating the ions according to m/z and measuring the ion abundances, a mass spectrum is obtained.

Quadrupole mass spectrometers (mass filters) allow ions at each m/z value to pass through sequentially; for example, ions at m/z 100, 101, 102 will pass one after the other through the quadrupole assembly so that first m/z 100 is transmitted, then m/z 101, then m/z 102 (or vice versa), and so on. Therefore, the ion collector (or detector) at the end of the quadrupole unit needs to cover only one point or focus in space (Figure 29.1a), and a complete mass spectrum is recorded over a period of time. The ions arrive at the collector sequentially, and ions are detected in a time domain, not in a spatial domain.

A magnetic-sector instrument separates ions according to their m/z values, but, unlike the quadrupole, it can also separate the ions by dispersing them in space (Figure 29.1b). The arrival of the dispersed ions can be recorded simultaneously in space (array or photographic plate focal-plane detection, Figure 29.1b), or, by manipulating the strength of the magnetic field, the ions can be brought sequentially to a focus at a point ion collector (Figure 29.1c). Other types of mass spectrometer can use point, array, or both types of ion detection. A time-of-flight (TOF) mass spectrometer collects ions sequentially and uses an array that is also a sequential detector (time-to-digital converter, TDC).

The original method for simultaneously recording a range of ions was to use a photographic plate that was placed in the focal plane (Figure 29.1b) such that all ions struck the photographic plate simultaneously at different positions along the plate. This method of detection is now rarely used because of the inconvenience of having to develop a photographic plate. Essentially, the array collector has taken its place. Ion-trap mass spectrometers can detect ions sequentially or simultaneously and, in some cases, may not use a formal ion collector at all; the ions can be detected by their different electric field frequencies in flight, according to m/z value.

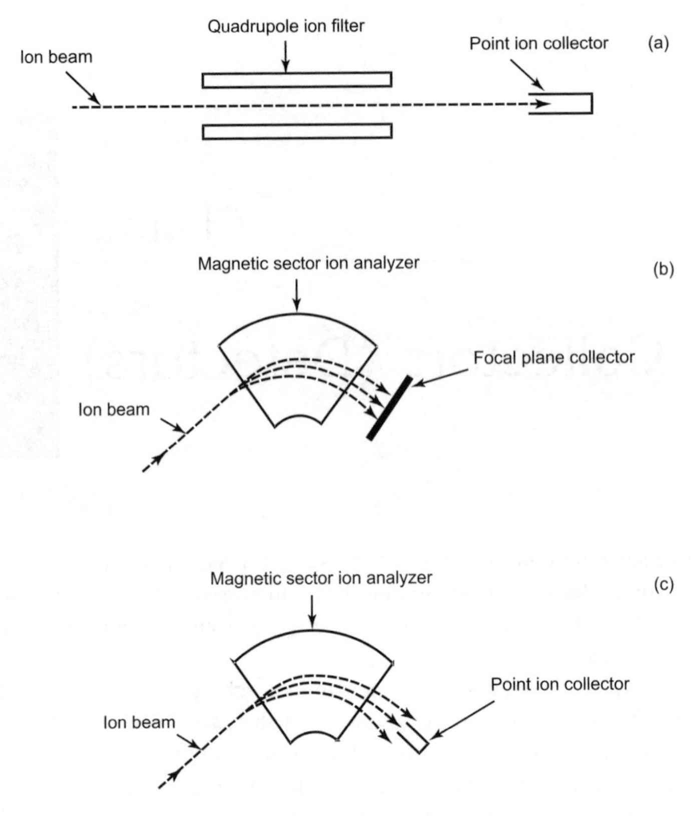

Figure 29.1
(a) Ions of different m/z values traverse a quadrupole mass filter one after another, and only a point ion collector is needed. (b) The ion beam is shown dispersed in space, each m/z value being brought to a focus at a point that lies in a focal plane. A photographic plate — or an array detector lying along the focal plane — can detect all ions at the same time. (c) The same dispersed ion beam as in (b) is brought to a focus at one point by changing the strength of the magnetic field, and in this case a point ion collector would be used.

Array Detection

The array detector (collector) consists of a number of ion-collection elements arranged in a line; each element of the array is an electron multiplier. Another type of array detector, the time-to-digital converter, is discussed in Chapter 31.

An Element of the Array

Where space is not a problem, a linear electron multiplier having separate dynodes to collect and amplify the electron current created each time an ion enters its open end can be used. (See Chapter 28 for details on electron multipliers.) For array detection, the individual electron multipliers must be very small, so they can be packed side by side into as small a space as possible. For this reason, the design of an element of an array is significantly different from that of a standard electron multiplier used for point ion collection, even though its method of working is similar. Figure 29.2a shows an electron multiplier (also known as a Channeltron®) that works without using separate dynodes. It can be used to replace a dynode-type multiplier for point ion collection but, because

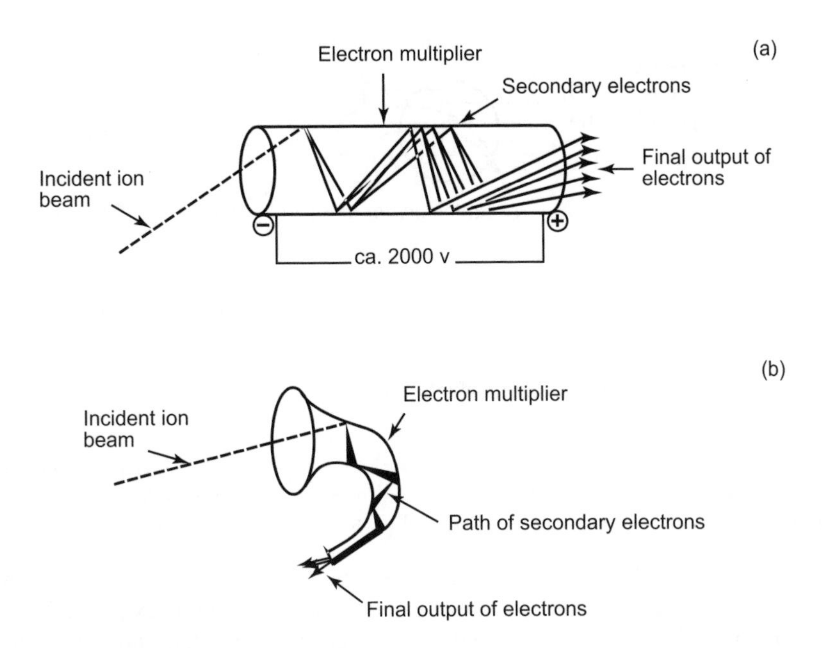

Figure 29.2
Diagram (a) represents a straight electron multiplier with a semiconducting inner surface. There is an electrical potential of about 1000 to 2000 V between the ends of the tube, which is shown set up for detection of positive ions. Incident ions striking the inner surface cause secondary electrons to be emitted, and these are accelerated by the high potential between the ends of the multiplier tube. In turn, the electrons strike the inner wall, and each causes more electrons to be emitted. Thus, as the electrons travel along the tube, striking the inner walls at intervals, more and more electrons are produced, so, for one incident ion, there will be a shower of electrons leaving the end of the multiplier. Diagram (b) shows a different shape of electron multiplier tube, designed to stop backscatter of the secondary electrons, which can cause excessive electrical noise. Otherwise, this design works on the same principle as the straight multiplier. This type of multiplier with no separate dynodes can be made in a wide variety of shapes and sizes, and, importantly for an array collector, can be made small.

the Channeltron® can be made very small, it is suited to array ion collection, where as many multipliers as possible, placed side by side, must be fitted into a small space. The Channeltron® can be made in a variety of shapes, one of which is shown in Figure 29.2b.

For operation of a Channeltron®, a large electrical potential is maintained between the open and closed ends of the element. An ion traveling at high speed enters the mouth of the Channeltron® and strikes the internal wall, causing a shower of electrons to be given off. These electrons are accelerated through the electric potential but travel more or less in straight lines, so they soon strike the internal wall again (Figure 29.2a, b). Each electron in the shower causes a second shower to be emitted. In this way, the electrons first produced travel down the Channeltron®, rebounding from the walls and increasing in number. Eventually, for the arrival of one ion, hundreds or thousands of electrons arrive at the closed end of the array element, where they constitute an electric current that can be measured and amplified. Often, this sort of electron multiplier is made in a curved shape (e.g., that illustrated in Figure 29.2b) to reduce backscattering of ions or electrons that would cause excessive electrical noise. Various operating modes can be used, such as analog, digital, light-coupled, etc., but there is neither the space nor the need to discuss them here.

Separation of Array Elements (Ion Mass Range)

Consider two array elements as illustrated in Figure 29.3, and suppose an ion beam has been dispersed to give ions of m/z values 100 and 101. If the dispersion is correct for the array size, the

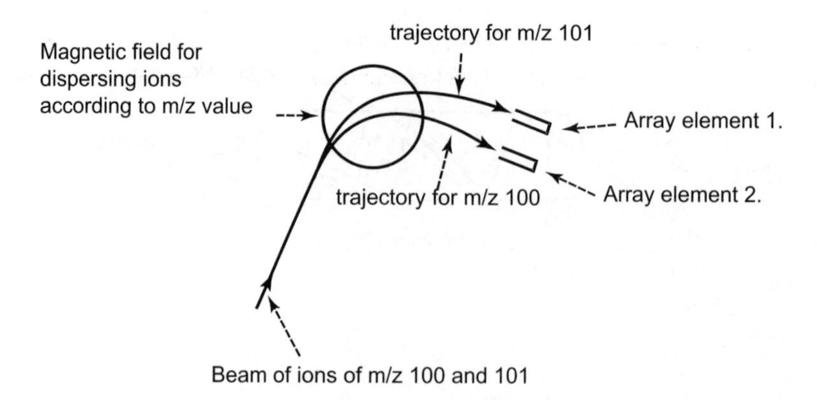

Figure 29.3
An ion beam containing just two types of ion of m/z values 100 and 101 dispersed in space on passing through a magnetic field. After dispersal, ions of individual m/z value 100 or 101 are focused at points close to the entries of two elements of an array collector. Each element of the array is a point ion collector.

ion of m/z 100 will enter the first element, and the ion of m/z 101 value will enter the second. Thus, at this level of dispersion, unit m/z values can be separated. Simple extrapolation to, say, ten ions of different m/z values (Figure 29.4) and ten array elements in a line shows that all of the ion m/z values and abundances can be measured simultaneously to give an instantaneous spectrum with integer m/z values. Further extrapolation indicates that more m/z values can be measured if there are more array elements. However, these additional measurements do not come without cost, and fitting a very large number of elements into a compact array becomes increasingly difficult. Therefore, a limited number of array elements is used, say 100, which means that sections of a mass spectrum can be measured instantaneously, but, if it is necessary to measure the whole spectrum, then it must be measured in sections at a time. Frequently, not all of a mass spectrum is needed, and it is unnecessary to measure more than one region of the spectrum. For example, maybe only the molecular ion region needs to be covered.

The array need not consist simply of one line of elements; there can be several such lines, one above the other. The single line is discussed here for the sake of simplicity.

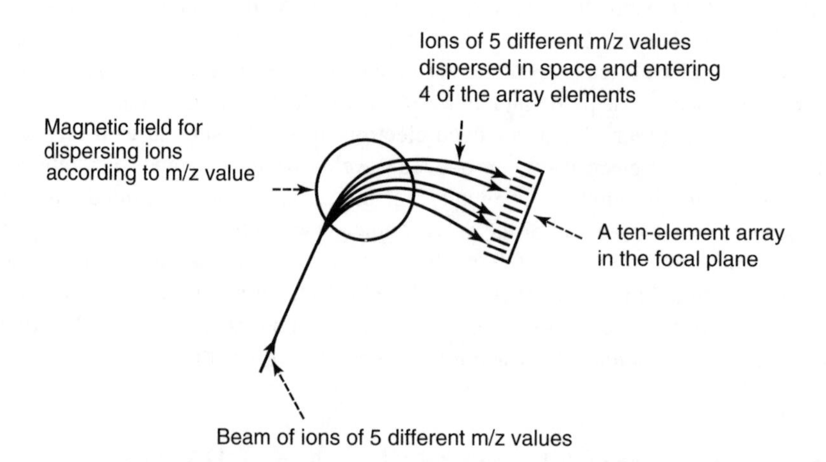

Figure 29.4
Ions of five different m/z values shown entering five elements of a ten-element array. All five ions would be recorded at the same time, with electrical outputs from the elements into which ions had entered and no output from those elements into which no ions had entered.

Now consider ions of m/z 100.1 and 100.2. If these are dispersed so as to enter the two elements of Figure 29.3, then the resolution improves to 0.1 mass units, but the number of m/z values that can be measured at the same time falls. In the ten-element array of Figure 29.4, to achieve a mass resolution of 0.1, only the mass range of m/z 100 to 101 would be covered, instead of the range 100 to 110 for unit mass resolution. Because the number of elements in the array does not change, increasing the resolution means that less of the spectrum can be covered. Conversely, the lower the resolution, the greater is the mass range that can be scanned. Thus, arrays can be used to cover wide regions of a mass spectrum at low resolution but only very small regions at high resolution.

Dynamic Range (Ion Abundance)

An electron multiplier used with point ion detection should ideally have a wide dynamic range of response to the number of ions arriving for collection; it should be almost equally efficient at detecting a few ions as detecting tens of thousands of ions (dynamic range), with a fair linear response between numbers of ions arriving and the final output from the multiplier. Therefore the sizes of peaks observed in a mass spectrum, no matter whether small or large, would relate reasonably well to the actual number of ions they are meant to represent (100 ions should be represented by a peak 100 times bigger than that for 1 ion).

The elements of an array detector are small and more easily saturated by the arrival of large numbers of ions than the much larger single-electron multiplier. Therefore the dynamic range of the array is generally worse than that of a single-point ion collector. For most practical situations this difference is not too important, but it should be borne in mind. It might also be worth noting that as the mass of an ion increases, the electrical response of any electron multiplier falls, and at high mass (say hundreds of mass units), the inherent sensitivity is very much less than for low mass. This effect is due to the fact that ions of large m/z values are usually traveling much slower than ions of small m/z values and therefore produce fewer electrons upon initial impact with the multiplier.

Uses of Array Collectors

The major advantage of array detectors over point ion detectors lies in their ability to measure a range of m/z values and the corresponding ion abundances all at one time, rather than sequentially. For example, suppose it takes 10 msec to measure one m/z value and the associated number of ions (abundance). To measure 100 such ions sequentially with a point ion detector would necessitate 1000 msec (1 sec); for the array detector, the time is still 10 msec because all ions arrive at the same time. Therefore, when it is important to be able to measure a range of ion m/z values in a short space of time, the array detector is advantageous.

There are two common occasions when rapid measurement is preferable. The first is with ionization sources using laser desorption or radionuclides. A pulse of ions is produced in a very short interval of time, often of the order of a few nanoseconds. If the mass spectrometer takes 1 sec to attempt to scan the range of ions produced, then clearly there will be no ions left by the time the scan has completed more than a few nanoseconds (ion traps excluded). If a point ion detector were to be used for this type of pulsed ionization, then after the beginning of the scan no more ions would reach the collector because there would not be any left! The array collector overcomes this difficulty by detecting the ions produced all at the same instant.

A second use of arrays arises in the detection of traces of material introduced into a mass spectrometer. For such very small quantities, it may well be that the tiny amount of substance will have disappeared by the time a scan has been has been completed by a mass spectrometer equipped with a point ion collector. An array collector overcomes this problem. Often, the problem of detecting trace amounts of a substance using a point ion collector is overcome by measuring not the whole mass spectrum but only one characteristic m/z value (single ion monitoring or single ion detection). However, unlike array detection, this single-ion detection method does not provide the whole spectrum, and an identification based on only one m/z value may well be open to misinterpretation and error.

Conclusion

An array ion collector (detector) consists of a large number of miniature electron multiplier elements arranged side by side along a plane. Point ion collectors gather and detect ions sequentially (all ions are focused at one point one after another), but array collectors gather and detect all ions simultaneously (all ions are focused onto the array elements at the same time). Array detectors are particularly useful for situations in which ionization occurs within a very short space of time, as with some ionization sources, or in which only trace quantities of a substance are available. For these very short time scales, only the array collector can measure a whole spectrum or part of a spectrum satisfactorily in the time available.

Chapter 30

Comparison of Multipoint Collectors (Detectors) of Ions: Arrays and Microchannel Plates

Introduction

All mass spectrometers analyze ions for their mass-to-charge ratios (m/z values) and simultaneously for the abundances of ions at any one m/z value. Once separated by m/z value, the ions must be detected (collected) and their numbers (abundances) measured for each m/z value. The resulting chart of m/z value versus abundance constitutes a mass spectrum.

In modern mass spectrometry, ion collectors (detectors) are generally based on the electron multiplier and can be separated into two classes: those that detect the arrival of all ions sequentially at a point (a single-point ion collector) and those that detect the arrival of all ions simultaneously (an array or multipoint collector). This chapter compares the uses of single- and multipoint ion collectors. For more detailed discussions of their construction and operation, see Chapter 28, "Point Ion Collectors (Detectors)," and Chapter 29, "Array Collectors (Detectors)." In some forms of mass spectrometry, other methods of ion detection can be used, as with ion cyclotron instruments, but these are not considered here.

Quadrupole mass spectrometers (mass filters) allow ions at each m/z value to pass through the analyzer sequentially. For example, ions at m/z 100, 101, and 102 are allowed to pass one after the other through the quadrupole assembly so that first m/z 100 is transmitted, then m/z 101, then m/z 102, and so on. Therefore, the ion collector at the end of the quadrupole unit needs to cover only one point or focus in space and can be placed immediately behind the analyzer (Figure 30.1). A complete mass spectrum is recorded over a period of time (temporally), which is set by the voltages on the quadrupole analyzer. In this mode of operation, the ions are said to be scanned sequentially. The resolution of m/z values is dependent solely on the analyzer and not on the detector. The single-point collector is discussed in detail in Chapter 28.

Magnetic-sector instruments separate ions according to their m/z values by dispersing them in space (Figure 30.2). By changing the strength of the magnetic field, the ions can also be focused one after another according to m/z values at one point in space and collected over a period of time (Figure 30.3); this mode of operation needs only a single-point detector, as for the quadrupole

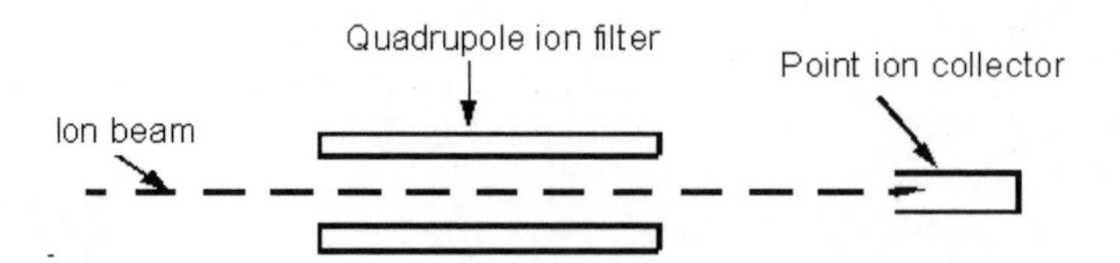

Figure 30.1
Ions of different m/z values pass sequentially in time through the quadrupole mass filter to reach an in-line, single-point ion collector.

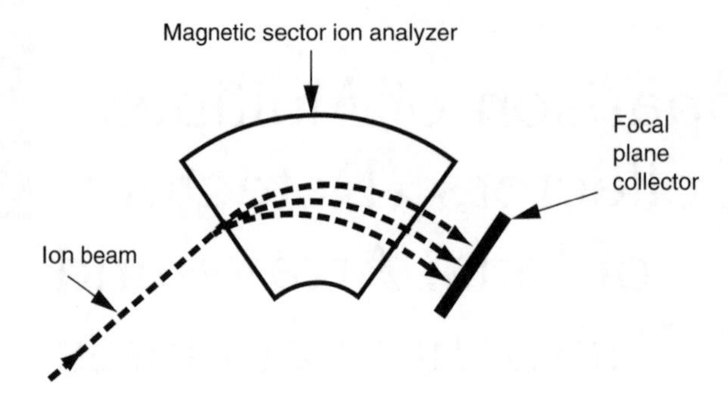

Figure 30.2
Ions of differing m/z values are dispersed by a magnetic sector and reach foci, which are distributed along a focal plane.

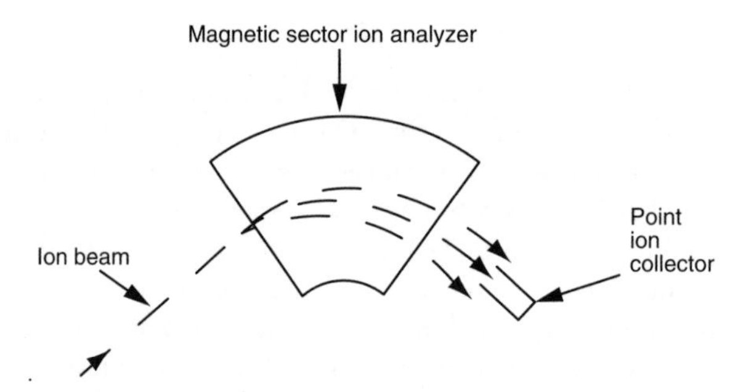

Figure 30.3
By adjusting the magnetic field, the dispersed ion beam in Figure 30.2 can be moved up or down so that ions of specific m/z values can be focused at a point ion collector.

instrument. For this last collection of ions in a time domain, resolution depends only on the analyzer. Alternatively, having dispersed the ions in space (resolved them according to m/z value), all can be detected at the same time over a section of space by using an array of single-point detectors (the focal plane collector in Figure 30.2).

The array system is discussed in Chapter 29. With array detection, resolution of m/z values depends both on the analyzer and the collector. Historically, the method for recording ions dispersed in space was to use a photographic plate, which was placed in the focal plane such that all ions struck the photographic plate simultaneously but at different positions along the plate, depending on m/z value. This method of detection is now rarely used because of the inconvenience of having to develop a photographic plate.

Other types of mass spectrometer can use point, array, or both types of ion detection. Ion trap mass spectrometers can detect ions sequentially or simultaneously and in some cases, as with ion cyclotron resonance (ICR), may not use a formal electron multiplier type of ion collector at all; the ions can be detected by their different electric field frequencies in flight.

Another form of array is called a microchannel plate detector. A time-of-flight (TOF) mass spectrometer collects ions sequentially in time and can use a point detector, but increasingly, the TOF instrument uses a microchannel plate, most particularly in an orthogonal TOF mode. Because the arrays and microchannel plates are both essentially arrays or assemblies of small electron multipliers, there may be confusion over their roles. This chapter illustrates the differences between the two arrays.

Arrays and Microchannel Plates

In both arrays and microchannel plates, the collector consists of a number of single-point ion detection elements, each of which is a very small electron multiplier. Each element is much smaller than the normal single-electron multiplier, and lots of them can be arranged close together as a planar array to cover a large area of space (Figure 30.4). The actual construction of the arrays is different as will be described below. The front face of the array contains the entrances or openings for each small detector, into which ions are deposited or collected. A fast-moving ion striking the entrance to an element starts a cascade of electrons that increases in size as the electrons bounce off the walls of the element during their passage to its back end. The back end of each element is either closed or open. In an array of closed elements, the end of each can be monitored for the arrival of cascading electrons, signaling the arrival of an ion or ions. In an array with open-ended elements, a cascade of electrons from any element is collected onto the same backing plate. Ions arriving anywhere in space over the face of the array are detected; viz., all of the elements are monitored as one.

There is potential confusion in the use of the word *array* in mass spectrometry. Historically, array has been used to describe an assemblage of small single-point ion detectors (elements), each of which acts as a separate ion current generator. Thus, arrival of ions in one of the array elements generates an ion current specifically from that element. An ion of any given m/z value is collected by one of the elements of the array. An ion of different m/z value is collected by another element. Ions of different m/z value are dispersed in space over the face of the array, and the ions are detected by m/z value at different elements (Figure 30.4).

An assemblage (array) of single-point electron multipliers in a microchannel plate is designed to detect all ions of any single m/z value as they arrive separated in time. Thus, it is not necessary for each element of the array to be monitored individually for the arrival of ions. Instead, all of

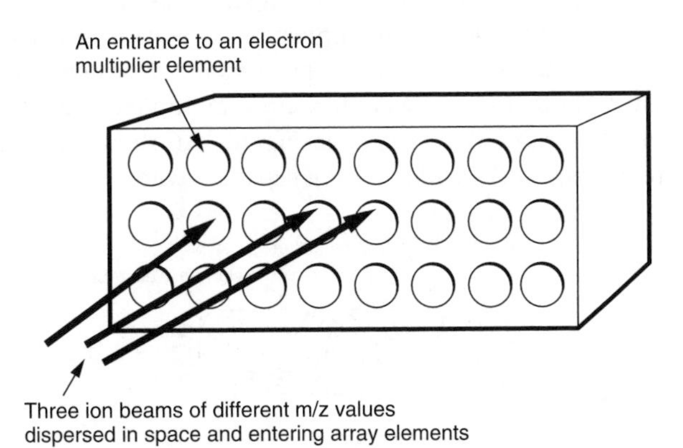

Figure 30.4
Idealized face view of a set of small electron multipliers arranged over a plane. Some typical individual multipliers are shown in later figures.

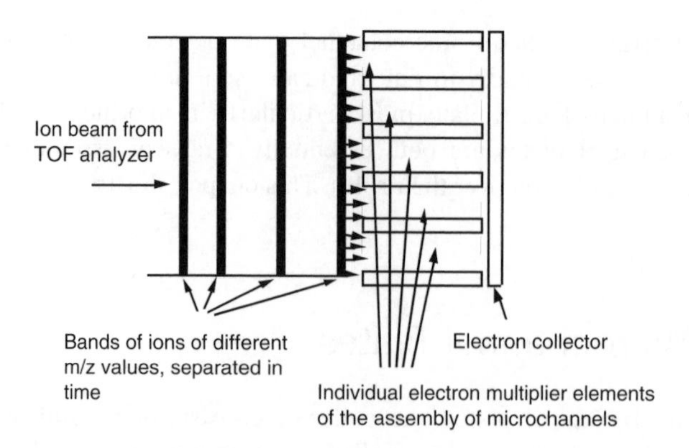

Figure 30.5
Bands of ions of different m/z values and separated in time in a broad ion beam traveling from left to right toward the front face of a microchannel assembly. The ions produce showers of electrons, and these are detected at the collector plate, which joins all the elements as one assemblage.

the back ends of the detection elements are connected together electronically. Thus, if ions of, for example, m/z 100 arrive at some time, t, but are spread spatially over the face of the array, they are all detected simultaneously, even though some may be collected by one element, others by another element, and so on. There is no discrimination in m/z value. That separation of ions by m/z must be effected in time by the analyzer, usually a TOF instrument (Figure 30.5).

To avoid confusion, the word *array* is now used to describe an assemblage of small single-point detectors that remain as individual ion monitoring elements, and the term *microchannel plate* is used to describe an assemblage of small single-point detectors, all of which are connected so as to act as a single large monitoring element. Because ion arrivals in the microchannel plate are recorded as instantaneous events (the ion current is digitized), this type of detector is called a time-to-digital converter (TDC).

The Elements of Array and Microchannel Plates

Where space is not a problem, it is possible to use a linear electron multiplier having separate dynodes to collect and amplify the electron current created each time an ion enters its open end. For array detection, the individual electron multipliers must be very small so they can be packed side by side into as small a space as possible. For this reason, the design of an element of an array is significantly different from that of a standard electron multiplier used for point ion collection, even though its method of working is very similar. Figure 30.6 shows an electron multiplier (also known as a Channeltron®) that works without using separate dynodes. Because each Channeltron®

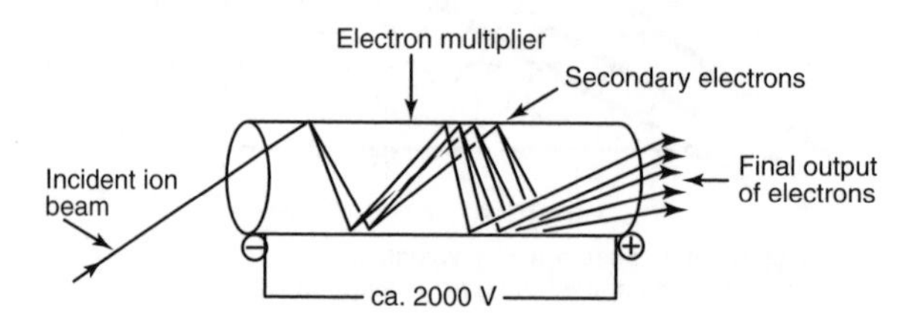

Figure 30.6
A typical single microchannel electron multiplier. Note how the primary ion beam causes a shower of electrons to form. The shower is accelerated toward the other end of the microchannel, causing the formation of more and more secondary electrons.

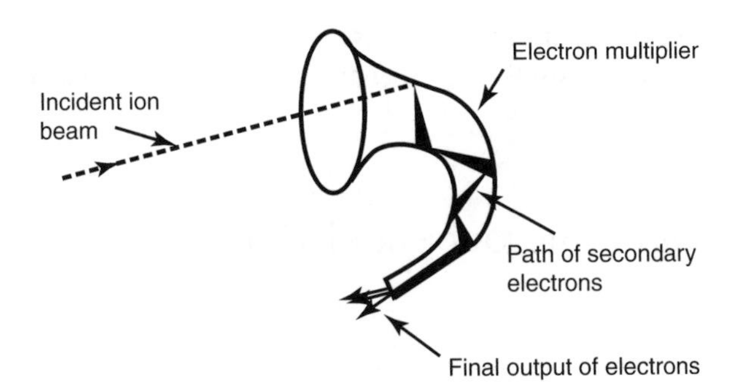

Figure 30.7
A different form of miniature electron multiplier. The curved shape is used to reduce backscattering of the electrons. The final output of electrons flows along a wire to an amplifier.

can be made very small, it is suited to array assemblages, in which as many multipliers as possible can be fitted side by side into a relatively small space. Each Channeltron® element can be made in a variety of shapes, one of which is shown in Figure 30.7.

For a microchannel plate, the back end of each element is left open, as shown in Figures 30.5 or 30.6, and forms a microchannel. Any electrons emerging from any element are all detected by the one collector plate.

Array Elements (Ion Mass Range)

Consider two array elements of the ones illustrated in Figure 30.4, and suppose an ion beam has been dispersed to give ions of m/z values 100 and 101. If the dispersion is correct for the array size, the ion of m/z 100 will enter one element and, at the same time, the ion of m/z 101 will enter a second adjacent element. Thus, at this level of dispersion, unit m/z values can be separated. Simple extrapolation to, say, five ions of different m/z values or ten array elements in a line shows that several ion m/z values and abundances can be measured simultaneously to give an instantaneous spectrum. Further extrapolation indicates that more m/z values can be measured if there are more array elements. However, these extra measurements do not come without cost, and fitting a very large number of elements into a compact array becomes increasingly difficult. Therefore, a limited number of array elements is used, say 100, which means that sections of a mass spectrum can be measured instantaneously, but, if it is required to measure a spectrum spread over several hundred or thousand mass units, then it must be measured in sections at a time. Frequently, not all of a mass spectrum is needed, and it may be unnecessary to measure more than one region of the spectrum, e.g., maybe only the molecular ion region needs to be covered.

Microchannel Elements (Ion Mass Range)

Consider again two detection elements, and suppose an ion beam has been dispersed in time such that ions of m/z 100 arrive at each of several elements (Figure 30.5). In this TOF mode, the next ion of m/z 101 has not yet arrived, and the ion of m/z 99 has arrived previously. Although the m/z ions are dispersed in time over a region of space and strike different elements of the detector, they are collected and monitored simultaneously because all of the microchannels are electronically connected. The operation of the microchannel plate is much easier than that of the array because all the elements are monitored as one at the plate, while each element must be monitored separately in the array. The microchannel plate detector is tremendously useful for those cases in which ions at each m/z value are separated in time by a mass analyzer but can be delivered to the collector spread out in space. No ions are lost through scanning, and the microchannel plate serves as a very sensitive detector.

Uses of Array and Microchannel Collectors

Array Collectors

The major advantage of array detectors over point ion detectors lies in their ability to measure both a range of m/z values and the corresponding ion abundances all at one time, rather than sequentially. For example, suppose it takes 10 msec to measure one m/z value and the associated number of ions (abundance). To measure 100 such ions of different m/z values with a point ion detector would require 1000 msec (1 sec). For the array detector, the time is still only 10 msec because all the ions of different m/z values arrive at the collector at the same time. Therefore, when it is important to be able to measure a range of ion m/z values in a short space of time, the array detector is advantageous.

There are two common occasions when instantaneous measurement of a range of m/z values is preferable. First, with ionization sources such as those using laser desorption or radionuclides, a pulse of ions is produced in a very short interval of time, often of the order of a few nanoseconds. If the mass spectrometer takes 1 sec to attempt to scan the range of ions produced, then clearly there will be no ions left by the time the scan has completed more than a few microseconds (ion traps excluded). The array collector overcomes this difficulty by detecting the ions produced all at the same instant.

A second use of arrays arises in the detection of trace components of material introduced into a mass spectrometer. For such very small quantities, it may well be that, by the time a scan has been carried out by a mass spectrometer with a point ion collector, the tiny amount of substance may have disappeared before the scan has been completed. An array collector overcomes this problem. Often, the problem of detecting trace amounts of a substance using a point ion collector is overcome by measuring not the whole mass spectrum but only one characteristic m/z value (single ion monitoring or single ion detection). However, unlike array detection, this single-ion detection method does not provide the whole spectrum, and an identification based on only one m/z value may well be open to misinterpretation and error.

Microchannel Plate Collectors

A major advantage of microchannel plate detectors over point ion detectors lies in their ability to measure the abundance of ions of a single m/z value, which are spread over a region of space. When used with a TOF analyzer, another major advantage appears. Typically in a TOF instrument, ions of adjacent m/z values in the ion beam are separated by about 20 to 30 nsec. Over a total mass range of 0–3000 mass units, all of the ions arrive at the collector within a period of about 30 μsec. Therefore, the microchannel plate acquires a mass spectrum in about 30 μsec, which on the human time scale appears to be instantaneous. Thus, like the array detector on a magnetic scanning instrument, the microchannel plate on a TOF instrument is capable of generating an almost instantaneous spectrum. Additionally, a full range of ions from, say, 0 to 3000 mass units can be pulsed into a TOF analyzer at the rate of about 30,000 times per second. The microchannel plate acquires all of these spectra, which can be summed. Thus, in 1 sec, a TOF/microchannel plate combination can sum about 30,000 spectra. This speed is a great advantage for examining spectra that contain spurious peaks of occasional electronic noise or that have been generated in a small period of time, as with laser-assisted ionization. Other uses of the microchannel plate are described in Chapter 20, "Hybrid Orthogonal Time-of-Flight Instruments."

Conclusion

A multipoint ion collector (also called the detector) consists of a large number of miniature electron multiplier elements assembled, or constructed, side by side over a plane. A multipoint collector can be an array, which detects a dispersed beam of ions simultaneously over a range of m/z values and is frequently used with a sector-type mass spectrometer. Alternatively, a microchannel plate collector detects all ions of one m/z value. When combined with a TOF analyzer, the microchannel plate affords an almost instantaneous mass spectrum. Because of their construction and operation, microchannel plate detectors are cheaper to fit and maintain. Multipoint detectors are particularly useful for situations in which ionization occurs within a very short space of time, as with some ionization sources, or in which only trace quantities of any substance are available. For such fleeting availability of ions, only multipoint collectors can measure a whole spectrum or part of a spectrum satisfactorily in the short time available.

Chapter 31

Time-to-Digital Converters (TDC)

Background

Point and array ion collectors are described in Chapters 28 and 29, respectively. The multipoint ion collector, in which many thousands of very-small-diameter microchannels are packed closely together, is a further development of the array. This arrangement gives a large number of ion collectors within an area large enough to encompass the cross-sectional area of a typical ion beam in a mass spectrometer. Unlike the array collectors, in which each Channeltron® is separately monitored so as to distinguish between m/z values, the microchannel plates simply record total ion arrivals and are particularly suited to time-of-flight (TOF) analyzers. Wherever an ion arrives at the front face of the microchannel plate, it results in a release of electrons to a positively charged electrode (back plate). The resulting electric current is passed to a suitable recording device. Ions in a TOF analyzer arrive sequentially at times that are proportional to m/z. The arrival events for each m/z (pulses) are recorded using electrical signals that are already digitized. Since the collector takes ion arrivals spread over time and converts these events into discrete (digital) electrical pulses, it is called a time-to-digital converter (TDC). This chapter focuses on the operation of a TDC.

Measurement of m/z Ratios by Time-of-Flight Instruments

Measurement of m/z ratios by TOF instruments is covered in detail in Chapter 26 and described only briefly here. After acceleration through an electric potential difference of V volts, ions reach a velocity v governed by the equation $v = (2zeV/m)^{0.5}$, in which e is the electronic charge. Note that the velocity is inversely proportional to the m/z value of an ion. If the ions travel a distance d before being detected, the time t needed to travel along the TOF flight tube is given by $t = d/v$ and m/z becomes proportional to t^2 (Figure 31.1). Thus, mass measurement resolves itself into the accurate measurement of the times needed for ions to travel along the flight tube. This timing must be done electronically because normal clocks cannot measure the times accurately enough for flight tubes of reasonable length (≈1.0 m). As an example, for a potential difference of 1000 Vs, a mass of 100, and a charge of z = 1, ions would take 22.8 μsec to travel the length of the TOF flight tube, and a mass of 101 would take 22.9 μsec. If mass 100 is to be differentiated (resolved) from 101, then the timer must be able to measure accurately time differences of at least 0.1 μsec. TDCs can effect such accurate

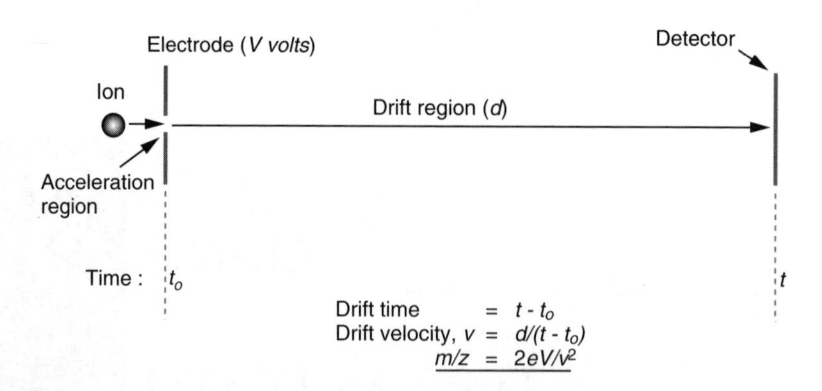

Figure 31.1
Upon acceleration through an electric potential of V volts, ions of unknown m/z value reach a velocity v (= $[2zeV/m]^{0.5}$). The ions continue at this velocity (drift) until they reach the detector. Since the start (t_0) and end (t) times are known, as is the length d of the drift region, the velocity can be calculated, and hence the m/z value can be calculated. In practice, an accurate measure of the distance d is not needed because it can be found by using ions of known m/z value to calibrate the system. Accurate measurement of the ion drift time is crucial.

time measurement and provide a very convenient method of recording mass spectra from a TOF instrument, particularly as the signal coming from the converter is not analog but is already digitized.

Ions in a TOF analyzer are temporally separated according to mass. Thus, at the detector all ions of any one mass arrive at one particular time, and all ions of other masses arrive at a different times. Apart from measuring times of arrival, the TDC device must be able to measure the numbers of ions at any one m/z value to obtain ion abundances. Generally, in TOF instruments, many pulses of ions are sent to the detector per second. It is not unusual to record 30,000 spectra per minute. Of course, each spectrum contains few ions, and a final mass spectrum requires addition of all 30,000 spectra to obtain a representative result.

Multichannel (Microchannel) Plate Array

The mode of action of a single Channeltron® (a miniaturized version used as one element of a microchannel plate array) is shown in Figure 31.2. A fast-moving ion striking the front end of a single microchannel causes a number of secondary electrons to be ejected. These electrons are accelerated by an electric field lying the length of the microchannel element and, after a short distance, strike its walls further along from their point of origin. Each electron impact on the Channeltron® walls causes several electrons to be ejected, with each initial secondary electron producing several more electrons. This process of producing more electrons continues along the length of the microchannel element, with further wall collisions leading to a cascade of many electrons, which emerge as a burst (pulse) from the other end of the microchannel. The pulse of electrons crosses to a positively charged metal plate collector and is detected as an electrical current. Thus, each time an ion arrives, an electrical pulse (electrical digit) is produced. A succession of ion arrivals yields a corresponding series of electrical pulses. As noted above, these ion arrival events are already digitized and can be stored directly in computer memory (discussed in the following section). Thus, each ion event is transformed into an electrical digital pulse, and this process gives rise to the term *time-to-digital converter* (TDC).

The front opening of such a microchannel element has a diameter of only a few microns, but it is only one element of a whole multichannel array (Figure 31.2). Whereas the orifice to one microchannel element covers an area of only a few square microns, an array of several thousand parallel elements covers a much larger area. In particular, the area covered by the array must be larger than

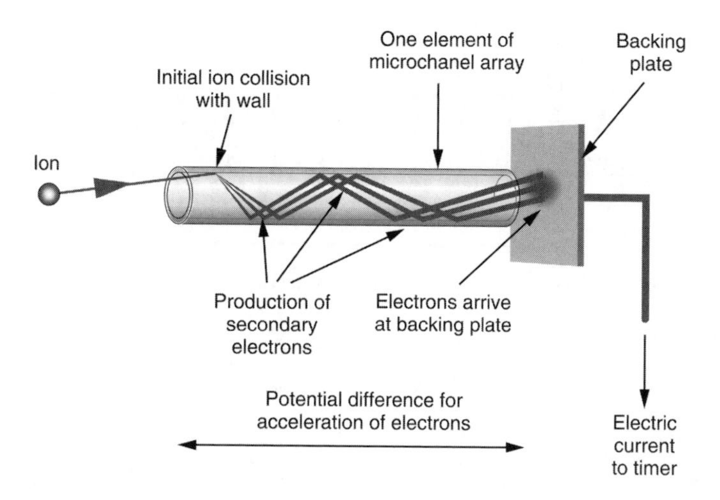

Figure 31.2
Diagram illustrating one element of a microchannel array. In a typical array, this Channeltron®-like tube is just one of thousands of similar ones, each only a few microns in diameter. All microchannel elements end a short distance from a common backing plate. An ion entering the front end of an array element and striking the wall produces secondary electrons. These in turn are accelerated along the tube and strike the wall. Each secondary electron produces several more secondary electrons. The process is repeated along the tube until a shower of electrons leaves its end and crosses to the backing plate. The arrival of these electrons constitutes a pulse of electric current, which is recorded by the TDC timer. The ion arrival time (an event) is recorded.

the cross-sectional area of the ion beam that is detected so that all ions are recorded. Figure 31.3 shows a small representative section of a multichannel array, together with the backing plate.

Timing of Electrical Pulses Resulting from Ion Arrivals at the Microchannel Plate Collector

As shown above, when an ion arrives at the microchannel array it releases a cascade of electrons onto the back plate. This cascade constitutes an electrical pulse from the microchannel plate, which

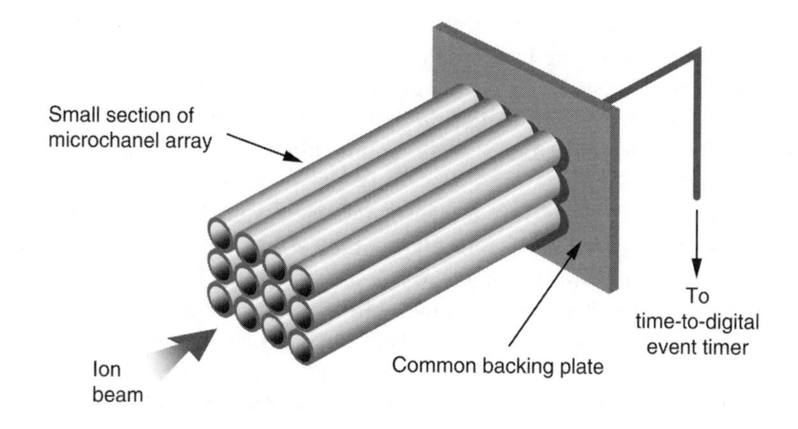

Figure 31.3
Diagrammatic representation of a small section of a microchannel array, with its backing plate. One element of the array is shown in greater detail in Figure 31.2. Arriving ions can enter any one of the array elements to produce an electric current at the backing plate. Since the latter is common to all of the array elements, it is immaterial which element of the array is involved because the signal is recorded by the common plate. Thus, arrival of an ion at any point on the array (an event) results in the sending of an electrical signal to the TDC timer (TDC converter). This point is the ion arrival time (t) shown in Figure 31.1.

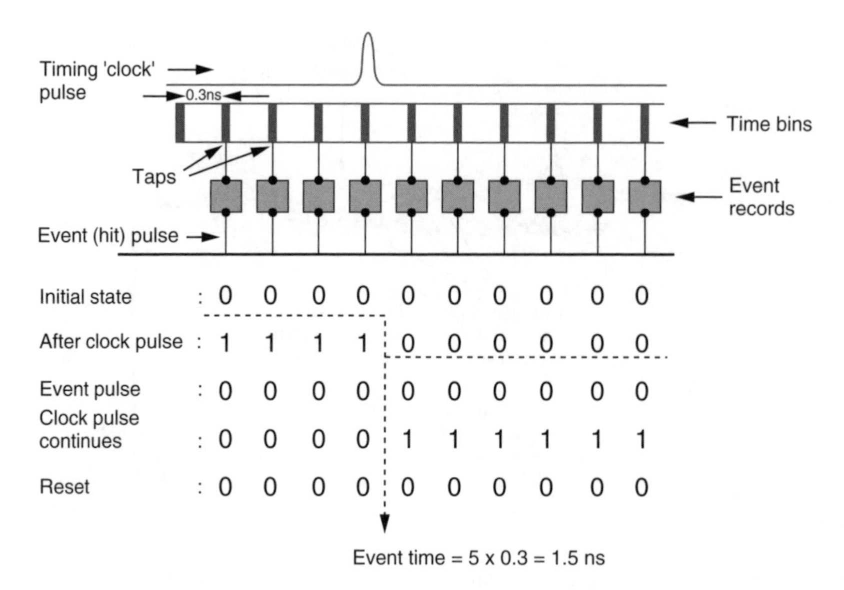

Figure 31.4
A clock pulse travels along a series of electronic gates (time bins). The time taken to cross each gate is about 0.3 nsec. Passage of the pulse through a gate is recorded by drawing off a part of the electrical signal from the clock pulse — viz., part of the signal is sent to the event recorders, which switch from a 0 state to a 1 state. In this example, after passing through five gates, the elapsed time from the start of the clock pulse is 1.5 nsec. Arrival of an ion (an event or hit) causes a new pulse to be sent to all of the event recorders which resets the bins affected by the clock pulse. The clock pulse continues to travel along the time bins, changing the event recorders from 0 to 1. Where the event has occurred is marked by a change of state of the event recorders. Before the timers are reset, the recorded event is stored in memory for later use.

is used for timing ion arrival events. Each electronic pulse resulting from an ion arrival travels to and is recorded by a time bin. A series of connected time bins can be regarded as a long chain of electronic gates through which a clock signal passes. The time for the signal to pass through each gate is about 3×10^{-10} sec (0.3 nsec). Thus, as shown in Figure 31.4, if a gate timing pulse is regarded as having started at time zero (t = 0), then the time taken for the pulse to pass through 100 bins is 30 nsec ($t_{100} = 3 \times 10^{-8}$ sec and $t_{100} - t_0 = 3 \times 10^{-8}$ sec).

Any one bin can be electronically distinguished from the next one, and therefore the bins can be used like the tick of a standard clock. Each bin serves as one tick, which lasts for only 0.3 nsec. By counting the ticks and knowing into which bin the ion pulse has gone, the time taken for the ion to arrive at the detector can be measured to an accuracy of 0.3 nsec, which is the basis for measuring very short ion arrival times after the ions have traveled along the TOF analyzer tube. Each ion arrival pulse (event) is extracted from its time bin and stored in an associated computer memory location.

An explanation of the scheme for storage of the event signals requires an extra degree of complexity. The bins have two states (on or off, digitally 0 or 1). Before the timing pulse begins, all bins are set to the 0 state, which changes to a 1 state when the timing pulse arrives. When an ion arrival event occurs, the clock timing signal will have already traveled through some of the bins; viz., it will have passed through some of them in sequence, setting their states from 0 to 1. When the ion signal (a hit) arrives, it is sent to all of the time bins simultaneously, and the bins will be affected in different ways (Figure 31.4). Those bins through which the timing signal has already passed will have been reset to the "1" state but, immediately in front of the timing signal, the next bin will still be set to the original "0" state. Those bins already affected by the timing signal will he set back to 0 when the "hit" signal arrives. Thus, all of the bins up to the event signal will be in the 0 state, but those after the event signal will be in the 1 state. After the timing signal has passed through all of the bins, they are examined. All those bins that have not recorded a hit are regarded as empty, but the first bin that was set to 1 by a combination of hit and timing pulses must be the one into which the "hit" signal went. Thus, an electronic inspection of the bins reveals that the "0/1" state of the bins changes at one of them, and this fact is stored in permanent memory. Since the position of any bin in the series is known relative to the starting bin, the time at which the "hit" signal from an ion arrival affected the bins can be measured to an accuracy of about 3×10^{-10} sec. After the event has been noted, all bins are reset to their 0 states before the next timing signal travels along them ready for the next ion arrival. A typical clock timer runs at about 3.6 GHz; viz., it ticks every 0.28×10^{-9} sec (every 0.28 nsec).

In a simplistic way, the bins can be regarded as a series of empty receptacles and an ion arrival (an event) results in one bin being filled. By looking into the line of bins, the full one can be found. Rather than discussing tiny fractions of a second, let it be supposed that it takes 1 second to pass from one bin to the next in a series of 1000 bins. The total time needed to pass through them all is 1000 sec. If only bin number 603 is found to be full, then the timing of the event that filled it must have occurred 603 seconds after the examination began. All empty bins are ignored, but the full bin is emptied and noted. Thus, the bins are returned to their original empty state, after which the timing begins all over again in preparation for the next ion event.

Each bin is connected to a memory location in a computer so that each event can be stored additively over a period of time. All the totaled events are used to produce a histogram, which records ion event times versus the number of times any one event occurs (Figure 31.5).With a sufficiently large number of events, these histograms can be rounded to give peaks, representing ion m/z values (from the arrival times) and ion abundances (from the number of events). As noted above, for TOF instruments, ion arrival times translate into m/z values, and, therefore, the time and abundance chart becomes mathematically an m/z and abundance chart; viz., a normal mass spectrum is produced.

Ion Abundances and Dead Time

A mass spectrum is a chart of ion abundances versus m/z values. It is shown above that the TDC measures ion arrival times, which are converted directly into m/z values. Notionally, the number of ions arriving at the detector at any one m/z value is equal to the number of events recorded (one

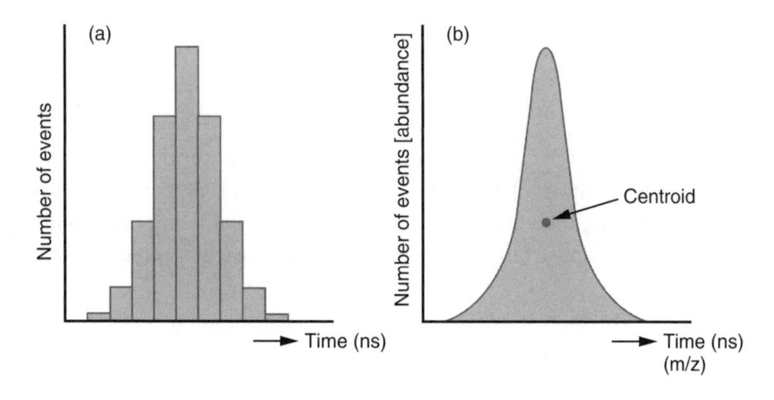

Figure 31.5
Ion arrival events are stored additively in memory locations (Figure 31.4), which also provide the times of arrival. After recording data for some time, the memories are examined to produce a histogram (a) of the number of events versus arrival times. The histogram shown relates to ions of one particular m/z value arriving at the detector at slightly different times. From the histogram, a peak shape is produced (b), the centroid of which gives the mean arrival time and, therefore, the m/z value (Figure 31.1). The area of the peak gives the total number of events and therefore the abundance of the ions. Similar procedures are used for all other ions of other m/z values in the mass spectrum.

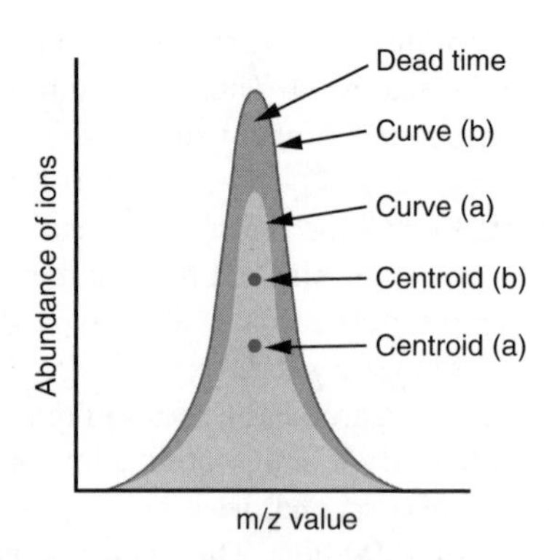

Figure 31.6
The curve (a) traces the outline of the peak obtained directly from the number of events recorded (Figure 31.5). The second curve (b) traces the outline of the peak obtained after correcting for coincidental events (dead time, shown by the shaded area). The centroids of peaks a and b are shown, and it can be seen that they occur at the same m/z value. Thus the dead-time correction alters only the abundances and not the m/z values of the ions.

ion in, one event recorded). With a recording time of only a few microseconds, there will not be many ions arriving together at the detector at exactly the same time, even when the ions are abundant. However, the more abundant the ions, the less likely it becomes that the relationship of one ion-one event will be true. Suppose for abundant ions at m/z 100 that, in one recording, two ions arrive at the microchannel plate array at the same time, even in different elements of the array. As shown above, the TDC system cannot differentiate one, two, or more ions arriving at the same time (within a time bin time of 0.3 nsec). Therefore, abundant ions are underrecorded. The difference between the true number of ions and the number suggested by the events counter can be substantial. The time bins are blind to multiple events occurring within a space of about 0.3 nsec. This "blindness" is known as dead time and must be corrected to give a more accurate assessment of true ion abundance. A mathematical algorithm is used to increase the apparent ion yield to a true ion yield before the mass spectrum is printed. The algorithm increases the number of events in proportion to the numbers of events recorded. Many events at one m/z value indicate that the correction needs to be larger and vice versa. Figure 31.6 illustrates the effect on ion abundances of the correction for dead time.

Conclusion

Time-to-digital converters (TDCs) are very convenient for measuring the times of arrival at a detector of ions in a TOF analyzer. The signal resulting from the detection of each ion is already digitized, and no analog-to-digital converter is needed prior to evaluating the signals by a computer. Before conversion to a mass spectrum, abundances of ions must be corrected for dead time.

Chapter 32

Origin and Uses of Metastable Ions

Introduction

A mass spectrum consists of peaks corresponding to ions. The position of a peak on the x-axis is proportional to its mass (strictly, its m/z value), while the height of the peak on the y-axis gives the number of ions (abundances) at a particular m/z. The ions giving rise to the spectrum are formed in an ion source and are passed through an analyzer for measurement of m/z and into a detector for measurement of abundance (Figure 32.1).

From molecules of a substance (M), electron ionization gives first the molecular ions (M^+), which remain in the ion source for several microseconds before traveling through the instrument to the detector. During their time in the source, some or many of the molecular ions that possess sufficient energy will decompose to give fragment ions F_1^+, F_2^+, … , F_n^+ (Figure 32.2) along various reaction (fragmentation) pathways; the fragment ions themselves decompose to other fragments until all excess of energy is used. The mass spectrometer draws out (samples) the assembly of ions (M^+, F_1^+, F_2^+, …) in the source and measures m/z and abundance values for each species to provide a spectrum of normal ions, viz., a spectrum of the ions in the source. For example, as shown in Figure 32.2, normal ions (F_1^+) will have m/z = x.

However, a small proportion of ions will have insufficient energy to fragment during the few microseconds spent in the source but will have just sufficient energy to fragment during the few microseconds of flight through the analyzer to the detector. Such ions, which start with a certain kinetic energy on leaving the source, must lose some of that energy on decomposing (conservation of momentum between the fragments). Therefore, as far as the spectrometer is concerned, these ions have the "wrong" kinetic energy for their mass, which will be measured as m/z = y (Figure 32.2; x > y). These are the so-called metastable ions. It is important to understand that the mass (m) of a normal fragment ion (F_1^+) is exactly the same as the mass (m) of the corresponding metastable ion (F_1^+). The only difference between them lies in their positions of formation (inside or outside the source). The spectrometer correctly measures m/z for F_1^+ (normal) ions formed in the source but gets the wrong m/z for F_1^+ (metastable) ions formed after the source. In other words, although normal and metastable F_1^+ ions have the same mass, the mass spectrometer interprets their m/z values differently because normal F_1^+ ions have greater momentum than the metastable ones. However, the spectrometer can be set to record the metastable ions correctly and not the normal

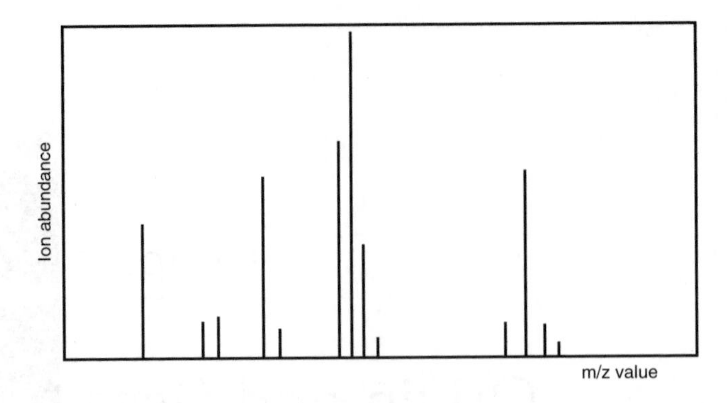

Figure 32.1
Example of a mass spectrum showing the peaks (or lines) corresponding to ions measured at various m/z values by the spectrometer; the heights of the peaks relate to the abundance of the ions.

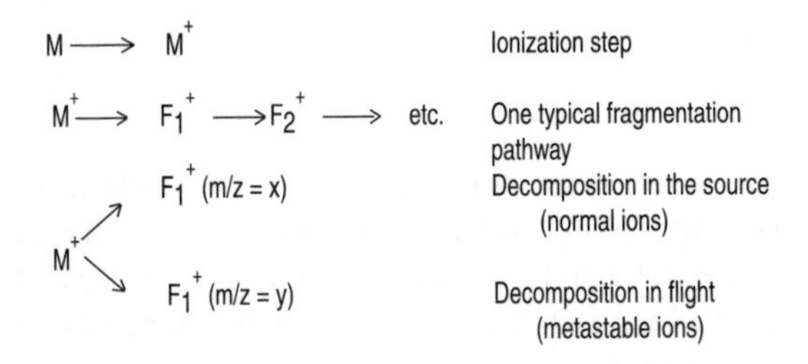

Figure 32.2
The first ionization step in the source converts neutral molecules (M) into ions (M^+). After a short time in the source, some molecular ions have fragmented to give smaller ions, F_1^+, F_2^+, … , F_n^+, along one or more reactions pathways. An ion (F_1^+) formed in the source and then analyzed will have an m/z = x; the same ion (F_1^+) formed outside the source will have an apparent m/z = y, where $y < x$.

ones. The required settings of the spectrometer to observe metastable ion processes are covered in Chapters 33 and 34, which examine the subject of linked scanning.

Perhaps it is worth emphasizing that the actual metastable ions are those that decompose, and the so-called metastable ions that are recorded are actually the products of decomposition, not the metastable ions themselves. It is more accurate to describe the recorded metastable ions as ions resulting from decomposition of metastable ions. The apparent m/z value of a metastable ion provides the link between an ion that fragments (the parent or precursor ion) and the ion that is formed by the fragmentation (the daughter or product ion).

Field Free Zones and the Formation of Metastable Ions

Certain regions of a mass spectrometer have no electric or magnetic fields to affect an ion trajectory (field-free regions). Figure 32.3 illustrates three such regions in a conventional double-focusing instrument.

Metastable ions formed in any of these field-free regions can be detected by specialized changes in the various electric and/or magnetic fields (linked scanning) without interference from normal ions formed in the source. In a one-sector magnetic instrument there are only two field-free regions. In tandem MS (mass spectrometry), there is always one more possible field-free region than there are sectors; for example, a four-sector instrument has five such regions although, in practice, only one or two are used for metastable ion observations. In a simple quadrupole mass spectrometer, differentiation between normal and metastable ions is not possible but can be carried out efficiently

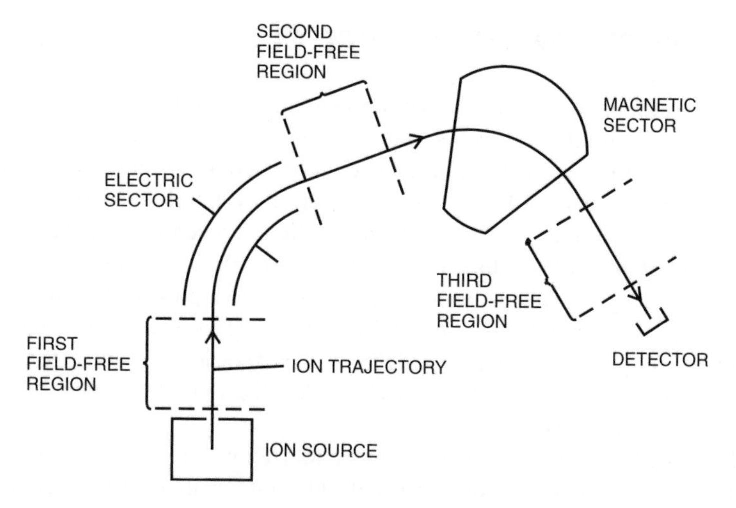

Figure 32.3
Ion trajectory through a conventional (EB) sector instrument, showing three field-free regions in relation to the sectors, the source, and the ion detector.

with a triple quadrupole instrument (see Chapter 33, " Linked Scanning and Metastable Ions in Quadrupole Mass Spectrometry").

Abundances of Metastable Ions

When molecules are ionized, particularly by electrons (EI), an excess of energy is usually left in the resulting molecular ions. The spread in excess of energy can be represented generally as shown in Figure 32.4. Only that small proportion of ions (metastable) indicated as region B (Figure 32.4) has just such an excess of energy that fragmentation is delayed and takes place not in the ion source but during the few microseconds of flight time between ion source and detector.

Ions with excesses of energy falling within regions A and C give rise to the normal molecular and fragment ions observed in a mass spectrum. In region C, all of the excess of energy is used in decomposition in the source, while in A there is not enough energy for decomposition. Only in the small region B do ions have just the right energy to fragment in flight. The low abundance of metastable ions makes their detection more difficult.

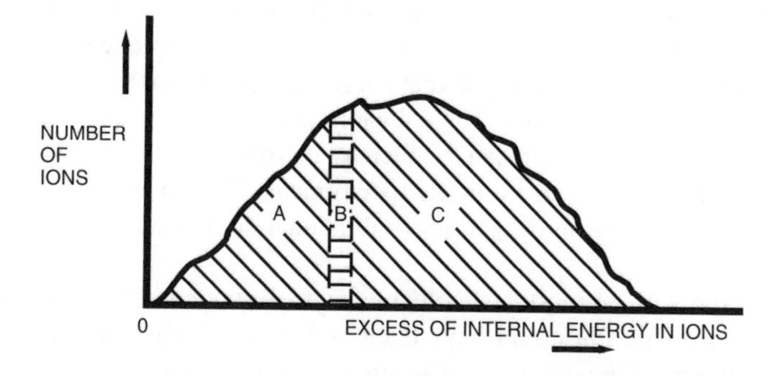

Figure 32.4
A correlation between the number of ions and their excess of energy. In region A, the molecular ions have insufficient energy to fragment, but in region C they have enough to cause decomposition in the ion source. Ions metastable in the narrow band, B, have just the amount of excess to cause fragmentation in flight.

Disadvantage of Low Abundance of Metastable Ions

The normal fragment ions in a mass spectrum potentially contain a wealth of information. Just as in a jigsaw puzzle, if the fragments (pieces) can be reassembled correctly, the original molecular structure (jigsaw picture) can be deduced (completed). Whereas in a jigsaw the relative positions of the pieces can be inferred from the way they interlock with neighboring pieces (shapes), in a conventional mass spectrum, connectivities between normal fragment ions are not obvious and may be impossible to discover for large molecules. The molecular jigsaw has to be completed without seeing the shapes of pieces and with knowing only their masses. Because the metastable ions appear differently from normal ions in a mass spectrometer and yet form a link between precursor and product ions, a metastable spectrum (scan) provides the required connectivities between the normal ions. It is unfortunate that the low abundance of metastable ions makes them difficult to observe.

Some mild methods of ionization (e.g., chemical ionization, CI; fast-atom bombardment, FAB; electrospray, ES) provide molecular or quasi-molecular ions with so little excess of energy that little or no fragmentation takes place. Thus, there are few, if any, normal fragment ions, and metastable ions are virtually nonexistent. Although these mild ionization techniques are ideal for yielding molecular mass information, they are almost useless for providing details of molecular structure, a decided disadvantage.

The absence or dearth of metastable ions can be counteracted by applying additional energy to normal ions after they have left the ion source and while they are in a field-free zone. The additional energy causes the normal ions to fragment in the field-free zone. This induced fragmentation is instrumentally indistinguishable from the decomposition giving rise to metastable ions, as discussed previously. Induced fragmentation increases the number of naturally occurring metastable transitions occurring in the electron ionization process and can produce metastable ions where none existed previously.

This enhanced fragmentation in field-free regions has led to big improvements in obtaining molecular structural information and in examining mixtures. Indeed, it has produced the whole new technique of tandem MS (or MS/MS). Along with the enhanced fragmentation, a wide variety of means for examining the fragments and their origins has developed (linked scanning).

Enhanced (Induced) Fragmentation

A number of techniques are available to increase the excess of energy within an ion so as to induce it to decompose. At the moment, only one of these has commercial importance: collisionally induced decomposition, also known as collisionally induced dissociation (CID). (This technique is also known as collisionally activated decomposition, CAD.) In this approach, normal ions in flight within a field-free region are given extra internal energy by causing them to collide with molecules of an inert gas, such as helium or argon. The collision takes place in a cell, a region of a field-free zone in which the inert gas pressure can be increased from zero to any desired value without decreasing the vacuum in the rest of the mass spectrometer. The collision process converts some of the kinetic energy of the ions into rotational, vibrational, and electronic energy which, in turn, leads to fragmentation.

Collision of an ion with an inert gas molecule leads to some deflection in the ion trajectory. After several collisions, the ion could have been deflected so much that it no longer reaches the detector. This effect attenuates the ion beam as it passes through the gas cell, leading to loss of instrumental sensitivity. An attenuation of 50 to 70% is acceptable and is not unusual in practice.

Conclusion

Metastable ions yield valuable information on fragmentation in mass spectrometry, providing insight into molecular structure. In electron ionization, metastable ions appear naturally along with the much more abundant normal ions. Abundances of metastable ions can be enhanced by collisionally induced decomposition.

Chapter 33

Linked Scanning and Metastable Ions in Quadrupole Mass Spectrometry

Introduction

The steps (reactions) by which normal ions fragment are important pieces of information that are lacking in a normal mass spectrum. These fragmentation reactions can be deduced by observations on metastable ions to obtain important data on molecular structure, the complexities of mixtures, and the presence of trace impurities.

For ordinary quadrupole instruments (Figure 33.1), it is not possible to study metastable ions. The field-free regions are usually very short, so, in the time taken to traverse them few ions dissociate. More importantly, a quadrupole is a mass filter. In the complex, crossed DC and RF (radio frequency) fields used for mass separation in a quadrupole assembly, there is no inherent force moving the ions along the central axis; i.e., there is nothing to drive the ions through the length of the assembly. Ions can only traverse the length of the quadrupole if they have been previously given some momentum by acceleration through a (small) electric potential (about 5 eV) upon leaving the ion source. Unlike magnetic/electric-sector instruments, in which the forward motion of ions is used to help achieve mass separation, the forward momentum of an ion in a quadrupole instrument is very much less and effectively plays no such part; the mass separation is achieved by the crossed DC and RF fields at right angles to ion motion.

Normal and Metastable Ions

Ions formed in the ion source are considered to be the normal ions in a mass spectrum. For example, as illustrated in Figure 33.2, some of the ions (m_1^+) dissociate to smaller mass ions (m_2^+) and neutrals (n_0); any unchanged ions (m_1^+) and the fragments (m_2^+) are drawn out of the source by an electric potential of V volts.

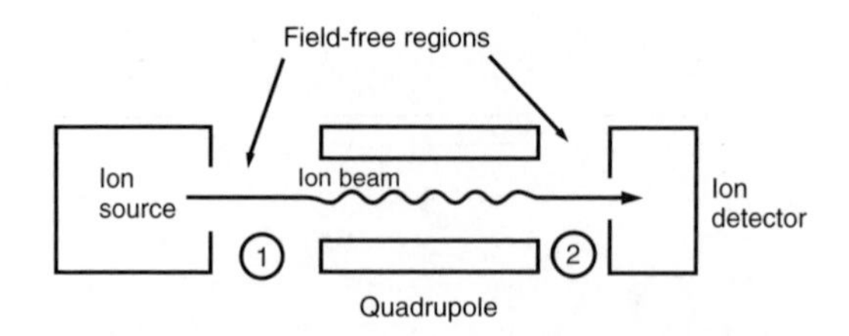

Figure 33.1
In a normal quadrupole instrument, the field-free regions are very short. Ions formed in region 1 will be transmitted by the quadrupole as normal ions. In region 2 there is no differentiation between metastable and normal ions.

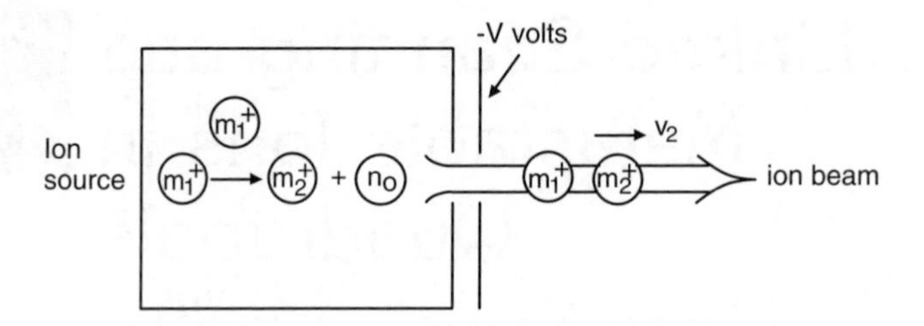

Figure 33.2
Formation of normal ions in an ion source. In this example, some initially formed ions (m_1^+) dissociate (fragment) to give smaller ions (m_2^+) and a neutral particle (n_0). Unchanged ions (m_1^+) and the fragment ions (m_2^+) are drawn out of the source as beams moving with velocities v_1, v_2, respectively.

The kinetic or translational energy of the ions is equal to the work done on moving the charged species through the potential, V, i.e., $1/2m_1v_1^2 = zV$ and $1/2m_2v_2^2 = zV$, where z is the charge on the ions and v_1, v_2 are their final velocities. From this, we obtain Equations 33.1 and 33.2.

$$V_1 = \sqrt{\frac{2zv}{m_1}} \tag{33.1}$$

$$V_2 = \sqrt{\frac{2zv}{m_1}} \tag{33.2}$$

Now consider the same fragmentation process, but this time the ion (m_1^+) dissociates outside the ion source. Initially, as for m_2^+, $m_1v_1^2 = 2zV$, and $V_1 = \sqrt{2zv/m_1}$. When fragmentation occurs, the total kinetic energy of m_1^+ is shared between the new (metastable) ion m_2^+ and its accompanying neutral n_0, but their velocities remain unchanged (v_1). Normal m_2^+ ions formed in the source have velocity v_2, but metastable m_2^+ ions formed outside the source have the velocity of m_1^+ (v_1; $v_1 < v_2$; Figure 33.3).

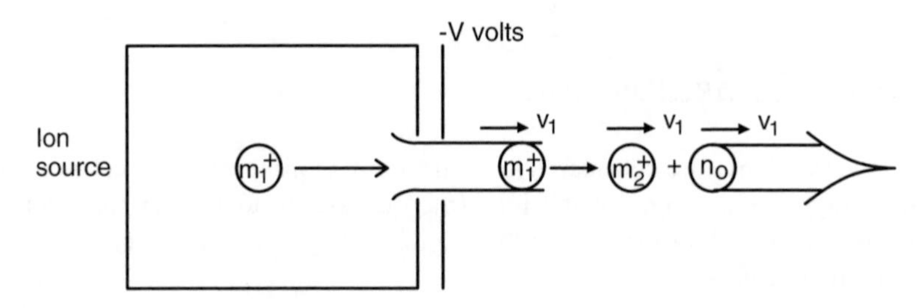

Figure 33.3
Ions (m_1^+) of velocity (v_1) that dissociate outside the ion source give ions (m_2^+) of the same velocity (v_1).

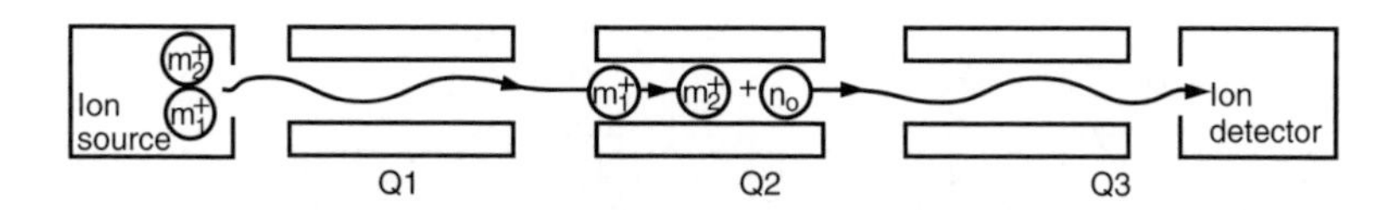

Figure 33.4
Normal ions (m_1^+ or m_2^+) formed in the ion source will pass (filter) through the first, second, and third quadrupoles (Q1, 2, 3) if these are set correctly. If Q1 is set to pass only m_1^+ ions, then normal m_2^+ ions cannot reach the detector, and if Q3 is set to pass only m_2^+ ions, then m_1^+ ions cannot reach the detector. Any m_2^+ ions that reach the detector must have been formed (metastable or induced by collision) by dissociation of m_1^+ ions in Q2.

In a sector instrument, which acts as a combined mass/velocity filter, this difference in forward velocity is used to effect a separation of normal and metastable m_2^+ ions (see Chapter 24, "Ion Optics of Magnetic/Electric-Sector Mass Spectrometers"). However, as discussed above, the velocity difference is of no consequence to the quadrupole instrument, which acts only as a mass filter, so the normal and metastable m_2^+ ions formed in the first field-free region (Figure 33.1) are not differentiated.

How Quadrupoles Can Be Used to Examine Metastable Ions

By using three successive in-line quadrupole assemblies, metastable ions can be examined (Figure 33.4). Ions from the source are mass separated in quadrupole Q1. For example, m_1^+ ions can be selected. The chosen ions are passed into quadrupole Q2, which has only RF voltages applied to it. In this mode, the ions from Q1 can pass through Q2 without further mass selection. If the m_1^+ ions dissociate in this region to give m_2^+, then these latter ions pass on into quadrupole Q3, which has its normal RF and DC voltages applied for mass resolution. Thus, m_1^+ ions are selected by Q1 but rejected by Q3, and m_2^+ ions are rejected by Q1 but selected by Q3. In this way, only metastable m_1^+ ions (passed by Q1 but rejected by Q3) that decompose to give m_2^+ in Q2 are detected in Q3. The triple quadrupole assembly is efficient for detection of metastable ions. The use of Q2 as an indiscriminate mass filter is sometimes indicated by the symbol, q, and the triple quadrupole can be referred to as QQQ or QqQ.

By introducing a collision gas into Q2, collision-induced dissociation (CID) can be used to cause more ions to fragment (Figure 33.4). For example, with a pressure of argon in Q2, normal ions (m_1^+) collide with gas molecules and dissociate to give m_2^+ ions. CID increases the yield of fragments compared with natural formation of metastable ions without induced decomposition.

Linked Scanning with Triple Quadrupole Analyzers

Linked scanning techniques by which fragmentation reactions can be examined are particularly easy to apply with QqQ instruments. The ease with which RF and DC voltages can be changed rapidly means that the scanning can be done very quickly. Three common and popular types of linked scan are briefly described here and serve to illustrate its principles.

Product Ion Scans

Precursor ions are selected by Q1 and passed into the collision cell (Q2 or q2 of Figure 33.5). Here, collision with an inert gas (argon or helium) causes dissociation to occur, and the resulting fragment (product) ions are detected by scanning Q3 (Figure 33.6).

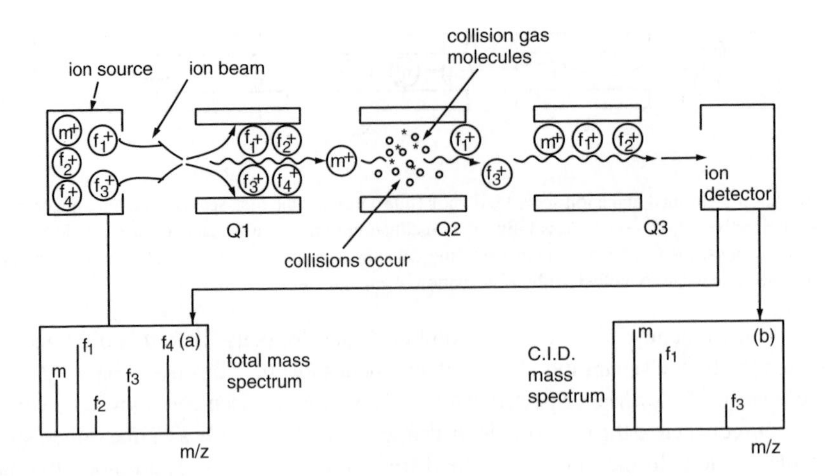

Figure 33.5
An example of linked scanning on a triple quadrupole instrument. A normal ion spectrum of all the ions in the ion source is obtained with no collision gas in Q2; all ions scanned by Q1 are simultaneously scanned by Q3 to give a total mass spectrum (a). With a collision gas in Q2 and with Q1 set to pass only m^+ ions in this example, fragment ions (f_1^+, f_2^+) are produced and detected by Q3 to give the spectrum (b). This CID spectrum indicates that both f_1^+ and f_2^+ are formed directly from m^+.

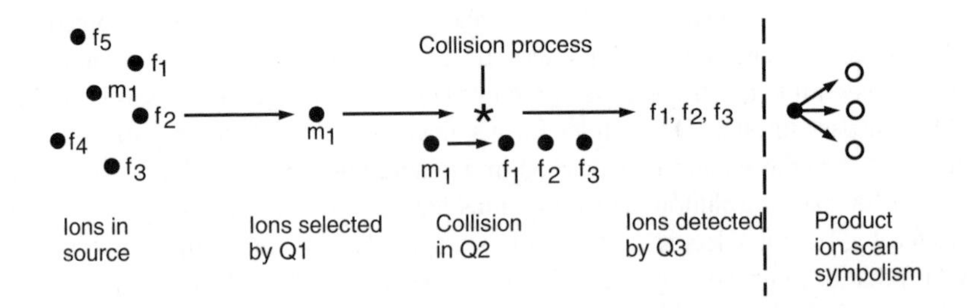

Figure 33.6
A product ion scan. Source ions (m_1^+, f_1^+, … , f_5^+) are selected by setting Q1, in this case, to pass only m_1^+. Collisional activation of these ions in Q2 induces dissociation to give fragment ions (f_1^+, f_2^+, f_3^+), which are detected by scanning Q3. The symbolism for this process is shown.

Precursor Ion Scans

Quadrupole, Q3, is set to pass only the product ions under investigation. All the ions from the ion source are scanned by Q1 and passed successively into Q2, where collisional activation occurs. Those ions that fragment to give product ions of interest are revealed by the appearance of the product ions in Q3 (Figure 33.7).

Constant Mass Difference Scans

For a process, $m_1^+ \rightarrow m_2^+ + n_0$, in which the difference in mass ($m_1 - m_2$) is a set value (Δm), the quadrupoles can be set to discriminate only that difference. Q1 scans the spectrum of ions from the source and Q3 scans the same masses but offset by Δm. For example, with a set mass difference of 28 amu, if Q1 is scanning 80, 81, 82, …, then Q3 is concurrently scanning 52, 53, 54, …, and so on. The ions from Q1 are collisionally activated in Q2, so, if they give ions in Q3 that are Δm mass units less, then the process, $m_1^+ \rightarrow m_2^+$ is established (Figure 33.8).

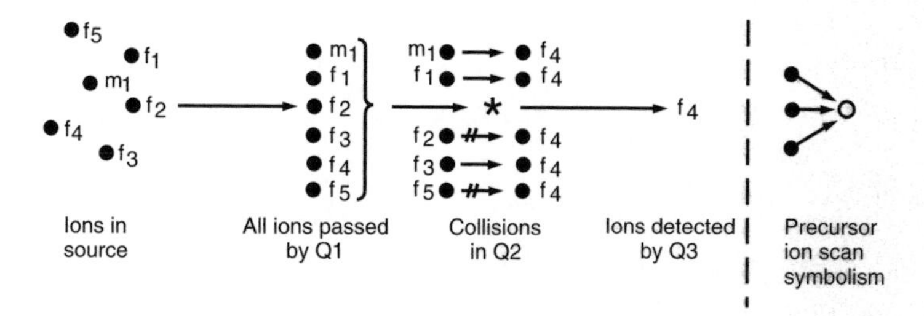

Figure 33.7
A precursor ion scan. Source ions (m_1^+, f_1^+, ... , f_5^+) are all passed successively by Q1 into the collision cell, Q2, where a selected fragment (f_4) is produced and detected by Q3. Only the ions (m_1, f_1, f_3) give f_4 fragment ions in this example.

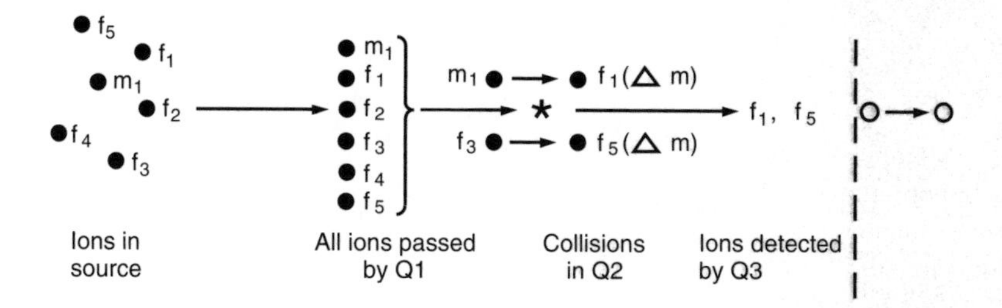

Figure 33.8
A constant-mass-difference scan. Source ions (m_1, f_1, ... , f_5) are passed successively by Q1 into Q2, where collisionally induced dissociation occurs. Q3 is set to pass only those ions produced in Q2 that have a predetermined mass difference (Δm) between the ions passed by Q1. In this example, they are m_1 - f_1 (= Δm) and $f_3 - f_5$ (= Δm), so, although all ions pass into Q2, only f_1, f_5 have a mass difference (Δm) equal to that selected for Q3.

Use of Metastable Ion and CID Data

The importance of linked scanning of metastable ions or of ions formed by induced decomposition is discussed in this chapter and in Chapter 34. Briefly, linked scanning provides information on which ions give which others in a normal mass spectrum. With this sort of information, it becomes possible to examine a complex mixture of substances without prior separation of its components. It is possible to look highly specifically for trace components in mixtures under circumstances in which other techniques could not succeed. Finally, it is possible to gain information on the molecular structures of unknown compounds, as in peptide and protein sequencing (see Chapter 40).

Conclusion

Metastable and collisionally induced fragment ions can be detected efficiently by a triple quadrupole instrument. By linking the scanning regions of the first and third quadrupoles, important information about molecular structure is easily obtained.

Chapter 34

Linked Scanning and Metastable Ions in Magnetic-Sector Mass Spectrometry

Introduction

The study of metastable ions concerns substances that have been ionized by electrons and have undergone fragmentation. The stable molecular ions that are formed by soft ionization methods (chemical ionization, CI; field ionization, FI) need a boost of extra energy to make them fragment, but in such cases other methods of investigation than linked scanning are generally used.

Chapter 32, "Origin and Uses of Metastable Ions," should be read in conjunction with this chapter. Briefly, normal ions are those that are formed in the ion source and are sufficiently stable to reach an ion collector without fragmenting further. Many ions formed in an ion source under electron ionization conditions have enough internal energy to decompose rapidly to give more-stable fragments. Once these stable states have been achieved, there is little excess of energy in the ions, and there is no further fragmentation before they reach an ion collector. These are the so-called normal ions (Figure 34.1a). However, some ions have just enough internal energy left over from their formation to fragment after leaving the ion source and before reaching the ion collector; these are the metastable ions. For example, as illustrated in Figure 34.1a, some ions m_1 dissociate in the ion source to give fragment ions m_2 and a neutral species n_0. All of these ions are extracted from the ion source by an electric potential (V) of a few thousand volts and are accelerated to a high velocity; each ion has kinetic energy equal to zeV. After being deflected in electric and magnetic sectors, the ions m_1, m_2 arrive at a collector to give a mass spectrum. Although in this instance it is obvious that m_2 must arise from m_1, in more complex fragmentations this information is missing. Since one of the attributes of a mass spectrum is the possibility of obtaining a chemical structure for the compound under investigation, the lack of connections between ions removes a vital step in knowing how the ions relate to each other. It is rather like trying to reassemble a clock knowing the masses of the bits and pieces but not how they relate to each other.

A small proportion of the ions m_1 in flight to the ion collector contain just enough energy to fragment to give m_2 ions after leaving the ion source. These are the metastable ions (Figure 34.1b). The main practical difference between ions m_1 dissociating in the ion source and those fragmenting outside it lies in the kinetic energies of the product ions. All normal ions, including m_2, formed in the ion source are accelerated out of it by the voltage V and achieve a kinetic energy zeV, but ions

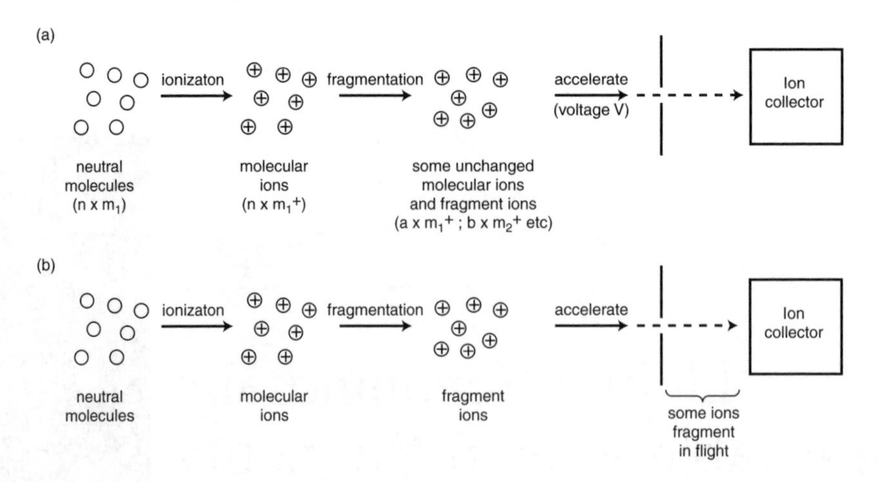

Figure 34.1
(a) Neutral molecules (m_1) enter the ion source and are ionized to give molecular ions (m_1^+). Within a few microseconds, many of the molecular ions have decomposed to give fragment ions (m_2^+, m_3^+, etc.). The ions are extracted from the source by an acceleration potential V and are directed by ion optics to the ion collector, where they are recorded as normal ions. (b) This process is initially the same as in (a). However, during flight (after acceleration but before ion collection) some of the ions fragment; these are the metastable ions.

m_2 formed in flight receive only a share of the kinetic energy of the fragmenting m_1 ions. Imagine a particle flying through the air and breaking into two pieces. The total kinetic energy remains the same before and after the ion breaks apart, but the individual kinetic energies of each fragment are less than the total. Therefore, m_2 ions formed in flight have kinetic energy less than zeV, and, consequently, the electric- and magnetic-sector optics treat the m_2 ion formed in flight as being different from m_2 ions formed in the ion source. By observing the ions m_2 formed in flight, it is possible to reveal the missing connection between m_1 and m_2. A mass spectrometer can be set to discover which product ions come from one precursor or which precursor ions give one product ion.

Scanning for metastable ions requires adjustment of electric and magnetic fields which can be adjusted individually or in conjunction with each other. If two fields are automatically adjusted at the same time, it is known as linked scanning. It is important to remember that metastable ions are not the product ions m_2 but are the ions undergoing fragmentation (precursor ions m_1).

Electric/Magnetic-Sector Geometry

Figure 34.2 illustrates the arrangement of electric and magnetic fields in a standard Nier-Johnson geometry. The electric sector focuses ions according to kinetic energy, the focal point lying between this sector and the magnetic. The magnetic sector focuses ions according to m/z value at the collector. Note the field-free regions between the ion source and the electric sector, between the electric and magnetic sectors, and between the magnetic field and the collector.

Metastable Ions Decomposing in the First Field-Free Region

An ion focused in the electric sector has a trajectory with a radius of curvature R = 2V/E. Note that m/z does not enter this equation and, for any given voltage E, focusing is determined by the accelerating voltage V and not by the mass of the ion. Any ion accelerated through V volts upon leaving the ion source and fragmenting in the first field-free region must share its kinetic energy between the two new particles, one being another ion and the other a neutral particle (see, for example, Figure 34.5). Such a product ion has insufficient kinetic energy ($m_2/m_1 \times zeV$) to be

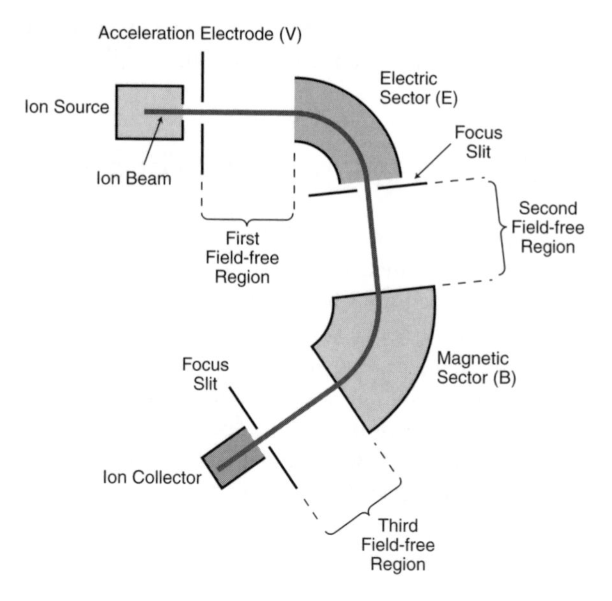

Figure 34.2
In a mass spectrometer of conventional Nier-Johnson geometry, the magnetic-mass-focusing (m/z) magnetic sector is placed after the kinetic-energy-focusing electric sector. The field-free regions lie between the acceleration electrode and the electric sector, between the electric and magnetic sectors, and between the magnetic sector and the ion collector. Two focus slits for the ion beam to pass through lie after the electric sector and before the ion collector.

focused properly by the electric sector. A solution to this problem is to increase the accelerating potential to V′, such that $V' = V\ (m_1/m_2)$. This solution defocuses all the normal ions but focuses the product ions of metastable decomposition. In use, normal ions (m_2) are recorded. Then, the voltage V is increased until the m_2 ions resulting from metastable ion decomposition appear in focus. The ratio V′/V is then known, m_2 is known, and, therefore, m_1 can be calculated. All precursor ions to any fragment ion can be found.

Alternatively, the electric focusing potential E can be changed, but this method needs another ion collector sited at the electric-sector focus, and it must be a collector that can be raised out of the ion beam when the mass of the ion being examined is required. This arrangement is not convenient. A better solution is obtained by linked scanning of the E/V voltages (see later discussion).

Metastable Ions Decomposing in the Second Field-Free Region

Ions m_1 fragmenting in the second field-free region give m_2 ions, which are focused by the magnetic field B but in the wrong place in the mass spectrum because of their reduced amount of kinetic energy. The mass spectrum shows normal ions at mass m_1 and m_2 but also another ion at apparent mass m*, which corresponds to the product ion m_2 deriving from metastable decomposition of m_1. The relationship is very simple: $m^* = m_2^2/m_1$. If a connection exists between m_1 and m_2 in the spectrum of an unknown substance, it is necessary to look only for another ion at the position m*. The method is very simple, but with modern instruments it is almost impossible to carry out because of the way that computer programs are written. For a computer, an ion peak is a sharp rise and fall on a baseline, and any slow rises and falls are ignored. Unfortunately, the metastable ions (m*) are low-abundance species, giving small, broad peaks that are invisible to the computer. They do not appear in a computer-generated mass spectrum. An older method of ion recording is needed to see them. Consequently, this method is no longer used and has been replaced with more complex linked-scanning methods, which involve simultaneous changes in the electric/magnetic field strengths (see discussion below) and are quicker to carry out.

Metastable Ions Decomposing in the Third Field-Free Region

In the third field-free region, all of the fields affecting the motion of an ion have been passed through on its way to the collector. A metastable ion m_1 fragmenting in this region gives an ion m_2 as before, but, because all ion separations have been done, the m_2 ions arrive at the same focus as their precursor ions m_1. There is only one way to differentiate them, and that is to make use of their kinetic energies. A normal ion m_1 arriving at the collector has kinetic energy zeV, but a product ion m_2 from decomposition of metastable m_1 ions has only kinetic energy $(m_2/m_1)zeV = zeV'$, in which $V' < V$. A special ion collector can be used to differentially reflect normal and metastable ions by control of an ion repeller voltage. Both normal and metastable ion spectra can be scanned by adjusting the repeller voltage. However, because this method requires a different ion collector than those normally used in mass spectrometry, it is a little-used technique, especially as more convenient linked scanning gives similar information on connections between ions.

Linked Scanning of V, E, and B Fields

There are a variety of possible linked scanning methods, but only those in more frequent use are discussed here. They differ from the linked scanning methods used in triple quadrupole instruments and ion traps in that two of the three fields (V, E, and B) are scanned simultaneously and automatically under computer control. The most common methods are listed in Table 34.1, which also defines the type of scanning with regard to precursor and product ions.

TABLE 34.1
Some Methods of Linked Scanning in the First Field-Free Region

Designation of scan type	Set search for precursor or product ions
E^2/V	Product ions
B/E	Product ions
B^2/E	Precursor ions
$(B/E)(1-E)^{1/2}$	Common neutral particles

V = accelerating voltage at the ion source; E = electric-sector voltage; B = magnetic-field strength.

E^2/V Scan

For ions to be focused as they pass through the electric sector, the ratio of V/E must be constant (the radius of curvature, R = 2V/E). The magnetic field is kept at a fixed value, but the accelerating potential V and the electric-sector voltage E are varied automatically in the ratio E^2/V. Using the terminology from above, a precursor ion m_1 having kinetic energy zeV is focused at the collector. Any metastable m_1 ions will give product ions m_2, which will have kinetic energy $(m_2/m_1)\cdot z\cdot e\cdot V$. If the accelerating voltage is changed to $(m_1/m_2)^2$, then these m_2 ions will have kinetic energy $(m_1/m_2)^2 \times (m_2/m_1)\cdot z\cdot e\cdot V = (m_1/m_2)\cdot z\cdot e\cdot V$. For these ions to pass through the electric sector with a trajectory of radius R, the electric field E must be increased simultaneously to $(m_1/m_2)E$. Thus, the two voltages V, E have been changed by $(m_1/m_2)^2$ and (m_1/m_2) and, for the ratio to remain constant, the voltages V/E^2 must be constant and hence the name of the scanning technique.

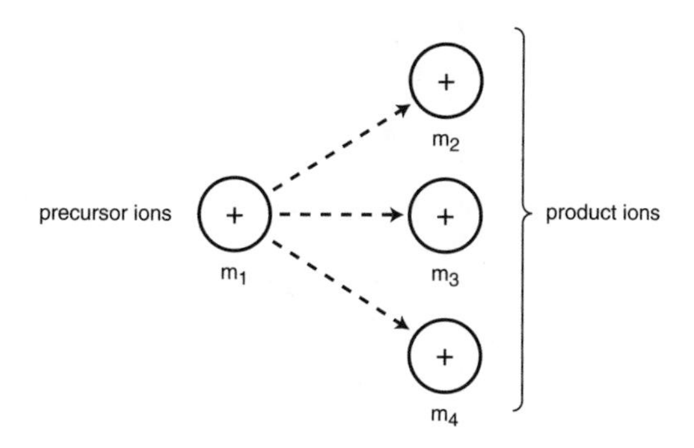

Figure 34.3
In an E^2/V- or a B/E-linked scan using an electric/magnetic-sector instrument, a precursor ion is selected. In this case it is m_1, which might be a molecular ion but equally could be any fragment ion. All product ions (m_2, m_3, m_4) from decomposition of m_1 in the first field-free region between the ion source and the ion collector are found, thereby giving connections m_1-m_2, m_1-m_3, m_1-m_4.

In use, a precursor ion is chosen, and then the V, E voltages are scanned such that E^2/V remains constant. Any fragment ions resulting from metastable decomposition of the precursor are automatically focused (Figure 34.3). The method provides information on which fragment ions arise from any given precursor. However, there is a limiting problem with this method of scanning. As the voltage V is varied, the sensitivity (efficiency) of detection at the ion collector changes, and the range of masses that can be examined at any one time is also limited on practical grounds for the values of E and V. To overcome this difficulty, the B/E scan was developed.

B/E Scan

To maintain ion detection efficiency, the accelerating voltage V is kept constant while the magnetic and electric fields are kept in a constant ratio. A metastable ion m_1 decomposing in the first field-free region to give a fragment ion m_2 will leave the m_2 ion with only $(m_2/m_1)zeV$ units of kinetic energy. For it to be transmitted through the electric sector, the voltage E must change by $(m_2/m_1)B$. The ion m_2 is then focused between the electric and magnetic sectors. However, the ion m_2 still has only $(m_2/m_1)zeV$ units of kinetic energy, and for it to be correctly focused by the magnetic field, the latter must decrease by (m_2/m_1). Both E and B have to be changed by the same amount. For any chosen precursor ion, the fields E and B are automatically adjusted so that the ratio B/E remains constant. The technique gives all product ions arising from any given precursor, as for the E^2/V scan (Figure 34.3). Ion resolution in the B/E scan remains very good.

B^2/E Scan

The B^2/E method is used to examine all precursors to any chosen product (fragment) ion. Suppose a normal fragment ion m_2 is selected. Since the accelerating voltage remains constant at V volts, the normal fragment ions m_2 have kinetic energy zeV. Any fragment ions m_2 arising from metastable decomposition must have kinetic energy $(m_2/m_1)zeV$. Therefore, for these ions to be transmitted through the electric sector, the sector voltage E must change to $(m_2/m_1)E$,

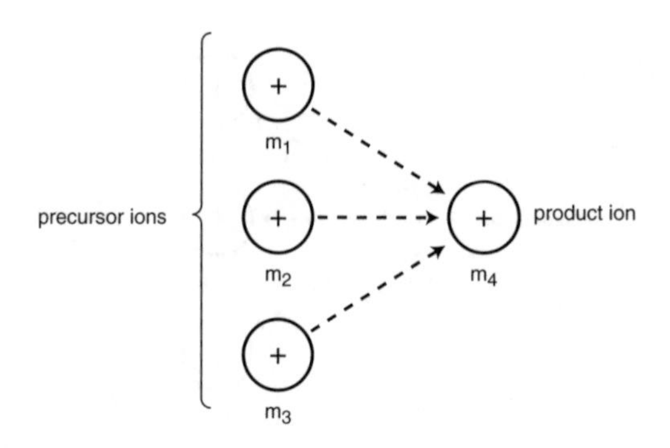

Figure 34.4
In a B^2/E-linked scan, a product ion (fragment ion) is selected. In this case it is m_4, which can be any fragment ion but not a molecular ion (connects with itself). All precursor ions (m_1, m_2, m_3), which decompose to give the product ion m_4, are found, giving connections m_1-m_4, m_2-m_4, m_3-m_4. Any one of the precursor ions could be a molecular ion.

and, for focusing by the magnetic field to the precursor m_1, the magnetic flux must increase by $(m_1/m_2)^2$. To find a precursor ion m_1, the square of the field B must be changed. In use, a fragment ion is chosen, and the two fields B and E are scanned in the ratio of B^2/E so that any precursor ions are focused by the magnetic field. The difference between the B/E and B^2/E scans lies in the directions of the connections. The former gives all product ions from a given precursor, while the latter gives all precursor ions to a given product ion (Figure 34.4). The mass resolution of a B^2/E scan is not as good as that of a B/E scan.

$(B/E)(1 - E)^{1/2}$ Scan

It sometimes happens that instead of finding which precursor ion gives which product ion (or vice versa), the object is to identify all pairs of precursor/product ions that show the loss of one particular mass. For example, it may be that a series of compounds contains some methyl esters (31 mass units). By looking for the loss of 31 mass units, viz., $m_1 - m_2 = 31$, methyl esters can be distinguished from other compounds (Figure 34.5).

Application of Linked Scanning

The electric and magnetic sectors may be of conventional or reversed geometry. In the latter situation, the magnetic sector precedes the electric one. For automation, the strength of the magnetic field must be determined, which is usually done by using a "Hall-effect" probe, which measures magnetic flux. The output from the probe (B) is used to set the value of the electric-sector voltage E in such a way that the required ratio of B and E is maintained. As an alternative, a calibration table relating scan time to mass can be used, with the value of E set in relation to time (and therefore mass and the magnetic field strength, B). When automated, complete metastable scanning can be accomplished within a few seconds. Since this time is less than that needed for most gas chromatography (GC) constituents to elute from a GC column, it is easily possible to carry out linked scanning in the GC/MS mode.

(B/E)(1 - E)$^{1/2}$ linked scan

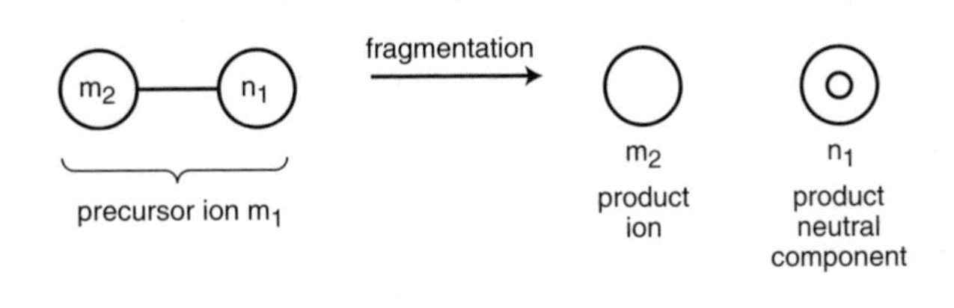

by mass, $n_1 = m_1 - m_2$

Figure 34.5
In a (B/E)(1 − E)$^{1/2}$-scanning mode, a mass difference is selected. For example, in this case a precursor ion m_1 is chosen (it is shown as being made up of two parts of mass: m_2, n_1). After fragmentation, the product ion is m_2 accompanied by a neutral particle of mass n_1. The mass difference ($n_1 = m_1 - m_2$) can be specified so only pairs of ions connected by this difference are found.

Linked Scanning, Ion Traps, and Hybrid Mass Spectrometers

The previous discussion of linked scanning has been concerned with only one process, where an ion m_1 fragments to give an ion m_2. However, a normal m_2 ion can also fragment to give another ion m_3, and this process can also be quantified by the previously discussed metastable-ion techniques. In general, these methods give either all product ions from one precursor or all precursor ions for one fragment ion. From this information, it is possible to extract links such as m_1 fragmenting to m_2 fragmenting to m_3. This same information can be obtained from ion traps. For example, an ion m_1 could be sped up so its collision with a neutral gas molecule gave a product ion m_2. This second ion would be retained in the trap and then sped up to cause a collision, making it fragment to m_3, with the latter retained in the trap. Thus a series of fragmentations can be examined on one sample. Such a succession of fragment-ion determinations have been termed MS^n, depending on how many fragmentation steps (n) are uncovered (Figure 34.6).

Besides obtaining this information from the electric/magnetic-sector methods described previously, this same information on ion connections can also be obtained quickly from collision-induced fragmentations in triple quadrupoles (see Chapter 33), in hybrid instruments having two mass analyzers — as in the Q/TOF (quadrupole and time-of-flight) instruments (see Chapter 23) — or in instruments having two successive magnetic sectors. These sorts of mass spectrometer are frequently used with ionization methods that give stable molecular ions with no tendency to fragment. If structural information is to be obtained in such cases, it is necessary to activate the molecular ions, which is usually done by colliding them with neutral gas molecules in a special

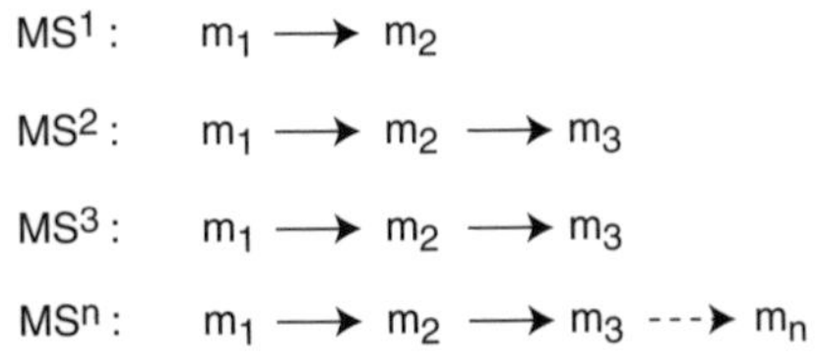

MS^1 : $m_1 \longrightarrow m_2$

MS^2 : $m_1 \longrightarrow m_2 \longrightarrow m_3$

MS^3 : $m_1 \longrightarrow m_2 \longrightarrow m_3$

MS^n : $m_1 \longrightarrow m_2 \longrightarrow m_3 \dashrightarrow m_n$

Figure 34.6
Diagram showing the relation between the designation MS^n and the number of sequential fragmentation steps involved.

collision cell. Triple quadrupole, Q/TOF, and double magnetic-sector instruments are used in this mode. Collision-induced decomposition produces significant fragmentation, thus providing a more sensitive technique for examining connections between ions rather than measuring the decomposition of natural metastable ions, which occur in low abundance.

Conclusion

In magnetic/electric-sector instruments, there are three major variables: the accelerating field of voltage V, an electric-sector field of voltage E, and the magnetic field of strength B. By linking two of these three fields (V with E or E with B), simple automated linked scanning can reveal the connections of molecular and subsequent fragment ions. These connections are important for obtaining structural information on the sample being investigated.

Chapter 35

Gas Chromatography (GC) and Liquid Chromatography (LC)

Introduction

Chromatography is a general term used to describe the separation of mixtures into their individual components by causing them to pass through a column of a solid or liquid. The name derives from the first attempts at such a separation when the green and yellow coloring matters of spinach were revealed by putting an extract of it onto a column of a solid such as chalk or sugar held inside a glass tube. The components showed up as colored bands observed at different positions along the column (Figure 35.1). The term *chromatography* is derived from chroma (color) and graph (writing).

Since then, chromatography has developed into a general technique for separating any kind of mixture, colored or colorless. It can be used to separate mixtures of gases, liquids, or solids and has achieved enormous importance. Two particular forms of chromatography, gas and liquid, are used more than any other and have been enhanced even more by coupling them to mass spectrometers, which act not only as a detection system (see below) but also provide valuable structural information about the separated components of mixtures. These coupled techniques of gas chromatography/mass spectrometry (GC/MS) and liquid chromatography/mass spectrometry (LC/MS) are described in Chapters 36 and 37, respectively. This chapter discusses the principles of gas and liquid chromatography and presents examples of their uses.

Principles of Gas and Liquid Chromatography

For chromatography to occur, a mobile phase and a stationary phase are needed. The mobile phase is a gas in gas chromatography and is a liquid in liquid chromatography. In GC, the stationary phase is almost always a column of liquid, gum, or elastomer but not a solid, while in LC the stationary phase is generally a column of a porous solid. Strictly, the names should be gas liquid chromatography (GLC) and liquid solid chromatography (LSC), as will be found in the early literature. The shortened names (GC, LC) came into use through widespread adoption of the methods. The flow of the mobile phase through the stationary phase forces a

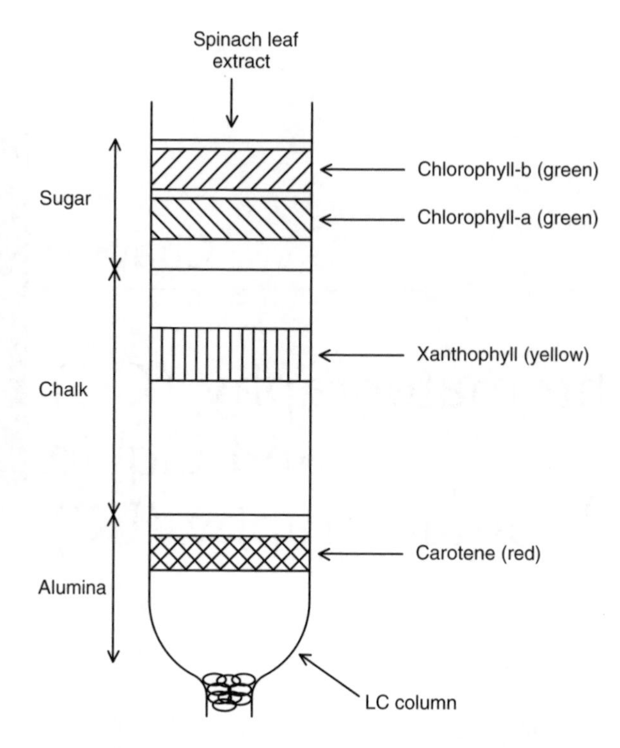

Figure 35.1
A chromatographic column filled in three sections with ground sugar, chalk, and alumina. When a petroleum extract of spinach leaves is run onto the top of the column, the extract spreads down the column, but not uniformly; bands of green chlorophylls stop near the top, yellow xanthophyll further down, and red carotene near the bottom.

mixture to pass through the chromatography column. During its passage through the column of stationary phase, a mixture gradually separates into its individual components, some passing through the column of stationary phase faster than others and emerging (eluting) earlier.

Requirements for Chromatographic Apparatus

In principle, a chromatographic instrument is very simple and is little more than an advanced plumbing system (Figure 35.2a, b). The mixture to be separated is put onto the start (or top) of the column using an injector. The gas or liquid mobile phase flows from a suitable store (typically a gas cylinder or bottle of solvent) past the injector and carries the injected mixture through the chromatographic column, which contains a stationary phase (a liquid for GC and a modified solid for LC). When the separated components emerge (elute) from the end (or bottom) of the column, they must be sensed by a detector. Finally, the detector must send an electrical signal, directly or indirectly, to a readout device or recorder. The time from injection of the mixture to recording the elution of individual components can be from a few seconds to an hour or more. The resulting chart from the recorder shows the amounts of separated components plotted against their times of elution and is called a chromatogram (Figure 35.3a, b).

Many kinds of detectors have been designed, ranging from the widely used, cheap but robust flame ionization (GC) or ultraviolet absorption type (LC) to the much more exciting and informative, if much more expensive, mass spectrometer.

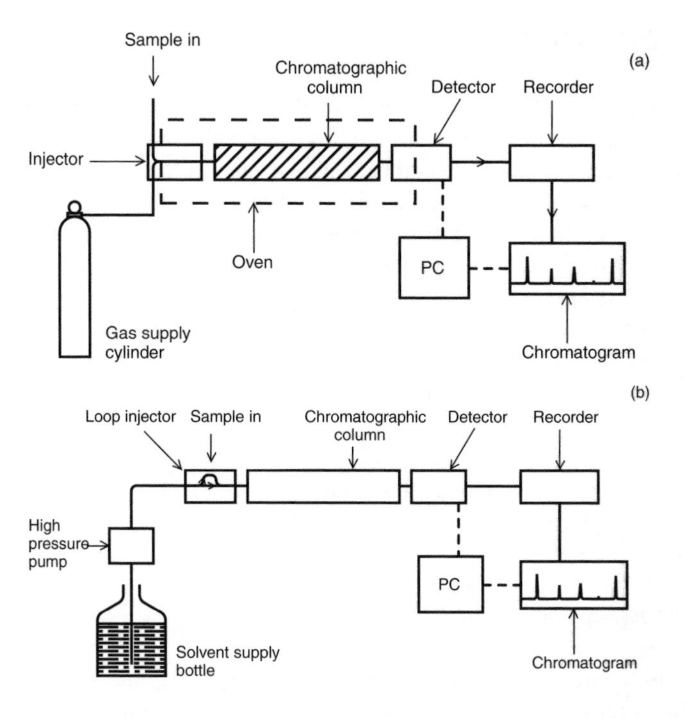

Figure 35.2
(a) Schematic layout of a gas chromatograph. A sample is vaporized in the injector, mixed with nitrogen, and carried by the gas stream through a column, where separation into components occurs. The emerging (eluting) components are detected and recorded to give a gas chromatogram. Alternatively, a personal computer can be used to acquire and display data. The injector, column, and the lower part of the detector are placed in an oven. (b) Schematic layout for a liquid chromatograph. Note the essential similarity to the gas chromatograph shown in Figure 35.2a; however, in detail, the injector, column, and detector are quite different. No oven is needed.

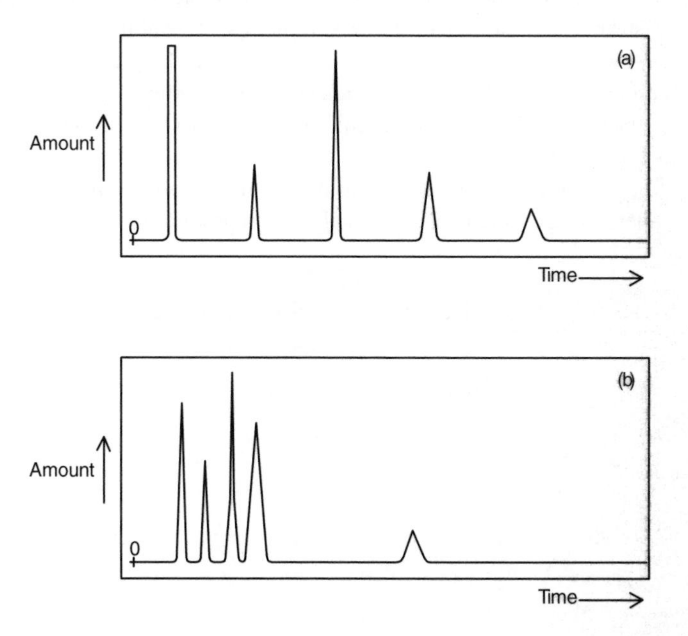

Figure 35.3
Typical (a) gas and (b) liquid chromatograms. The charts show amounts (y-axis) of substance emerging from a column versus time (x-axis). The time taken (measured at the top of a peak) for a substance to elute is called a retention time.

The Chromatographic Process

The principles behind separation of mixtures by GC or LC are more or less the same, but the experimental procedures are different. During chromatography, any one component of a mixture is partitioned between the mobile and stationary phases such that, at any one instant, some of the component is in one phase and the remainder is in the other. In this dynamic equilibrium system, progress of a component along the chromatographic column is effected by the flow of the mobile phase in a process of continual exchange between the stationary and mobile phases. The process can be likened simplistically to the movement of a bus along its route. For some of the time, the bus moves (molecules in the mobile phase), but, at other times, the bus is stopped (molecules in the stationary phase). The time taken by a bus to cover its route (retention time for molecules in chromatography) depends on the number and the length of time of the stops it makes and its speed between stops.

Similarly, with molecules, their speed of movement through the chromatographic column depends on the time spent in the mobile phase compared with that in the stationary one and on the flow rate of the mobile phase.

Gas Chromatographic Phases

For GC, the mobile phase is a gas (usually nitrogen or helium), and the movement of mixture components along the column will depend on their volatilities; the more volatile a component, the more time is spent in the gas phase and the faster it will be swept along. This factor is a major one, but not the only, controlling the rate of movement of a substance through a gas chromatographic column. The various components of a mixture will have different affinities for the stationary phase, a nonvolatile liquid over which the mobile gas phase flows. Thus, for two components having the same volatilities but differing polarities, their rates of passage will be different. A polar component will tend to stick on a polar stationary phase but not on a nonpolar one, while nonpolar compounds will be held more strongly by nonpolar stationary phases. The ability of a stationary phase to separate mixtures in GC depends on its chemical composition. Many types of stationary phase are used, a few of which are listed in Table 35.1. The choice of stationary phase affects GC operation. An example of the change in gas chromatographic behavior for two different columns of the same length is shown in Figure 35.4.

Because volatility is such an important factor in GC, the chromatographic column is contained in an oven, the temperature of which can be closely and reproducibly controlled. For very volatile

TABLE 35.1
Some Typical Stationary Phases for Use in GC

Name	Approximate		
	Type	Polarity	(McReynolds Number)*
Nujol	Mineral oil	Nonpolar	11
Silicone SE-30	Elastomer	Nonpolar	41
Apiezon L	Grease	Nonpolar	42
Silicone OV-17	Elastomer	Intermediate	202
Didecyl phthalate	Pure liquid	Intermediate	235
Carbowax 20M	Polymer	Polar	510
DEGS	Polymer	Very polar	835

* The McReynolds number gives an approximate indication of polarity on a scale of 0 (nonpolar) to about 1000 (extremely polar).

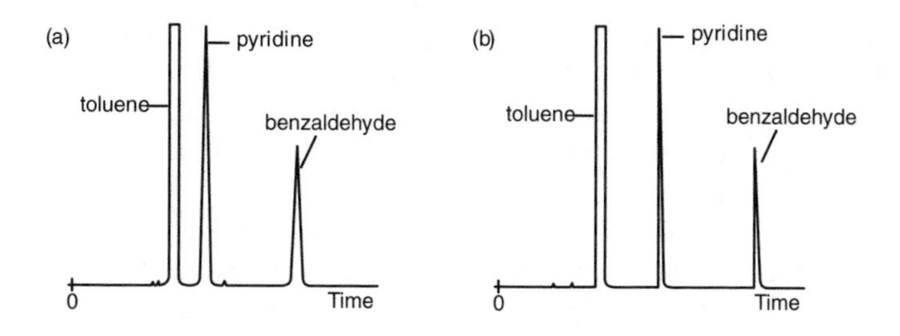

Figure 35.4
Two gas chromatograms showing the effect of polarity of the stationary phase on the separation efficiency for three substances of increasing polarity: toluene, pyridine, and benzaldehyde. (a) Separation on silicone SE-30, a nonpolar phase, and (b) separation on elastomer OV-351, a more polar phase. Note the greatly changed absolute and relative retention times; the more polar pyridine and benzaldehyde are affected most by the move to a more polar stationary phase.

compounds, the oven may be operated at only 30–40°C, but for nonvolatile substances, temperatures of 200–250°C may be needed. For GC analysis, the column can be operated at one fixed temperature (isothermally) or over a temperature range (temperature programming).

In general, the longer a chromatographic column, the better will be the separation of mixture components. In modern gas chromatography, columns are usually made from quartz and tend to be very long (coiled), often 10–50 m, and narrow (0.1–1.0 mm, internal diameter) — hence their common name of capillary columns. The stationary phase is coated very thinly on the whole length of the inside wall of the capillary column. Typically, the mobile gas phase flows over the stationary phase in the column at a rate of about 1–2 ml/min.

Liquid Chromatographic Phases

For LC, temperature is not as important as in GC because volatility is not important. The columns are usually metal, and they are operated at or near ambient temperatures, so the temperature-controlled oven used for GC is unnecessary. An LC mobile phase is a solvent such as water, methanol, or acetonitrile, and, if only a single solvent is used for analysis, the chromatography is said to be isocratic. Alternatively, mixtures of solvents can be employed. In fact, chromatography may start with one single solvent or mixture of solvents and gradually change to a different mix of solvents as analysis proceeds (gradient elution).

The stationary phase in LC is a fine granular solid such as silica gel. It can be used as such (mainly for nonpolar compounds), or the granules can be modified by a surface-bonded coating that changes (reverses) the polarity of the gel. A very small selection of stationary phases is listed in Table 35.2.

TABLE 35.2
Some Solid Stationary Phases for Use in LC

Name	Polarity
Bondapak C18	Non-polar
Zorbax ODS	Non-polar
Carbowax 400	Intermediate
Micropak CN	Intermediate
Corasil II	Polar
Silica gel Very	Polar

Reversed-phase columns are used to separate polar substances. Although in LC the stationary phase is a solid, it is necessary to bear in mind that there may be a thin film of liquid (e.g., water) held on its surface, and this film will modify the behavior of sample components equilibrating between the mobile and stationary phases. A textbook on LC should be consulted for deeper discussion on such aspects.

In LC, because the mobile phase is a liquid and the stationary one is a granular solid, viscosity limitations rule out the simple use of the long capillary columns found in GC. Short columns of 10–25 cm and 2–4 mm internal diameter are more usual in LC. It is difficult to force the mobile solvent phase through such columns, and high-pressure pumps must be used to get a reasonable flow rate — hence the name high-pressure liquid chromatography (HPLC). More recently, very narrow LC columns have become available, but these nanocolumns must be operated at very high pressure to force the mobile liquid phase through them. Typically, a liquid phase flow of some 1 to 2 ml/min at a pressure of 20 to 200 bar (300 to 3000 lb/in^2) is used. Very often the high-pressure part of the terminology is omitted, as in LC/MS rather than HPLC/MS

Injectors

For GC, the injector is most frequently a small heated space attached to the start of the column. A sample of the mixture to be analyzed is injected into this space by use of a syringe, which pierces a rubber septum. The injector needs to be hot enough to immediately vaporize the sample, which is then swept onto the head of the column by the mobile gas phase. Generally, the injector is kept at a temperature 50°C higher than is the column oven. Variants on this principle are in use, in particular the split/splitless injector. This injector can be used in a splitless mode, in which the entire injected sample goes onto the column, or in a split mode, in which only part of the sample goes onto the column, the remainder vented to atmosphere. For other less usual forms of injector, a specialist book on GC should be consulted.

For liquid chromatography, a sample of the mixture solution is injected through a loop injector which allows a quantity of the solution to be placed in a small tubular loop at atmospheric pressure. By manipulating a valve, the high-pressure flow of solvent to the column is diverted through the loop, carrying the sample with it (Figure 35.5).

Detectors

Specialized detectors and inlet systems for GC/MS and LC/MS are described in Chapters 36 and 37, respectively.

The effluent from the end of a GC column is usually nitrogen or helium and contain a very small proportion of organics as they emerge (elute) from the column. The most widely used detector is one in which the eluate is burned continuously in a flame after admixing it with hydrogen and air (flame ionization detector). Ions and electrons formed when emerging organics burn in the flame are monitored electronically, and the resulting electric current is used to drive a chart recorder (Figure 35.6).

In LC, the most common means for monitoring the eluant is to pass it through a cell connected into an ultraviolet spectrometer. As substances elute from the column, their ultraviolet absorption is measured and recorded. Alternatively, the refractive index of the eluant is monitored since it varies from the value for a pure solvent when it contains organics from the column.

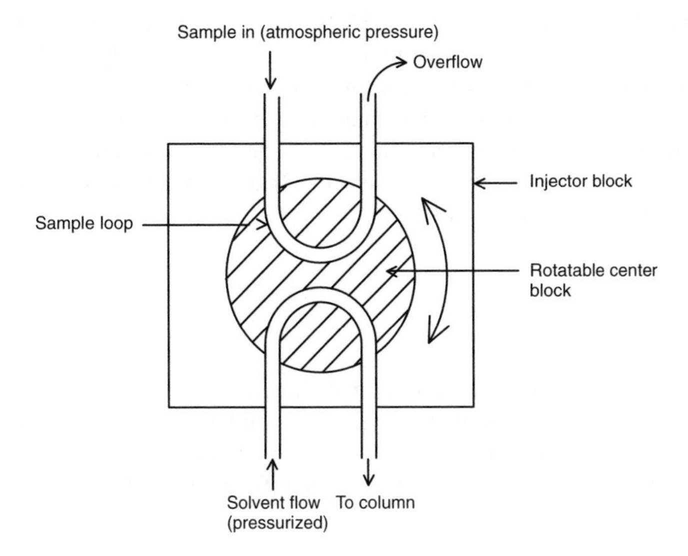

Figure 35.5
A typical loop injector showing the sampling position with pressurized solvent flowing through one loop onto the column and the sample solution placed in the other loop at atmospheric pressure. Rotation of the loop carrier through 180° puts the sample into the liquid flow at high pressure with only momentary change in pressure in the system.

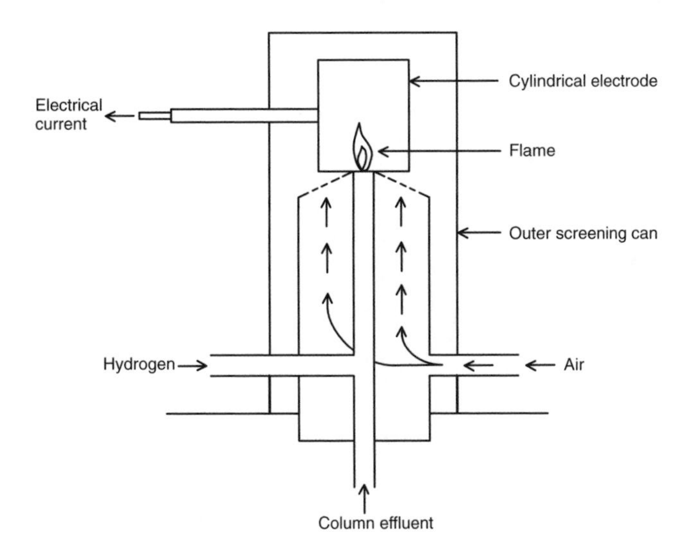

Figure 35.6
Schematic diagram of a flame ionization detector. Ions and electrons formed in the flame provide an electrically conducting path between the flame at earth potential and an insulated cylindrical metal electrode at high potential surrounding the flame; the flow of current is monitored, amplified, and passed to the recording system.

Uses of GC and LC

As a rule of thumb, one can say that the efficiency of separation of mixtures and the simplicity of operating and maintaining apparatus are much greater for GC than for LC. Hence, other things being equal, GC is most often the technique of first choice and can be used with a very wide variety of compound types. However, for nonvolatile or thermally labile substances like peptides, proteins, nucleotides, sugars, carbohydrates, and many organometallics, GC may be ruled out completely and LC comes into its own. Apart from availability of instrumentation, in choosing between GC and LC, it is necessary to consider the nature of the sample to be analyzed. Whereas GC cannot be used with nonvolatile or thermally labile compounds, LC can be used for almost all the types of substance routinely analyzed by GC. The major reasons for not simply using LC for all types of compound lie mostly in the far greater achievable resolution of mixtures into their components by GC, the generally greater sensitivity of GC detectors (nanogram amounts can be analyzed easily), and the easier operation of GC instruments.

Conclusion

Mixtures passed through special columns (chromatography) in the gas phase (GC) or liquid phase (LC) can be separated into their individual components and analyzed qualitatively and/or quantitatively. Both GC and LC analyzers can be directly coupled to mass spectrometers, a powerful combination that can simultaneously separate and identify components of mixtures.

Chapter 36

Gas Chromatography/Mass Spectrometry (GC/MS)

Introduction

As its name implies, this important analytical technique combines two separate procedures: gas chromatography (GC) and mass spectrometry (MS). Both individual techniques are quite old. GC developed as a means of separating volatile mixtures into their component substances and provided a big step forward in the analysis of mixtures. The method is described fully in Chapter 35, but it can be summarized as follows (Figure 36.1): by passing a mixture in a gas stream (the gas phase) through a long capillary column, the inside walls of which are thinly coated with a liquid (the liquid phase), the components of the mixture become separated and emerge (elute) one after another from the end of the column. In a simple GC instrument, the emerging components are either burnt in a flame for detection (the popular flame ionization detector) or passed to atmosphere after traversing some other kind of detector. The detected components are recorded as peaks on a chart (the gas chromatogram). The area of a peak correlates with the amount of a component, and the time taken to pass through the column (the retention time, i.e., the time to the peak maximum) gives some information on the possible identity of the component. However, the identification is seldom absolutely certain and is often either vague or impossible to determine.

In complete contrast to a GC apparatus, a mass spectrometer is generally not useful for dealing with mixtures. If a single substance is put into a mass spectrometer, its mass spectrum (Figure 36.2) can be obtained with a variety of ionization methods that are described in Chapters 1 through 5. Once the spectrum is obtained, it is often possible to make a positive identification of the substance or to confirm its molecular structure. Clearly, if a mixture of substances were put into the MS, the resulting mass spectrum would be a summation of the spectra of all the components (Figure 36.3). This spectrum would be extremely complex, and it would be impossible to identify positively the various components. (Note that some MS instruments can deal with mixtures, and these are described in Chapters 20 through 23, which deal with hybrid MS/MS systems, and Chapters 33 and 34, which deal with linked scanning.)

Thus, there is one instrument (GC) that is highly efficient for separating mixtures into their components but is not good on identification, and another instrument (MS) that is efficient at identifying single substances but is not good with mixtures. It is not surprising to find that early efforts were made to combine the two instruments into one system (GC/MS) capable of separating, positively identifying, and quantifying complex mixtures, provided these could be vaporized. Com-

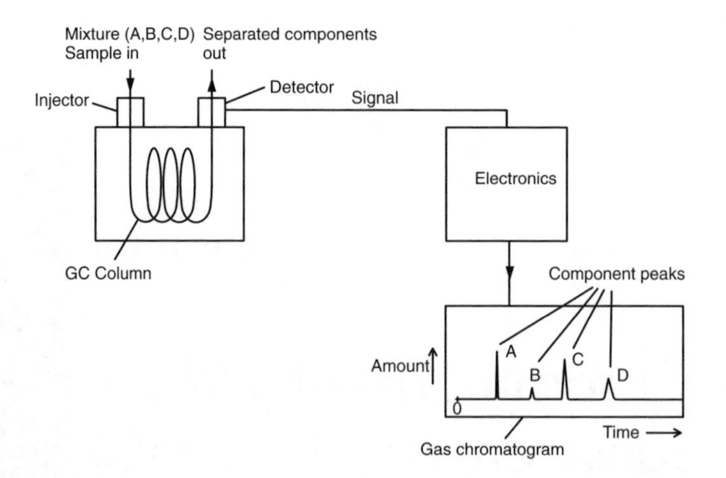

Figure 36.1
Schematic diagram showing the injection of a mixture of four substances (A, B, C, D) onto a GC column, followed by their separation into individual components, their detection, and the display (gas chromatogram) of the separated materials emerging at different times from the column.

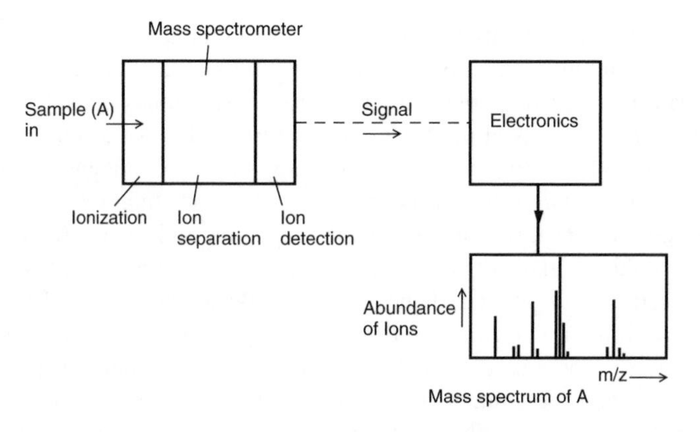

Figure 36.2
Schematic diagram of a mass spectrometer. After insertion of a sample (A), it is ionized, the ions are separated according to m/z value, and the numbers of ions (abundances) at each m/z value are plotted against m/z to give the mass spectrum of A. By studying the mass spectrum, A can be identified.

bining GC and MS was not without its problems (see below), but modern GC/MS is now a routinely used methodology in many areas, ranging from interplanetary probes to examination of environmental dust samples to determine dioxin levels. Further, the addition of GC to MS does not simply give a sum of the two alone; the information provided by combined GC/MS yields information that could not be extracted from either technique alone, an aspect that is discussed below.

Connection between GC and MS

As described above, the mobile phase carrying mixture components along a gas chromatographic column is a gas, usually nitrogen or helium. This gas flows at or near atmospheric pressure at a rate generally about 0.5 to 3.0 ml/min and eventually flows out of the end of the capillary column into the ion source of the mass spectrometer. The ion sources in GC/MS systems normally operate at about 10^{-5} mbar for electron ionization to about 10^{-3} mbar for chemical ionization. This large pressure

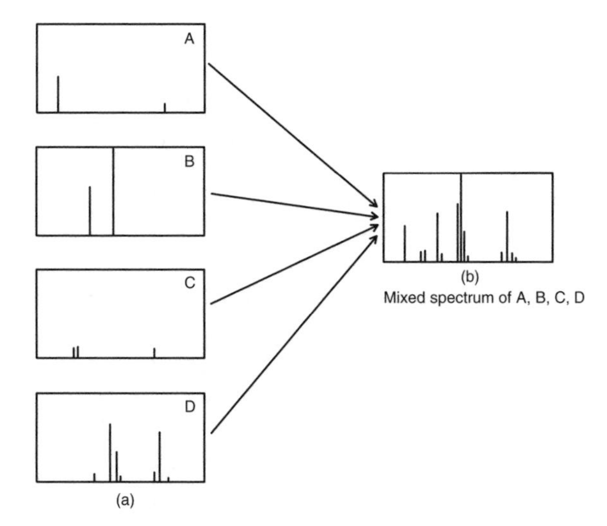

Figure 36.3
By way of illustration, very simple spectra for four substances (A, B, C, D) are shown: (a) separately and (b) mixed in unequal proportions. The mixture spectrum is virtually impossible to decode if A, B, C, D are not known beforehand to be present.

change between the end of the chromatographic column and the inside of the ion source causes the gas to expand and reach a flow equivalent to several liters per minute. Therefore, large pumps are required to remove the excess gas while maintaining the vacuum inside the source near the level optimum for ionization. In modern GC/MS installations, the use of capillary chromatographic columns and high-speed pumps means that the end of the column can be literally right inside the ion source. In older GC/MS instruments, gas flow rates were much greater, and it was necessary to have an interface between the end of the column and the ion source. This interface or separator removed much of the GC carrier gas without also removing the eluting mixture components, which traveled into the ion source. The jet separator was a popular form that can still be found occasionally.

Recording Mass Spectra

As each mixture component elutes and appears in the ion source, it is normally ionized either by an electron beam (see Chapter 3, "Electron Ionization") or by a reagent gas (see Chapter 1, "Chemical Ionization"), and the resulting ions are analyzed by the mass spectrometer to give a mass spectrum (Figure 36.4).

For capillary GC, separated mixture components elute in a short time interval, often lasting only a few seconds. Thus, the amount of any one component in the ion source is not constant as its mass spectrum is being obtained. Rather, it starts at zero, rises rapidly to a maximum, and drops rapidly back to zero. If this passage through the ion source is faster than the mass spectrometer's ability to scan the spectrum, then a true spectrum will not be found because the start and end of the scan will show less compound than at the middle of the scan. This changing concentration of eluting component results in a distorted mass spectrum that might not be recognizable (Figure 36.5). The answer to this problem is to scan the spectrum so fast that, in effect, the concentration of the eluting component scarcely changes during the time needed to acquire a spectrum.

For a quadrupole mass spectrometer, this high rate of scanning is not difficult because it requires only simple changes in some electrical voltages, and these changes can be made electronically at very high speed, which is why quadrupoles are popular in GC/MS combinations. In the early days of magnetic-sector mass spectrometers, the required scanning speed was not possible because of serious hysteresis effects in the magnets. With modern magnet technology, scanning can be done

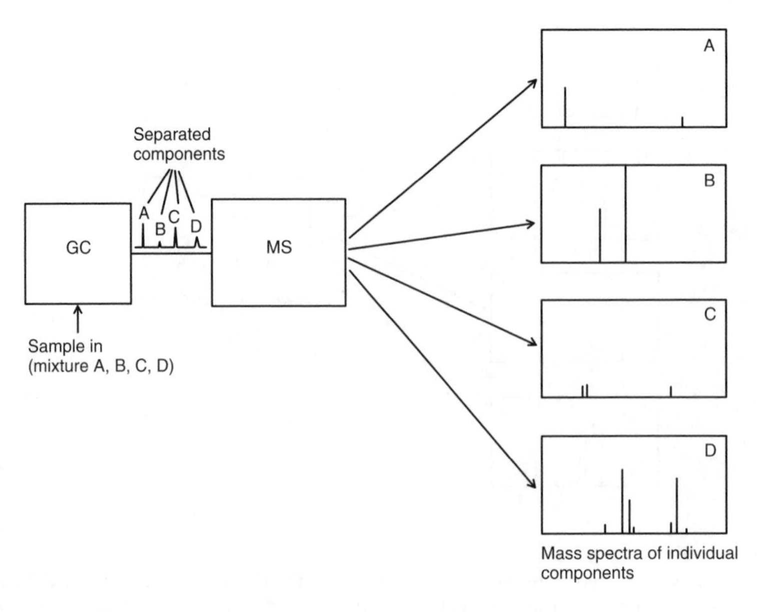

Figure 36.4
In a GC/MS combination, passage of the separated components (A, B, C, D) successively into the mass spectrometer yields their individual spectra.

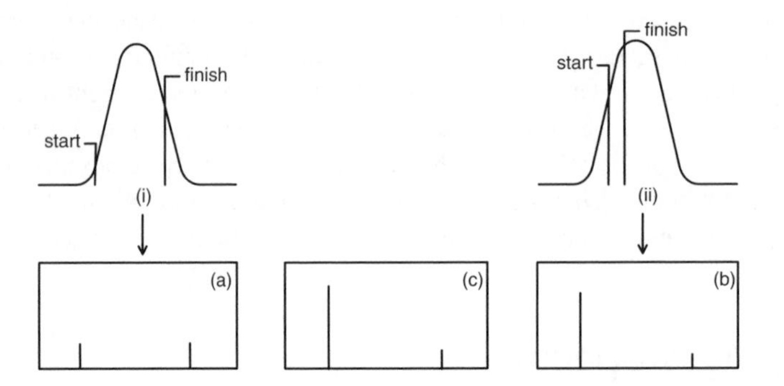

Figure 36.5
Slow scanning (i) of the mass spectrum over a GC peak for substance A gives spectrum (a), but rapid scanning (ii) gives spectrum (b), which is much closer to the true spectrum (c).

at high speed with insignificant hysteresis, and magnetic-sector instruments can now compete with quadrupoles. While ultimate scan speed for a magnetic instrument is not as good as the quadrupole's, the former does have the advantage of providing greater mass resolution at higher mass. However, in GC/MS, where the highest analyzed mass is likely to be less than about 600 Da, this advantage of the magnetic instrument is of little consequence.

As described above, the concentration of an eluting component in the ion source goes from zero to zero through a maximum. Where should the scan be taken? Usually, the greater the amount of a substance in an ion source, the better will be the resulting mass spectrum (within reason!). This situation suggests that the best time for a scan will be near the maximum concentration (the top of a GC peak) and that the instrument operator must watch the developing chromatogram continually, trying to judge the best moment to measure a spectrum. Since a gas chromatogram can routinely take 20 to 50 min to obtain, such a watching brief is labor intensive (and deadly dull!). The simple answer is to set the mass spectrometer scanning continuously. Therefore, as the mixture is injected onto the chromatographic column, repetitive scanning is instituted over a preset mass range (e.g., 50 to 500 mass units) at a preset interval (e.g., every

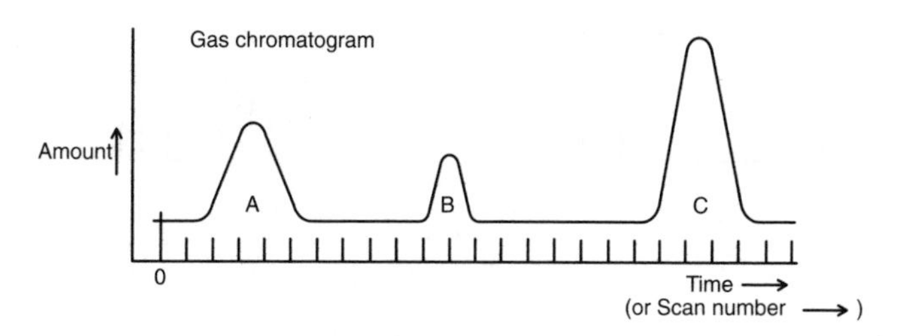

Figure 36.6
Typical gas chromatogram showing three components (A, B, C) emerging at different times. Continuous scanning for mass spectra is started at zero, and scans are repeated regularly (small tick marks). Thus peak A is scanned five times during its passage through the ion source. Because scanning is regular, the time axis can be replaced by a scan-number axis.

10 sec). Thus, scan follows scan right through the chromatogram, and hundreds of mass spectra can be recorded in a routine GC/MS experiment (Figure 36.6).

Data flow at such high speeds and in such copious quantities from the mass spectrometer that a microprocessor/computer is necessary. At the end of a GC/MS run, all of the data are in storage, usually on a hard disc in the computer, in the form of a mass spectrum for every scan that was done. These spectra can then be manipulated. For example, the scan corresponding to the top of a chromatographic peak can be selected (scan 3 or 4 in Figure 36.6 for example) and the mass spectrum displayed or printed out. As a routine, the computer adds all the ion peaks in each mass spectrum (actually summing electrical currents) to give a total ion current for the mass spectrum. These total ion currents are plotted along an x-axis (time for elution) and a y-axis (amount of total ion current) to give a total ion current (TIC) chromatogram showing the elution of all the components of a mixture. Storage of all these data has other major advantages.

Manipulation of Scan Data

Some ways in which data can be utilized are described briefly below.

Background Subtraction with Library Search

Gas chromatograms are often obtained by running the column at temperatures that are sufficiently high to volatilize the stationary liquid phase on the column walls so that it, too, begins to elute in the mobile gas stream from the end of the column. This effect is called column bleed. Clearly, if the liquid phase is eluting all the time, it must reach the ion source when there is (and even when there is not) an eluting mixture component, and, since the spectrometer scans continuously, the mass spectrum of the column bleed is recorded continuously. Therefore, the recorded spectrum for an eluting component will not be pure but, rather, will be mixed with the column bleed. Fortunately, it is a simple matter to get the computer to subtract the (known) column bleed spectrum from the mixed spectrum to obtain the desired pure spectrum of the eluting component. This process is called background subtraction and is carried out routinely before a library search (next item) or before the spectrum is printed out (Figure 36.7).

Once a mass spectrum from an eluting component has been acquired, the next step is to try to identify the component either through the skill of the mass spectroscopist or by resorting to a library search. Most modern GC/MS systems with an attached data station include a large library of spectra from known compounds (e.g., the NIST library). There may be as many as 50,000 to 60,000 stored spectra covering most of the known simple volatile compounds likely to be met in analytical work. Using special search routines under the control of the computer, one can examine

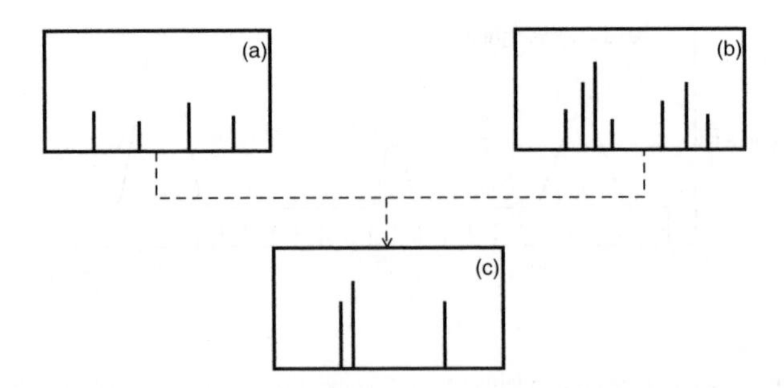

Figure 36.7
Column bleed gives a mass spectrum (a) that is mixed with an eluting component to give a complex spectrum (b). By subtracting (a) from (b), the true spectrum (c) of the eluting component is obtained.

this huge database very quickly, comparing the mass spectrum of the eluted mixture component with each of the library spectra. The computer then provides a short list of the best matches between the library spectra and the measured one. Often, from the closeness of the match or fit and its chromatographic retention time, the eluted component can be identified positively.

Resolution Enhancement

Sometimes (not infrequently) a single peak from a gas chromatogram reflects the elution of two, three, or more substances. By examining the scans across the peak and using a simple mathematical process, the computer can reveal the existence of more than one eluting component in a single peak and print out the mass spectra of the discovered components. This process is a form of resolution enhancement; the GC/MS combination can do what neither the GC nor the MS alone could do. In effect, the mass spectrometer has improved (enhanced) the resolution of the gas chromatograph (Figure 36.8).

Mass Chromatogram

There could be a series of compounds, designated RX, in which the R part was the same for all, but X was different. In such a series, the R part can give a characteristic ion — the mass corresponding to R is the same for every member of the series. When one member of such a series elutes from the chromatographic column, its resulting mass spectrum will always contain one mass common to the whole series. Therefore, even if the series is mixed with other compounds, its members can be recognized from the characteristic ion. It is a simple process to get the computer to print out a chromatogram in which only those scans containing the characteristic ion are used to draw the chromatogram. Basically, the output is blind to any component other than those containing an R group. Such a selected chromatogram is called a mass chromatogram and is useful for pinpointing where certain compounds elute without the need for examining all the spectra (Figure 36.9).

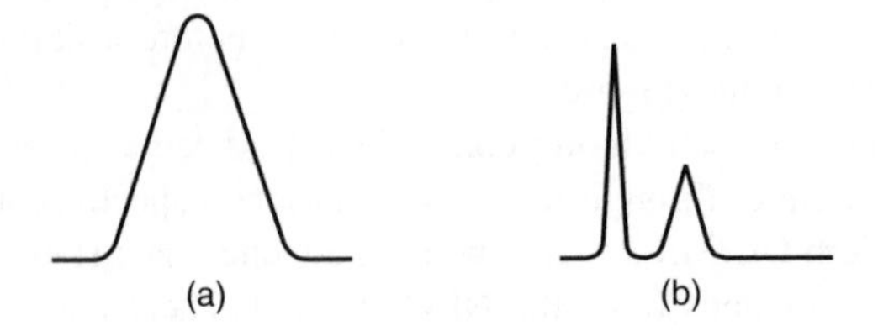

Figure 36.8
A single peak from an ordinary gas chromatogram (a) is revealed as two closely separated peaks by resolution enhancement (b).

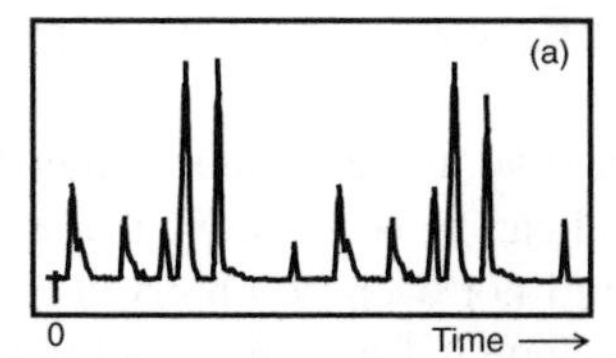

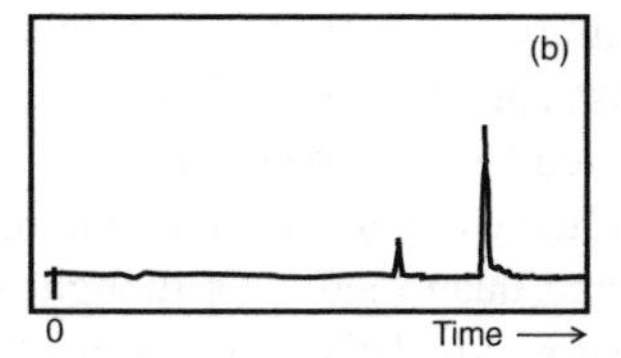

Figure 36.9
(a) A typical total-ion-current gas chromatogram showing many components, some of which have a part R (e.g., of characteristic mass 91). (b) A mass chromatogram based on m/z 91. The same data as used in (a) are plotted, but they are manipulated so only ion currents corresponding to m/z 91 are shown. Note the decrease in complexity, making the desired identification much easier.

Selected Ion Recording

In a process similar to that described in the previous item, the stored data can be used to identify not just a series of compounds but specific ones. For example, any compound containing a chlorine atom is obvious from its mass spectrum, since natural chlorine occurs as two isotopes, ^{35}Cl and ^{37}Cl, in a ratio of 3:1. Thus its mass spectrum will have two molecular ions separated by two mass units (35 + 2 = 37) in an abundance ratio of 3:1. It becomes a trivial exercise for the computer to print out only those scans in which two ions are found separated by two mass units in the abundance ratio of 3:1 (Figure 36.10). This selection of only certain ion masses is called selected ion recording (SIR) or, sometimes, selected ion monitoring (SIM, an unfortunate

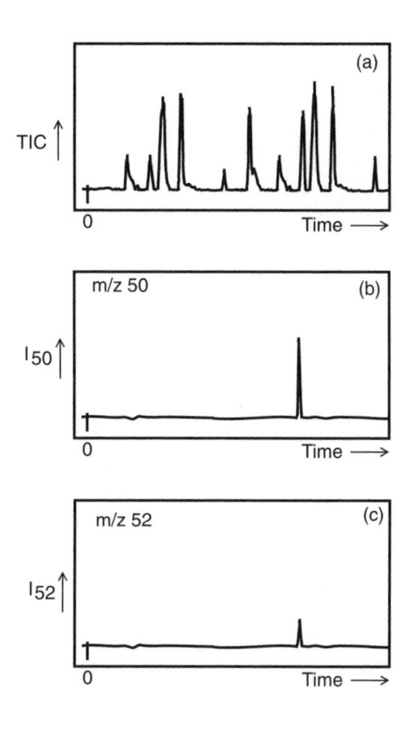

Figure 36.10
(a) A typical total-ion-current chromatogram showing many components of a mixture. The aim is to find if any methyl chloride (CH_3Cl; RMM = 50, 52) is present. To this end, the data (or a new chromatogram) are scanned at two specific positions: (b) at m/z 50 and (c) at m/z 52. Note how m/z 50, 52 both reach a maximum at the same scan and have a ratio in peak heights of 3:1 (^{35}Cl:^{37}Cl = 3:1). Such an experiment in selected-ion recording identifies the suspected component against a complex background and can be made very sensitive.

choice of terminology, since there is another well-known mass spectrometric technique called secondary ion mass spectrometry, SIMS).

The ions selected for such recording can be one, two, three, or more (multiple ion recording). In fact, through judicious choice of ions, the method can be so selective that a chosen component can be identified and quantified even though it could not even be observed in the original total-ion-current chromatogram. This very powerful technique is frequently used to examine extremely complex mixtures when the goal is to identify small amounts of a particular substance in a mass of other things, as with detection of banned drugs in the body fluids of athletes or racehorses.

Conclusion

By connecting a gas chromatograph to a suitable mass spectrometer and including a data system, the combined method of GC/MS can be used routinely to separate complex mixtures into their individual components, identify the components, and estimate their amounts. The technique is widely used.

Chapter 37

Liquid Chromatography/Mass Spectrometry (LC/MS)

Introduction

As its name implies, this important analytical technique combines two separate procedures: liquid chromatography (LC) and mass spectrometry (MS). Both individual techniques are quite old. LC developed as a means of separating nonvolatile mixtures into their component substances and provided a big step forward in revealing their complexities and analyzing them. The method is described fully in Chapter 35, but it can be summarized as follows (Figure 37.1): by passing a mixture in a liquid stream (the mobile or liquid phase) through a long column packed with a stationary phase (particles of a special solid), the components of the mixture become separated and emerge (elute) one after another from the end of the column. In a simple LC instrument, the emerging components dissolved in the liquid mobile phase are measured by passing the liquid stream through either an ultraviolet (UV) or a refractive-index detector. The detected components are recorded as peaks on a chart (the liquid chromatogram). The area of a peak correlates with the amount of a component, and the time taken to pass through the column (the retention time, i.e., the time to the peak maximum) gives some information on the possible identity of the component. However, the identification is seldom absolutely certain and is often either vague or impossible to determine.

In complete contrast to an LC apparatus, a mass spectrometer is generally not useful for dealing with mixtures. If a single substance is put into a mass spectrometer, its mass spectrum (Figure 37.2) can be obtained with a variety of ionization methods that are described in Chapters 1 through 5. Once the spectrum is obtained, it is often possible to make a positive identification of the substance or to confirm its molecular structure. Clearly, if a mixture of substances were put into the MS, the resulting mass spectrum would be a summation of the spectra of all the components (Figure 37.3). This spectrum would be extremely complex, and it would be impossible to identify positively the various components. (Note some MS instruments can deal with mixtures and are described in Chapters 20 through 23, which deal with hybrid MS/MS systems, and Chapters 33 and 34, which deal with linked scanning.)

Thus, there is one instrument (LC) that is highly efficient for separating mixtures into their components but is not good on identification, and another instrument (MS) that is efficient at identifying single substances but is not good with mixtures. It is not surprising to find that early efforts were made to combine the two instruments into one system (LC/MS) capable of separating,

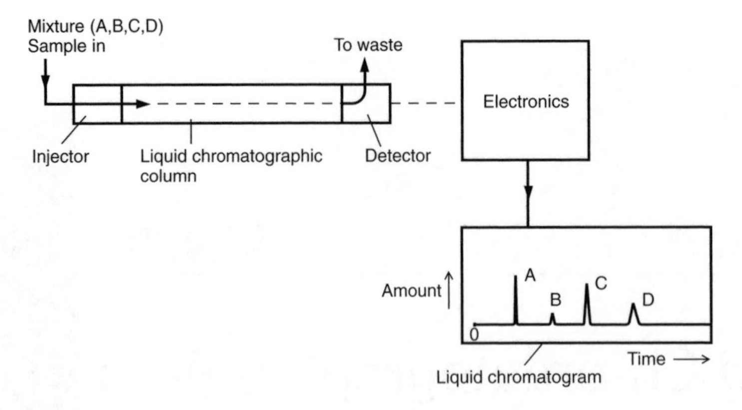

Figure 37.1
Schematic diagram showing injection of a mixture of four substances (A, B, C, D) onto an LC column, followed by their separation into individual components, their detection, and the display (chromatogram) of the separated materials emerging at different times from the column.

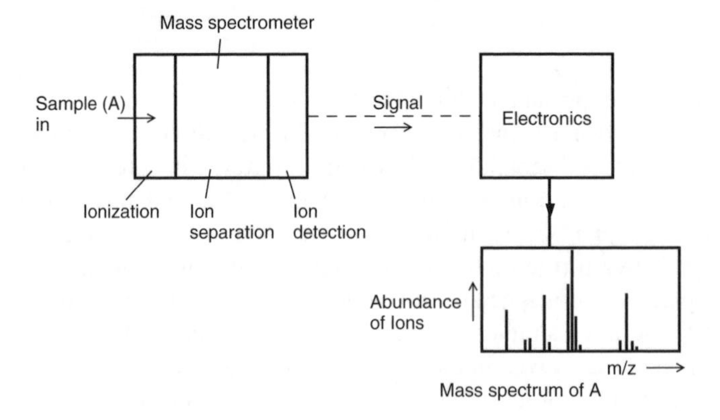

Figure 37.2
Schematic diagram of a mass spectrometer. After insertion of a sample (A), it is ionized, the ions are separated according to m/z value, and the numbers of ions (abundances) at each m/z value are plotted against m/z to give the mass spectrum of A. By studying the mass spectrum, A can be identified.

positively identifying, and quantifying complex mixtures. Combining LC and MS is much more difficult than was the case with GC and MS, but modern LC/MS is now an established routine in many different kinds of laboratories. Further, the addition of LC to MS does not simply give a sum of the two alone; the information provided by combined LC/MS yields information that could not be extracted from either technique alone, an aspect that is discussed below.

Connection Between LC and MS

As described above, the mobile phase carrying mixture components along a liquid chromatographic column is a liquid, usually one of a range of solvents of which water, acetonitrile, and methanol are common. This liquid flows from the end of the column at a rate of generally about 0.5 to 3.0 ml/min. In very narrow nanocolumns, it is even less. This liquid flow must be transferred into the ion source of the mass spectrometer, which is under high vacuum. The large difference in pressure between the end of the chromatographic column and the inside of the ion source would cause enormous problems for the mass spectrometer vacuum system if there were no interface, since a

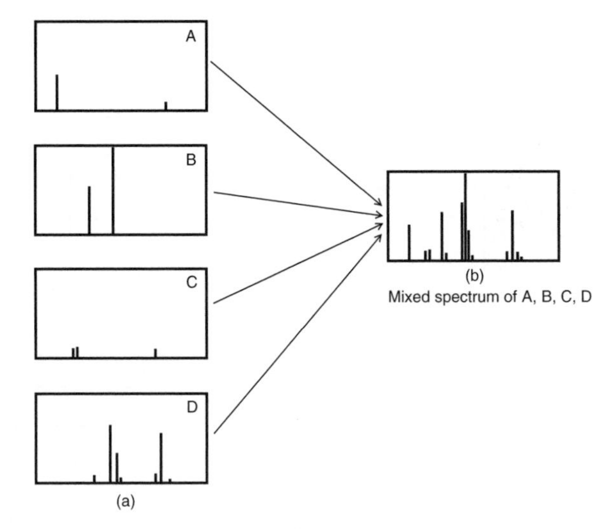

Figure 37.3
By way of illustration, very simple spectra for four substances (A, B, C, D) are shown: (a) separately and (b) mixed in unequal proportions. The mixture spectrum is virtually impossible to decode if A, B, C, D are not known beforehand to be present.

liquid stream of 0.5 to 3.0 ml/min from the LC would almost immediately become a gas stream of hundreds of liters per minute in the MS. Further, in this expansion, the mixture components would become diluted, so, even if the MS system could deal with the expanding solvent vapor, the sensitivity of detection of mixture components would be reduced catastrophically. Some method needed to be found for removing most of the liquid flowing from the column without losing the dissolved mixture components. Several methods (interfaces or separators) for overcoming this problem have been advanced, all with advantages and disadvantages that need to be considered.

In the earliest interface, a continuous moving belt (loop) was used onto which the liquid emerging from the chromatographic column was placed as a succession of drops. As the belt moved along, the drops were heated at a low temperature to evaporate the solvent and leave behind any mixture components. Finally, the dried components were carried into the ion source, where they were heated strongly to volatilize them, after which they were ionized.

This method is still in use but is not described in this book because it has been superseded by more recent developments, such as particle beam and electrospray. These newer techniques have no moving parts, are quite robust, and can handle a wide variety of compound types. Chapters 8 through 13 describe these newer ionization techniques, including electrospray, atmospheric pressure ionization, plasmaspray, thermospray, dynamic fast-atom bombardment (FAB), and particle beam.

It is worth noting that some of these methods are both an inlet system to the mass spectrometer and an ion source at the same time and are not used with conventional ion sources. Thus, with electrospray, the process of removing the liquid phase from the column eluant also produces ions of any emerging mixture components, and these are passed straight to the mass spectrometer analyzer; no separate ion source is needed. The particle beam method is different in that the liquid phase is removed, and any residual mixture components are passed into a conventional ion source (often electron ionization).

Finally, note that the ions produced by the combined inlet and ion sources, such as electrospray, plasmaspray, and dynamic FAB, are normally molecular or quasi-molecular ions, and there is little or none of the fragmentation that is so useful for structural work and for identifying compounds through a library search. While production of only a single type of molecular ion may be useful for obtaining the relative molecular mass of a substance or for revealing the complexity of a mixture, it is often not useful when identification needs to be done, as with most general analyses. Therefore,

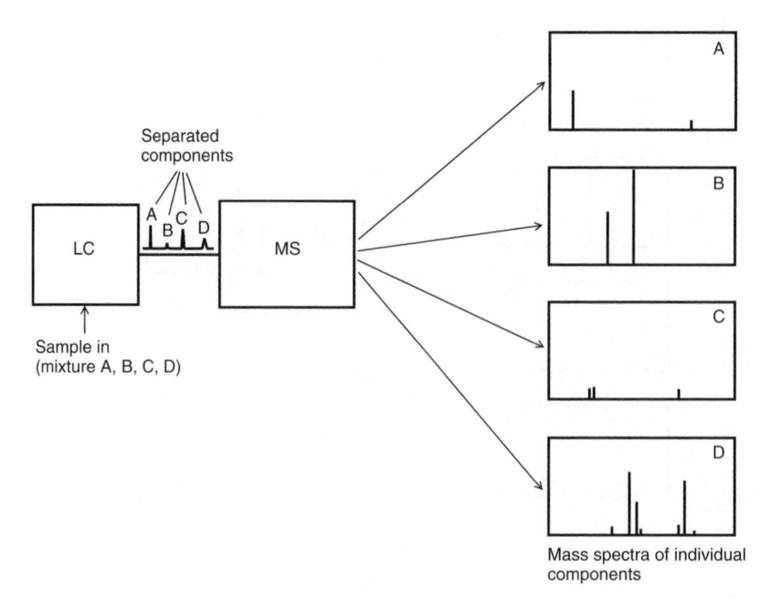

Figure 37.4
In an LC/MS combination, passage of the separated components (A, B, C, D) successively into the mass spectrometer yields their individual spectra.

the ions resulting from these combined inlet/ion source interfaces must be given extra energy to induce fragmentation (see Chapters 32 through 34, which deal with metastable ions and linked scanning). The particle beam system does not suffer from this problem.

Recording Mass Spectra

As each mixture component elutes, the resulting ions are analyzed by the mass spectrometer to give a mass spectrum (Figure 37.4).

Separated mixture components elute in a short time interval, often lasting only a few seconds. Thus, the amount of any one component in the ion source is not constant as its mass spectrum is being obtained. Rather, it starts at zero, rises to a maximum, and drops back to zero. If this passage through the ion source is faster than the mass spectrometer's ability to scan the spectrum, then a true spectrum will not be found because the start and end of the scan will show less compound than at the middle of the scan. This changing concentration of eluting component results in a distorted mass spectrum that might not be recognizable (Figure 37.5). The answer to this problem is to scan the spectrum so fast that, in effect, the concentration of the eluting component has scarcely changed during the time needed to acquire a spectrum.

For a quadrupole mass spectrometer, this high rate of scanning is not difficult because it requires only simple changes in some electrical voltages, and these changes can be made electronically at very high speed, which is one reason why quadrupoles are popular in LC/MS combinations. In the early days of magnetic-sector mass spectrometers, the required scanning speed was not possible because of serious hysteresis effects in the magnets. With modern magnet technology, scanning can be done at high speed with insignificant hysteresis, and magnetic-sector instruments can now compete with quadrupoles. While ultimate scan speed for a magnetic instrument is not as good as the quadrupole's, the former does have the advantage of providing greater mass resolution at higher mass.

As described above, the concentration of an eluting component in the ion source goes from zero to zero through a maximum. Where should the scan be taken? Usually, the greater the amount of a

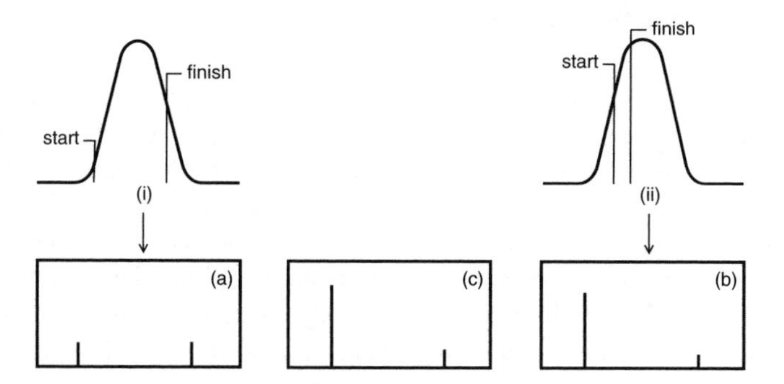

Figure 37.5
Slow scanning (i) of the mass spectrum over an LC peak for substance A gives spectrum (a), but rapid scanning (ii) gives spectrum (b), which is much closer to the true spectrum (c).

substance in an ion source, the better will be the resulting mass spectrum (within reason!). This fact suggests that the best time for a scan will be near the maximum concentration (the top of an LC peak) and that the instrument operator must watch the developing chromatogram continually, trying to judge the best moment to measure a spectrum. Since a liquid chromatogram can routinely take 20 to 50 min to obtain, such a watching brief is labor intensive (and deadly dull!). The simple answer is to set the mass spectrometer scanning continuously. Therefore, as the mixture is injected onto the chromatographic column, mass spectrometer scanning is instituted over a preset mass range (e.g., 50 to 500 mass units) at a preset interval (e.g., every 10 sec). Thus, scan follows scan right through the chromatogram, and hundreds of mass spectra can be recorded in a routine LC/MS experiment (Figure 37.6).

Data flow at such high speeds and in such copious quantities from the mass spectrometer that a microprocessor/computer is necessary. At the end of an LC/MS run, all of the data are in storage, usually on a hard disc in the computer, in the form of a mass spectrum for every scan that was done. These spectra can then be manipulated. For example, the scan corresponding to the top of a chromatographic peak can be selected (scan 3 or 4 in Figure 37.6 for example) and the mass spectrum displayed or printed out. As a routine, the computer adds all the ion peaks in each mass spectrum (actually summing electrical currents) to give a total ion current for the mass spectrum. These total ion currents are plotted along an x-axis (time for elution) and a y-axis (amount of total ion current) to give a total ion current (TIC) chromatogram showing the elution of all the components of a mixture. Storage of all these data has other major advantages.

Manipulation of Scan Data

Some ways in which data can be utilized are described briefly below.

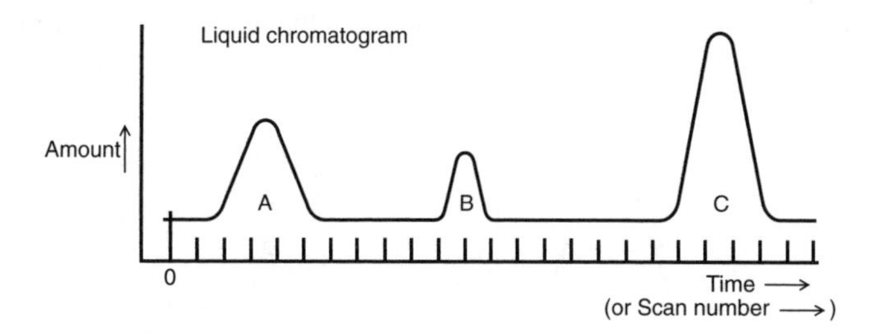

Figure 37.6
Typical liquid chromatogram showing three components (A, B, C) emerging at different times. Continuous scanning for mass spectra is started at zero, and scans are repeated regularly (small tick marks). Thus peak A is scanned five times during its passage through the ion source. Because scanning is regular, the time axis can be replaced by a scan-number axis.

Background Subtraction with Library Search

Once a mass spectrum from an eluting component has been acquired, the next step is to try to identify the component either through the skill of the mass spectroscopist or by resorting to a library search. If only molecular or quasi-molecular ions have been produced, they will not be of much use for identification purposes. If the ions are given extra energy so that they fragment — or if the particle beam or moving-belt inlets are in place so that electron ionization can be used — then the resulting mass spectra can be compared with those held in a large library of spectra from known compounds (e.g., the NIST library). There may be as many as 50,000 to 60,000 stored spectra covering most of the known simple volatile compounds likely to be met in analytical work. Using special search routines under the control of the computer, one can examine this huge database very quickly, comparing the mass spectrum of an eluted mixture component with each of the library spectra. The computer then provides a short list of the best matches between the library spectra and the measured one. Often, from the closeness of the match or fit and its chromatographic retention time, the eluted component can be identified positively.

Resolution Enhancement

Sometimes (not infrequently) a single peak from a liquid chromatogram reflects the elution of two, three, or more substances. By examining the scans across the peak and using a simple mathematical process, the computer can reveal the existence of more than one eluting component in a single peak and print out the mass spectra of the discovered components. This process is a form of resolution enhancement — the LC/MS combination can do what neither the LC nor the MS alone could do. In effect, the mass spectrometer has improved (enhanced) the resolution of the liquid chromatograph (Figure 37.7).

Mass Chromatogram

There could be a series of compounds, designated RX, in which the R part was the same for all, but X was different. In such a series, the R part can give a characteristic ion; viz., the mass corresponding to R is the same for every member of the series. When one member of such a series elutes from the chromatographic column, its resulting mass spectrum will always contain one mass common to the whole series. Therefore, even if the series is mixed with other compounds, its members can be recognized from the characteristic ion. It is a simple process to get the computer to print out a chromatogram in which only those scans containing the characteristic ion are used to draw the chromatogram. Basically, the output is blind to any component other than those containing an R group. Such a selected chromatogram is called a mass chromatogram and is useful for pinpointing where certain compounds elute without the need for examining all the spectra (Figure 37.8).

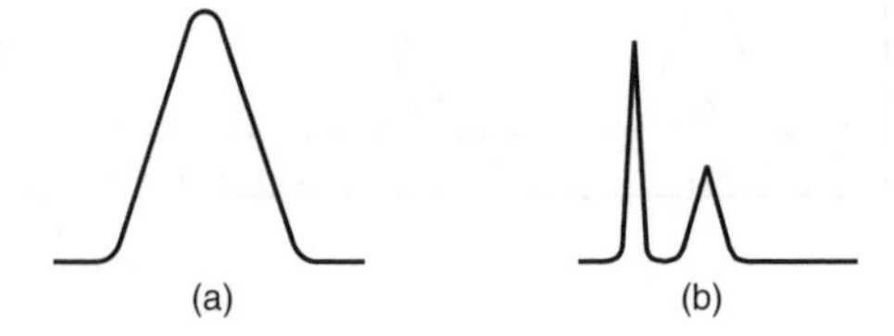

Figure 37.7
A single peak from an ordinary liquid chromatogram (a) is revealed as two closely separated peaks by resolution enhancement (b).

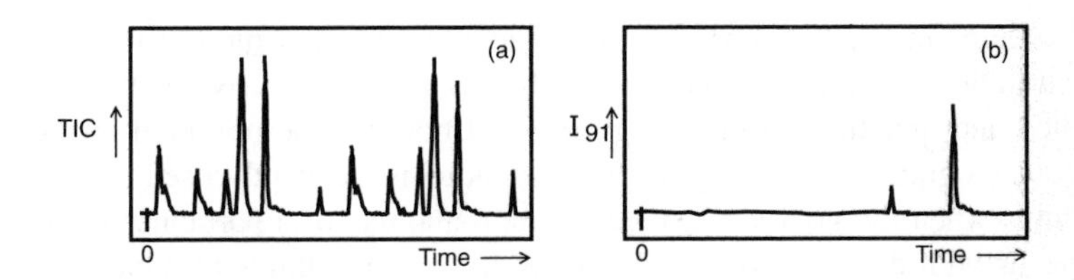

Figure 37.8
(a) A typical total-ion-current chromatogram showing many components, some of which have a part R (e.g., of characteristic mass 91). (b) A mass chromatogram based on m/z 91. The same data as used in (a) are plotted, but they are manipulated so that only ion currents corresponding to m/z 91 are shown. Note the decrease in complexity, making the desired identification much easier.

Selected Ion Recording

In a process similar to that described in the previous item, the stored data can be used to identify not just a series of compounds but specific ones. For example, any compound containing a chlorine atom is obvious from its mass spectrum, since natural chlorine occurs as two isotopes, ^{35}Cl and ^{37}Cl, in a ratio of 3:1. Thus its mass spectrum will have two molecular ions separated by two mass units (35 + 2 = 37) in an abundance ratio of 3:1. It becomes a trivial exercise for the computer to print out only those scans in which two ions are found separated by two mass units in the abundance ratio of 3:1 (Figure 37.9). This selection of only certain ion masses is similar to selected ion recording (SIR) or, sometimes, selected ion monitoring (SIM).

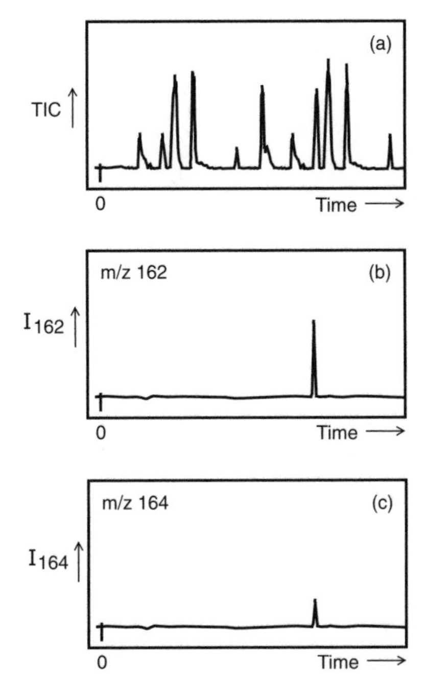

Figure 37.9
(a) A typical total-ion-current chromatogram showing many components of a mixture. The aim is to find if any chloronaphthalene ($C_{10}H_7Cl$; RMM = 162, 164) is present. To this end, the data (or a new chromatogram) are scanned at two specific positions: (b) at m/z 162 and (c) at m/z 164. Note how m/z 162, 164 both reach a maximum at the same scan and have a ratio in peak heights of 3:1 (^{35}Cl:^{37}Cl = 3:1). Such an experiment in selected-ion recording identifies the suspected component against a complex background and can be made very sensitive.

The ions selected for such recording can be one, two, three, or more (multiple ion recording). In fact, through judicious choice of ions, the method can be so selective that a chosen component can be identified and quantified even though it could not even be observed in the original total-ion-current chromatogram. This very powerful technique is frequently used to examine extremely complex mixtures when the goal is to identify small amounts of a particular substance in a mass of other things, as with detection of banned drugs in the body fluids of athletes or racehorses or dioxins in industrial waste.

Conclusion

By connecting a liquid chromatograph to a suitable mass spectrometer through an interface and including a data system, the combined method of LC/MS (sometimes written HPLC/MS) can be used routinely to separate complex mixtures into their individual components, identify the components, and estimate their amounts. The technique is widely used.

Chapter 38

High-Resolution, Accurate Mass Measurement: Elemental Compositions

Introduction

In many areas of science (chemistry, biochemistry, environmental, etc.), there is a need to determine the molecular formula of a substance. For instance, natural gas (methane) has the formula CH_4, which indicates that each carbon atom (C) has four hydrogen atoms (H_4) attached; similarly, water has the formula H_2O, indicating two hydrogens (H_2) attached to each oxygen. For more complex molecules, the formulae are correspondingly more complex; for example, the natural product, adenosine, has the formula $C_{10}H_{13}N_5O_4$. In the past, the only way of obtaining such formulae was to break down a known weight of the molecule into its constituent elements and weigh them, the resulting proportions giving the required formula. This was a painstakingly slow and often inaccurate process for complex molecules. The use of mass spectrometry for accurate mass measurement has transformed the situation such that, in modern science, molecular formulae for a vast range of substances are obtained using this technique.

Atomic and Molecular Mass: Fragment Ion Mass

The actual mass of an atom is very small indeed. For example, a hydrogen atom weighs something like 10^{-24} g. Instead of using such an absolute scale, a relative integer mass scale is easier to handle. On this relative scale, all atomic masses have values near integers. Thus, the absolute masses for hydrogen (1×10^{-24} g) and deuterium (2×10^{-24} g) are relatively 1 and 2, respectively. Table 38.1 gives relative integer masses for some common elements. Although this relative integer scale is acceptable in many circumstances where only approximate values are needed, for some applications, accurate relative masses are needed. By definition, on this accurate mass scale, carbon (^{12}C) is given the value of exactly 12, i.e., 12.00000. Table 38.1 indicates the accurate masses for some elements compared with their integer values.

Atoms combine in definite proportions to give molecules. For example, natural gas is mostly composed of methane, a substance in which four hydrogen atoms (H) are combined with one carbon (C); the molecular formula is written as CH_4. Similarly, water, ammonia, ethanol, and glucose have

TABLE 38.1
Relative Integer and Accurate Atomic Masses for Some Common Elements.

	Integer Mass		Accurate Mass	
Hydrogen	H	^{1}H	1	1.00783
(Deuterium)	(D)	^{2}H	2	2.01410
Carbon	C	^{12}C	12	12.00000
	^{13}C	13	13.00335	
Nitrogen	N	^{14}N	14	14.00307
Oxygen	O	^{16}O	16	15.99491
Chlorine	Cl	^{35}Cl	35	34.96885
	^{37}Cl	37	36.96590	
Silicon	Si	^{28}Si	28	27.97693

Methane (CH_4) : 1 x carbon + 4 x hydrogen = 1 x 12 + 4 x 1 = 16 (16.031)

Water (H_2O) : 2 x hydrogen + 1 x oxygen = 2 x 1 + 1 x 16 = 18 (18.01057)

Ammonia (NH_3) : 1 x nitrogen + 3 x hydrogen = 1 x 14 + 3 x 1 = 17.0265)

Ethanol (C_2H_6O) : 2 x carbon + 6 x hydrogen
+ 1 x oxygen = 2 x 12 + 6 x 1
+ 1 x 16 = 46 (46.0418)

Glucose ($C_6H_{12}O$): 6 x carbon + 12 x hydrogen
+ 1 x oxygen = 6 x 12 + 12 x 1
+ 6 x 16 = 192 (192.1634)

Figure 38.1
Calculation of molecular mass from a molecular formula for several simple substances. Accurate masses are shown in parentheses.

the respective molecular formulae: H_2O, NH_3, C_2H_6O, and $C_6H_{12}O_6$. For each of these substances a molecular mass can be computed, as detailed in Figure 38.1. Since these masses are calculated from *relative* atomic masses, they should be referred strictly as relative molecular mass (RMM). Recent publications use RMM, but older publications used molecular weight (MW or M.Wt). Values for the accurate relative molecular masses are also given in Figure 38.1.

In a mass spectrum, removal of an electron from a molecule (M) gives a molecular ion (Equation 38.1). The mass of an electron is very small compared with the mass of even the lightest element, and for all practical purposes, the mass of $M^{\bullet+}$ is the same as that of M. Therefore, mass measurement of a molecular ion gives the original relative molecular mass of the molecule.

$$M + e^- \rightarrow M^{\bullet+} + 2e^- \qquad (38.1)$$

Decomposition of ions gives fragment ions that can also be mass measured. Figure 38.2 illustrates a simple example in which the molecular ion of iodobenzene cleaves to give a fragment ion, $C_6H_5^+$, and an iodine atom, I. The molecular mass at 203.94381 corresponds to the molecular formula (or elemental composition) of C_6H_5I, while the fragment ion mass at 77.03915 corresponds to the elemental composition, C_6H_5; the difference in mass equates to the mass of the ejected neutral iodine atom.

A suitable mass spectrometer can be set to measure relative atomic, molecular, or fragment ion mass.

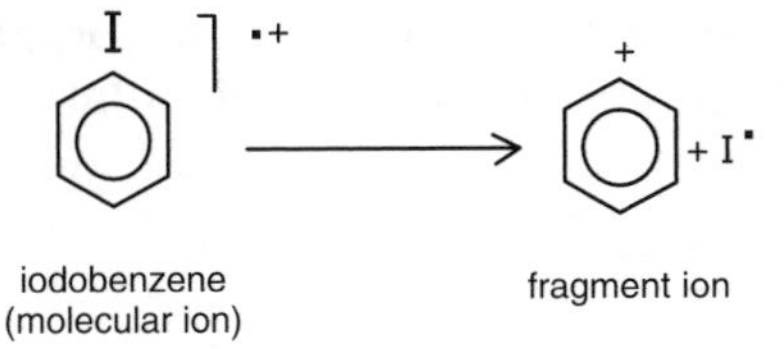

Formula or composition:	C_6H_5I (molecular)	C_6H_5 (elemental)
Integer mass:	204	77
Accurate mass:	203.94 381	77.039 15

Figure 38.2
Simple fragmentation of the molecular ion of iodobenzene gives a fragment ion, $C_6H_5^+$. The difference in measured masses between the molecular and fragment ions gives the mass of the ejected neutral iodine atom.

The Value of Accurate Mass Measurement

A mass spectrometer can measure integer relative mass with high accuracy, but the result is not nearly so informative as measurement of accurate relative mass. An example illustrates the reason.

Three substances — carbon monoxide (CO), ethene (C_2H_4), and nitrogen (N_2) — each have an integer molecular mass of 28 (Figure 38.3). Their occurrence in a mass spectrometer would give molecular ions at integer masses of 28, and they could not be distinguished from each other. However, their respective accurate masses of 27.99491, 28.03132, and 28.00614 are different, and a mass spectrometer capable of accurate mass measurement would be able to distinguish them even if all three occurred in the same sample. Thus, accurate mass measurement is useful for confirming the molecular composition of a tentatively identified material.

There is a more important use. Suppose a mass spectrometer has accurately measured the molecular mass of an unknown substance as 58.04189. Reference to tables of molecular mass vs. elemental composition will reveal that the molecular formula is C_3H_6O (see Table 38.2). The molecular formula for an unknown substance can be determined which is enormously helpful in identifying it.

Finally, accurate mass measurement can be used to help unravel fragmentation mechanisms. A very simple example is given in Figure 38.2. If it is supposed that accurate mass measurements were made on the two ions at 203.94381 and 77.03915, then their difference in mass (126.90466) corresponds exactly to the atomic mass of iodine, showing that this atom must have been eliminated in the fragmentation reaction.

Resolution of Mass Spectrometers

The ability of a mass spectrometer to separate two masses (M_1, M_2) is termed resolution (R). The most common definition of R is given by Equation 38.2, in which $\Delta M = M_1 - M_2$ and $M = M_1 \cong$

Chemical name	Formula	Integer Mass	Accurate Mass
Carbon monoxide	CO	28	27.99491
Ethene	C_2H_4	28	28.03132
Nitrogen	N_2	28	28.00614

Figure 38.3
Integer and accurate masses for three different gases, each having the same integer relative molecular mass (RMM = 28).

TABLE 8.2
A Listing of Elemental Compositions vs. Accurate Mass at Nominal Integer Mass of 58

Integer Mass = 58 *				
C	H	N	O	Accurate Mass
1	—	1	2	57.992902
1	2	2	1	58.016711
1	4	3	—	58.040520
2	2	—	2	58.005478
2	4	1	1	58.053096
2	6	2	—	58.053096
3	6	—	1	58.041862
3	8	1	—	58.065671
4	10	—	—	58.078247

* Compositions are read from left to right. Thus, the fifth entry in the table would be C_2H_4NO, with an accurate mass of 58.053096.

M_2. Thus, if a mass spectrometer can separate two masses (100, 101), then $\Delta M = 1$, $M = 100$, and $R = 100$. For conventional accurate mass measurement, R needs to be as large as possible, typically having a value of 20,000. At this sort of resolution, a mass of 100.000 can be separated from a second mass at 100.0050 ($\Delta M = M/R = 100/20{,}000 = 0.005$)

$$R = M/\Delta M \qquad (38.2)$$

To use the formula in Equation 38.2, it is necessary to define at what stage the two peaks representing the two masses are actually separate (Figure 38.4). The depth of the valley between the two peaks serves this purpose, with valley definitions of 5, 10, or 50%. A 5% valley definition is a much stricter criterion of separation efficiency than the 50% definition.

Measurement of Accurate Mass

By Automated Methods

All routine mass spectrometers measure accurate mass by reference to standard substances. High-resolution, accurate mass measurement requires that a standard substance and the sample under investigation be in the mass spectrometer at the same time. Ions from the standard have known mass, and ions from the unknown (sample) are mass measured by interpolation between successive masses due to the standard (Figure 38.5). High resolution is needed to ensure that there will be mass separation between ions due to the standard and those due to the sample. For this reason, the standard substances are usually perfluorinated hydrocarbons, since ions from these substances have masses somewhat less than an integer value, whereas most other (organic) compounds have masses somewhat greater than an integer value. For example, at integer mass 124, the fluorocarbon C_4F_4 has an accurate mass of 123.9936 while a hydrocarbon, C_9H_{16}, has an accurate mass of 124.12528. A mass spectrometer resolution of 4000 is sufficient to separate these masses. The whole operation is usually automated by a data system, but it can be done manually by a system of peak matching.

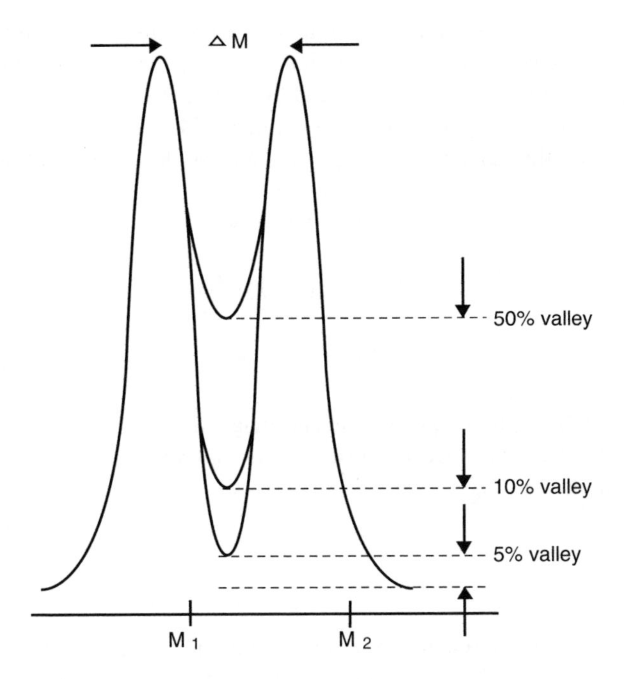

Figure 38.4
The separation (ΔM) of two peaks representing masses (M_1, M_2) can be defined as having a 5, 10, or 50% valley — the depth of the valley is 5, 10, or 50% of peak height. The 5% definition is a more severe test of instrument performance.

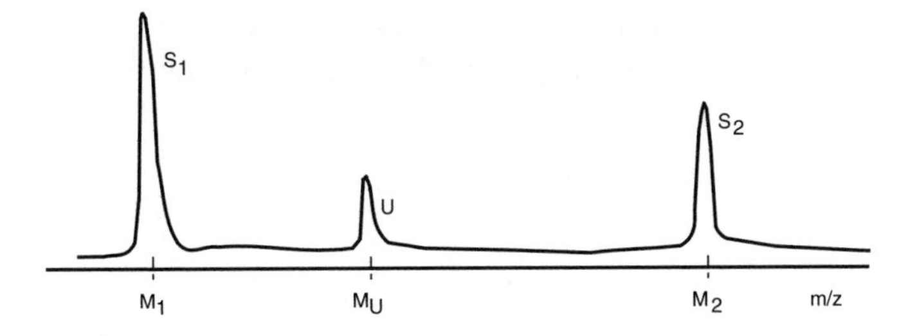

Figure 38.5
Part of a high-resolution mass spectrum showing two peaks (S_1, S_2) due to ions of mass (M_1, M_2) from a standard substance and one peak (U) due to an ion of unknown mass (M_u) from a sample substance. The difference between M_1 and M_2 is accurately known, and therefore the mass, M_u, can be obtained by interpolation (the separation of M_u from M_1 or M_2). If the separation of M_1, M_2 is a time (T) and that of M_1, M_u is time (t), then for a linear mass scale $M_u = [t\,(M_2 - M_1)/T] + M_1$.

By Peak-Matching Methods

In the example of Figure 38.5, for a magnetic-sector instrument, two masses (M_1, M_u) will follow different trajectories in the mass spectrometer, defined by radii of curvature (r_1, r_u; see Chapter 24, "Ion Optics of Magnetic/Electric-Sector Mass Spectrometers"). If the magnetic-field strength is B, and the ions are accelerated from the ion source through an electric potential of V volts, then Equation 38.3 follows. If the voltage (V) is changed to V* so that $r_1 = r_u$, then Equation 38.4 shows that the unknown mass (M_u) can be calculated if mass (M_1) from the standard is known and the ratio of V*/V is measured.

$$M_1/z = B^2r_1^2/2V$$

$$M_u/z = B^2r_u^2/2V \qquad (38.3)$$

Therefore, for accurate mass measurement, a standard mass peak (M_1) is selected, and the accelerating voltage (V) is changed until the sample ion peak (M_u) exactly coincides with the position of M_1. This technique is called peak matching, and the ratio between the original and new voltages (V/V*) multiplied by mass (M_1) gives the unknown mass, M_u.

For greatest accuracy, the standard mass (M_1) should be as close as possible to the unknown (M_u).

$$M_1/z = B_2 r_1^2/2V$$

$$M_u/z = B_2\, r_1^2/2V^* \qquad (38.4)$$

$$\therefore\ M_u = M_1(V/V^*)$$

Peak matching can be done on quadrupole and magnetic-sector mass spectrometers, but only the latter, particularly as double-focusing instruments, have sufficiently high resolution for the technique to be useful at high mass.

Other Methods

Other techniques for mass measurement are available, but they are not as popular as those outlined above. These other methods include mass measurements on a standard substance to calibrate the instrument. The standard is then withdrawn, and the unknown is let into the instrument to obtain a new spectrum that is compared with that of the standard. It is assumed that there are no instrumental variations during this changeover. Generally, this technique is less reliable than when the standard and unknown are in the instrument together. Fourier-transform techniques are used with ion cyclotron mass spectrometers and give excellent mass accuracy at lower mass but not at higher.

Conclusion

By high-resolution mass spectrometry, ions of known mass from a standard substance can be separated from ions of unknown mass derived from a sample substance. By measuring the unknown mass relative to the known ones through interpolation or peak matching, the unknown can be measured. An accurate mass can be used to obtain an elemental composition for an ion. If the latter is the molecular ion, the composition is the molecular formula.

Chapter 39

Choice of Mass Spectrometer

Introduction

Early in the history of mass spectrometry, instruments were built by individual scientists and often could not be used or understood by anyone but the inventor. Now, very few scientists build their own mass spectrometers, and many commercial types are available. Although commercial mass spectrometers can be adapted to specialized needs, in most cases they are used by people who want fast, accurate answers to a wide range of questions, such as, "Is this painting genuinely old or is it a modern fake?" or "Was this horse doped when it won the race?" or "What is the structure of this new protein?" These questions can be broadly classified under the heading of analysis, a topic that embraces most uses of commercial mass spectrometers. Given this wide usage and the range of mass spectrometers that are commercially available, it can be difficult to choose the right instrument. This chapter reviews some of the factors that should be considered in selecting a mass spectrometer system. Of course, any decision is likely to be a compromise between what is desirable and what is available within a given price range.

Note that many of the terms mentioned in this chapter are discussed in detail elsewhere in this book. For example, the theory and practical uses of electron ionization (EI) are fully discussed in Chapter 3.

Objectives in Buying a Mass Spectrometer

Deciding on the main objective for buying a mass spectrometer is an obvious first step, but it is essential in achieving a satisfactory result. Before approaching suppliers of commercial mass spectrometers, it is wise to set out on paper the exact analytical requirements for both the immediate and near future. The speed of advance in science — especially in analysis and mass spectrometry — means that long-range prognostications of future requirements are likely to be highly speculative and therefore of little relevance.

Once basic requirements and secondary objectives have been established, the prospective purchaser will find it easier to discuss details with sales representatives. From the latter's viewpoint, it is easier to talk to a potential customer who knows what he needs from a mass spectrometer system rather than to a customer who has only a vague idea of what is required. In fact, an uninformed customer can end up purchasing an expensive instrument that is far too good for the analyses required or, at the other extreme, a cheap instrument that is inadequate for immediate needs, let alone ones that might arise in the near future.

Future usage of a new instrument is worth considering, even if long-range forecasting is unlikely to be useful. It is almost inevitable that, once a new instrument has been installed and is operating routinely, people or groups within an organization that has just acquired a new mass spectrometer begin to hear about what it can do. This news leads to them to devise new ways of carrying out analyses with the help of the new instrument. Such healthy developments often lead to pressure on available time on the new machine, and an instrument that cannot accommodate some increase in output leads to disappointment and, often, a need to make another purchase sooner than expected. Therefore, when contemplating the purchase of a new instrument, it is useful to explore the possibility that the machine may need to be adapted to increase throughput or to change the way it operates. For example, it may be desirable to fit an alternative type of ion source, and an instrument that cannot be adapted easily could be a drawback. These are relevant issues to raise with a sales representative, if only to set boundaries in the putative purchaser's own mind. Some instruments come as a complete system aimed at one primary objective and are almost impossible to alter, even modestly (unless expense is no object!). Other instruments may place an unacceptably low limit on the number of samples that can be examined per hour or per day.

Setting out the objectives in a detailed fashion and assigning priority to those objectives is an essential first step and is time well spent. Once carefully considered, it could become clear that the objectives fall into two different categories and that it might be cheaper to purchase two dedicated instruments rather than one large, all-encompassing mass spectrometer.

Types of Sample

Samples can be single substances; complex mixtures of well-known, relatively simple substances; complex mixtures of substances of totally unknown structure; or combinations of such analyses. It is impossible to generalize in such situations, but it is possible to offer guidelines on some of the important issues.

For the simpler analyses, it may be necessary to look only for the presence of impurities in a single substance that is volatile, thermally stable, and easy to handle. Such analyses can be tackled with most kinds of mass spectrometer having simple inlet systems. At the other end of the scale of difficulty, there are several major analytical pathways. The sample can be a single substance and still be very intractable. For example, ceramics or bone are often considered single entities, whereas each is chemically complex, nonvolatile, and difficult to break down into simpler identifiable species, posing severe analytical problems. The sample could be a mixture of polar, thermally unstable proteins or peptides from a biological sample, with many of the constituents of unknown structure. In such a case, the whole sample needs chromatography, a specialized inlet system, and possible MS/MS (two MS systems used in tandem) facilities. As in the case of toxic substances, the important components of a sample may be present only as traces, and high sensitivity of detection is required.

Complexity of Sample

A sample can be complex in one of two ways. It might be a single substance with a very complex chemical structure, or it might contain several substances of varying polarity, volatility, and thermal stability.

If samples are largely pure, single substances, then the sample inlet can be quite simple, as with a direct insertion probe or a gas inlet. However, most analyses require assessment of the number of components, their relative proportions, and their chemical structures. This level of complexity will normally need some degree of separation of the components before examination by mass spectrometry. Isolation and concentration of the components that need to be examined by mass spectrometry is usually a first step, and, when properly designed, this sort of preanalytical step can lead to big improvements in overall sensitivity and to less demand on the mass spectrometer. For example, volatile organic components from soil samples are usually removed by vaporization before analysis rather than trying to put the soil sample into the mass spectrometer.

After this preseparation stage (sometimes called sample cleanup), it may still be necessary to effect some separation of the required sample components. For example, with regard to the soil sample mentioned above, the isolated volatile organics may consist of literally hundreds of components present in widely varying amounts. In these circumstances, it is necessary to use a chromatographic procedure to separate the components before they are passed into the mass spectrometer. This separation can be done manually, in that each component can be trapped as it emerges from a gas chromatograph (GC) or liquid chromatograph (LC) column and then transferred to a simple inlet system on the mass spectrometer (MS). It is much more likely that the GC or LC apparatus will be linked directly to the mass spectrometer so that, as each component emerges from the chromatographic column, it is transferred immediately into the mass spectrometer. Therefore, combined GC/MS, LC/MS, or CE/MS (capillary electrophoresis and mass spectrometry) may be needed, depending on the nature of the substances to be analyzed. While there is some overlap, in that some substances could be examined by either GC/MS or LC/MS, usually analyses are required using only one type of inlet. This consideration is important because inlets for GC/MS, LC/MS, and CE/MS are different, and the addition of a second or third inlet will substantially increase the cost of the mass spectrometer system. If changing from one inlet system to another is time consuming, the process will lead to time when the mass spectrometer cannot be used, which can be costly for a busy analytical laboratory. This factor is another reason for carefully distinguishing between essential inlet systems and those that would merely be useful.

It might be noted at this stage that some mass spectrometer inlets are also ionization sources. For example, with electrospray ionization (ES) and atmospheric pressure chemical ionization (APCI), the inlet systems themselves also provide the ions needed for mass spectrometry. In these cases, the method of introducing the sample becomes the method of ionization, and the two are not independent. This consideration can be important . For example, electrospray produces abundant protonated molecular ions but no fragment ions. While this is extremely important for accurate mass information and for dealing with mixtures by MS/MS, the lack of fragment ions gives almost no chemical structure information. If such information is needed, then the mass spectrometer will have to be capable of fragmenting the protonated molecular ions through incorporation of a collision gas cell or other means, again adding to cost.

Prior separation of mixtures into individual components may not be needed. If the mass spectrometer is capable of MS/MS operation, one of the mass spectrometers is used to isolate individual ions according to m/z value (mass-to-charge ratio), and the other is used to examine their fragmentation products to obtain structural information.

For analytical laboratories with high throughputs of samples, it is usually necessary to have automatic samplers, so that the mass spectrometer can work 24 hours per day, even in the absence of an operator. Therefore, one consideration that may be important in deciding on an instrument could be as simple as asking how long it takes to set the operating characteristics of the mass spectrometer. How easy is it to calibrate? Most modern systems use computer programs that automatically check and adjust voltages, peak shapes, and so on at the touch of a button. Calibration for accuracy of m/z value may be trivial, but it can also be complex, especially at high resolution.

Table 39.1 indicates very broadly which arrangement of mass spectrometer might be used for various sample types.

TABLE 39.1
Complexity of Samples to Be Examined and Mass Spectrometer Type

Complexity of Sample	Mass	GC/MS	LC/MS	CE/MS	MS/MS
Single substances	√	√	√	√	√
Mixture of volatile substances	√	√	—	—	√
Mixture of solids in solution	—	—	√	√	√
Mixture of insoluble solids	—	—	—	—	√

Sample Volatility, Polarity, and Thermal Stability

Gases

The ease of vaporization of a sample can be an overriding factor in choice of mass spectrometer. Generally, the three phases — gas, liquid, and solid — need to be considered. By definition, a gas is volatile, and inserting a gas into a mass spectrometer is easy. A simple system of valves and filters is sufficient for transferring a gas into the vacuum of a mass spectrometer, and often an EI source is all that is needed. Additionally, gases tend to be of low molecular mass, so ion analyzers for gases need not be very sophisticated or have more than a modest resolving power to cover the range needed (often less than an upper limit of m/z 100–150). Mass spectrometers for such purposes can be very small and light, and they are used on space probes to other planets. Similarly, small mass spectrometers (usually quadrupole instruments) are used to monitor atmospheres on earth in places where noxious substances may be present. These small mass spectrometers can be used in and transported by small vans or cars.

Apart from substances that are gaseous at normal ambient temperatures, other materials cross the divide between gas and liquid when they have boiling points in the range from –10 to +40°C. Such substances are so volatile that their mass spectrometric analysis becomes just as easy as that for gases. For example, butane has a boiling point just below normal ambient temperatures, and diethyl ether has a boiling point not much above most ambient temperatures. This intermediate volatility produces a gray area from the viewpoint of mass spectrometric analysis, since the cutoff between a gas (such as argon) and a highly volatile liquid (such as petrol) is not sharp and is highly dependent on ambient or operating temperatures. Even some solids produce significant vapor pressures, so they could be analyzed by simple gas mass spectrometers if the mass range is sufficiently large. Camphor is a solid at normal ambient temperatures, but it volatilizes (sublimes) very easily. It is worth recalling in this context that, usually, whatever the ambient atmospheric temperature, the mass spectrometer itself will not normally operate outside certain limits. Above about 35°C, the electronic components in computer-controlled systems become increasingly unstable, and frequently it is necessary to install suitable air-conditioning. Below about 10°C, condensation of atmospheric moisture onto a mass spectrometer can lead to electrical problems through short circuits. Therefore, in considering whether or not a mass spectrometer would be useful for volatile liquids, it is worth considering the temperature of the operating mass spectrometer and not simply outside ambient temperatures. A mass spectrometer operating in the Arctic or Antarctic requires air-conditioning just as much as one operating near the equator.

Of course, some substances are sufficiently volatile that a heated inlet line can be used to get them into a mass spectrometer. Even here, there are practical problems. Suppose a liquid or solid is sufficiently volatile, that heating it to 50°C is enough to get the vapor into the mass spectrometer through a heated inlet line. If the mass spectrometer analyzer is at 30°C, there is a significant possibility that some of the sample will condense onto the inner walls of the spectrometer and slowly vaporize from there. If the vacuum pumps cannot remove this vapor quickly, then the mass spectrometer will produce a background spectrum of the substance added through the heated inlet line throughout successive analyses of other substances. This well-known memory effect can be tedious to remove and can slow the throughput of analyses. Thus it is probably best to consider that any substance that remains a liquid above about 50 to 70°C should not be considered for analysis in a gas mass spectrometer. Similarly, it would probably not be wise to insert sublimable solids into such instruments.

Liquids

As with gases, there are no sharply defined limits for what should be considered a liquid under MS operating conditions, and the best guide with regard to use of mass spectrometers probably comes from the operating temperature range (10 to 30°C) of the instrument or any associated apparatus. A mass spectrometer inlet may be at atmospheric pressure or it may be under a high vacuum (10^{-5} to 10^{-6} mm of mercury). Clearly, introduction of a low-boiling liquid into the inlet of a system under high vacuum will lead to its rapid volatilization. It may even be that the pressure rise resulting from volatilization of a liquid becomes so great that the instrument shuts itself down to safeguard its vacuum gauges and ion detectors. It is worth recalling that even substances with a boiling point of 100°C or more will evaporate rapidly in a vacuum of 10^{-5} mm of mercury. The higher the boiling point, the less it is a problem. Liquids and even solids may well arrive at the mass spectrometer inlet in vapor form, as when they come from a gas chromatograph. Generally, the amounts of such emerging substances are very low, so, if the vacuum is high, transfer from the chromatographic column is easy through a heated line to the inlet or ion source of a mass spectrometer.

Liquids that are sufficiently volatile to be treated as gases (as in GC) are usually not very polar and have little or no hydrogen bonding between molecules. As molecular mass increases and as polar and hydrogen-bonding forces increase, it becomes increasingly difficult to treat a sample as a liquid with inlet systems such as EI and chemical ionization (CI), which require the sample to be in vapor form. Therefore, there is a transition from volatile to nonvolatile liquids, and different inlet systems may be needed. At this point, LC begins to become important for sample preparation and connection to a mass spectrometer.

To achieve sufficient vapor pressure for EI and CI, a nonvolatile liquid will have to be heated strongly, but this heating may lead to its thermal degradation. If thermal instability is a problem, then inlet/ionization systems need to be considered, since these do not require prevolatilization of the sample before mass spectrometric analysis. This problem has led to the development of inlet/ionization systems that can operate at atmospheric pressure and ambient temperatures. Successive developments have led to the introduction of techniques such as fast-atom bombardment (FAB), fast-ion bombardment (FIB), dynamic FAB, thermospray, plasmaspray, electrospray, and APCI. Only the last two techniques are in common use. Further aspects of liquids in their role as solvents for samples are considered below.

Solids

Substances that are solid at ambient temperatures are likely to have strong polar, electrostatic, and sometimes hydrogen-bonding forces. In some solids, such as sterols, these forces are sufficiently weak that they can be vaporized easily and analyzed by GC or GC/MS. They are also fairly stable to heat. However, many other solids, such as proteins, carbohydrates, or oligonucleotides, also have very strong electrostatic, polar, and hydrogen-bonding forces, and efforts to volatilize these substances using heat simply leads to their decomposition. Such materials need special methods to get them into a mass spectrometer, especially if they are in solution in aqueous solvent, as is often the case. Therefore, while some solids can be examined by GC/MS methods or even by simple introduction probes by EI or CI (perhaps after suitable derivatization), many other solids need to be inserted into a mass spectrometer in a different manner, such as by ES or APCI.

Solutions of solids may need to be converted into aerosols by pneumatic or sonic-spraying techniques. After solvent has evaporated from the aerosol droplets, the residual particulate solid matter can be ionized by a plasma torch.

Some solid materials are very intractable to analysis by standard methods and cannot be easily vaporized or dissolved in common solvents. Glass, bone, dried paint, and archaeological samples are common examples. These materials would now be examined by laser ablation, a technique that produces an aerosol of particulate matter. The laser can be used in its defocused mode for surface profiling or in its focused mode for depth profiling. Interestingly, lasers can be used to vaporize even thermally labile materials through use of the matrix-assisted laser desorption ionization (MALDI) method variant.

For solids, there is now a very wide range of inlet and ionization opportunities, so most types of solids can be examined, either neat or in solution. However, the inlet/ionization methods are often not simply interchangeable, even if they use the same mass analyzer. Thus a direct-insertion probe will normally be used with EI or CI (and desorption chemical ionization, DCI) methods of ionization. An LC is used with ES or APCI for solutions, and nebulizers can be used with plasma torches for other solutions. MALDI or laser ablation are used for direct analysis of solids.

Table 39.2 shows the ionization modes that are suitable for different physical properties of sample substances.

Mass Analyzers

Commercial mass analyzers are based almost entirely on quadrupoles, magnetic sectors (with or without an added electric sector for high-resolution work), and time-of-flight (TOF) configurations or a combination of these. There are also ion traps and ion cyclotron resonance instruments. These are discussed as single use and combined (hybrid) use.

Single Analyzers

The advantages and disadvantages of single analyzers need to be examined within the context for which they will be used rather than for their overall attributes. For example, a simple quadrupole,

TABLE 39.2
Sample Type and Mode of Ionization[a]

Sample[b]	EI	CI	ES	APCI	MALDI	PT	TI
Gas	√	√	—	—	—	√	—
Liquid (volatile)	√	√	—	—	—	√	—
Liquid (nonvolatile)	—	—	√	√	—	√	√
Solid(volatile)	√	√	—	—	√	√	—
Solid (nonvolatile)[c]	—	—	√	√	√	√	√
Solution (direct insertion)	—	—	√	√	—	√	√

[a] EI = electron ionization; CI = chemical ionization; ES = electrospray; APCI = atmospheric-pressure chemical ionization; MALDI = matrix-assisted laser desorption ionization; PT = plasma torch (isotope ratios); TI = thermal (surface) ionization (isotope ratios).

[b] These are only approximate guides.

[c] Solids must be in solution.

which is robust and can easily examine substances at unit mass resolution up to m/z 200, would be ample for most gas analyzers but would be totally unsuitable for accurate mass measurement in most applications. Conversely, a magnetic-sector instrument of 50,000 resolution and capable of measurement to m/z 100,000 would be totally wasted for the previously mentioned gas analysis. Since the difference in cost would be very large, there would be no difficulty in making a choice. However, just as defining boundaries between gases, liquids, and solids is a tenuous exercise in the face of considerations of vacuum, volatility, and thermal stability, so the choice of mass analyzer can become a difficult one.

Resolving Power and Mass Range

The ability of an analyzer to effect unit resolution of m/z values is important. It is useful to differentiate a working resolution from a best resolution. For obvious reasons, a mass spectrometer manufacturer will want to quote the best measured resolution attainable on any one instrument, but the purchaser needs to remember that this measurement will have been obtained when everything is working perfectly. Under everyday conditions, the mass spectrometer is unlikely to be operating even close to this best limit; therefore effective resolving power is likely to be much lower than the best as stated in a brochure. As a rough guide, it is probably reasonable to subtract 10–15% from best-resolution figures to get some idea of the effective resolution obtainable under normal working conditions. Even then, choice of analyzer is not necessarily easy. As a general guide, up to m/z 600–1000 with unit mass resolution, all types of analyzer will be sufficient, but the quadrupole or ion trap is likely to be cheapest. As the working range increases, the quadrupoles and ion traps begin to drop out of consideration, and TOF or sector instruments come to the fore. Of these, the TOF is good up to m/z 3000–5000, although the higher mass ranges would probably need to include the use of a reflectron. As with the quadrupoles and ion traps, the TOF instruments are robust and easy to operate. Ion cyclotron resonance instruments may also be used at high m/z values if operated in Fourier-transform mode.

An added consideration is that the TOF instruments are easily and quickly calibrated. As the mass range increases again (m/z 5,000–50,000), magnetic-sector instruments (with added electric sector) and ion cyclotron resonance instruments are very effective, but their prices tend to match the increases in resolving powers. At the top end of these ranges, masses of several million have been analyzed by using Fourier-transform ion cyclotron resonance (FTICR) instruments, but such measurements tend to be isolated rather than targets that can be achieved in everyday use.

Simple considerations of achievable mass resolution and measurement become unrealistic when using electrospray methods of ionization because m/z values change markedly. A substance of mass 10,000 with a single positive charge has an m/z value of 10,000, which is easily measurable with a magnetic/electric sector combination but not with a quadrupole or a TOF instrument. However, a substance of mass 10,000 but with 10 positive charges has an m/z ratio of only 1000, which can be measured with a quadrupole, ion trap, or TOF instrument. A simple change from one ionization method to another can reduce or increase the demands on mass resolving power, with a significant effect on the final cost of the system.

Accurate mass measurement requires high resolving power. The difference in degrees of difficulty between measuring an m/z of 28 and one of 28.000 is likely to be large. Table 39.3 shows the broad mass ranges achievable with various analyzers.

Combined Analyzers (Hybrid Instruments)

For some kinds of analyses, it is convenient to have two combined mass spectrometers. This combination naturally increases costs; therefore purchase of such hybrid instruments tends to require much

TABLE 39.3
Typical Mass Ranges Achievable with Various Analyzers

M/z range[a]	Quadrupole	TOF	Sector	Ion trap	ICR
0–1500	√	√	√	√	√
1500–3000	—	√	√	—	√
3000–30,000	—	√	√	—	√
30,000–100,000	—	—	√	—	√
> 100,000	—	—	√	—	√

[a] Most ion sources produce singly charged ions, i.e., $z = 1$ and the ranges shown here apply to such ions. Matrix assisted methods may produce ions with $z > 1$. When $z = 1$, $m/z = m$, *viz.*, mass can be measured directly. An ES ion source produces ions with $z > 1$ and this effectively extends the mass ranges that can be examined. For example, with $z = 1$ and $m = 10{,}000$, the m/z value is 10,000 and this would be beyond the capabilities of a quadrupole instrument. However, with $z = 10$ and $m = 10{,}000$, $m/z = 1000$ and this value is within the range of a quadrupole analyzer.

more careful consideration of needs. Hybrid instrumentation is necessary for MS/MS purposes, for which the two mass spectrometers may be quadrupoles, magnetic sectors, or a combination of these with each other or with TOF analyzers. Often the combination is linked via a hexapole or quadrupole bridge, which serves to channel all ions from one MS to the other, or the quadrupole may serve as a gas collision cell. For MS/MS purposes, ions selected by m/z value in one mass spectrometer are passed through a collision cell to induce fragmentation, and the fragment ions are examined in the second mass spectrometer. Thus in the triple quadrupole (QqQ) analyzer, the central quadrupole is operated in such a way that no mass separation occurs, but collisions of ions with gas can take place. Analysis by MS/MS is particularly easy in this combination, as is selected-reaction monitoring, a technique for seeking out small traces of substances from other (mass spectrometrically) overlying (masking) substances. Two sector instruments linked together give a wide range of MS/MS techniques but at greater cost. The TOF instrument is frequently linked to a quadrupole (Q/TOF) for MS/MS purposes.

A major divergence appears for ion traps and ion cyclotron resonance (ICR) mass spectrometers. In both of these, MS/MS can be carried out without the need for a second analyzer. The differentiation is made possible by the length of time ion traps or ICR instruments can hold selected ions in their mass analyzers. For quadrupoles, magnetic sectors, and TOF analyzers, the ions generated in an ion source pass once through the analyzer, usually within a few microseconds. For ion traps and ICR instruments, ions can be retained in the trap or resonance cell for periods of milliseconds. In the latter case, it becomes possible to select ions at low background gas pressure, to collisionally activate these ions by increasing their velocity relative to background "bath" gas for a short time, and then to examine the resulting fragment ions. It is even possible to carry out MS^n analyses, where n can be from two to about five. Ion traps, like quadrupoles, are limited in ultimate mass that can be measured, but they provide a relatively cheap introduction to MS/MS. The ICR instruments tend to be as expensive as bigger hybrid instruments, but they can also be used in MS^n measurements.

Ionization Methods

When mass spectrometry was first used as a routine analytical tool, EI was the only commercial ion source. As needs have increased, more ionization methods have appeared. Many different types of ionization source have been described, and several of these have been produced commercially. The present situation is such that there is now only a limited range of ion sources. For vacuum ion sources, EI is still widely used, frequently in conjunction with CI. For atmospheric pressure ion sources, the most frequently used are ES, APCI, MALDI (lasers), and plasma torches.

Electron Ionization (EI) and Chemical Ionization (CI)

For gases and volatile liquids or solids, these methods are generally useful, giving both molecular and fragment ions with relatively high efficiency. EI often results in low yields of molecular ions compared with fragment ions. In contrast, CI gives high proportions of quasi-molecular ions to fragment ions, so it is convenient to have combined EI/CI sources to allow easy, rapid switching between the two ionization modes. Then, the molecular mass of a sample substance becomes clearly defined (CI), and its chemical structure can be inferred from the fragment ions (EI). Most manufacturers of mass spectrometers offer a combined EI/CI source. Switching between the two modes occupies only a second or two, and the alternate switching from EI to CI and back can be accomplished within the time taken for a substance to emerge from a GC column.

The poor yield of molecular ions during EI is partly due to the inherent instability of some cation radicals formed upon ionization. Since CI gives quasi-molecular ions that have been largely thermally equilibrated, there is little fragmentation. However, both CI and EI require the sample to be in the gas phase, which means that thermal instability of the sample can be a serious problem. Thus a substance such as toluene can be easily vaporized under the vacuum conditions inside a mass spectrometer, but sugars, proteins, or oligonucleotides are nonvolatile and degrade extensively when heated to modest temperatures. Obtaining an EI or CI spectrum of toluene is easy, but polar, high-molecular-mass compounds will give only spectra characteristic of breakdown products. (There are some instruments that make use of this thermal effect to investigate complex mixtures, as in pyrolysis/MS or even pyrolysis/GC/MS.)

There are ill-defined limits on EI/CI usage, based mostly on these issues of volatility and thermal stability. Sometimes these limits can be extended by preparation of a suitable chemical derivative. For example, polar carboxylic acids generally give either no or only a poor yield of molecular ions, but their conversion into methyl esters affords less polar, more volatile materials that can be examined easily by EI. In the absence of an alternative method of ionization, EI/CI can still be used with clever manipulation of chemical derivatization techniques.

There are methods for vaporizing solids of low volatility by placing them on a thin wire, which is then raised to a high temperature within a fraction of a second (direct chemical ionization, DCI). This rapid heating allows some vaporization without decomposition, but with the development of later ionization methods, it is now rarely used.

Electrospray Ionization (ES) and Atmospheric Pressure Chemical Ionization (APCI)

ES and APCI are ionization methods that were initially developed to enable the effluent from an LC column to be passed directly to a mass spectrometer. It is necessary to remove the solvent from the effluent without removing too much of the solute itself and without applying too much heat. For ES, the effluent is sprayed at atmospheric pressure and temperature into a desolvation region, where most of the solvent evaporates, and the residual sample (solute) molecules pass into a mass analyzer, usually as protonated molecules. The sample does not need to be heated, and the protonated (quasi-) molecular ions are thermally equilibrated by collisional processes during the evaporation of solvent. No fragment ions are formed. APCI is similar, except that the method of spraying is different. Additionally, APCI uses another ionization region (corona discharge) to enhance the yield of protonated molecular ions.

Both ES and APCI provide good yields of stable, protonated molecular ions. Because there is no fragmentation and therefore little structural information, it becomes necessary to add some excess of internal energy to the molecular ions to cause them to decompose. (This addition is usually done by collisionally induced decomposition or activation.) This extra internal energy can be added simply by making the ion beam pass through a region having a modest density of collision gas present (e.g., helium or argon). Alternatively, the molecular ions can be sped up in the desolvation region so they collide with evaporating neutral solvent molecules. Both ES and APCI serve dual roles as an ion source and an inlet system for whatever mass analyzer is used (quadrupole, ion trap, magnetic sector, TOF, ion cyclotron resonance).

Apart from ES and APCI being excellent ion sources/inlet systems for polar, thermally unstable, high-molecular-mass substances eluting from an LC or a CE column, they can also be used for stand-alone solutions of substances of high to low molecular mass. In these cases, a solution of the sample substance is placed in a short length of capillary tubing and is then sprayed from there into the mass spectrometer.

There is an added attribute of ES. Because the spraying process involves the application of a high positive or negative electric charge, the resulting protonated and other ions have more than one electric charge which facilitates measurement of high molecular mass, as indicated earlier. Accurate measurement of large m/z values is much more demanding, and ES provides a means for measuring large molecular masses with high accuracy.

Mass-Analyzed Laser Desorption Ionization (MALDI)

A laser pulse lasting a few pico- or nanoseconds and focused into a small area can deposit a large amount of energy into a sample so that it is rapidly vaporized to form an aerosol. Unfortunately, most of the ablated material has no excess of electric charge, and, unless a secondary ionization source is used (see later), there would be very low sensitivity because of the low yields of ions. If the sample substance is intimately mixed with a suitable matrix material, the ion yield can be greatly enhanced. The matrix is usually a volatile acidic substance that can absorb laser light efficiently. During vaporization following absorption of a laser pulse, the entrained sample substance is also vaporized. Since the sample is usually less ionizable than the matrix, proton transfer occurs between them to give protonated molecular ions of the sample. As these ions are produced by a low-energy proton-transfer step, there is little fragmentation. This method is very good for producing molecular mass information from polar, thermally unstable, high-molecular-mass biochemicals such as oligonucleotides, carbohydrates, and proteins. These sorts of samples often need to be examined in large numbers and special auto-inlets (auto-samplers) are available. Once inserted into the instrument, the autosampler allows large numbers of samples to be examined routinely. Various mass analyzers can be used with the MALDI ion source. If structural information is needed, there must be some induced fragmentation of the protonated molecular ions, and this fragmentation is mostly provided by the use of gas collision cells, as with ES and APCI, or by ion/molecule collision in an ion trap or ICR instrument.

Plasma Torch (PT) and Thermal (Surface) Ionization (TI)

The previous discussion has centered on how to obtain as much molecular mass and chemical structure information as possible from a given sample. However, there are many uses of mass spectrometry where precise isotope ratios are needed and total molecular mass information is unimportant. For accurate measurement of isotope ratio, the sample can be vaporized and then directed into a plasma torch. The sample can be a gas or a solution that is vaporized to form an aerosol, or it can be a solid that is vaporized to an aerosol by laser ablation. Whatever method is used to vaporize the sample, it is then swept into the flame of a plasma torch. Operating at temperatures of about 5000 K and containing large numbers of gas ions and electrons, the plasma completely fragments all substances into ionized atoms within a few milliseconds. The ionized atoms are then passed into a mass analyzer for measurement of their atomic mass and abundance of isotopes. Even intractable substances such as glass, ceramics, rock, and bone can be examined directly by this technique.

TABLE 39.4
Types of Ions Formed by Various Ion Sources[a]

Ionization Method	Type of Molecular Ion Formed	Good Molecular Mass Information	Abundant Fragment Ions	MS/MS Needed for Structural Information	Accurate Values for Isotope Ratios
EI	M^+, M^-	—[b]	√	—	—
CI	$[M + H]^+$, $[M + X]^+$, $[M - H]^-$	√	—	√	—
ES	$[M + nH]^{n+}$, $[M - nH]^{n-}$	√	—	√	—
MALDI	$[M + H]^+$, $[M + X]^+$	√	—	—	√
APCI	$[M + H]^+$, $[M + X]^+$, $[M - H]^-$	—	√[c]	√[c]	√
PT	A^+, A^-)	√	—[c]	—[c]	√
TI	A^+, (A^-)	√[c]	—[c]	—[c]	√

[a] Indications in this table are generalizations.

[b] Electron ionization usually gives some molecular ions and sometimes these are abundant. Often it is difficult to be certain which ion, if any, is the molecular ion. For this reason, it is advantageous to obtain both EI and CI spectra, the first giving good structural information and the second good molecular mass information.

[c] Plasma torches and thermal ionization sources break down the substances into atoms and ionized atoms. Both are used for measurement of accurate isotope ratios. In the breakdown process, all structural information is lost, other than an identification of elements present (e.g., as in inductively coupled mass spectrometry, ICP/MS).

Since detailed chemical structure information is not usually required from isotope ratio measurements, it is possible to vaporize samples by simply pyrolyzing them. For this purpose, the sample can be placed on a tungsten, rhenium, or platinum wire and heated strongly in vacuum by passing an electric current through the wire. This is thermal or surface ionization (TI). Alternatively, a small electric furnace can be used when removal of solvent from a dilute solution is desirable before vaporization of residual solute. Again, a wide variety of mass analyzers can be used to measure m/z values of atomic ions and their relative abundances.

Table 39.4 lists the suitable ionization sources that are commonly available for various classes of substance.

Overall View of Choices

The previous discussion has concentrated on major factors likely to be important in choosing the best mass spectrometer for a given defined purpose. Clearly, there are likely to be other issues that need to be considered, and instrument price will be one of these. The major considerations are listed in Tables 39.1–4.

Conclusion

The choice of a mass spectrometer to fulfill any particular task must take into account the nature of the substances to be examined, the degree of separation required for mixtures, the types of ion source and inlet systems, and the types of mass analyzer. Once these individual requirements have been defined, it is much easier to discriminate among the numerous commercial instruments that are available. Once suitable mass spectrometers have been identified, it is then often a case of balancing capital and running costs, reliability, ease of routine use, after-sales service, and manufacturer reputation.

Chapter 40

Analysis of Peptides and Proteins by Mass Spectrometry

Introduction

Until 1981, mass spectrometry was limited, generally, to the analysis of volatile, relatively low-molecular-mass samples and was difficult to apply to nonvolatile peptides and proteins without first cutting them chemically into smaller volatile segments. During the past decade, the situation has changed radically with the advent of new ionization techniques and the development of tandem mass spectrometry. Now, the mass spectrometer has a well-deserved place in any laboratory interested in the analysis of peptides and proteins.

Fast-Atom Bombardment (FAB)

The first breakthrough came with the development of fast-atom bombardment (FAB), which enabled polar compounds of high molecular mass to be ionized without application of heat for volatilization of the sample. In FAB (see Chapter 4, "Fast-Atom Bombardment and Liquid-Phase Secondary Ion Mass Spectrometry Ionization"), the sample is dissolved in a suitable solvent (also called a matrix) of low volatility (e.g., glycerol, thioglycerol, m-nitrobenzyl alcohol) and is bombarded by a beam of fast xenon or argon atoms. Ionization produces protonated $[M + H^+]$ or deprotonated $[M - H]^-$ molecular ions, sometimes accompanied by a little fragmentation. The matrix reproducibly gives background ion peaks, but these can interfere with sample ion peaks and usually dominate the low-mass end of the spectrum (Figure 40.1).

Different samples exhibit different levels of response to FAB, and, with a mixture of components, it is feasible that not all will be detected. In some cases, the minor components of a sample appear more prominently in the mass spectrum than the major ones. Despite these limitations, FAB is in widespread use and is an excellent technique for determining the molecular masses of peptides up to 10,000 Daltons, with an accuracy of 0.5 Da.

FAB has evolved, and fast atoms are being replaced by fast ions, such as cesium (Cs^+). This variation is called liquid secondary ion mass spectrometry (LSIMS) because the sample solution affords the secondary ion beam while the bombarding ions constitute the primary beam. Spectra

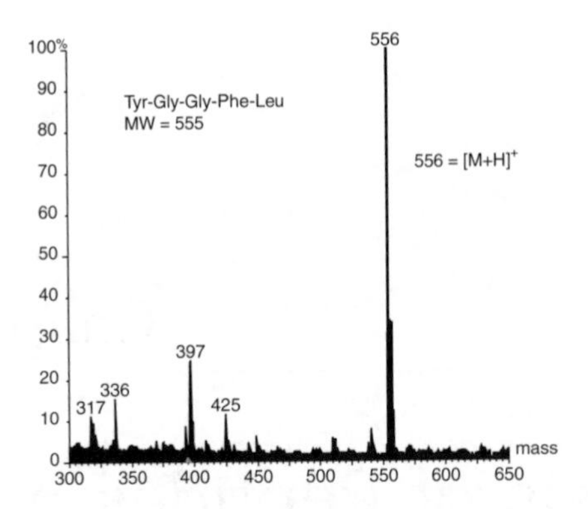

Figure 40.1
For this FAB experiment, a sample of the pentapeptidic enkephalin, Tyr.Gly.Gly.Phe.Leu., dissolved in glycerol was bombarded by xenon atoms. The resulting mass spectrum shows abundant protonated molecular ions at m/z 556.

produced by FAB and LSIMS are virtually identical, although higher sensitivity at high mass (10 kDa) is claimed for the latter.

Dynamic FAB

Another development arising from FAB has been its transformation from a static to a dynamic technique, with a continuous flow of a solution traveling from a reservoir through a capillary to the probe tip. Samples are injected either directly or through a liquid chromatography (LC) column. The technique is known as dynamic or continuous flow FAB/LSIMS and provides a convenient direct LC/MS coupling for the on-line analysis of mixtures (Figure 40.2).

Mixtures, as with the acyl coenzyme A factors shown in Figure 40.2, can be separated and analyzed on-line. In peptide and protein work, the peptidic substance is often reacted (digested) with enzymes that cleave the peptide or protein at places along its backbone to give smaller peptides. This digest (mixture of peptides) must be separated into its components and the newly formed peptides identified so that the original structure can be deduced. (This is called mapping and is something like assembling a linear jigsaw puzzle.) LC/FAB/MS is well suited to the separation of such mixtures and the identification of components through their molecular masses. However, not only the molecular mass of a peptide is important. The actual sequence (order) of amino acid residues making up the peptide chain is also important, and FAB, which gives predominantly molecular mass information and few structural pointers, must be supplemented by another technique such as MS/MS, also called tandem MS.

Mass Spectrometry/Mass Spectrometry (MS/MS)

Typically, a sample is analyzed by FAB/MS to obtain a relative molecular mass and then by FAB/MS/MS to obtain structural information by fragmenting the molecular ion and examining the fragment ions. This analysis is achieved by passing the molecular ion from a first mass analyzer into a collision cell (Figure 40.3), where collision gas (e.g., argon) is used to fragment the ion. The fragment ions produced are analyzed by a second mass spectrometer. The fragments can arise only directly from the molecular ion and so provide useful sequence information for peptides. Peptides have been found to fragment by predictable pathways along their backbone from the C-terminus and/or the N-terminus, as shown by the example of angiotensin III (Figure 40.4).

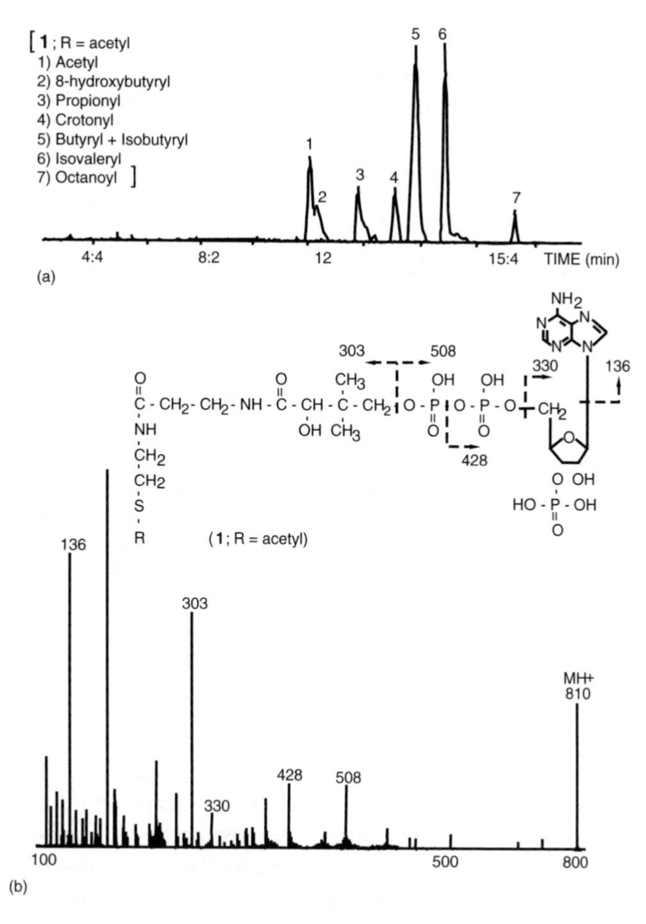

Figure 40.2
(a) LC/FAB/MS analysis of short- and medium-chain acyl coenzyme A compounds (1, = 1–7; 0.5 nmol of each). These compounds carry acyl groups for enzyme reactions, and a number of metabolic diseases can be traced to enzyme deficiencies that result in the accumulation of toxic coenzyme A thioesters. (b) FAB mass spectrum of acetyl coenzyme A [component 1.1 from the trace shown in (a)]. The likely origin of major fragment ions is indicated.

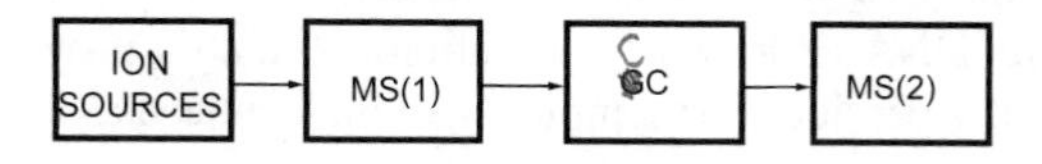

Figure 40.3
Typical MS/MS configuration. Ions produced from a source (e.g., dynamic FAB) are analyzed by MS(1). Molecular ions (M^+ or $[M + H]^+$ or $[M - H]^-$, etc.) are selected in MS(1) and passed through a collision cell (CC), where they are activated by collision with a neutral gas. The activation causes some of the molecular ions to break up, and the resulting fragment ions provide evidence of the original molecular structure. The spectrum of fragment ions is mass analyzed in the second mass spectrometer, MS(2).

Tandem mass spectrometers most commonly used for MS/MS studies include the following analyzer combinations, although many others are possible:

1. Quadrupole–collision cell–quadrupole
2. Magnetic/electrostatic analyzer–collision cell–quadrupole
3. Magnetic/electrostatic analyzer–collision cell–magnetic/electrostatic analyzer

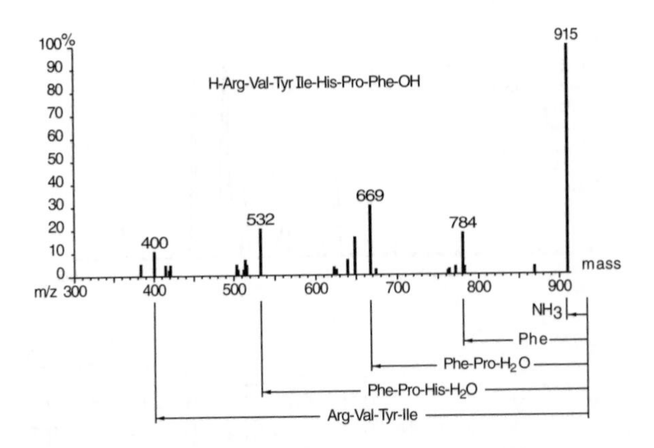

Figure 40.4
Typical FAB/MS/MS experiment on a peptide. The FAB/MS spectrum of angiotensin III is dominated by the protonated molecular ion $[M + H]^+$ at m/z 932, confirming the molecular mass of 931. In an MS/MS experiment, only the ion at m/z 932 was allowed to pass through the first mass analyzer into the collision cell (Figure 40.3). When the ion at m/z 932 passes through the collision cell, several fragment ions were produced, and all were analyzed by the second mass analyzer to produce the spectrum shown here. The ion at m/z 915 arises by loss of NH_3 from 932. The ions at m/z 784, 669, 532, and 400 arise, respectively, by loss of a phenylalanine residue; phenylalanine, proline, and a water molecule; phenylalanine, proline, histidine, and a water molecule; and arginine, valine, tyrosine, and isoleucine, from 932. These fragments verify the expected sequence of amino acid residues in angiotensin III.

The collision cell used with the first two types (1, 2) is an RF-only quadrupole or a hexapole. This type of cell adds only a small amount of energy to an ion during its collision with the cell gas. The third type uses a high-energy collision cell that has the advantage of producing fragmentation of amino acid side chains as well as the backbone fragmentation shown in Figure 40.4. This extra fragmentation gives information that permits differentiation of the two isomeric amino acid residues, leucine and isoleucine. Sequence information has been obtained by MS/MS on samples up to 2500 Da, often at low picomole levels, and usually in just a few minutes.

Other Ion Sources

The techniques described thus far cope well with samples up to 10 kDa. Molecular mass determinations on peptides can be used to identify modifications occurring after the protein has been assembled according to its DNA code (post-translation), to map a protein structure, or simply to confirm the composition of a peptide. For samples with molecular masses in excess of 10 kDa, the sensitivity of FAB is quite low, and such analyses are far from routine. Two new developments have extended the scope of mass spectrometry even further to the analysis of peptides and proteins of high mass.

Laser Desorption Mass Spectrometry (LDMS)

The LDMS technique uses solid matrices (e.g., cinnamic acid derivatives) to absorb energy from a laser pulse, which volatilizes and ionizes proteins premixed within the matrix. Mass analysis is achieved by a time-of-flight (TOF) analyzer which, as the name suggests, measures precisely the time taken for the ions to travel from the source through the flight tube to the detector. The heavier the ion, the longer is the flight time. The spectrum generally contains a protonated $[M + H^+]$ or deprotonated $[M - H]^-$ molecular ion cluster, together with doubly charged and perhaps multicharged ions. Fragmentation, hence sequence information, is usually absent.

In principle, there is no upper limit to the mass range, and proteins as large as 200 kDa have been measured using as little as 1 pmol of material, making LDMS one of the most sensitive techniques available. However, the resolution of this technique is low compared with other mass spectrometric methods, and the ions constituting the molecular mass cluster are unresolved. Heterogeneous proteins can present a problem, as the mass accuracy of 0.1% (e.g., 50 Da at 50 kDa) means that some post-translational modifications go undetected, and mass changes associated with a single amino acid substitution would be unnoticed.

Electrospray (ES)

The development of ES has radically increased the use of mass spectrometry in biotechnology by providing an ionization technique capable of analyzing large, polar, thermally sensitive biomolecules with unprecedented mass accuracy and good sensitivity. In ES, the sample, in solution, is passed through a narrow capillary and reaches an atmospheric-pressure ionization source as a liquid spray. The voltage at the end of the capillary is significantly higher (3 kV) than that of the mass analyzer, so the sample emerging is dispersed into an aerosol of highly charged droplets known as the electrospray. Evaporation of solvent, aided by a stream of nitrogen, causes a decrease in the size of the droplets until eventually multicharged ions from individual protein molecules, free from solvent, are released and can be mass analyzed.

The ions so produced are separated by their mass-to-charge (m/z) ratios. For peptides and proteins, the intact molecules become protonated with a number (n) of protons (H^+). Thus, instead of the true molecular mass (M), molecular ions have a mass of [M + nH]. More importantly, the ion has n positive charges resulting from addition of the n protons $[M + nH]^{n+}$. Since the mass spectrometer does not measure mass directly but, rather, mass-to-charge (m/z) ratio, the measured m/z value is $[M + nH]/n$. This last value is less than the true molecular mass, depending on the value of n. If the ion of true mass 20,000 Da carries 10 protons, for example, then the m/z value measured would be (20,000 + 10)/10 = 2001.

This last m/z value is easy to measure accurately, and, if its relationship to the true mass is known (n = 10), then the true mass can be measured very accurately. The multicharged ions have typical m/z values of <3000 Da, which means that conventional quadrupole or magnetic-sector analyzers can be used for mass measurement. Actually, the spectrum consists of a series of multicharged protonated molecular ions $[M + nH]^{n+}$ for each component present in the sample. Each ion in the series differs by plus and minus one charge from adjacent ions ($[M + nH]^{n+}$; n = an integer series for example, 1, 2, 3, …, etc.). Mathematical transformation of the spectrum produces a true molecular mass profile of the sample (Figure 40.5).

Proteins of Large Molecular Mass

While electrospray is used for molecules of all molecular masses, it has had an especially marked impact on the measurement of accurate molecular mass for proteins. Traditionally, direct measurement of molecular mass on proteins has been difficult, with the obtained values accurate to only tens or even hundreds of Daltons. The advent of electrospray means that molecular masses of 20,000 Da and more can be measured with unprecedented accuracy (Figure 40.6). This level of accuracy means that it is also possible to identify post-translational modifications of proteins (e.g., glycosylation, acetylation, methylation, hydroxylation, etc.) and to detect mass changes associated with substitution or deletion of a single amino acid.

A typical electrospray analysis can be completed in 15 min with as little as 1 pmol of protein. An analysis of the cord blood of a baby (Figure 40.6) showed quite clearly that five globins were present, viz., the normal ones (α, β, Gγ, and Aγ) and a sickle-cell variant (sickle β). The last one is easily revealed in the mass spectrum, even at a level of only 4% in the blood analyzed.

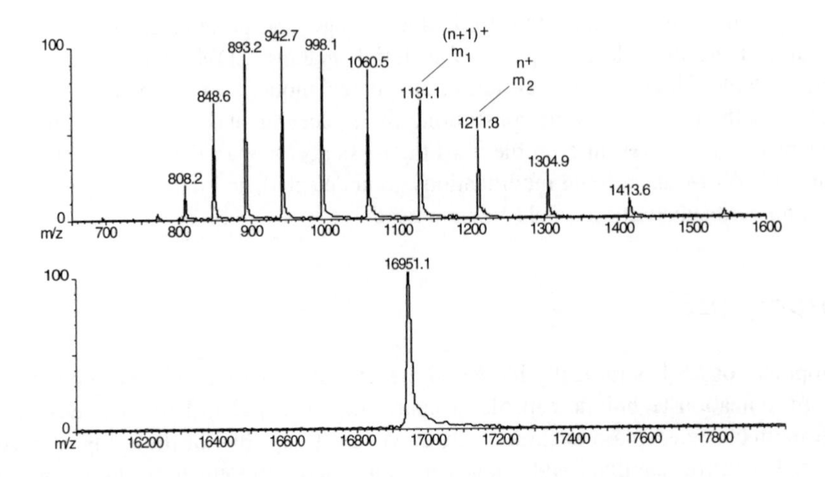

Mass to Charge Ratio (m/z)	No. of Charges (n)	Molecular Mass (RMM)
1542.04	11	16951.40
1413.59	12	16950.94
1304.93	13	16950.95
1211.80	14	16951.11
1131.12	15	16951.62
1060.46	16	16951.26
998.11	17	16950.67
942.75	18	16951.30
893.15	19	16950.71
848.57	20	16951.25
808.21	21	16951.14
771.49	22	16950.72
	Mean	16951.09
	S.D.	±0.30

Figure 40.5A
A sample of the protein, horse heart myoglobin, was dissolved in acidified aqueous acetonitrile (1% formic acid in H_2O/CH_3CN, 1:1 v/v) at a concentration of 20 pmol/l. This sample was injected into a flow of the same solvent passing at 5 µl/min into the electrospray source to give the mass spectrum of protonated molecular ions $[M + nH]^+$ shown in (a). The measured m/z values are given in the table (b), along with the number of protons (charges; *n*) associated with each. The mean relative molecular mass (RMM) is 16,951.09 ± 0.3 Da. Finally, the transformed spectrum, corresponding to the true relative molecular mass, is shown in (c); the observed value is close to that calculated (16,951.4), an error of only 0.002%.

While this example shows that small differences in large masses are easily discerned by electrospray methodology, note that absolute accuracy of mass measurement is unprecedentedly high. The accuracy is sufficiently high that substitution of one amino acid by another can be detected with ease. Figure 40.7 illustrates this point following the discovery of a new variant of β-globin, called Hb Montreal-Chori, in which a threonine residue has been substituted by isoleucine at position 87 in the β-chain. The substitution caused a change in mass from 15,837.2 Da to 15,879.3 Da.

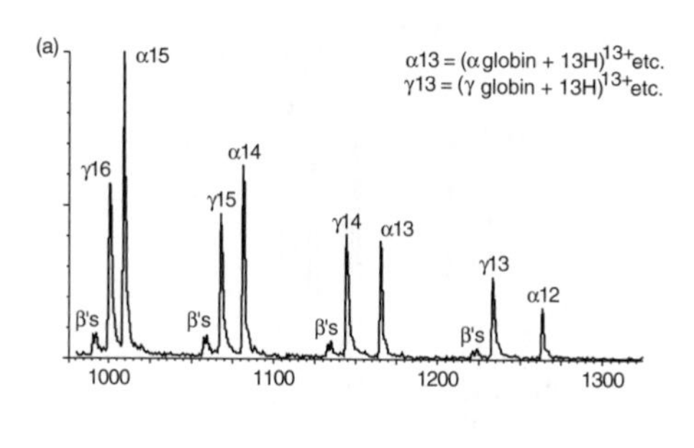

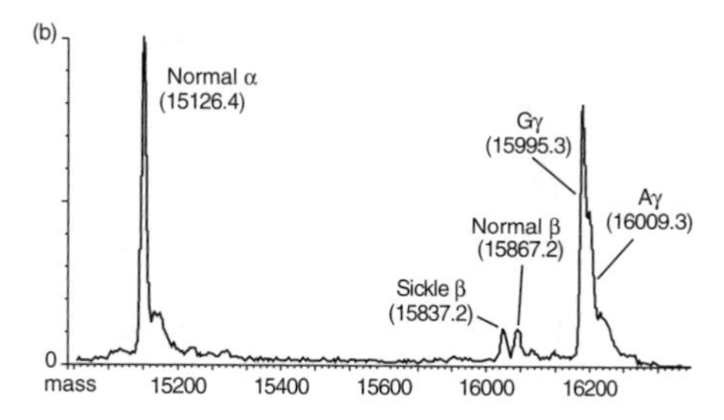

Figure 40.6
(a) Part of an electrospray spectrum of globins from the cord blood of a sickle-cell carrier and (b) the same after computer transformation onto a true molecular mass scale. Normal α, β, and γ globins are clearly visible, along with the sickle-cell variant (sickle β). The two γ-globins (A, G), although having a mass difference of only 14 Da in 16,000 Da, are easily distinguished from each other.

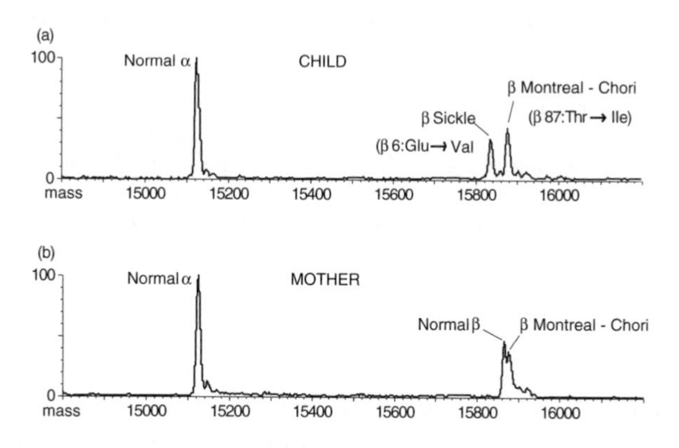

Figure 40.7
Electrospray mass spectra of globins from the blood of (a) a child diagnosed as having the sickle-cell anemia trait and (b) of its mother. As well as the usual β-globin sickle-cell variant at m/z 15,837.2, a new variant (β-Montreal-Chori) appears at m/z 15,879.3 and is observed in both the child and the mother.

Analysis of Peptide Mixtures

The electrospray source can be coupled directly to a liquid chromatographic (LC) column so that, as components of a mixture emerge from the column, they are passed through the source to give accurate mass data. As an example, a mixture of the peptides shown in Figure 40.8(a) was separated by LC and accurately mass-analyzed by ES.

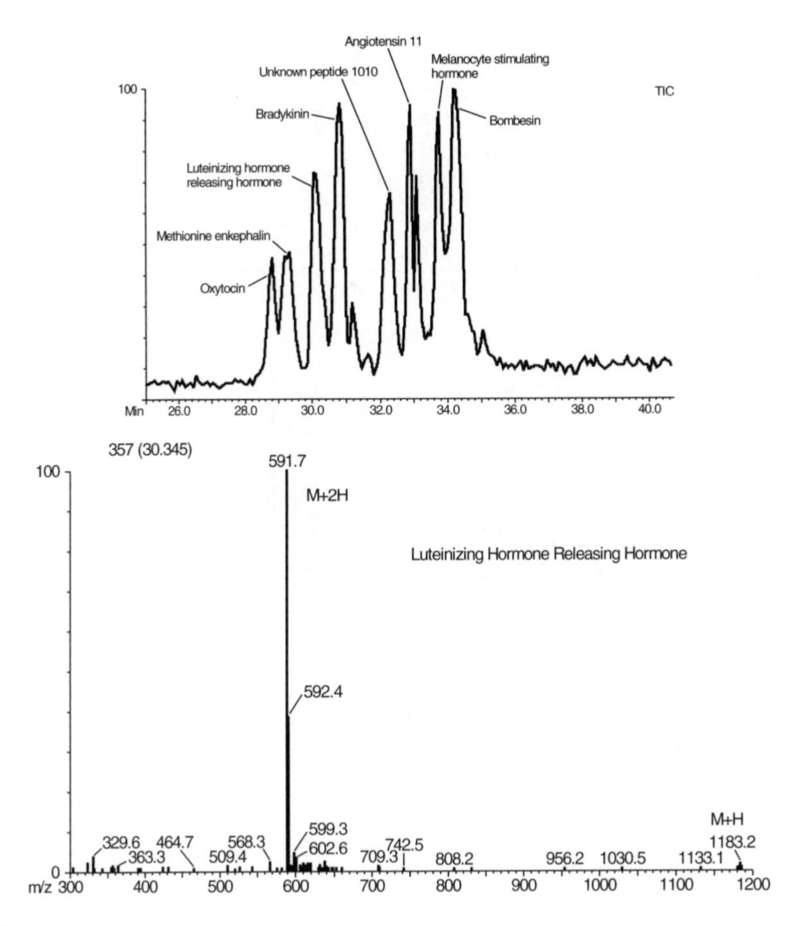

Figure 40.8

(a) A mixture of peptides was separated by LC, and the eluant was passed into an electrospray source. The total ion current trace (a) reveals the individual component peptides, each of which was identified through its measured accurate mass, an illustration of which is shown in (b) for luteinizing hormone-releasing hormone. (b) Mass spectrum of one of these peptides (luteinizing hormone-releasing hormone), which gave an abundant ion representing a doubly protonated molecule $[M + 2H]^+$ at m/z 592.4 and, therefore, with a true relative molecular mass (M) of 1182.8.

Conclusion

Intact peptides and proteins can be examined by a variety of new techniques, including MS/MS, dynamic FAB, APCI, and electrospray. Large masses of tens of thousands of Daltons can be accurately measured with unprecedented accuracy by electrospray.

Chapter 41

Environmental Protection Agency Protocols

Introduction

The Environmental Protection Agency of the United States (U.S. EPA) exerts strict control over analysis procedures for monitoring, quantifying, and recording a wide range of organic compounds that could cause environmental hazards if present at levels higher than those recommended. The stringent control exerted by the EPA covers many application areas, with gas chromatography (GC) featuring prominently. The availability of computer-supported quadrupole mass spectrometry, its high intrinsic sensitivity, and its compatibility with gas chromatographs led to its adoption in the late 1970s as the primary tool for monitoring trace organics.

Environmental Laws

Complex environmental samples originate from diverse matrices (the predominant material of which the sample to be analyzed is composed). These matrices, usually either water or soil/sediment, can contain as many as 50 to 100 organic components at widely varying concentrations. The EPA approach to the analysis of these samples involves the analysis of specific (or target) compounds and the use of authentic standards for quality control. The current number of standards in the EPA repository is about 1500, and their analysis is covered by various approved methods.

The EPA Contract Laboratory Program (CLP) has responsibility for managing the analysis programs required under the U.S. Comprehensive Environmental Response, Compensation, and Liability Act (CERCLA). The approved analytical methods are designed to analyze water, soil, and sediment from potentially hazardous waste sites to determine the presence or absence of certain environmentally harmful organic compounds. The methods described here all require the use of GC/MS.

The Contract Laboratory Program

Certain laboratories can, after a contract has been awarded, register under the Contract Laboratory Program (CLP) of the EPA. To earn a contract, one or more specifically prepared samples must be analyzed under very similar conditions to those used in standard protocols. Only if the data are deemed satisfactory will a contract be awarded. Further evaluation samples must be analyzed at three-month intervals afterward to ensure that performance is being maintained.

Protocols

The EPA publishes Series Methods that describe the exact procedures to be followed with respect to sample receipt and handling, analytical methods, data reporting, and document control. These guidelines must be followed closely to ensure accuracy, reproducibility, and reliability within and among the contract laboratories.

500 Series Methods

The Safe Drinking Water Act (1974) controls the monitoring of finished drinking water, raw source water, or drinking water in any treatment stage. Method 524 monitors 60 purgeable (volatile) organic compounds in drinking water by the use of GC/MS. A purge-and-trap device is used to isolate the organic compounds by stripping them from water using a stream of an inert gas. The purged components are then trapped onto a porous polymer from which they can be thermally desorbed directly onto a GC column. Method 525 covers the analysis of 86 basic, neutral, and acidic compounds in drinking water.

600 Series Methods

The Clean Water Act (1972) requires discharge limits to be set on industrial and municipal wastewater, and these analyses are outlined in the National Pollution Discharge Elimination System for the 600 Series Methods. Method 624 covers the analysis of purgeable organic compounds; Method 625 covers the analysis of 81 bases, neutrals, and acids; Method 613 describes the analysis of dioxins and furans.

8000 Series Methods

To satisfy the Resource Conservation and Recovery Act (1977) and its amendment for hazardous and solid waste (1984), the 8000 Series Methods have been designed to analyze solid waste, soils, and groundwater. In particular, methods 8240/8260 require the use of a purge-and-trap device in conjunction with packed or capillary GC/MS, respectively, for the analysis of purgeable organic compounds. Methods 8250/8270 concern analyses for the less-volatile bases, neutrals, and acids by GC/MS after extraction from the matrix by an organic solvent.

Sample Analysis by GC/MS

Calibration

Before sample preparation, the laboratory must demonstrate that the mass spectrometer is operating satisfactorily. First, the instrument must be tuned by calibration using one of two compounds.

Br—(benzene ring)—F C_6H_4BrF (a)

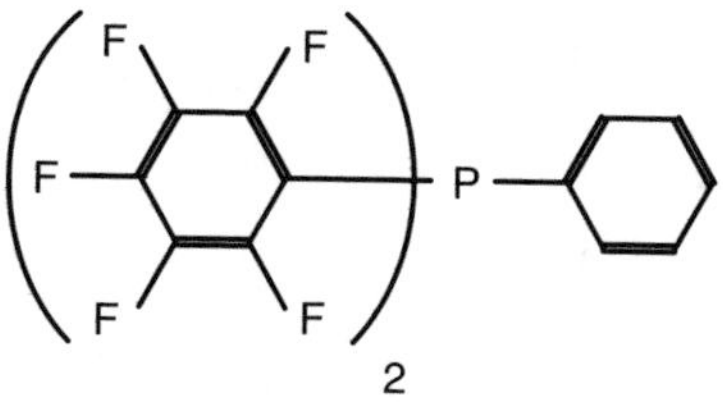

(F, F, F, F, F substituted ring)$_2$—P—(benzene ring) $C_{18}H_5F_{10}P$ (b)

Figure 41.1
Chemical structures for BFB and DFTPP and their relative molecular masses (RMM; molecular weight): (a) 4-bromofluorobenzene (BFB); RMM = 174 and 176; (b) decafluorotriphenylphosphine (DFTPP); RMM = 442.

TABLE 41.1
Ion Abundance Criteria

m/z	Ion Abundance Criteria for BFB	Relative Abundance (%)
Bromofluorobenzene (BFB)		
50	15.0–40.0% of base peak	24.1
75	30.0–60.0% of base peak	50.5
95	Base peak, 100% relative abundance	100.0
96	5.0–9.0% of base peak	7.4
173	Less than 2.0% of mass 174	0.3 (7.1)[a]
174	Greater than 50.0% of base peak	50.6
175	5.0–9.0% of mass 174	3.6 (7.1)[a]
176	Greater than 95.0% but less than 101.0% of mass 174	50.1 (99.1)[a]
177	5.0–9.0% of mass 176	3.5 (7.0)[b]
Decafluorotriphenylphosphine (DFTPP)		
m/z	Ion abundance criteria for DFTPP	Relative Abundance (%)
51	30.0–60.0% of mass 198	44.3
68	Less than 2.0% of mass 69	0.6 (1.3)[c]
69	Mass 69 relative abundance	43.3
70	Less than 2.0% of mass 69	0.6 (1.5)[c]
127	40.0–60.0% of mass 198	57.0
197	Less than 1.0% of mass 198	0.0
198	Base peak, 100% relative abundance	100
199	5.0–9.0% of mass 198	7.0
275	10.0–30.0% of mass 198	24.7
365	Greater than 1.00% of mass 198	2.5
441	Present, but less than mass 443	10.3
442	Greater than 40% of mass 198	69.5
443	17.0–23.0% of mass 442	13.7 (19.8)[d]

[a] Value in parentheses is percent of mass 174.
[b] Value in parentheses is percent of mass 176.
[c] Value in parentheses is percent of mass 69.
[d] Value in parentheses is percent of mass 442.

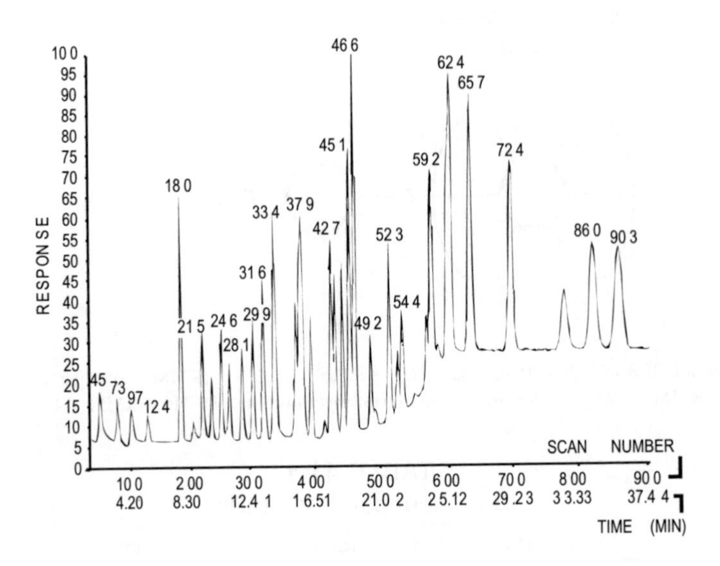

No.	Rank	Spectrum Match	Spectrum Fit	Scan Diff	Peak Area	Figs	Scan Found	Scan Pred	Quan m/z	Compound Name
1	99	60	99	0	2404000	bb	260	260	128	Bromochloromethane
2	86	44	93	0	7130000	bb	332	332	65	1, 2-Dichloroethane-d4
3	89	44	99	0	7178000	bb	45	45	50	Chloromethane
4	94	51	99	0	3706000	bb	73	73	94	Bromomethane
5	93	50	99	0	6428000	bb	97	97	62	Vinyl Chloride
6	83	34	99	0	3418000	bb	124	124	64	Chloroethane
7	100	79	99	0	11963000	bb	180	180	84	Methylene Chloride
8	72	33	99	0	2778500	vv	200	200	43	Acetone
	52	12	99	8	495000	?v	208			
	47	4	98	-9	131000	??	191			
9	100	78	99	0	17782000	bb	215	215	76	Carbon Disulfide
10	100	69	99	0	4551000	bb	246	246	96	1, 1-Dichloroethane
11	98	58	99	0	11006500	bb	281	281	63	1, 1-Dichloroethane
12	100	68	99	0	5588500	bv	299	299	96	1, 2-Dichloroethene (to
13	100	76	99	0	13830000	bb	316	316	83	Chloroform
14	91	49	96	0	10453000	bb	335	335	62	1, 2-Dichloroethane
15	90	45	99	0	16378000	bb	523	523	114	1, 4-Difuorobenzene
16	78	25	99	0	2427652	bv	334	334	72	2-Butanone
17	100	69	99	0	10077000	bv	368	368	97	1, 1, 1-Trichloroethane
18	92	48	99	0	9371000	vb	378	378	117	Carbon Tetrachlor
19	80	46	82	0	21057960	vb	380	380	43	Vinyl Acetate
20	100	64	99	0	10350460	bv	393	393	83	Bromodichloromethane
21	100	72	98	0	9990000	bb	427	427	63	1, 2-Dichloropropane
22	100	64	98	33	11815000	bb	466	466	75	cis-1, 3-dichloropropen
23	100	66	98	0	6840000	bb	446	446	130	Trichloroethene
24	79	30	96	0	6787000	bb	465	465	129	Dibromochloromethane
25	73	41	96	0	6394530	bv	467	467	97	1, 1, 2-Trichloroethane
26	100	74	97	0	28387000	bb	458	458	78	Benzene
27	83	35	97	-33	11249710	vb	433	433	75	trans-1, 3-dichloroprop
28	91	47	99	0	4062000	bb	536	536	173	Bromoform
29	67	2	47	0	4509000	?b	630	630	117	Chlorobenzene-d5
	28	3	46	2	29000	??	632			
30	76	23	98	0	8146502	bb	812	812	95	4-Bromofluorobenzene
31	88	41	99	0	18205590	bv	619	619	98	Toluene-d8
32	98	57	99	40	10291170	vv	584	584	43	4-Methyl-2-Pentanone
33	88	42	98	-40	7731600	vb	544	544	43	2-Hexanone
34	94	53	98	0	5249000	bb	591	591	164	Tetrachloroethene
35	88	41	99	0	9299502	vv	594	594	83	1, 1, 2, 2-Tetrachloroeth
36	96	55	98	0	16024560	bv	624	624	92	Toluene
37	88	41	99	0	17310850	bv	659	659	112	Chlorobenzene
38	89	45	98	0	10301000	bb	724	724	106	Ethylbenzene
39	83	34	98	0	19074000	bb	860	860	104	Styrene
40	82	32	99	0	11223000	bb	902	902	106	m-Xylene
41	57	32	99	4	10456000	bb	906	902	106	o-/p-Xylene

Figure 41.2
(a) GC mass chromatogram showing the separation of a mixture of standards during initial calibration. The compounds corresponding to the peaks are identified in Figure 41.2(b). (b) An interim report on the analysis of the mixture of standards illustrated in Figure 41.2(a).

No.	Rank	Spectrum Match		Scan FitDiff	Peak Area	Figs	Scan Found	Scan Pred	Quan m/z	Compound Name
1	100	79	99	1	2005000	bb	210	209	128	Bromochloromethane
2	83	41	97	-1	5365000	bb	267	268	65	1, 2-Dichloroethane-d4
3	100	248	99	-4	2262000	bb	32	36	50	Chloromethane
4	óóóó	No Trace Found			óóó-			59	94	Bromomethane
5	óóóó	No Trace Found			óóó-			78	62	Vinyl Chloride
6	óóóó	No Trace Found			óóó-			100	64	Chloroethane
7	92	84	99	-5	17767000	bb	140	145	84	Methylene Chloride
8	óóóó	No Trace Found			óóó-			161	43	Acetone
9	óóóó	No Trace Found			óóó-			174	76	Carbon Disulfide
10	óóóó	No Trace Found			óóó-			199	96	1, 1-Dichloroethene
11	100	76	99	-1	36246000	bb	226	227	63	1, 1-Dichloroethane
12	óóóó	No Trace Found			óóó-			242	96	1, 2-Dichloroethene (t
13	óóóó	No Trace Found			óóó-			255	83	Chloroform
14	óóóó	No Trace Found			óóó-			270	62	1, 2-Dichloroethane
15	87	59	99	3	14924000	bb	426	423	114	1, 4-Difluorobenzene
16	óóóó	No Trace Found			óóó-	*		272	72	2-Butanone
17	óóóó	No Trace Found			óóó-			300	97	1, 1, 1-Trichloroethane
18	óóóó	No Trace Found			óóó-			308	117	Carbon Tetrachloride
19	86	63	81	-1	23897000	bv	309	310	43	Vinyl Acetate
20	óóóó	No Trace Found			óóó-			320	83	Bromodichloromethane
21	100	85	99	-3	33719000	bb	345	348	63	1, 2-Dichloropropane
22	óóóó	No Trace Found			óóó-			353	75	cis-1, 3-Dichloropropen
23	óóóó	No Trace Found			óóó-			363	130	Trichloroethene
24	óóóó	No Trace Found			óóó-			379	129	Dibromochloromethane
25	óóóó	No Trace Found			óóó-			380	97	1, 1, 2-Trichloroethane
26	óóóó	No Trace Found			óóó-			373	78	Benzene
27	óóóó	No Trace Found			óóó-			380	75	trans-1, 3-Dichloroprop
28	óóóó	No Trace Found			óóó-			445	173	Bromoform
29	36	0	57	0	20730000	bb	482	482	117	Chlorobenzene-d5
30	100	65	99	0	14070500	bb	646	646	95	4-Bromofluorobenzene
31	97	62	99	1	27585000	bb	504	503	98	Toluene-d8
32	óóóó	No Trace Found			óóó-			447	43	4-Methyl-2-Pentanone
33	óóóó	No Trace Found			óóó-			416	43	2-Hexanone
34	óóóó	No Trace Found			óóó-			452	164	Tetrachloroethene
35	óóóó	No Trace Found			óóó-			455	83	1, 1, 2, 2-Tetrachloroethene
36	óóóó	No Trace Found			óóó-			477	92	Toluene
37	óóóó	No Trace Found			óóó-			504	112	Chlorobenzene
38	óóóó	No Trace Found			óóó-			554	106	Ethylbenzene
39	óóóó	No Trace Found			óóó-			658	104	Styrene
40	óóóó	No Trace Found			óóó-			690	106	m-Xylene
41	óóóó	No Trace Found			óóó-			693	106	o-/p-Xylene

Figure 41.3
Identification of some target compounds in an environmental sample. The report lists the closeness of match and the estimated quantities for the listed compounds.

4-Bromofluorobenzene (BFB) is used to establish tuning performance prior to the analysis of purgeable organic compounds, and decafluorotriphenylphosphine (DFTPP) is used prior to the analysis of bases, neutrals, and acids (Figure 41.1).

The positive-ion electron-ionization spectra of BFB and DFTPP must exhibit molecular and specified fragment ions, the relative abundances of which must fall within a predefined range. Ion abundance criteria for BFB and DFTPP are shown in Table 41.1.

This calibration procedure, which must be demonstrated at the start of each working period (or 12-hour shift) ensures that all samples are analyzed with respect to a known reference point of the mass spectrometer.

The next step is to show that the response for the analysis of any target compound is linear. This step is known as the initial calibration and is achieved by the analysis of standards for a series of specified concentrations to produce a five-point calibration curve (Figure 41.2a, b). On subsequent days, a continuing calibration must be performed on calibration check compounds to evaluate the calibration precision of the GC/MS system.

Sample Analysis

Before sample preparation, surrogate compounds must be added to the matrix. These are used to evaluate the efficiency of recovery of sample for any analytical method. Surrogate standards are often brominated, fluorinated, or isotopically labeled compounds that are not expected to be present in environmental media. If the surrogates are detected by GC/MS within the specified range, it is

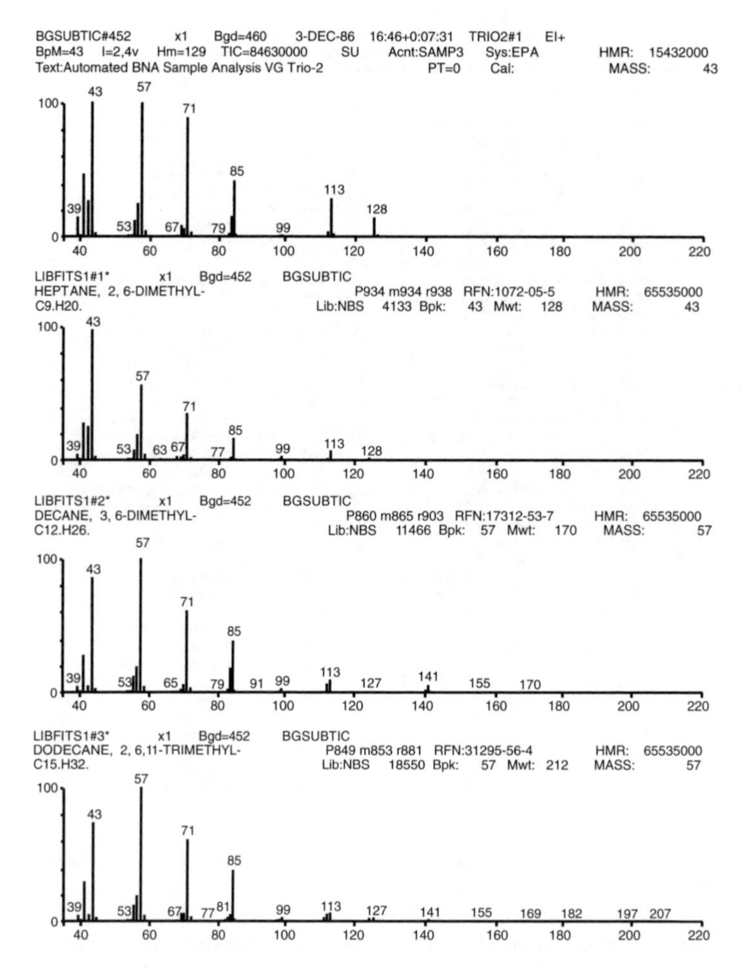

Figure 41.4
Comparison of the mass spectrum from a target compound (top), with the three best fits from the library of standard spectra (lower three traces). The closeness of fit of the mass spectra and the chromatographic retention time lead to a positive identification of 2, 6-dimethylheptane.

taken to be an indication that sample preparation and analysis have been performed satisfactorily. If the surrogates are out of specification, reanalysis or even repreparation of the sample is required.

Internal standards at a known concentration are added to the sample after its preparation but prior to analysis to check for GC retention-time accuracy and response stability. If the internal standard responses are in error by more than a factor of two, the analysis must be stopped and the initial calibration repeated. Only if all the criteria have been met can sample analysis begin.

A considerable amount of time is necessary to reach the point at which sample analyses can commence, and it is essential that the stability and reliability of the mass spectrometer be high to ensure maximum sample throughput during the limited time available between calibration checks.

Data Processing

After analysis, target compounds (those compounds that the analysis is aiming to detect) are recorded by the data processing facilities of the mass spectrometer (Figure 41.3). The correct location of target peaks in a gas chromatogram is verified by the use of target-compound databases, stored by the data system of the mass spectrometer. These databases hold information pertaining to the target compounds: name, retention time, search window with respect to the retention time, and so on. Once chromatographic peaks have been assigned to target compounds, the various mass spectra from the sample are compared with library spectra of the target compounds; a match between an observed and a library spectrum serves to verify the identification of a component (Figure 41.4). The closeness of the match is given a score, with a higher score indicating better identification. The National Institutes of Health (NIH)–EPA mass spectral library contains over 44,000 entries.

Chromatographic peak areas are calculated automatically by the data system by reference to the response obtained from certain specified, compound-dependent ions. From the peak areas of the target compounds, quantification is achieved by comparison with the internal standards, which are present in known concentration. The laboratory responsible for the analysis must report the target compounds and all tentatively identified (nontarget) compounds. Standard EPA forms must be completed and submitted. A laboratory is said to be in compliance when it has satisfied all aspects of its CLP contract.

Conclusion

The U.S. Environmental Protection Agency publishes sets of Series Methods that describe procedures for detecting and estimating the quantity of environmentally hazardous substances. There are strict requirements for accuracy, reproducibility, and for calibration of mass spectrometers.

Chapter 42

Computers and Transputers in Mass Spectrometers, Part A

Introduction

An understanding of computer and transputer operation requires some basic knowledge of the working of a microprocessor. This simplified account of the use of computers and transputers begins with some discussion of the fundamentals of computing. It must be emphasized that the following account is very elementary, and books on computing should be consulted for a deeper understanding.

The account begins with binary arithmetic, moves on to on-off (flip-flop) electronic switches, then to serial and parallel processing, and finally to computers/transputers.

Multibase Arithmetic

Much of modern life revolves around the use of the decimal system of numbers, although for many purposes it is far from ideal. This system (dec = ten) takes 10 as the basic unit of operation, and we say that we are working to *base 10*. However, the base 10 is not the only one possible; the previous British and many other countries' coinage was founded on a base of 12. The meaning of a base for purposes of calculation is illustrated in Figure 42.1.

If two numbers (17 and 24) are to be added together, then the calculation proceeds as follows: 7 + 4 = 11, but 11 is 10 + 1, and so a 1 is written down and a ten is added to the next column. In the next column, there is one ten from the 17, two tens from the 24, and one ten carried over, and the total number of tens is 1 + 2 + 1 = 4. Therefore, the answer is 41 — but only in the decimal system! Suppose the calculation is done on a base of seven — a heptimal system. Then, repeating the above calculation, 7 + 4 = 11 as before, but now 11 is *one* 7 with 4 left over. Therefore, one 7 is carried to the left (Figure 42.2), and the 4 left over is written down.

In the next column, there are one seven (not one ten, as with decimal 17) and two sevens (from the 24), making a total of three sevens; the total number of sevens is then 3 + 1 (carried from the previous column) = 4, and this is written down. Thus, the total obtained from the addition of 17 and 24 in the two systems is 41 in one and 44 in the other! Which is correct? In fact, we have cheated a little because 17 (decimal) written in the heptimal system should be *two* sevens and *a*

1 7	1 7
2 4	2 4
1	4 1

1 ⟶ ⟶

Figure 42.1
Addition of two decimal numbers. Note the carryover of one ten from the rightmost column into the left one.

1 7	1 7
2 4	2 4
4	4 4

1 ⟶ ⟶

Figure 42.2
Addition in the heptimal system. Now, one seven is carried from the rightmost column into the leftmost one.

three (23) and the 24 should be *three* sevens and *a* three (33). It is 23 and 33, which must be added: 3 + 3 = 6 (less than 7, so 6 is written down, with no seven to carry into the next column) and the next column is 2 + 3 = 5, giving an answer of 56 in a fully heptimal system.

The above results appear most confusing, but only if the bases are mixed. Working with one base all the time (e.g., decimal), there is no confusion.

Binary Arithmetic

Two systems have been illustrated above, one in use in everyday life and the other just a mathematical curiosity employed to emphasize that it is not necessary to work always to the base of ten. To use the decimal base, ten numbers are needed: 1, 2, 3, 4, 5, 6, 7, 8, 9, and 0. All other numbers are made up from these. In the binary system (bini = two) there are only two numbers, viz., 0 and 1, and all numbers can be made up from just these two. We need to know that $2^0 = 1$, $2^1 = 2$, $2^2 = 4$, $2^3 = 8$, and so on. Just as decimal addition was arranged in ones and tens and hundreds ($10^0 = 1$, $10^1 = 10$, $10^2 = 100$, etc.; Figure 42.3), so the binary system can be arranged in the same way. In binary code, the number 3 is a 1 ($1 \times 2^0 = 1$) plus a 2 ($1 \times 2^1 = 2$) and is written 11 (Figure 42.3).

Similarly, the number 17 in binary, must be written 2^0 (= 1) plus 2^4 (= 16) or, in binary code, 10001 ($[1 \times 2^4] + [0 \times 2^3] + [0 \times 2^2] + [0 \times 2^1] + [1 \times 2^0]$; Figure 42.4).

Simple arithmetical operations in binary code are easy in that only two numbers are used (0,1). An example of adding 17 and 24 in binary code is shown in Figure 42.5. The answer is 101001, and the other decimal numbers (2 to 9) are unnecessary. The penalty lies in the complexity! Imagine remembering that the decimal numbers 1, 2, 3, 4, 5, 6, 7, 8, 9 are 1, 10, 11, 100, 101, 110, 111, 1000, 1001 and trying to work out your change after purchasing something, even ignoring all the other numbers. A 999 call becomes 1111100111. However, it turns out that for machines, a base of two is highly desirable because only two fundamental operations are needed, viz., *off* or *on* corresponding to 0 and 1. As far as a machine is concerned, the number 3 (decimal), which in binary is 11, means setting two switches, both *on*. If the machine can sense that a switch is off or on, then it can carry out arithmetical operations in binary language.

(a)	1000	100	10	1
or,	10^3	10^2	10^1	10^0
thus,	3(1000)	2(100)	6(10)	5(1) = 3265 (decimal)
(b)	8	4	2	1
or,	2^3	2^2	2^1	2^0
thus,	1(8)	0(4)	0(2)	1(1) = 1001 (binary)

Note: The sequence of columns in binary runs 1, 2, 4, 8, 16, 32, 64, 128, 256, 512, 1024, 2048, 4096 and so on, each column being two times greater than the previous one. Compare this with the decimal system where each column is 10 times greater than the previous one.

To convert 1001 into a decimal number we have to take 1 x 8 plus 0 x 4 plus 0 x 2 plus 1 x 1 which is 9 (decimal). Conversion of 3265 into binary is a little more difficult: the largest binary number near 3265 is 2048 (or 2^{11}) and there is only one of them with 1217 left over; the next largest binary number is 1024 (or 2^{10}), leaving 193; the next binary number is 128 (or 2^7) and then 64 (2^6) and 1 (2^0). Thus, the binary equivalent of 3265 is 110011000001.

3265		
-2048	$(1\text{x}2^{11})$ 1	
1217		
-1024	$(1\text{x}2^{10})$	1 00 (2^9, 2^8)
193		
-128	$(1\text{x}2^7)$	1
65		
-64	$(1\text{x}2^6)$	1 00000 (2^5 to 2^1)
1	$(1\text{x}2^0)$	1

Figure 42.3
(a) In the decimal system, the number 3265 means (3 × 1000) + (2 × 100) + (6 × 10) + (5 × 1). (b) In the binary system, the number 1001 means (1 × 8) + (0 × 4) + (0 × 2) + (1 × 1). Note how the columns are arranged in powers of the base 10 for decimal and 2 for binary.

2^0	2^1	2^2	2^3	2^4	2^5	2^6	2^7	2^8	2^9
1	2	4	8	16	32	64	128	256	512

What is the binary equivalent of decimal 5? The largest power of 2 which fits 5 is $2^2 = 4$. Therefore there is 1 x 2^2 in 5 with 1 left over; $2^1 = 2$ which is too big and so we write 0 x 21; finally we see that 1 x $2^0 = 1$. The number 5 is made up from a 1 x 2^2 and a 1 x $2^0 = 1$ but in binary we must not forget to put 0 x 2^1; in decimal we may write 304 meaning three hundred and four and if the 0 had been omitted then the number would have been thirty four (34). Thus, the binary equivalent of 5 is 101 and not 11. Other numbers may be converted into binary in an exactly similar fashion. Decimal 39 is:
32 + (0 x 16) + (0 x 8) + 4 + 2 + 1 which is 100111 in binary.

Figure 42.4
To convert a decimal number into binary, it is necessary to look at the powers of 2 as described in Figure 42.3. Decimal numbers corresponding to ascending powers of 2 are shown here, up to 2^9. Each number is just twice the previous one.

Addition of 17 and 24:

17 (decimal) =	1 0 0 0 1 (binary)
24 (decimal) =	1 1 0 0 0 (binary)
41 (decimal) =	1 0 1 0 0 1 (binary)

Figure 42.5
Addition of two numbers in binary code. Note the carryover in the left-most column caused by adding two ones.

Electronic Switching and Binary Code

There is an electronic circuit called a *flip-flop*. It consists of two transistors connected in such a way that, if a voltage is applied, one side of the circuit becomes active and the other side not; if a second voltage is applied, the circuit flips so that the active side becomes inactive and vice versa. Thus, just as with a conventional switch for which one touch puts it on and a second touch turns it off, one touch of the flip-flop turns it *on* and a second touch turns it *off*. Addition of two binary numbers now becomes possible. Suppose we want to add 2 + 1 (= 3; decimal). First, the numbers must be converted into binary code (10 and 01) and these become switch settings in the machine, but we need four switches so that 10 becomes on, off and 01 becomes off, on (Figure 42.6).

If the first pair of switches is examined, one is off and the other on, and the result of *touching* each must be a resulting *on* (off–on and on–off, giving a *total* of *on*). For the other pair, exactly the opposite sequence is present but the net result is *on*. As far as the machine is concerned, the result is "on, on," which in binary code is 11 and in decimal code is 3, the correct answer. Therefore, to get the machine to add in binary, it is necessary to have a switch for each power of two that we want. The number 2^6 is 64 (decimal) and, to represent any number up to 63, we must have seven switches (seven flip-flop circuits), viz., 2^5, 2^4, 2^3, 2^2, 2^1 2^0, and zero. In computer jargon, these switches are called *bits*. Normally, a rack of switches comes in multiples of 8 or 16 or 32 or 64. A *byte* refers to the whole rack. Thus, 8 bits make a byte or 16 bits make a byte, and so on (Figure 42.7). A megabyte of computer memory in 8-bit binary code will have 8,000,000 switches!

A consequence of this mode of handling is that the electronic signals into, out of, and inside the computer flow as series of tiny pulses with gaps between, corresponding to the on/off switching of the bits. For this reason they are known as *digital* devices because of their relation to the two binary numbers used (two digits). For it to be usable, a continuous (analog) electrical signal must be *digitized* before it can be used; viz., its continuous nature must be split up into a series of discrete pulses.

Registers

The various parts of a computer have different functions to deal with peripherals (the viewing screen, the mouse, the keyboard, and inputs and outputs for transmission of data), and all of these

Switches (1,2)	On Off	equivalent to	1	0
Switches (3,4)	Off On	equivalent to	0	1
Add	On On	equivalent to	1	1

Figure 42.6
An electronic switching circuit can be *on* or *off*, and these positions are used to represent the two basic binary numbers 1 and 0, respectively. Decimal 2 is 10 in binary (switch settings on, off), and decimal 1 is 01 in binary (switch settings off, on).

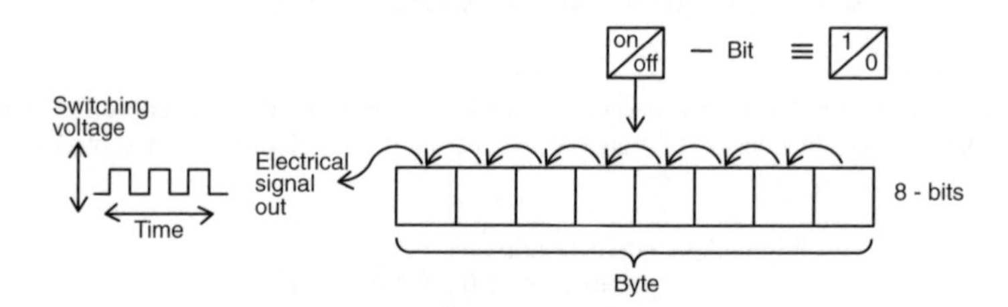

Figure 42.7
Bits are put together as bytes. This example is an 8-bit byte. Faster, more powerful computers have more bits to the byte (16, 32, 64). In reading a byte, the bits flow one after the other out of the byte as electronic pulses (a positive voltage for *on* and zero for *off*).

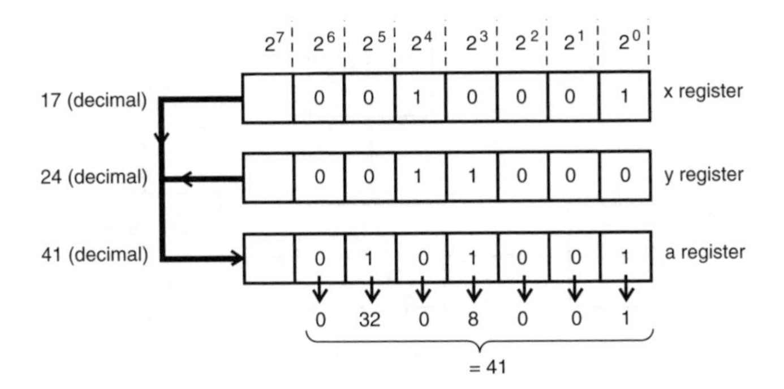

Figure 42.8
Two decimal numbers (17, 24) represented in binary bits in an x-register (0010001) and a y-register (0011000). Addition into the a-register gives 0101001, which translates as 41 (decimal).

must be controlled by a central processor. These functions are not considered in the present discussion. The processor also supports an arithmetic unit that enables addition, subtraction, multiplication, and division to be carried out. To do this processing, the processor has a number of registers that are racks of bits. A typical 8-bit register is shown in Figure 42.8.

Let this register in Figure 42.8 be called the x-register. Underneath is a similar y-register and then an a-register. The x- and y-registers are now each filled with a number, 17 and 24 (decimal), respectively, that we want to add. Seven of the bits are used to represent 2^0 to 2^6, viz., to represent the numbers 1, 2, 4, 8, 16, 32, 64; the last bit is for "carryover." Number 17 (decimal) is made up of 1×16 plus 1×1, and in our 7-bit binary code becomes 0010001 (binary). The switches or bits in the x-register are set to off off on off off off off on. Similarly, 24 (decimal) becomes 1×16 plus 1×4 or 0011000 (binary) and is entered into the y-register. To add them is a simple matter: two zeros (two offs) are still zero; a zero and a one (off and on) are one; two ones become zero and carry one to the left (on plus on is zero but turns on the next bit along). The sum of this addition is shown in Figure 42.8, where we see the result 0101001. (Note that there has been a carryover from bit 5 to bit 6.) In decimal numbers, 0101001 means $(0 \times 64) + (1 \times 32) + (0 \times 16) + (1 \times 8) + (0 \times 4) + (0 \times 2) + (1 \times 1) = 41$ (which is correct for 17 + 24!).

Operations such as the above are carried out very rapidly by the computer through voltage switching, each switch lasting only a few nanoseconds. Therefore, although it is clumsier to represent numbers in binary for the human mind, and instead we use ten symbols (0, 1, … , 9) to help us with complicated arithmetic, the speed with which we can do this arithmetic is nothing like the speed of the computer. Computer addition seems instantaneous, whereas human response to addition takes a finite time.

Most of the previous discussion has concerned addition. Subtraction in binary is very similar, but multiplication is awkward (try it!). For this reason it is quicker for a computer to multiply by carrying out a series of additions. Multiplying 3×5 becomes adding $5 + 5 + 5$. Because each addition is very fast, the time taken for even a large multiplication is very little and still appears instantaneous to us. Only with very large computations does this speed become obvious enough to merit special computers, more powerful than the ones being considered here for use in mass spectrometry. Finally, division is very similar to multiplication, except that a series of subtractions is carried out instead of additions.

Other Registers

The previous discussion concentrated on arithmetical operations by computing in binary numbers represented as bits and bytes. However, other computer functions also use bytes of information.

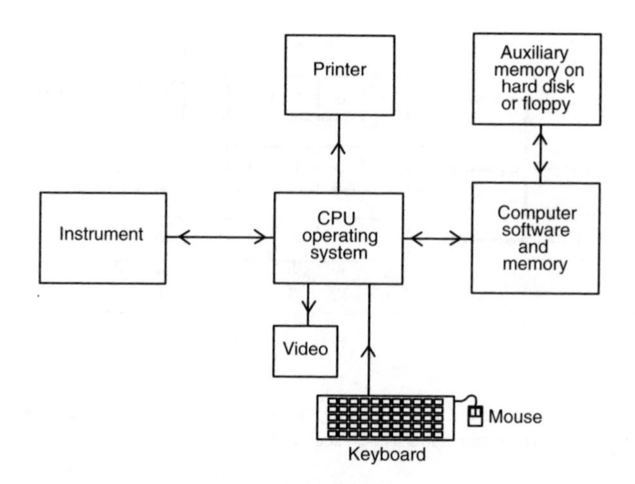

Figure 42.9
A typical layout controlled by the central microprocessor (CPU). Electrical inputs are received from the keyboard, mouse, or instrument. Outputs go to the video screen, printer, and the instrument. Memory and software are utilized by the CPU on command.

For example, in word processing, all the letters, symbols, punctuation marks, and so on must be coded to reside in memory, and the memory itself must be organized. Most characters are represented by a standard coding (ASCII code) in which each is given a byte value. For example, the binary code for a capital letter *M* is 1001101. The computer must be able to differentiate between this value meaning capital *M* and its arithmetic value (77 decimal), and does so with *flags*. The flags are simply reserved bits that the computer uses as a guide. Similarly, there must be flags to tell the computer to stop or start and other flags so that the computer knows where to look in its memory for various items. Each memory location is a byte for which there is a flag or addressing pointer.

All operations of the computer processor or memory banks revolve around the use of bytes of information in which some of the bits carry the essentials and one or two other bits in each byte carry directional or instructional flags. A central processor of a computer is a piece of hardware that has been preprogrammed with its own memory (not accessible to the user — the so-called ROM or read-only memory) and is used to organize the reception of information (input), initial processing of the information, sending the initial information to a store (memory) or a software program for further processing, and finally sending results or instructions (output).

The input could be from a PC (personal computer), as in word processing, but could equally well be from an instrument; the output could be to a video screen, a printer, or to the same or another instrument (Figure 42.9). All these functions are organized by the central processor in so-called real time, i.e., virtually immediately.

Bits and Bytes

A simple 8-bit device has been described above. These are very limited in what they can handle. For example, in terms of numbers, an 8-bit device (1 byte) alone can deal with numbers only up to 127 or 255, depending on how the eighth bit is used. This device would not be much use for general work, although it might be sufficient for simple instrument control. To get to larger numbers, other bytes must be used in conjunction with each other. Use of several bytes together necessarily slows computation. To get around this problem, larger sized bytes are used. The first major step was the introduction of 16-bit devices, followed soon thereafter by the now-common 32-bit bytes.

There are even available 64-bit and greater bytes. The latter devices can deal directly with numbers up to about 1000 million (10^9) and are very much faster than having to string together four or five 8-bit bytes. However, there is a price to be paid. Constructing 8-bit byte devices (chips) is now straightforward, but, even so, the yield of perfectly functioning chips is not good, and many have to be thrown away. For the 64-bit chips, the engineering complexities are enormous and the yield of perfect devices is quite low, hence the high cost. Except for very large number-crunching computers, most general-purpose computers, such as the PC, work with 16- or 32-bit chips.

Languages

A computer must be instructed in the operations that it has to carry out. Although some instruction is built into the computer and appears as hardware (the physical construction), most of the instruction set is written in a suitable language, and then the computer is programmed to do its job. At the lowest level of language, the so-called machine code, instructions are written in a very elementary fashion, and literally each step of the instruction sequence must be spelled out. For example, in everyday life one might say something like, "Draw a square." This statement would be high-level language. In machine code, the same instruction might be, "Put a dot on the video screen at location x, y; move horizontally to the right for z squares; stop; move vertically down for z squares; stop," and so on.

Programming in machine code is slow, skilled work but has the advantage that any instruction is carried out at the fastest possible speed. Thus, where speed is important, as in the central processor unit (CPU), all instructions are built permanently in machine code and usually cannot be altered by the person using the computer. The CPU has its own memory, which carries instructions on what to do in the event of a command (e.g., pressing a key on the keyboard is a command) and directions for dealing with a variety of signals (as from an external instrument). For many other peripheral purposes, instructions are written in high-level languages. The instruction "draw a square" could be a statement to the computer in some high-level instruction set. Actually, this statement would be translated into low-level machine code automatically. Although on the face of it only a simple, readily understood (by humans) instruction has been given, in fact, this instruction has been quietly converted (translated) into machine code by the programming language so the computer can understand it. All high-level languages are actually indirect means of providing programmers with a less tedious way of instructing a computer than through the direct use of machine code. Therefore, once a basic set of instructions has been built into a central processor to enable it to deal quickly with electronic impulses (signals) from keyboards, instruments, printers, and so on, other instructions can be written as a program (software) in a convenient high-level language that can be loaded into the computer's main memory and which the computer will obey once started. Clearly, if a single instruction in a high-level language is made up of, say, 20 machine-level instructions (and usually many more), it will take 20 times longer to carry out a simple instruction like "draw a square." For this reason, it takes a computer much longer to process data when its instructions (the program) are written in high-level languages. Nevertheless, this disadvantage is more than made up for by the relative simplicity of writing instruction sets in a high-level language. Typical high-level languages are Fortran, Pascal, C, and Occam.

Computer Memory

As set out above, certain parts of computer memory are reserved exclusively for the central processor and other parts for driving any peripherals, such as a keyboard. Although this memory can be accessed (read), it cannot be changed, hence its name *read-only memory* (ROM). The user has access to another type of memory that can be changed by either the user directly or indirectly via a program in the computer. These memory locations can be changed at will, hence the name *random-access memory* (RAM). The amount of RAM in a computer is usually quoted in megabytes (1 Mbyte = 1 million locations or memory registers), and the more RAM there is available, the more complicated the tasks that can be tackled. A typical PC might have 200 Mbytes of memory available and be able to support several software programs. Simpler computers, such as those aimed mainly at the games market, have more modest amounts of memory.

The Clock

All the instruction sets in the computer could not operate without some form of timing device available since the length of time needed by the various instructions changes with the instruction. The problem could be exemplified by reference to control systems such as traffic lights. These are operated on a timer basis and, without them, traffic would soon jam at busy periods as each vehicle (instruction) tried to get to its destination (memory location) across the paths of others. Of course the clocks used with computers need to tick at a very fast rate if the instructions are to be carried out quickly. For example, a clock ticking at the rate of 20 million times per second (20 MHz) can deal with each basic instruction in about 0.00000005 seconds. These clocks are not the usual kind found with clockwork mechanisms; rather, they consist of special crystals that oscillate constantly at these high rates when an alternating electrical voltage is applied to them, rather like the modern quartz watches. The ticks of these crystal clocks are not mechanical ones but consist of electrical impulses that allow something to proceed during the *on* time and stop anything from happening in the *off* period. Computer instructions go by the clock.

Conclusions

By electronic engineering, a system of interconnected switching devices is able to respond in one of only two modes (on or off), and these modes can be controlled at the basic level of a bit. Bits are assembled into bytes, as with an 8-bit device, and through programming of the bytes a computer central processor can be made to follow sets of instructions (programs) written in special languages, either at a direct level (machine code) that can be acted upon immediately by a computer or at a high level that is translated for the user into machine code.

Chapter 43

Computers and Transputers in Mass Spectrometers, Part B

Introduction

Many of the terms used in this chapter are introduced in Part A (Chapter 42), and it is assumed that the reader has read Part A or is familiar with its contents.

Serial and parallel processing are two different ways in which information is passed around a computer system. Normal computers rely on serial methods, where each snippet of information follows another in a logical fashion until the result is achieved (Figure 43.1). Most software programs developed up until now rely on a serial approach in which the total set of instructions is laid out one after the other, and no part of the program can proceed until its turn arrives. A great deal of effort, time, and money has gone into production of processors, construction of software programs, and development of hardware for serial computing, and these will not be lightly jettisoned. Nevertheless, when it comes to speed of processing information, the parallel mode has many advantages and has given rise to a new type of computer called a transputer (Figure 43.2).

This chapter briefly discusses the advantages to be gained from the use of transputers in acquiring and processing data from an instrument like a mass spectrometer, which routinely deals with large-scale input and output at high speed.

Basic Speed Differential between Parallel and Serial Modes

Movement of information in a computer could be likened to a railway system. Carriers of information (bits or bytes) move together (like a train and wagons) from one location to another along electronic tracks. It is important that no two bits of information are mixed up, and therefore all the moves must be carefully synchronized with a clock. This situation resembles the movement of trains on a railway; many trains use the same track but are not all in the same place at the same time. The railways run to a timetable. Similarly, information is moved around the computer under the control of the central processor unit (CPU).

Consider a process that requires the movement of 100 pieces of information — this could be part of a calculation, a bit of word processing, direction of data inputs, and so on. Suppose the

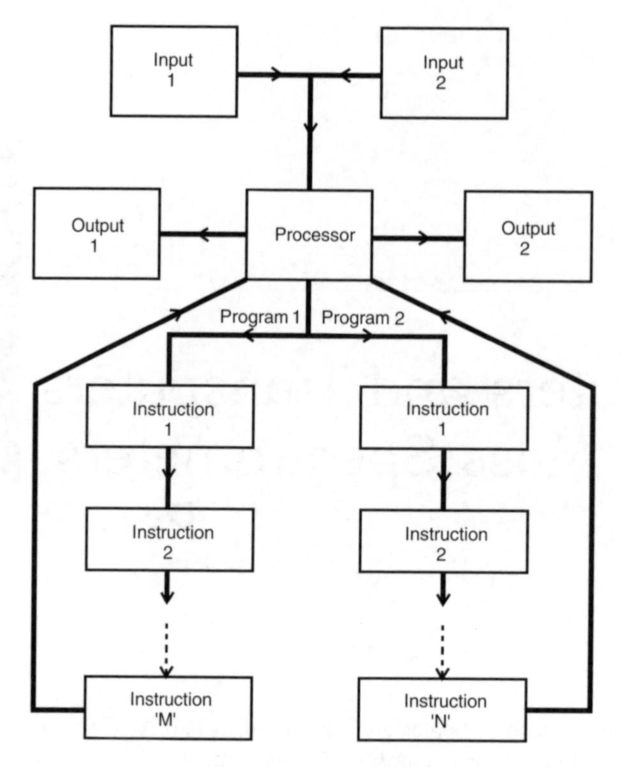

Figure 43.1
A very simple illustration in which information from two inputs is dealt with sequentially (serially) by a microprocessor. Input 1 is accepted, and the left-hand series of instructions (program 1) are carried out. Then, Input 2 is examined, and the right-hand set of instructions is followed through. The processes are iterated. If each program (1, 2) takes 1 msec, the total time for one iteration is 2 msec.

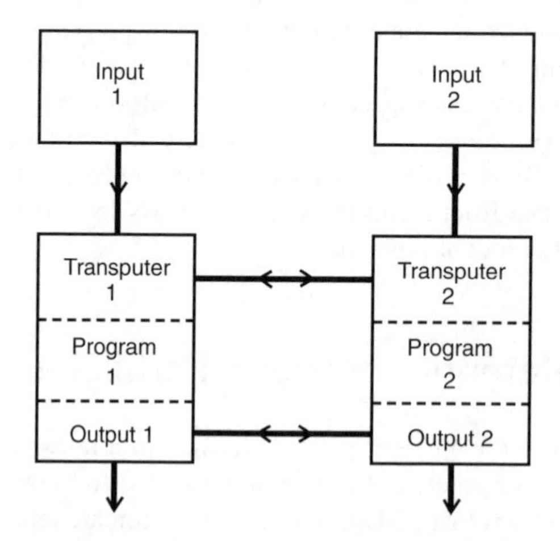

Figure 43.2
In contrast with Figure 43.1, there are two transputers, each dealing with a set of instructions. If the processing time for each is 1 msec, the total time required is still about 1 msec. Communication links are needed between the transputers.

average time taken to process each piece of the information is 100 nsec; then the total time taken for 100 pieces of information is 10,000 nsec or 10 msec. This process speed is very fast compared with the railway and seems instantaneous to the human brain. However, 100 pieces of information is not much and might be equivalent only to putting 10 characters onto a screen. When it comes to dealing with information entering the computer at a high rate, as when a mass spectrum is being recorded, this rate of information transfer is not fast enough, as described in Part C (Chapter 44). The major problem arises from the need to move all the information sequentially on the same electronic track (called a *data bus*). The speed with which the information moves is one factor; the other is the need to control its flow, which all goes through the CPU along the bus, which turns out to be a bottleneck. One way out of this bottleneck problem is to provide more routes for the flow of information (parallel processing), just as one way (very expensive) of speeding up a railway would be to have many more lines. In a computer, these extra lines can be provided, but a different way of timetabling the movement of information is needed which is where the transputer comes in.

Consider the above example of 100 pieces of information, but instead of being processed serially, the information is carried over a network of ten electronic tracks. Now the time taken to move this information is only 1 msec; viz., the system is ten times faster than the serial one. However, simply providing ten new tracks is not an answer in itself. Each track must be provided with its own microprocessor to deal with the information, and the processors must be able to communicate with each other so the information flows in an orderly fashion. (It is no use having two different train tracks if each train arrives at the same platform at exactly the same time!) Thus, the transputer is a microprocessor that has its own memory bank and input and output lines enabling it to communicate with other transputers in a parallel fashion. For example, one transputer could deal with data input into a mass spectrometer (recording a mass spectrum), while another could be controlling the actual scanning of the mass spectrometer. The two would work in conjunction in that, once scanning had been instituted by one transputer, the other would record information until told to stop by the first.

Parallel processing of information is a technique that speeds computer operations without the need to push the limits of existing technology by attempting to build increasingly faster processors. A typical outline of transputer architecture compared with that of a standard computer is shown in Figure 43.3.

Because the transputer has a 32-bit processor and fast access to considerable quantities of on-chip RAM, it has been called a computer on a chip. Transputers are inherently faster than microprocessors, which have to refer to RAM outside the chip on which they reside. Thus the 100-nsec cycle time used in the above illustration may be only 50 nsec when carried out on the transputer chip.

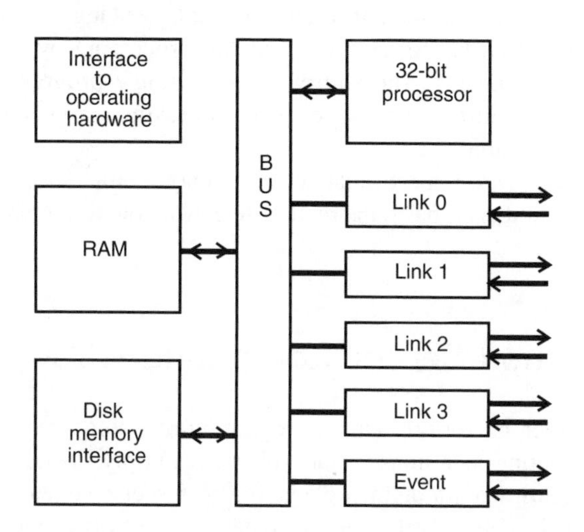

Figure 43.3
A typical transputer architecture. The transputer (sometimes referred to as a computer on a chip) has four input/output links (0, 1, 2, 3) to other transputers, a channel for inputting/requesting data (event link), some built-in random-access memory, an interface to the main operating system (clock, boot, etc.), and an external memory interface. Internal communication is via a bus.

This big increase in speed has not been without cost. The everyday machine codes and high-level languages (Fortran, Pascal, C, etc.) used to control operations in a standard computer are inappropriate for parallel processing, which needs its own instruction set and has led to the development of special languages for use with the transputer.

Occam

The most highly developed language for controlling transputer operations is called *Occam* in honor of the philosopher William of Occam who, along with other minimalists, maintained that there is no point in trying to describe anything with more than the bare minimum of facts. In Occam's view, added suppositions are superfluous and should be chopped out (hence the term Occam's Razor). The Occam language enables programs to be written in parallel mode in which instructions can be passed from transputer to transputer. Thus, one transputer can acquire and process information but will then wait for a signal from another transputer before passing on the result to possibly a third transputer, to memory or to a peripheral. Each transputer has four inputs and four output lines, so it can communicate directly with up to four other transputers. Occam facilitates control of these inputs and outputs.

This Occam language allows groups of transputers to operate together, each carrying out a specific task and with access to its own memory bank, and then sending forward the processed information in harmony with the operations of the other transputers. This timetabling avoids the difficulty of the two trains arriving at the same platform at exactly the same time — one would be held back slightly until the other had cleared the platform.

Because of the past effort that has been invested in other high-level serial languages and the large number of useful software programs written in these languages, it is unlikely that transputers and Occam will simply take their place. However, in applications where there is a clear advantage in being able to handle high rates of information flow and process it at high speed — as in mass spectrometry — transputers will be used more and more. In applications that do not need the high speeds attainable with the transputer (word processing, for example), it is unlikely that transputers will be used. Because most computer programmers are not conversant with parallel programming and are likely to be reluctant to learn yet another language, there have been some efforts to adapt Fortran and Pascal to deal with parallel processing by adding some additional parallel-compatible instruction sets. The ability to use a single programming language would also ease the marriage of two different approaches to processing (serial and parallel) where both are used in the same application.

The transputer's advantage in speed relative to common computer operations has also been boosted by reducing the number of basic instruction sets available to the programmer. This aspect is discussed next.

Reduced Instruction Set for Computing (RISC)

A high-level language, such as Fortran, makes programming a computer much easier than using machine code directly. Routine operations — add, subtract, multiply, divide, logarithm, sine, cosine, go to a defined place, return from a defined place, repeat a sequence, etc. — are all available as simple written statements in a high-level language. What is not readily apparent is that the processor in the computer must translate these statements into machine code before the computer can carry them out. A simple "add" instruction to a computer may require 10–20 machine-code operations.

Although there is not much difference in the execution time for carrying out an instruction directly in machine code or indirectly after translation, there is a definite delay involved while the computer looks for the translation. If "add" is one instruction among a compilation of perhaps thousands of others, then each time the computer carries out this instruction it must search through this large set to find the one it wants. If there is only a small set in the first place, then the search time is greatly reduced. (At the same time, the high-level language becomes more unwieldy, in a general sense, in that the programmer has to formulate specific operations, e.g., sine.) For many computer operations, the provision of a large instruction set is overkill, since many of them will rarely be used by the application to which the language is being applied. This circumstance especially applies to most data-acquisition and -processing applications for instruments like mass spectrometers. In such applications, it makes sense to reduce the number of instructions and simplify the computer search, which has led to the design of processors with a reduced instruction set (RISC) that facilitates faster transfer of information. The transputer was designed from the start with a RISC vocabulary, making it inherently faster than processors having a full instruction set, with or without parallel processing.

Conclusion

Microprocessors (transputers), with the help of a special language (Occam), can handle flows of information in a parallel fashion instead of sequentially (serially), thereby greatly increasing the speed of operation. Transputers also control the flow of information by communicating with each other.

Chapter 44

Computers and Transputers in Mass Spectrometers, Part C

Introduction

A brief outline of the workings of computers and transputers has been presented in Parts A and B of this discussion (see Chapters 42 and 43), and both should be read before reading this chapter unless the reader is already familiar with the basics of computing. Additional details on some of the functions discussed here are available in other chapters of this book, and cross-references are given where relevant.

From the point of view of computation, a mass spectrometer is a source of data that must be acquired, processed, stored, and printed, but it is also an instrument that must be tuned and controlled in its operations. This chapter discusses both types of operation separately. For either type of operation, it is generally true that most electrical signals reaching the computer input are analog (continuous), as seen in Figure 44.1a. (The time-to-digital converter, which has a digital output, is a significant exception to this generalization.) However, the computer or transputer is a digital device — its electrical signals are in the form of pulses (Figure 44.1b). Therefore, before the analog signal can be accepted by the computer it must be digitized, for which an analog-to-digital converter (ADC) is used. Similarly, any signal coming from the computer is in digital form and must be turned into an analog signal by a digital-to-analog converter (DAC). Interconversion of analog to digital and vice versa is exemplified in Figure 44.2 by reference to a voltage varying with time. In the following discussion in this chapter it will be assumed that the electrical signals are in their correct form through use of ADCs or DACs as necessary.

Data Processing

A mass spectrum consists of a series of peaks at different m/z values, with the height of the peak proportional to the number of ions. A partial mass spectrum is shown in Figure 44.3 and is seen to be an analog signal that varies as the peaks rise from and fall to the baseline. Between the peaks are relatively long intervals when there is only the baseline. As described above, the signal is first digitized.

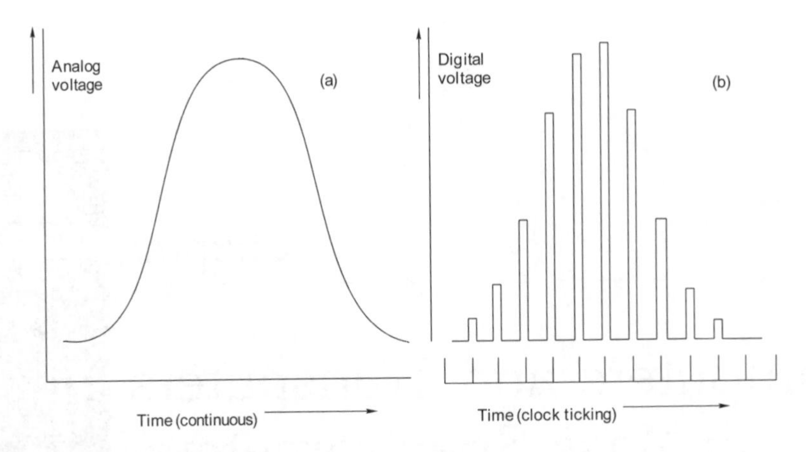

Figure 44.1.
(a) An analog voltage signal representing a mass spectral peak varies continuously with time, rising through a maximum and back to the baseline. (b) In a digitized form, a clock ticks, and on each tick the analog voltage is read out. Thus, the peak now appears outlined as a series of discrete digital readings rather than in continuous form.

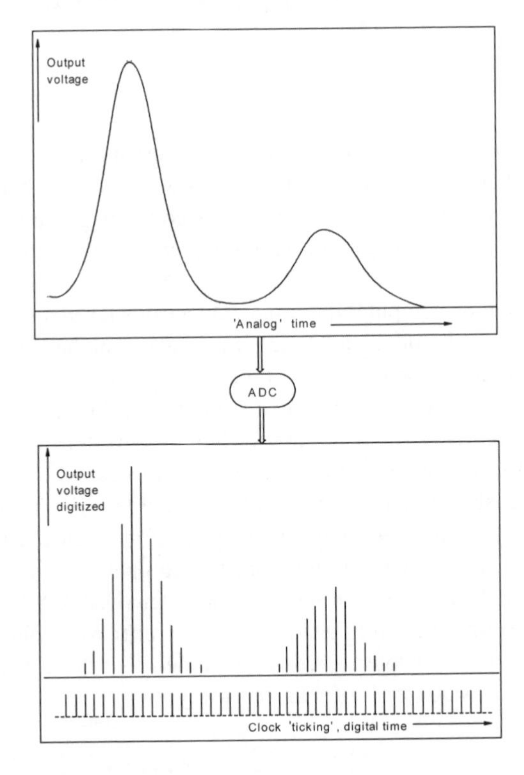

Figure 44.2
An analog-to-digital converter (ADC) converts a continuous signal into a series of digital pulses in which the voltage represents snapshots of the analog signal taken at regular time intervals.

If one imagines a ticking clock, then each time it ticks a measure of the peak height is read (actually as a voltage). Figure 44.4 gives a representation of this process. The clock ticks very fast, at something like 100,000 times per second, and therefore a stream of voltage readings flows to the computer. These large amounts of data are indigestible and would rapidly fill the available memory. Accordingly, there is a preprocessor to reduce the amount of information. First, an artificial baseline is set (Figure 44.5) to mitigate the effects of natural noise on the true baseline, which would appear as additional small peaks.

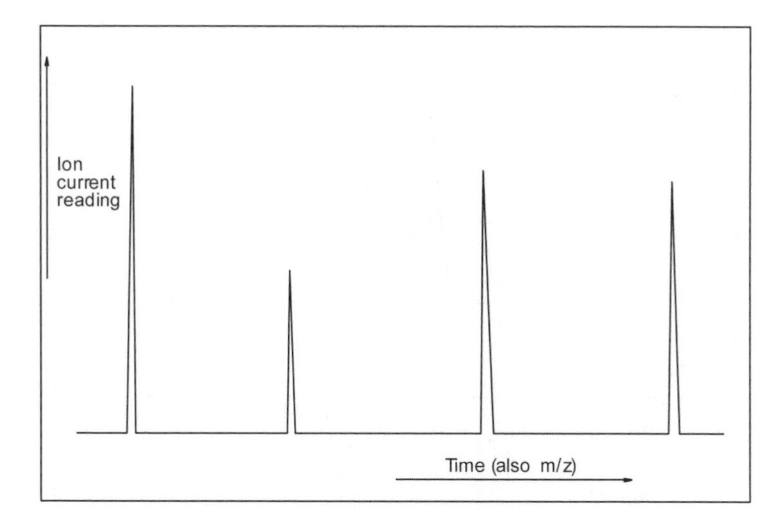

Figure 44.3
The voltage from the ion detector varies as ions are brought to a focus. As each m/z value is focused, a peak ion current leads to the voltage change. Between the individual m/z values (between the peaks), no voltage change is observed because no ions arrive at the detector. The time between individual peaks is very much longer than the time representing peak width.

Clock 'ticks' (sec)	Detector output (volts)	
5.00001	3.5	peak
5.00002	5.0	peak
5.00003	15.1	peak
5.00004	6.2	peak
5.00005	4.3	peak
5.00006	3.5	
5.21001	3.5	baseline
5.21002	3.5	baseline
5.21003	3.5	baseline
5.21004	3.5	baseline
5.21005	3.5	

Figure 44.4
A typical flow of digital data for a small part of a mass spectrum. An ion peak appears after 5 sec of scanning; the digital voltages are read from the analog signal at intervals of 0.0001 sec. At other times only a steady baseline voltage reading is observed.

With the artificial baseline, the voltage reading between peaks is constant, and the true peaks clearly rise up from the baseline and then fall back below it again (Figure 44.5). Note that the artificial baseline should not be set at such a high value that it covers most of the peaks, because such a setting would lead to false estimates of the peak heights. As a peak starts to form, the change in voltage is noted by the preprocessor, which then collects the digitized data until the peak disappears (the voltage returns to its steady value). From the digital voltage readings at known time intervals, the preprocessor calculates the peak centroid (center of gravity) as a measure of position

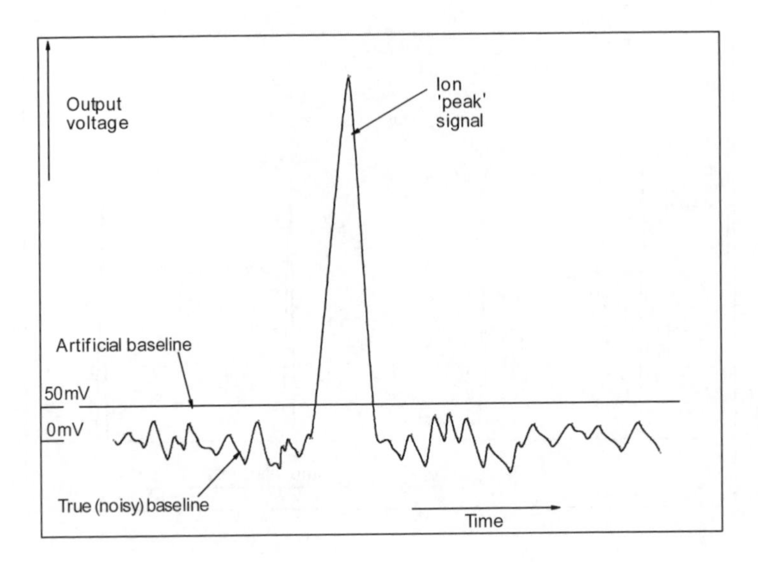

Figure 44.5
A true baseline output from an ion detector is electrically noisy and, if recorded as such, the noise would appear as a great many small (unwanted) peaks. By creating an artificial baseline at a voltage just above the noise, the small peaks are eliminated and only the desired signal is recorded. It is important not to set the artificial baseline voltage too high, since this would eliminate too much of the required peak.

(m/z value) and peak area (a measure of ion abundance) by simple mathematical algorithms illustrated in Figure 44.6. These two bits of information (peak area and arrival time), together with a flag to enable the computer to keep track of the peak data, are sent to the computer memory, which stores the information until the mass spectrometer stops scanning.

The spaces (time) between peaks is large compared with the peak width (measured as a time interval) and, if the above reduction in data did not take place, then something like 1 million bits of information would have to be stored per spectrum. In a typical spectrum there may be 100 peaks, and, after passing through the preprocessor, these correspond to only 300 pieces of information, which are much easier to deal with than a million. Also, because the computer does not have to collect data when there is only the steady baseline (between peaks), it can perform other tasks. Thus, as each peak arrives (is detected), there is a rush of activity as the digitized signals are processed, but between the peaks there are no data arriving and the computer can be programmed to use this time for other tasks (Figure 44.3).

Once the peaks have been collected and stored, the computer can be asked to work on the data to produce a mass spectrum and print it out, or it can be asked to carry out other operations such as library searching, producing a mass chromatogram, and making an accurate mass measurement on each peak. Many other examples of the use of computers to process mass data are presented in other chapters of this book.

Instrument Control

Peak Shape

In a well-tuned (adjusted) instrument, the shape of a mass spectral peak is approximately triangular (Figure 44.7a), but, in an instrument that is poorly tuned the peak will appear misshapen (Figure 44.7b). Usually, the cause of the skewing of the peak arises from incorrectly adjusted

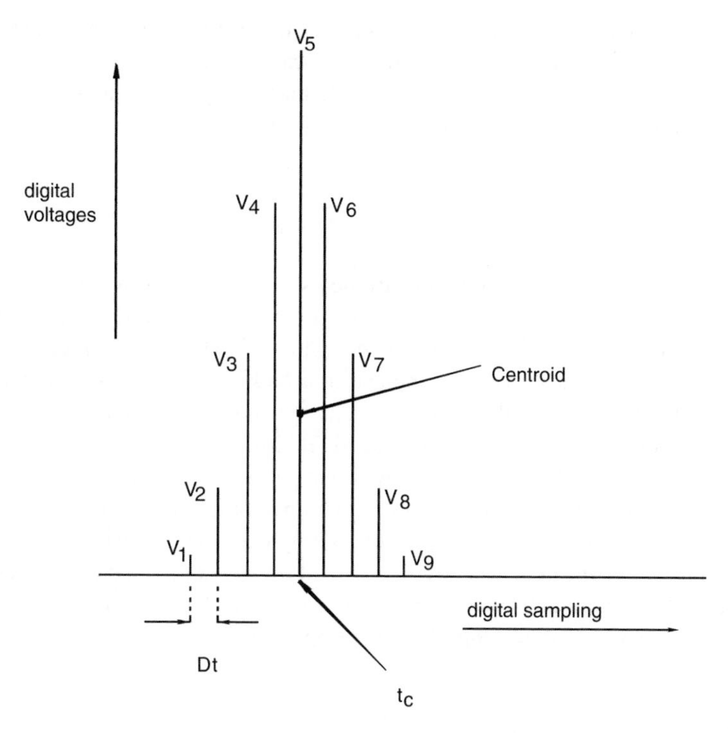

Figure 44.6
If digital voltage readings (V1–V9) are taken at time intervals (Δt = 0.0001 sec in the example of Figure 44.4), then the area of the true peak (dotted) can be (mathematically) closely approximated to give ion abundance and, similarly, the time (t_c) to the center of gravity (centroid) of the peak can be determined, thereby giving the m/z value.

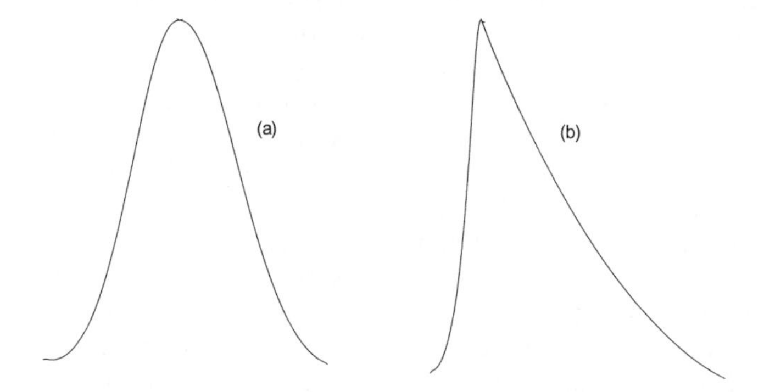

Figure 44.7
(a) In a well-tuned instrument, the ion beam has only a small spread of translational energies, causing the ions to arrive at the collector evenly and closely grouped around an average time of arrival. The resulting analog signal (peak) is approximately an isosceles triangle. (b) In a poorly tuned instrument, the peak is distorted, with more ions on one side of the beam than the other. The mean arrival time in such an instance is no longer the same and leads to error in the recorded m/z value, particularly in accurate mass work.

voltages in the various electric or magnetic lenses used to align the ions in flight, so they are focused at a collector (see Chapters 24 and 25 for further discussion of ion optics for sector and quadrupole mass spectrometers). Maladjusted voltages can cause some ions to have too much or too little momentum, or they can even rotate the beam somewhat so that it does not pass truly through the defining slits. The misshapen peak can be detected by the computer. The digital voltage samples obtained across the peak (as described above) are compared with those expected for a perfectly shaped peak. Then, the computer is programmed to send out signals to the various ion optic lenses to increase or decrease the voltages on them to correct the astigmatism in the ion beam. Each time an adjustment is made, the computer checks the new peak shape. This process continues until the shape is right. The process has been completely automated, and in the latest mass spectrometers getting the peak shape correct is all part of an automatic setup process.

Voltage Checking

A mass spectrometer has several integrated instruments that provide feedback information on how it is functioning. For example, the vacuum system is monitored by an ion gauge that gives out an electrical signal proportional to the vacuum being achieved. Thus, the computer can check this voltage routinely, and, if it finds a reading indicating a fault, it can draw it to the attention of the operator or, if the fault is serious, the computer can shut down the electronic components of the spectrometer to prevent them from being damaged. If, for instance, the pressure in an electron ionization source rises too high, the filaments will burn out; early detection warning of the imminence of such a state is invaluable in avoiding time-consuming repairs. This type of checking is performed routinely, and, because of the speed of a computer, it can be carried out on a very short cycle time (e.g., once a second). Similarly, other parts of a mass spectrometer are routinely checked on a cyclical basis. For many faults, programming has reached a stage where the computer, having detected a fault, can diagnose it and report it to the operator.

Different Scanning Modes

The primary objective of data processing is to obtain the routine mass spectrum, viz., to produce a chart showing the number of ions (ion current or ion abundance) at each m/z value within a given preset range of m/z values. To this end, the computer controls electrical or magnetic fields, so scanning of one or the other starts at some predetermined value and changes uniformly until a second predetermined point is reached. The changing fields allow ions of steadily increasing (or decreasing) m/z values to reach the ion detector, where they are recorded by the computer. However, there are more complicated scanning modes required for *metastable* ion work and for MS/MS. The actual nature of these modes is set out in Chapters 32–34. Here, it is sufficient to note that the scans are electrically and/or magnetically quite complicated, requiring the fields to be scanned together in predetermined fashioned. For example, a B^2/E scan needs the square of a magnetic field strength to be varied with a first-order change in the electric field so that the ratio of B^2 to E remains constant as a mass spectrum is obtained. Since the computer can monitor the field strengths (described in the previous section, "Voltage Checking"), it is able to effect the required variation while acquiring the incoming mass data. This dual working function could be accomplished using two transputers working in parallel (one for acquiring and processing data and the other for checking and controlling field strengths). Alternatively, a single microprocessor deals with both functions at the same time by switching extremely quickly from one function to the other iteratively.

Manipulation of Mass Spectral Data

Apart from the actual acquisition of the mass spectrum and its subsequent display or printout, the raw mass spectral data can be processed in other ways, many of which have been touched on in other chapters in this book. Some of the more important aspects of this sort of data manipulation are explained in greater detail below.

Library Searching

Most mass spectrometers for analytical work have access to a large library of mass spectra of known compounds. These libraries are in a form that can be read immediately by a computer; viz., the data corresponding to each spectrum have been compressed into digital form and stored permanently in memory. Each spectrum is stored as a list of m/z values for all peaks that are at least 5% of the height of the largest peak. To speed the search process, a much shorter version of the spectrum is normally examined (e.g., only one peak in every fourteen mass units).

When a mass spectrum has been acquired by the spectrometer/computer system, it is already in digital form as m/z values versus peak heights (ion abundances), and it is a simple matter for the computer to compare each spectrum in the library with that of the unknown until it finds a match. The shortened search is carried out first, and the computer reports the best fits or matches between the unknown and spectra in the library. A search of even 60,000 to 70,000 spectra takes only a few seconds, particularly if transputers are used, thus saving the operator a great deal of time. Even a partial match can be valuable because, although the required structure may not have been found in the library, it is more than likely that some of the library compounds will have structural pieces that can be recognized from a partial fit and so provide information on at least part of the structure of the unknown.

Presenting Data in Different Ways

Once mass spectral data have been transformed into a set of m/z values together with ion abundances and put into memory, the computer can be programmed to manipulate the information in different ways. For example, instead of a straightforward mass spectrum, it can produce a normalized one in which all the peak heights have been recalculated with respect to the largest one, which is made equal to 100%. In GC/MS and LC/MS operations (Chapters 36 and 37, respectively), the computer will present a total ion current (TIC) chromatogram or a mass chromatogram (Figure 44.8a), or it can be used for selected ion monitoring (Figure 44.8b).

All of these sorts of computer operations are done independently of data acquisition and spectrometer control functions. Rather, they are done through specially written software programs acting on the acquired and preprocessed data. In fact, once the data have been processed into m/z values and ion abundances, any number of calculations can be performed if the relevant program is available. Manufacturers usually supply a suite of such programs, and, depending on the way the computer/transputer has been set up, these additional software programs might run only when the computer is not acquiring data and controlling the spectrometer (foreground working), or they might be capable of running at the same time as data acquisition (foreground/background working). The latter mode is more powerful but requires enhanced computer processor capacity, often provided by the use of one or more transputers.

Accurate Mass Measurement

Assuming that the mass spectrometer has sufficient mass resolution, the computer can prepare accurate mass data on the m/z values from an unknown substance. To prepare that data, the system must acquire the mass spectrum of a known reference substance for which accurate masses for its ions are already known, and the computer must have a stored table of these reference masses. The computer is programmed first to inspect the newly acquired data from the reference compound in comparison with its stored reference spectrum; if all is well, the system then acquires data from the unknown substance. By comparison and interpolation techniques using the known reference

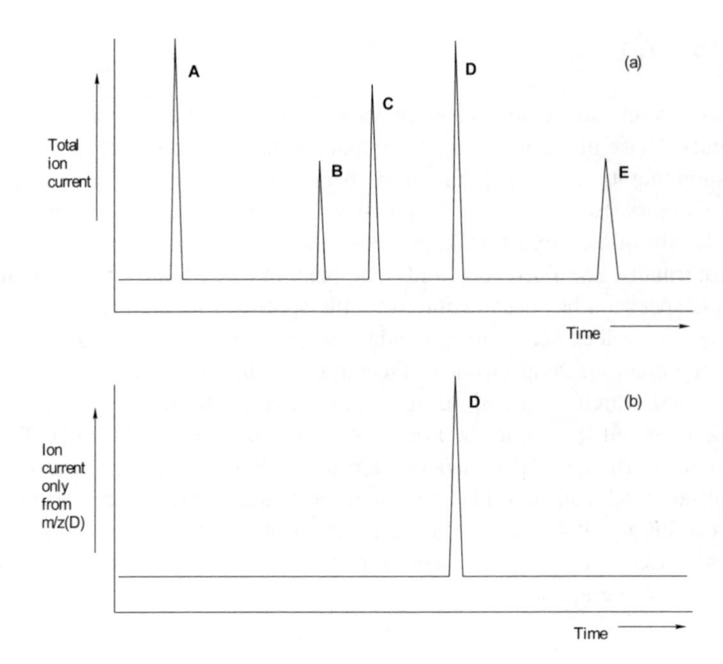

Figure 44.8
(a) The output from GC/MS or LC/MS instruments can be a total ion current (TIC) chromatogram showing the detection of all the substances (A–E) eluting from the chromatographic column. (b) Alternatively, by setting the computer to ignore all ions except those characteristic of, say, compound D, only that one peak will appear in the resulting mass chromatogram.

spectrum (Figure 44.9), the m/z values from the spectrum of the unknown are calculated accurately (see Chapter 38).

Interestingly, if the original comparison of the spectrum from the reference compound with stored m/z data for that compound reveals discrepancies, the stored reference data are updated before the computer goes on to acquire data from the unknown compound. In this mode, the computer is not used simply to acquire and manipulate data but is also used to make decisions

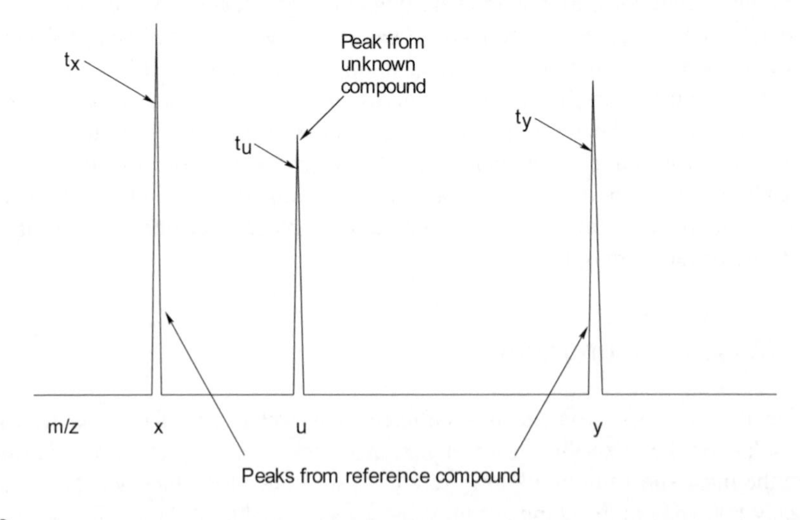

Figure 44.9
If the m/z values (x, y) for two peaks in a reference compound are known, the m/z value (u) for an unknown lying between them can be calculated accurately by interpolation. The m/z values are actually peak arrival times at the ion collector (tx, tu, ty). Thus m/z(u) = m/z(x) + {[(tx – tu)/(tx – ty)] [m/z(x) – m/z(y)]}. Since all the data on the right of this equation are known, m/z(u) is easily calculated.

regarding the accuracy of the figures it holds about the known reference, with any discrepancy leading the computer to refresh its reference table.

Conclusion

Computers, often combined with transputers, are used for three main functions when connected to a mass spectrometer. The foremost requirements involve the acquisition and preprocessing of basic data and the control of the instrument's scanning operations. Additional software programs are available to manipulate the preprocessed data in a wide variety of ways depending on what is required, e.g., a mass spectrum or a total ion chromatogram.

Chapter 45

Introduction to Biotechnology

Genetics

Chromosomes, DNA, and Protein Synthesis

All living matter — humans, animals, and plants — is made up of cells, which are small bags of saline water and a rich variety of chemicals. The cells are fundamental units of life. Cells must be able to make the peptides, proteins, and enzymes that are required to build an organism and to utilize raw material (food) to keep the organism alive. However, the most important property of all living cells is their ability to produce replicas of themselves without any external instructions. The study of the information content of cells, or heredity, is called genetics, and the genetic information of a cell determines the distinguishing features of the organism.

The nucleus of the cell contains chromosomes, which appear in pairs. When cells divide, each new cell receives a complete set of chromosomes by uncoupling of the pairs. Chromosomes contain both proteins and deoxyribonucleic acids (DNA), the latter being the genetic material. DNA strand is a long, linear molecule with a backbone of alternating sugar (deoxyribose) and phosphate units; in addition, each sugar is also attached to one of four possible bases — adenine (A), thymine (T), guanine (G), and cytosine (C) — to form nucleosides and nucleotides (Figure 45.1). Two strands of DNA run in parallel but opposite directions in a helical configuration with relatively weak hydrogen bonds between the two strands.

As shown in Figure 45.1, the bases appear in complementary pairs, A with T and G with C; in this particular example, the sequence for one strand of DNA is A-T-C-G-T- while the other strand is -T-A-G-C-A-. The sequences of the bases attached to the sugar-phosphate backbone direct the production of proteins from amino acids. Along each strand, groups of three bases, called codons, correspond to individual amino acids. For example, in Figure 45.1, the triplet CGT, acting as a codon, would correspond to the amino acid serine. One codon, TAC, indicates where synthesis should begin in the DNA strand, and other codons, such as ATT, indicate where synthesis should stop.

When the cell requires instructions for protein production, part of the code on DNA, starting at an initiator and ending at a stop codon, is converted into a more mobile form by transferring the DNA code into a matching RNA code on a messenger ribonucleic acid (mRNA), a process known as transcription. The decoding, or translation, of mRNA then takes place by special transfer ribonucleic acids (tRNA), which recognize individual codons as amino acids. The sequence of amino acids is assembled into a protein (see "Proteins" section). In summary, the codes on DNA

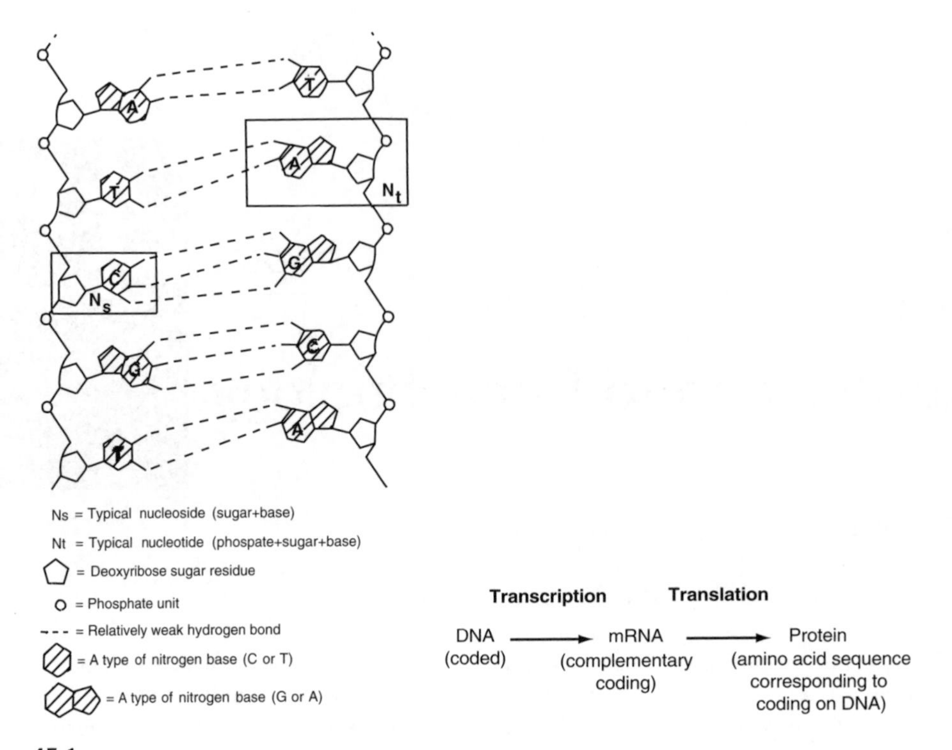

Figure 45.1
Schematic diagram of deoxyribonucleic acid (DNA) showing the pairing of nitrogen bases (A-T; C-G) between parts of two strands of DNA, the backbone of each strand being composed of alternately linked sugar and phosphate units.

are used to synthesize proteins, as seen in Figure 45.1. This whole process can be likened to the translation of one language into another by means of two interpreters, one of which (mRNA) writes down the words of the first language (DNA) while the second (tRNA) translates the words into the second language (protein).

Genetic Engineering and Gene Cloning

The term *genetic engineering* is applied to methods of altering an organism's characteristics by direct manipulation of its constituent DNA. For example, genetic recombination is the process in which a new DNA molecule is formed by chemically cutting and joining DNA strands using special enzymes. *Transposition* refers to the transfer of a gene from one chromosome to another or from one site to a different one on the same chromosome, using this recombinant technology. Much recombinant work is done with natural gene (DNA) material, but synthetic DNA can be used. Synthesis of specified lengths of DNA (or RNA) from constituent nucleotides can be carried out using special automatic apparatus, a gene machine. These artificial lengths of DNA, sometimes called oligonucleotides, can be spliced into chromosomal DNA by recombinant technology just as natural DNA can.

Gene cloning, or exact copying, is achieved by the use of microorganisms. Fragments of DNA containing just one or a few genes can be taken from any source and placed in one of the nucleic acids of a microorganism such as *Escherichia coli,* which then treats the new DNA as if it were its own and produces millions of exact copies (clones).

The basic steps of gene cloning first involve cutting a precise DNA segment (gene) from a donor source DNA by use of a restriction enzyme (Figure 45.2). At the same time, a small looped

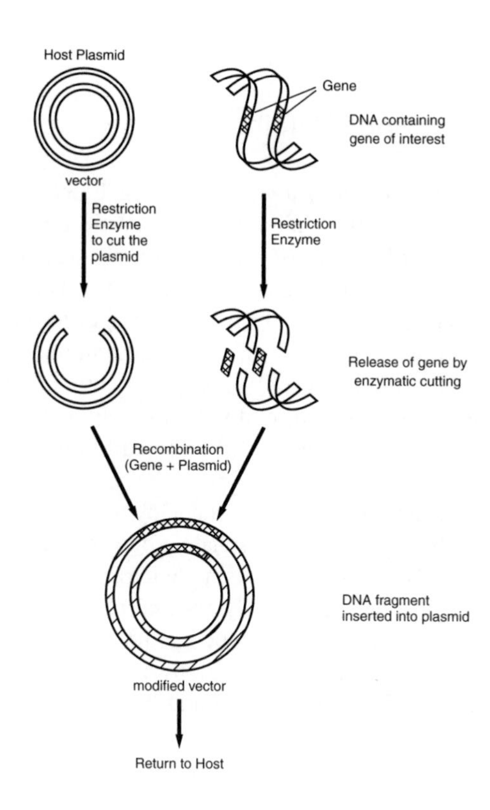

Figure 45.2
Gene cloning.

DNA (plasmid), a vector from a host microorganism, is snipped to open it up (Figure 45.2). The gene released from the donor is inserted (ligated) into the plasmid by a ligase, thereby closing the loop, so the plasmid is returned to its original state, except for the newly inserted gene (Figure 45.2). The modified vector is reinserted into the host organism, which replicates it (cloning). Large quantities of otherwise scarce proteins, corresponding to the newly inserted gene, can be synthesized by such cloning techniques simply by growing and harvesting the modified microorganisms.

DNA Sequencing

DNA from a gene contains hundreds to thousands of nucleotide units for which the sequence is needed in order to interpret its code. Sequencing methods require only small amounts (5 μg) of purified DNA, which can be produced by cloning. Automated sequencers are available that can daily sequence DNA containing hundreds of nucleotide units.

In the human cell there are 23 pairs of chromosomes containing approximately 3000 million base pairs of DNA. Short sequences of DNA, perhaps with as few as 20 nucleotide units and sometimes radiolabeled, can be obtained either by chemical synthesis (gene machine) or from cloning. These short sequences can be used to "probe" for a complementary sequence by looking for the position to which they bind to any DNA sample under investigation, from blood for example. Such probes can detect as little as 100 fg of DNA and are the basis of forensic genetic fingerprinting tests.

DNA sequence data have been used to investigate inherited diseases such as hemophilia and muscular dystrophy, and also in cancer research.

Proteins

Amino Acids, Peptides, and Proteins

Amino acids (or strictly α-amino acids) are the building blocks of peptides, proteins, and enzymes. Some examples (alanine, phenylalanine, and aspartic acid) are shown in Figure 45.3. Peptides and proteins play key roles in most biological processes. Besides forming structural materials in animate species (hair, muscle, ligaments, etc.), other peptides and proteins and especially enzymes determine the pattern of chemical reactions in cells and mediate many other functions, such as transport and storage of nutrients, immune protection, and the control of growth. For example, hemoglobin transports oxygen in blood, and the related myoglobin transports oxygen in muscle; iron is carried in blood plasma by yet another protein, transferrin; ovulation is controlled by simple hormonal peptides; carbohydrates are broken down into sugars by enzymes.

Peptides and proteins are chain-like molecules made by the sequential linking of amino acids. Sometimes sugar molecules (glycosylates) or phosphate groups are attached at various points along the chain. For example, Figure 45.3 shows how a simple tetrapeptide can be constructed formally from four constituent amino acids, one each of glycine and phenylalanine and two of alanine, by elimination of water. The resulting tetrapeptide contains three amide linkages and can be named H(gly.ala.ala.phe)OH, using the three-letter abbreviations in place of the full name for each amino acid (or GAAF if the one letter code is used). Note that the amino acid residues in the chain are

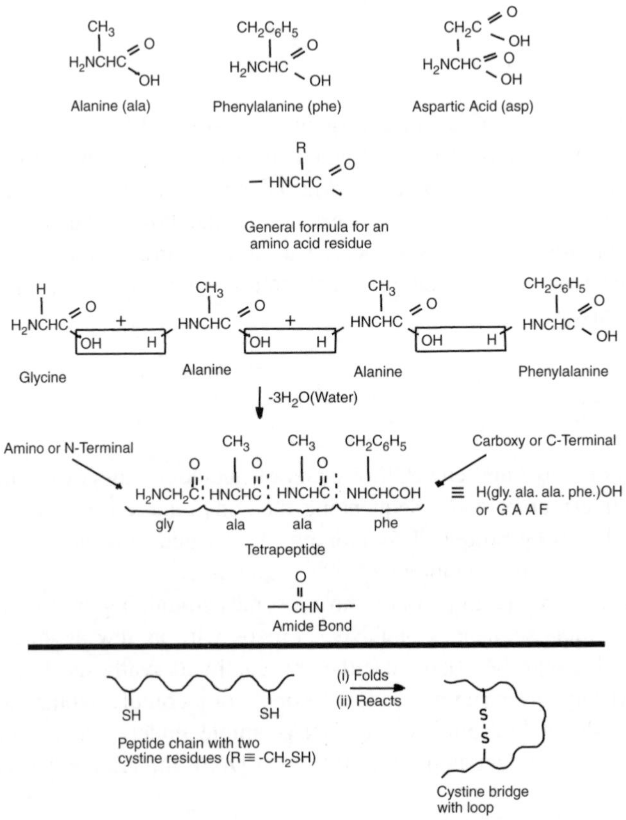

Figure 45.3
Formation of peptides and proteins.

joined by amide bonds, which give the peptide and protein chains much physical and chemical strength (Figure 45.3).

Short chains of amino acid residues are known as di-, tri-, tetrapeptide, and so on, but as the number of residues increases the general names oligopeptide and polypeptide are used. When the number of chains grow to hundreds, the name protein is used. There is no definite point at which the name polypeptide is dropped for protein. Twenty common amino acids appear regularly in peptides and proteins of all species. Each has a distinctive side chain (R in Figure 45.3) varying in size, charge, and chemical reactivity.

The sequence of amino acids in a peptide can be written using the three-letter code shown in Figure 45.3 or a one-letter code, both in common use. For example, the tripeptide, ala.ala.phe, could be abbreviated further to AAF. Although peptides and proteins have chain-like structures, they seldom produce a simple linear system; rather, the chains fold and wrap around each other to give complex shapes. The chemical nature of the various amino acid side groups dictates the way in which the chains fold to arrive at a thermodynamically most-favored state.

In enzymes, this folding process is crucial to their activity as catalysts, with part of the structure as the center of reactivity. Heating enzymes (or other treatments) destroys their three-dimensional structure so stops further action. For example, in winemaking, the rising alcohol content eventually denatures the enzymes responsible for turning sugar into alcohol, and fermentation stops.

Peptides and proteins can be purified by multiple techniques. They can be separated from small molecules by dialysis through semipermeable membranes or by gel-filtration chromatography. They can be separated from each other by ion-exchange chromatography or by electrophoresis and its variants. The high affinity of many proteins for specific chemical groups is used to advantage for their separation and purification in affinity chromatography.

Sequencing Methods

The sequence of each different peptide or protein is important for understanding the activity of peptides and proteins and for enabling their independent synthesis, since the natural ones may be difficult to obtain in small quantities. To obtain the sequence, the numbers of each type of amino acid are determined by breaking down the protein into its individual amino acids using concentrated acid (hydrolysis). For example, hydrolysis of the tetrapeptide shown in Figure 45.3 would give one unit of glycine, two units of alanine, and one unit of phenylalanine. Of course, information as to which amino acid was linked to which others is lost.

The N-terminal residue of a peptide (gly in the tetrapeptide of Figure 45.3) can be identified by bonding another flag molecule at this position and then removing the modified residue from the peptide by hydrolysis (Figure 45.4). Unfortunately, this technique destroys the rest of the sequence, so the Edman technique was developed to remove sequentially one amino acid at a time while leaving the remainder of the chain intact. Reaction with phenylisothiocyanate generates a phenylthiohydantoin (PTH) derived from the N-terminal amino acid (Figure 45.5). The terminal residue can be identified, leaving the remainder of the peptide intact, so the procedure can be iterated until the entire sequence has been determined.

Such chemical approaches to obtaining amino acid sequences have been fully automated and can be used with extremely small quantities of peptide or protein. They are frequently referred to as *wet* methods because they are done in solution. Although a sensitive technique, the method begins to fail after about 20 to 40 residues have been determined, so it is necessary to break proteins into smaller peptide units. The Edman degradation is then used on the smaller fragments of the original chain. Although the amino acid sequences of these peptides are obtained, the order in which these smaller segments occur in the original protein is not defined. (Compare the result of cutting a ribbon without identifying which cut end goes with which.) The necessary additional information is obtained by creating "overlap" peptides from treatment of the original protein with

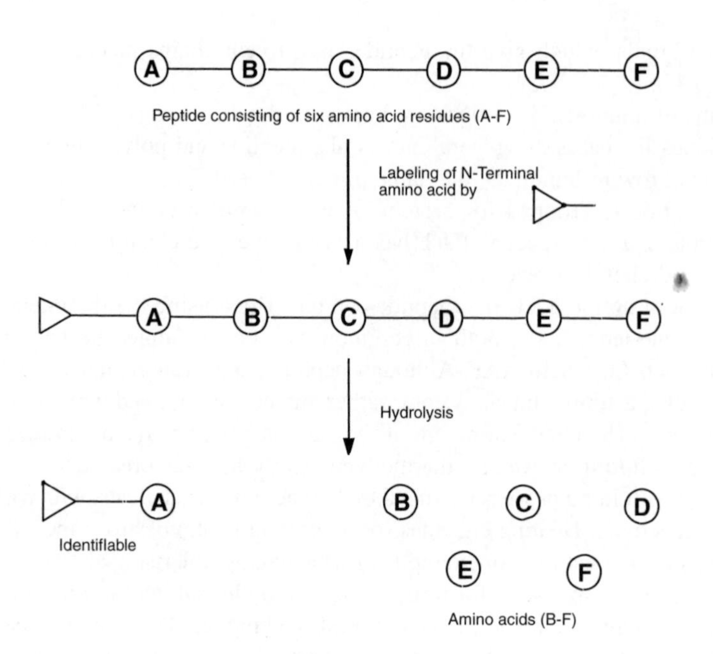

Figure 45.4
Identification of the amino-terminal residue of a peptide.

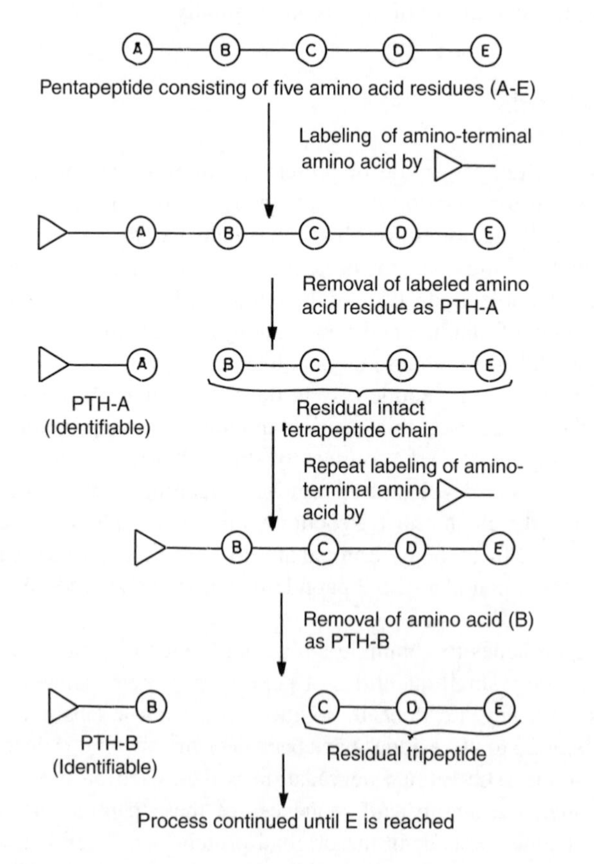

Figure 45.5
The Edman sequential degradation of a peptide or protein.

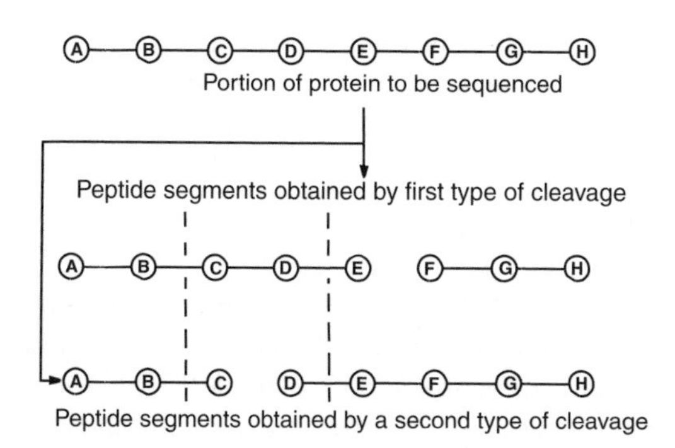

Figure 45.6
Protein sequencing by creation of overlap peptides (mapping).

different chemicals or enzymes (cutting the ribbon at different points). This peptide-mapping technique can be used to determine the whole of a protein sequence containing hundreds of amino acids (Figure 45.6). As mentioned above, automated sequencers are available and can carry out complete Edman degradations on polypeptides containing tens of amino acids in less than one day.

An alternative approach to peptide sequencing uses a *dry* method in which the whole sequence is obtained from a mass spectrum, thereby obviating the need for multiple reactions. Mass spectrometrically, a chain of amino acids breaks down predominantly through cleavage of the amide bonds, similar to the result of chemical hydrolysis. From the mass spectrum, identification of the molecular ion, which gives the total molecular mass, followed by examination of the spectrum for characteristic fragment ions representing successive amino acid residues allows the sequence to be read off in the most favorable cases.

However, interpretation of, or even obtaining, the mass spectrum of a peptide can be difficult, and many techniques have been introduced to overcome such difficulties. These techniques include modifying the side chains in the peptide and protecting the N- and C-terminals by special groups. Despite many advances made by these approaches, it is not always easy to read the sequence from the mass spectrum because some amide bond cleavages are less easy than others and give little information. To overcome this problem, tandem mass spectrometry has been applied to this dry approach to peptide sequencing with considerable success. Further, electrospray ionization has been used to determine the molecular masses of proteins and peptides with unprecedented accuracy.

Enzymes

Enzymes have been touched upon already in various sections; of their uses outside of natural metabolism are discussed here.

Enzymes are proteins that catalyze specific chemical reactions in biological systems by binding to specific substrates. Chemical modifications of the substrate take place on the enzyme, and the products are released. Because of their specificity in reaction, enzymes are being used increasingly in the chemical industry to catalyze chemical reactions, often to produce chiral compounds that are difficult to isolate by conventional methods. Other uses of enzymes are developing, and the technique of gene cloning allows large quantities of enzymes to be produced by utilizing harmless bacteria as biochemical factories. As a final note, if the gene corresponding to a particular enzyme can be identified, it is no longer necessary to sequence the enzyme chain by identifying the amino acids directly. Instead, the sequence of nucleotides in the DNA is found, and then, from the succession of codons, the complete sequence of the enzyme can be read off.

Conclusion

Deoxyribonucleic acids are the chemical codes for genes, which are grouped together in chromosomes. The coding directs cellular synthesis in an organism to give a wide variety of peptides, proteins, and enzymes that, in turn, form structural materials (hair, tissue, etc.), other chemicals for breaking down food (needed for growth and energy), and still other chemicals for protecting the organism (immune response). All these processes can be interfered with naturally (mutation, disease) or artificially (genetic engineering).

Chapter 46

Isotopes and Mass Spectrometry

Introduction

The concept that all substances are composed of elements and atoms goes back at least 2000 years. Originally, only four elements were recognized: air, earth, fire, and water. Each substance was thought to consist of very small particles, called atoms, that could not be subdivided any further. This early mental concept of the nature of matter was extremely prescient, considering there were no experimental results to indicate that matter should be so and none to verify that it was so. Modern atomic theory is much more rigorously based, and we even have the ability to see atoms with special tunneling microscopes. All of chemistry is based on how atoms react with each other.

The idea that air, earth, fire, and water are elements — and that there are only these four elements — has long gone, but the basic idea of atoms as the simplest building blocks of matter is still accepted, although with a proviso. In chemistry, reactions occur between atoms, and, in that sense, atoms can be regarded as the simplest building blocks. However, the inner structures of atoms have important consequences, and under special conditions atoms are not regarded as the simplest building blocks of matter that can exist alone. At this next level, atoms are seen to be composed of three entities: electrons, protons, and neutrons. In particular, the numbers and masses of protons and neutrons determine the character of each element. The ratio of protons and neutrons in an atomic nucleus is important and gives rise to the existence of isotopes. Mass spectrometers are particularly effective general instruments for exploring the existence and abundance ratios of isotopes. The next sections explain the structure of atoms and then show how isotopes arise.

Atomic Structure and the Elements

Many elements are familiar to us in everyday life. Iron is an element used for making ships, cars, spades, etc. There are about 90 such familiar elements, including helium, oxygen, nitrogen, mercury, platinum, and gold. As an element, iron consists of atoms of iron, the smallest building blocks, each of which is indivisible by chemical means. A lump of iron comprises millions, trillions, and zillions of atoms, and the mass of each atom of iron is very small, about 10^{-22} g! In a piece of iron weighing 50 g, there are about 10^{23} atoms.

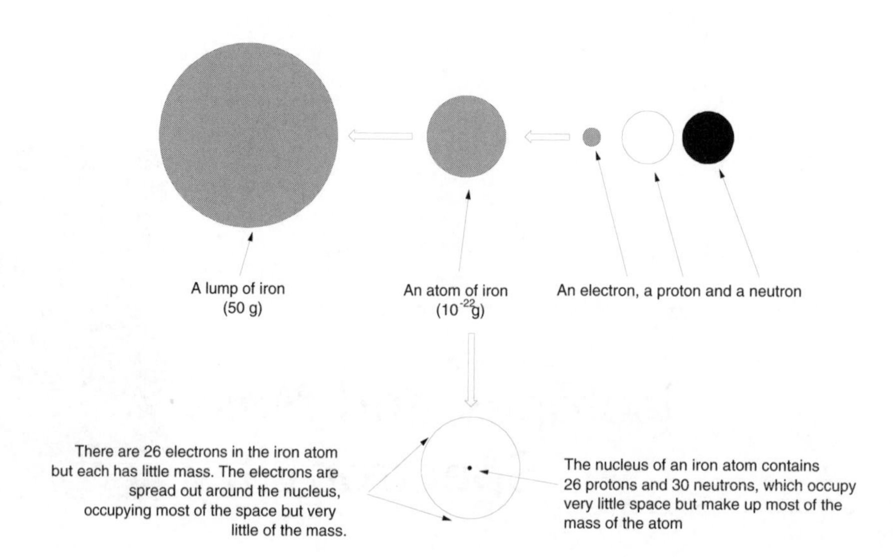

Figure 46.1
A representation of atomic structure. The various spheres are not drawn to scale. The lump of iron on the left would contain almost a million million million million (10^{24}) atoms, one of which is represented by the sphere in the top center of the page. In turn, each atom is composed of a number of electrons, protons, and neutrons. For example, an atom of the element iron contains 26 electrons, 26 protons, and 30 neutrons. The physical size of the atom is determined mainly by the number of electrons, but almost all of its mass is determined by the number of protons and neutrons in its dense core or nucleus (lower part of figure). The electrons are spread out around the nucleus, and their number determines atomic size; but the protons and neutrons compose a very dense, small core, and their number determines atomic mass.

With only 90 elements, one might assume that there could be only about 90 different substances possible, but everyday experience shows that there are millions of different substances, such as water, brick, wood, plastics, etc. Indeed, elements can combine with each other, and the complexity of these possible combinations gives rise to the myriad substances found naturally or produced artificially. These combinations of elemental atoms are called compounds. Since atoms of an element can combine with themselves or with those of other elements to form molecules, there is a wide diversity of possible combinations to make all of the known substances, naturally or synthetically. Therefore, atoms are the simplest chemical building blocks. However, to understand atoms, it is necessary to examine the structure of a typical atom or, in other words, to examine the building blocks of the atoms themselves. The building blocks of atoms are called electrons, protons, and neutrons (Figure 46.1).

It might be noted here in passing that, in high-energy physics, even these simplest building blocks comprise even smaller units such as quarks and gluons. However, these much smaller building blocks are not concerned in chemical reactions or chemistry and will not be considered further here. For mass spectrometry, only the structure of the atomic building blocks (electrons, protons, and neutrons) is of importance.

Electrons

An electron carries one unit of negative electrical charge (Figure 46.2). Its mass is about 1/2000 that of a proton or neutron. Therefore, very little of the mass of an atom is made from the masses of the electrons it contains, and generally the total mass of the electrons is ignored. For example, an atom of iron has a mass of 56 atomic units (au; also called Daltons), of which only about 0.02% is due to the 26 electrons. Thus an iron atom (Fe^0) is considered to have the same mass as a doubly charged cation of iron (Fe^{2+}), even though there is a small mass difference.

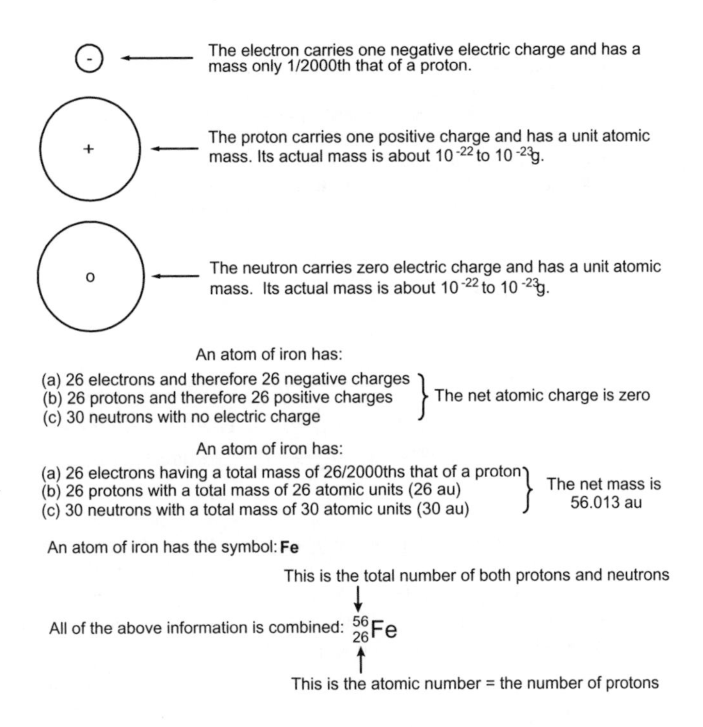

Figure 46.2
The top part of the figure gives an indication of the electric charges and relative sizes of the three building blocks of atoms (electron, protons, and neutrons) that are relevant for the purposes of mass spectrometry. The relative sizes as drawn are not meant to be to scale. The middle part of the figure shows how the electrons in an atom of iron are mostly responsible for its size and how the nucleus is responsible for most of the mass. In this simplified discussion, there is no consideration of packing fractions, which are partly responsible for atomic masses not being whole numbers. The bottom part of the figure illustrates (by the example of iron) how an atom of any element is designated so as to show the numbers of protons, neutrons, and electrons it contains.

Although each electron is very small, all of the electrons in an atom move around the nucleus, sometimes being close to the nucleus and at other times being quite far away. On average, the distance of the electrons from the nucleus is many times the diameter of the nucleus. Therefore, although the electrons add very little to the mass of the atom, they do determine the overall size of the atom (Figure 46.1). Imagine an atom as big as a football, with a small pellet of lead at its center. The air (electrons) surrounding the lead pellet (nucleus) represents the volume occupied by the electrons, but the small pellet constitutes most of the mass.

From a chemical and mass spectrometric viewpoint, the other major property of an electron is its electric charge of one unit. Removal of one electron from an atom or molecule [M] gives a singly charged cation [M^+], and addition of an electron gives a singly charged anion [M^-]. The electric charge is necessary for the mass spectrometer to be able to measure a mass-to-charge (m/z) ratio. Because m/z gives the mass of M^+ or M^- — and the difference between these masses is very small compared with the mass of the neutral species M — it can be said that the m/z value gives the mass M. Although not strictly true, for most purposes it is a close enough approximation (Figure 46.3).

Protons

Each proton is about 2000 times heavier than an electron, and its mass is one atomic unit. Importantly, it also carries one unit of positive electric charge (Figure 46.2). The proton is very

The mass of an atom or ion resides mostly in the nucleus

Atom of iron: Fe^0, mass approximately 56.013 from 26 protons, 30 neutrons and 26 electrons

A cation of iron: Fe^{++}, mass approximately 56.012 from 26 protons, 30 neutrons and 24 electrons

The m/z value for Fe^{++} is 56.012/2 = 28.006

Some naturally occurring multi-isotopic elements (approximate % natural abundances are shown in brackets). Note that only one carbon isotope is radioactive

Carbon	$^{12}_{6}C$ (98.9)	$^{13}_{6}C$ (1.1)	$^{14}_{6}C$ (very small)*				
Oxygen	$^{16}_{8}O$ (99.8)	$^{17}_{8}O$(0.04)	$^{18}_{8}O$ (0.2)				
Sulphur	$^{32}_{16}S$ (95.0)	$^{33}_{16}S$ (0.8)	$^{34}_{16}S$ (4.2)				
Bromine	$^{79}_{35}Br$ (50.5)	$^{79}_{35}Br$ (49.5)					
Molybdenum	$^{92}_{42}Mo$ (15.8)	$^{94}_{42}Mo$ (9.0)	$^{95}_{42}Mo$ (15.7)	$^{96}_{42}Mo$ (16.5)	$^{97}_{42}Mo$ (9.5)	$^{98}_{42}Mo$ (23.8)	$^{100}_{42}Mo$ (9.6)
Mercury	$^{196}_{80}Hg$ (0.2)	$^{198}_{80}Hg$(10.0)	$^{199}_{80}Hg$ (16.8)	$^{200}_{80}Hg$ (23.1)	$^{201}_{80}Hg$ (13.2)	$^{202}_{80}Hg$ (29.8)	$^{204}_{80}Hg$ (6.9)

Figure 46.3

The upper part of the figure illustrates why the small difference in mass between an ion and its neutral molecule is ignored for the purposes of mass spectrometry. In mass measurement, ^{12}C has been assigned arbitrarily to have a mass of 12.00000. All other atomic masses are referred to this standard. In the lower part of the figure, there is a small selection of elements with their naturally occurring isotopes and their natural abundances. At one extreme, xenon has nine naturally occurring isotopes, whereas, at the other, some elements such as fluorine have only one.

small and is confined to the nucleus of the atom (along with any neutrons), and the mass of the atom resides almost entirely in this very dense nucleus. The nucleus is not the most important factor in determining atomic size (the electrons do that), but it does determine where most of the mass of the atom resides.

The unit positive charge on the proton balances the unit negative charge on the electron. In neutral atoms, the number of electrons is exactly equal to the number of protons. In an iron atom (Fe^0), there are 26 electrons and just 26 protons. A cation is formed by removing electrons not by adding protons. An ion M^+ has one electron less than the neutral atom M^0. Similarly, an anion M^- is formed by adding an electron and not by subtracting a proton from M^0.

The number of protons in an atom determines which atomic species is present. The simplest element hydrogen has two atoms, each of which has just one proton. No element other than hydrogen has only one proton. The next element (helium) has two protons in each atom, and so on through all of the known elements. Iron, as has been shown, has 26 protons, and that number of protons is the reason it is iron and not, say, lead. In fact, each element is characterized by its atomic number, which is the number of protons in an atom of the element (Figure 46.2). Hence, an atom of iron is represented by the symbol for iron (Fe) with a number written as a subscript in front of it; $_{26}Fe$ signifies iron of atomic number 26. The designation $_{17}Cl$ signifies chlorine of atomic number 17 (17 protons in the nucleus).

Neutrons

A neutron is characterized by having no electrical charge but has one unit of atomic mass, the same as that of a proton (Figure 46.2). Neutrons, like protons, reside in the atomic nucleus and contribute to the mass of the atom. The chemistry of an atom, like its size, is determined by the electrons in the atom. The mass of the atom is characterized mainly by the total number of neutrons and protons in the nucleus (atomic binding energies are ignored in this discussion). For mass spectrometric purposes of measurement, it is the mass that is important in establishing m/z values.

Atomic Nucleus and Isotopes

Consider a nucleus of the simplest element, hydrogen, of atomic number 1; it is designated ${}^{1}_{1}H$. There is one proton (and of course one electron), and the atomic mass is 1 Da (mostly from the one proton). The total number of protons plus neutrons is then indicated by a another (superscripted) prefix. In this present case it is ${}^{1}_{1}H$ (Figure 46.2). The chemistry of hydrogen is mostly determined by that one electron, but its mass is determined mostly by the one proton. There is another type of hydrogen atom called deuterium, which includes a neutron in its nucleus and has its own symbol, D. However, it is less confusing to write it as ${}^{2}_{1}H$. This type of hydrogen atom also has one electron and one proton, and its chemistry is the same as that of ${}^{1}_{1}H$. Because the nucleus of ${}^{2}_{1}H$ contains one neutron and one proton, its atomic mass is 2. There are two types of hydrogen atom, one about twice as heavy as the other, but the general chemistry of the two is identical for most purposes. These two kinds of hydrogen atom are called isotopes. The predominantly abundant isotope is ${}^{1}_{1}H$, and the isotope in much lesser abundance is ${}^{2}_{1}H$. These isotopes are not radioactive. However, there is even a third isotope of hydrogen called tritium (${}^{3}_{1}H$), which has two neutrons and one proton in its nucleus and is three times heavier than the first hydrogen isotope, ${}^{1}_{1}H$. All of these hydrogen isotopes react chemically in the same way, and only the ${}^{3}_{1}H$ isotope is radioactive.

Some elements in their natural state have only one isotope, as with fluorine, phosphorus, and rhodium but others have several isotopes, as with carbon (three), oxygen (three), and molybdenum (seven). A few examples are given in Figure 46.3. A small proportion of naturally occurring isotopes is radioactive, but most isotopes are not. The radioactivity results from the nucleus being unstable, which happens if the numbers of protons and neutrons in a nucleus become seriously unbalanced. Nuclei of carbon having six protons and six neutrons (${}^{12}_{6}C$) or six protons and seven neutrons (${}^{13}_{6}C$) are stable and not radioactive, but the carbon isotope with six protons and eight neutrons (${}^{14}_{6}C$) is unstable and is radioactive (Figure 46.3).

Recently, it has become possible to create isotopes that do not exist naturally. These are the artificial isotopes, and all are radioactive. For example, 13 artificially created isotopes of iodine are known, as well as its naturally occurring monoisotopic form of mass 127. Mass spectrometry is able to measure m/z values for both natural and artificial isotopes.

Isotope Ratios

Approximate Abundance Ratios

Naturally occurring isotopes of any element are present in unequal amounts. For example, chlorine exists in two isotopic forms, one with 17 protons and 18 neutrons (${}^{35}Cl$) and the other with 17 protons and 20 neutrons (${}^{37}Cl$). The isotopes are not radioactive, and they occur, respectively, in a ratio of nearly 3:1. In a mass spectrum, any compound containing one chlorine atom will have two different molecular masses (m/z values). For example, methyl chloride (CH_3Cl) has masses of 15 (for the CH_3) plus 35 (total = 50) for one isotope of chlorine and 15 plus 37 (total = 52) for the other isotope. Since the isotopes occur in the ratio of 3:1, molecular ions of methyl chloride will show two molecular-mass peaks at m/z values of 50 and 52, with the heights of the peaks in the ratio of 3:1 (Figure 46.4).

This example can be used in reverse to show the usefulness of looking for such isotopes. Suppose there were an unknown sample that had two molecular ion peaks in the ratio of 3:1 that were two mass units apart; then it could reasonably be deduced that it was highly likely the unknown contained chlorine. In this case, the isotope ratio has been used to identify a chlorine-containing compound. This use of mass spectrometry is widespread in general analysis of materials, and it

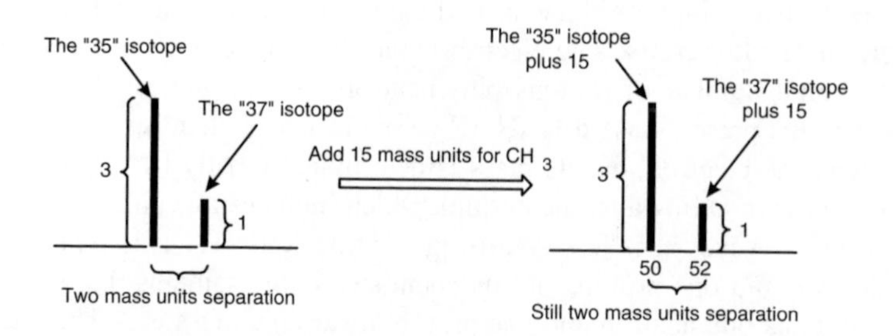

Figure 46.4
A diagrammatic illustration of the effect of an isotope pattern on a mass spectrum. The two naturally occurring isotopes of chlorine combine with a methyl group to give methyl chloride. Statistically, because their abundance ratio is 3:1, three ^{35}Cl isotope atoms combine for each ^{37}Cl atom. Thus, the ratio of the molecular ion peaks at m/z 50, 52 found for methyl chloride in its mass spectrum will also be in the ratio of 3:1. If nothing had been known about the structure of this compound, the appearance in its mass spectrum of two peaks at m/z 50, 52 (two mass units apart) in a ratio of 3:1 would immediately identify the compound as containing chlorine.

makes use of only approximate ratios of isotopes because that is all that is necessary for identification. It is not usually necessary to know if the ratio is 3.001:1.00 or 3.002:1.00; all that is needed is a ratio near 3:1. Where there are several isotopes of an element, the actual pattern of masses and abundances is enough for identification. It would be very difficult to miss the evidence for mercury or molybdenum in a mass spectrum, since there are seven isotopes with distinctive patterns of abundances and mass differences (Figure 46.3). These uses of isotopes are discussed in Chapter 47. However, there are other uses of isotopes that require very accurate determinations of their ratios of abundances.

Accurate Abundance Ratios

The use of accurate isotope ratio measurement is exemplified here by a method used to determine the temperature of the Mediterranean Sea 10,000 years ago. It is known that the relative solubility of the two isotopic forms of carbon dioxide ($^{12}CO_2$, $^{13}CO_2$) in sea water depends on temperature (Figure 46.5).

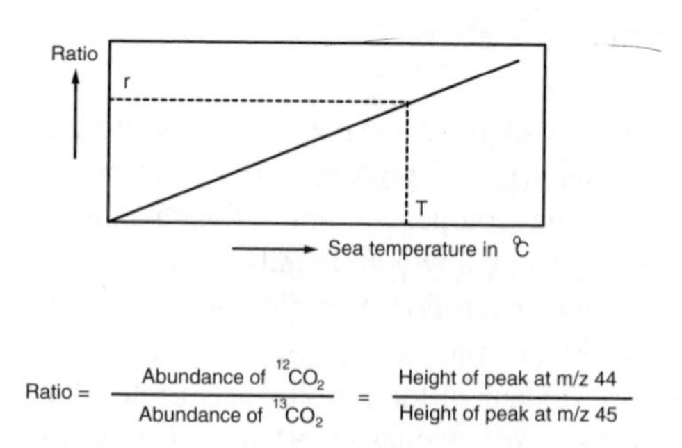

Figure 46.5
By experimentally determining the ratio of abundances of ^{12}C and ^{14}C isotope peaks for CO_2 dissolved in sea water at various temperatures, a graph can be drawn relating the solubility of $^{12}CO_2$ compared with that of $^{13}CO_2$ (the ratio described above). On extracting the CO_2 from sediment containing the shells (calcium carbonate) of dead sea creatures by addition of acid, a ratio (R) of abundances of $^{12}CO_2$ to $^{13}CO_2$ can be measured. If this value is read from the graph, a temperature T is extrapolated, indicating the temperature of the sea at the time the sediment was laid down. Such experiments have shown that 10,000 years ago the temperature of the Mediterranean was much as it is now.

One method for measuring the temperature of the sea is to measure this ratio. Of course, if you were to do it now, you would take a thermometer and not a mass spectrometer. But how do you determine the temperature of the sea as it was 10,000 years ago? The answer lies with tiny sea creatures called diatoms. These have shells made from calcium carbonate, itself derived from carbon dioxide in sea water. As the diatoms die, they fall to the sea floor and build a sediment of calcium carbonate. If a sample is taken from a layer of sediment 10,000 years old, the carbon dioxide can be released by addition of acid. If this carbon dioxide is put into a suitable mass spectrometer, the ratio of carbon isotopes can be measured accurately. From this value and the graph of solubilities of isotopic forms of carbon dioxide with temperature (Figure 46.5), a temperature can be extrapolated. This is the temperature of the sea during the time the diatoms were alive. To conduct such experiments in a significant manner, it is essential that the isotope abundance ratios be measured very accurately.

This accurate measurement of the ratio of abundances of isotopes is used for geological dating, estimation of the ages of antiquities, testing athletes for the use of banned steroids, examining fine details of chemical reaction pathways, and so on. These uses are discussed in this book under various headings concerned with isotope ratio mass spectrometry (see Chapters 7, 14, 15, 16, 17, 47, and 48).

Conclusion

Isotopes of an element are formed by the protons in its nucleus combining with various numbers of neutrons. Most natural isotopes are not radioactive, and the approximate pattern of peaks they give in a mass spectrum can be used to identify the presence of many elements. The ratio of abundances of isotopes for any one element, when measured accurately, can be used for a variety of analytical purposes, such as dating geological samples or gaining insights into chemical reaction mechanisms.

Chapter 47

Uses of Isotope Ratios

Introduction

A general discussion of isotopes appears in Chapter 46. For naturally occurring elements, the ratios of abundances of isotopes are well known and, for convenience, are listed in Tables 47.1 and 47.2. These ratios can be used in routine mass spectrometry to identify various elements. For example, molybdenum has the seven isotopes listed here (the percent abundances are given in parentheses): ^{92}Mo (14.8), ^{94}Mo (9.25), ^{95}Mo (15.9), ^{96}Mo (16.7), ^{97}Mo (9.55), ^{98}Mo (24.1), ^{100}Mo (9.63); it would not be very difficult to identify the presence of molybdenum in a sample from the pattern of isotopic peaks in a mass spectrum. Although listed in Tables 47.1 and 47.2 as fixed ratios, the actual ratios of isotopes differ slightly, depending on the source of the elements being investigated. Thus, although ^{12}C and ^{13}C are listed as occurring in a ratio of 98.882 to 1.108, if this ratio is measured in metabolic products, as in drug testing, the ratio will be found to be slightly different from that given in Tables 47.1 and 47.2 because metabolic reactions differ somewhat between ^{12}C and ^{13}C. Such accurate measurement of precise isotope ratios is important in many areas, including chemistry, geology, environmental science, the nuclear industry, and medicine. Special instruments are needed to measure these precise levels. It also becomes necessary to stipulate that certain materials should be regarded as having standard isotope ratios, and these are used to gauge any changes appearing in nonstandard substances.

For the naturally occurring elements, many new artificial isotopes have been made, and these are radioactive. Although these new isotopes can be measured in a mass spectrometer, this process could lead to unacceptable radioactive contamination of the instrument. This practical consideration needs to be considered carefully before using mass spectrometers for radioactive isotope analysis.

Apart from naturally occurring elements, there are now newly made elements beyond uranium. These constitute the transuranic series. All the elements in this series are radioactive.

Isotope Ratios in Routine Mass Spectrometry

General Chemistry

The occurrence of the elements carbon, nitrogen, and oxygen manifests itself in the isotope patterns occurring for all molecular or fragment ions. For small numbers of carbon atoms, the

TABLE 47.1
Relative Abundances of Naturally Occurring Isotopes

IE (eV)	Da/e	1	2	3	4	5	6	7	8	9	10	11	12	13	14	15	16	17	18	19	20	Da/e
13.6	H	99.9	0.02																			H
24.6	He				100																	He
5.3	Li						7.50	92.5														Li
9.3	Be									100												Be
8.3	B										19.9	80.1										B
11.3	C												98.9	1.10								C
14.5	N														99.6	0.37						N
13.6	O																99.8	0.04	0.20			O
17.4	F																			100		F
21.6	Ne																				90.5	Ne
	Da/e	21	22	23	24	25	26	27	28	29	30	31	32	33	34	35	36	37	38	39	40	Da/e
21.6	(Ne)	0.27	9.22																			(Ne)
5.1	Na			100																		Na
7.6	Mg				79.0	10.0	11.0															Mg
6.0	Al							100														Al
8.2	Si								92.2	4.67	3.10											Si
10.5	P											100										P
10.4	S												95.0	0.75	4.21		0.02					S
13.0	Cl															75.8		24.2				Cl
15.8	Ar																0.34		0.06		99.6	Ar
4.3	K																			93.3	0.01	K
6.1	Ca																				96.9	Ca
	Da/e	41	42	43	44	45	46	47	48	49	50	51	52	53	54	55	56	57	58	59	60	Da/e
4.3	(K)	6.73																				(K)
6.1	(Ca)		0.65	0.14	2.09		.004		0.19													(Ca)
6.5	Sc					100																Sc
6.8	Ti						8.00	7.30	73.8	5.50	5.40											Ti
6.7	V										0.25	99.8										V
6.8	Cr										4.35		83.8	9.50	2.37							Cr
7.4	Mn															100						Mn
7.9	Fe														5.80		91.7	2.20	0.28			Fe
7.9	Co																			100		Co
7.6	Ni																		68.3		26.1	Ni
	Da/e	61	62	63	64	65	66	67	68	69	70	71	72	73	74	75	76	77	78	79	80	Da/e
7.6	(Ni)	1.13	3.59		0.91																	(Ni)
7.7	Cu			69.2		30.8																Cu
9.4	Zn				48.6		27.9	4.10	18.8		0.60											Zn
6.0	Ga					60.1		39.9														Ga
7.9	Ge										20.5		27.4	7.80	36.5		7.80					Ge
9.8	As															100						As
9.8	Se														0.90		9.00	7.60	23.6		49.7	Se
11.8	Br																			50.7		Br
14.0	Kr																		0.35		2.25	Kr
	Da/e	81	82	83	84	85	86	87	88	89	90	91	92	93	94	95	96	97	98	99	100	Da/e
9.8	(Se)		9.20																			(Se)
11.8	(Br)	49.3																				(Br)
14.0	(Kr)		11.6	11.5	57.0		17.3															(Kr)
4.2	Rb					72.2		27.8														Rb
5.7	Sr				0.56		9.86	7.00	82.6													Sr
6.4	Y									100												Y
6.8	Zr										51.5	11.2	17.2		17.4		2.80					Zr
6.9	Nb												100									Nb
7.1	Mo											14.8		9.25	15.9	16.7	9.55	24.1		9.63		Mo
7.3	Tc																					Tc
7.4	Ru																5.52		1.88	12.7	12.6	Ru

most abundant ions are those containing the ^{12}C isotope because it is nearly 100 times more abundant than the ^{13}C isotope. However, the relative abundance of the ^{13}C isotopic peak in a mass spectrum increases as the number of carbon atoms increases. Thus, for *n* carbon atoms, the abundance of the ions containing one ^{13}C isotope of carbon increases relative to those ions containing only the ^{12}C isotope by about $n \times 1.1\%$. The presence of isotopes from other elements can make the isotopic pattern quite complex. For the compound fluorodecane, the fluorine is monoisotopic (^{19}F), and the natural abundance of ^{2}H is very low compared with ^{1}H. Figure 47.1a shows the mass spectrum of the molecular ion region for fluorodecane, from which it is seen that the dominant peak is that corresponding to the ^{12}C isotope, and those

TABLE 47.1 (CONTINUED)
Relative Abundances of Naturally Occurring Isotopes

IE (eV)	Da/e	101	102	103	104	105	106	107	108	109	110	111	112	113	114	115	116	117	118	119	120	Da/e
7.4	(Ru)	17.0	31.6		18.7																	(Ru)
7.5	Rh			100																		Rh
8.3	Pd		1.02		11.1	22.3	27.3		26.5		11.7											Pd
7.6	Ag							51.8		48.2												Ag
9.0	Cd						1.2		0.89		12.5	12.8	24.1	12.2	28.7		7.49					Cd
5.8	In													4.30		95.7						In
7.3	Sn												0.97		0.65	0.36	14.5	7.68	24.2	8.58	32.6	Sn
8.6	Sb																					Sb
9.0	Te																				0.10	Te
	Da/e	121	122	123	124	125	126	127	128	129	130	131	132	133	134	135	136	137	138	139	140	Da/e
7.3	(Sn)		4.63		5.79																	(Sn)
8.6	(Sb)	57.3		42.7																		(Sb)
9.0	(Te)	2.60		0.91	4.82	7.14	19.0		31.7		33.8											(Te)
10.5	I							100														I
12.1	Xe				0.10		0.09		1.91	26.4	4.10	21.2	26.9		10.4		8.90					Xe
3.9	Cs													100								Cs
5.2	Ba										0.11		0.10		2.42	6.59	7.85	11.2	71.7			Ba
5.6	La																		0.09	99.9		La
5.5	Ce																0.19		0.25		88.5	Ce
	Da/e	141	142	143	144	145	146	147	148	149	150	151	152	153	154	155	156	157	158	159	160	Da/e
5.5	(Ce)		11.1																			(Ce)
5.4	Pr	100																				Pr
5.5	Nd		27.1	12.2	23.8	8.30	17.2		5.76		5.64											Nd
5.6	Pm																					Pm
5.6	Sm					3.10		15.0	11.3	13.8	7.40		26.7		22.7							Sm
5.7	Eu											47.8		52.2								Eu
6.1	Gd												0.20		2.18	14.8	20.5	15.7	24.8		21.9	Gd
5.8	Tb																			100		Tb
5.9	Dy																0.06		0.10		2.34	Dy
	Da/e	161	162	163	164	165	166	167	168	169	170	171	172	173	174	175	176	177	178	179	180	Da/e
5.9	(Dy)	18.9	25.5	24.9	28.2																	(Dy)
6.0	Ho					100																Ho
6.1	Er		0.14		1.61		33.6	23.0	26.8		14.9											Er
6.2	Tm									100												Tm
6.3	Yb								0.13		3.05	14.3	21.9	16.1	31.8		12.7					Yb
5.4	Lu															97.4	2.59					Lu
6.7	Hf														0.16		5.21	18.6	27.3	13.6	35.1	Hf
7.9	Ta																				0.01	Ta
8.0	W																				0.13	W
	Da/e	181	182	183	184	185	186	187	188	189	190	191	192	193	194	195	196	197	198	199	200	Da/e
7.9	(Ta)	99.9																				(Ta)
8.0	(W)		26.3	14.3	30.7		28.6															(W)
7.9	Re					37.4		62.6														Re
8.7	Os				0.02		1.58	1.60	13.3	16.1	26.4		41.0									Os
9.1	Ir											37.3		62.7								Ir
9.0	Pt										0.01		0.79		32.9	33.8	25.3		7.20			Pt
9.2	Au																	100				Au
10.4	Hg																0.14		10.0	16.8	23.1	Hg
	Da/e	201	202	203	204	205	206	207	208	209	210	211	212	213	214	215	216	217	218	219	220	Da/e
10.4	(Hg)	13.2	29.8		6.85																	(Hg)
6.1	Ti			29.5		70.5																Ti
7.4	Pb				1.40		24.1	22.1	52.4													Pb
7.3	Bi									100												Bi
	Da/e	221	222	223	224	225	226	227	228	229	230	231	232	233	234	235	236	237	238	239	240	Da/e
6.1	Th												100									Th
5.9	Pa																					Pa
6.1	U														.006	0.72			99.3			U

H e l

ions containing one ^{13}C atom constitute about 11% of this peak height at an m/z value one unit greater. Similarly, the peak at two mass units greater is due to those ions that have two ^{13}C atoms, the remainder being ^{12}C. The pattern of isotopic peaks is simply a distribution of the probabilities of finding different combinations of ^{12}C and ^{13}C atoms in the ten carbon atoms of fluorodecane. If there are ten carbon atoms in the molecule, then each of the ten has a 1.1% chance of being ^{13}C, so the chance that one ^{13}C will turn up is $10 \times 1.1\%$.

TABLE 47.2
Atomic Weights of the Elements Based on the Carbon 12 Standard

Symbol	Amu	Abundance	Symbol	Amu	Abundance
H 1	1.007825037	99.985	Zn 64	63.9291454	48.89
H 2	2.014101787	0.015	Zn 66	65.9260352	27.81
He 3	3.016029297	0.00013	Zn 67	66.9271289	4.11
He 4	4.00260325	100.00	Zn 68	67.9248458	18.56
Li 6	6.0151232	7.52	Zn 70	69.9253249	0.62
Li 7	7.0160045	92.48	Ga 69	68.9255809	60.2
Be 9	9.0121825	100.00	Ga 71	70.9247006	39.8
B 10	10.0129380	18.98	Ge 70	69.9242498	20.52
B 11	11.0093053	81.02	Ge 72	71.9220800	27.43
C 12	12.00000000	98.892	Ge 73	72.9234639	7.76
C 13	13.003354839	1.108	Ge 74	73.9211788	36.54
N 14	14.003074008	99.635	Ge 76	75.9214027	7.76
N 15	15.000108978	0.365	As 75	74.9215955	100.00
O 16	15.99491464	99.759	Se 74	73.9224771	0.96
O 17	16.9991306	0.037	Se 76	75.9192066	9.12
O 18	17.99915939	0.204	Se 77	76.9199077	7.50
F 19	18.99840325	100.00	Se 78	77.9173040	23.61
Ne 20	19.9924391	90.92	Se 80	79.9165205	49.96
Ne 21	20.9938453	0.257	Se 22	81.916709	8.84
Ne 22	21.9913837	8.82	Br 79	78.9183361	50.57
Na 23	22.9897697	100.00	Br 81	80.916290	49.43
Mg 24	23.9850450	78.60	Kr 78	77.920397	0.354
Mg 25	24.9858392	10.11	Kr 80	79.916375	2.27
Mg 26	25.9825954	11.29	Kr 82	81.913483	11.56
Al 27	26.9815413	100.00	Kr 83	82.914134	11.55
Si 28	27.9769284	92.18	Kr 84	83.9115064	56.90
Si 29	28.9764964	4.71	Kr 86	85.910614	17.37
Si 30	29.9737717	3.12	Rb 85	84.9117996	72.15
P 31	30.9737634	100.00	Rb 87	86.9091836	27.85
S 32	31.9720718	95.018	Sr 84	83.913428	0.56
S 33	32.9714591	0.750	Sr 86	85.9092732	9.86
S 34	33.96786774	4.215	Sr 87	86.9088902	7.02
S 36	35.9670790	0.107	Sr 88	87.9056249	82.56
Cl 35	34.968852729	75.4	Y 89	88.9058560	100.00
Cl 37	36.965902624	24.6	Zr 90	89.9047080	51.46
Ar 36	35.967545605	0.337	Zr 91	90.9056442	11.23
Ar 38	37.9627322	0.063	Zr 92	91.9050392	17.11
Ar 40	39.9623831	99.600	Zr 94	93.9063191	17.40
K 39	38.9637079	93.08	Zr 96	95.908272	2.80
K 40	39.9639988	0.012	Nb 93	92.9063780	100.00
K 41	40.9618254	6.91	Mo 92	91.906809	15.05
Ca 40	39.9625907	96.92	Mo 94	93.9050862	9.35
Ca 42	41.9586218	0.64	Mo 95	94.9058379	14.78
Ca 43	42.9587704	0.13	Mo 96	95.9046755	16.56
Ca 44	43.9554848	2.13	Mo 97	96.9060179	9.60
Ca 46	45.953689	0.0032	Mo 98	97.9054050	24.00
Ca 48	47.952532	0.179	Mo 100	99.907473	9.68
Sc 45	44.9559136	100.00	Ru 96	95.907596	5.68
Ti 46	45.9526327	7.95	Ru 98	97.905287	2.22
Ti 47	46.9517649	7.75	Ru 99	98.9059371	12.81
Ti 48	47.9479467	73.45	Ru 100	99.9042175	12.70
Ti 49	48.9478705	5.51	Ru 101	100.9055808	16.98
Ti 50	49.9447858	5.34	Ru 102	101.9043475	31.34
V 50	49.9471613	0.24	RU 104	103.905422	18.27
V 51	50.9439625	99.76	Rh 103	102.905503	100.00
Cr 50	49.946463	4.31	Pd 102	101.905609	0.80
Cr 52	51.9405097	83.76	Pd 104	103.904026	9.30
Cr 53	52.9406510	9.55	Pd 105	104.905075	22.60
Cr 54	53.9388822	2.38	Pd 106	105.903475	27.10
Mn 55	54.9380463	100.00	Pd 108	107.903894	26.70
Fe 54	53.9396121	5.90	Pd 110	109.905169	13.50
Fe 56	55.9349393	91.52	Ag 107	106.905095	51.35
Fe 57	56.9353957	2.25	Ag 109	108.904754	48.65
Fe 58	57.9332778	0.33	Cd 106	105.906461	1.22
Co 59	58.9331978	100.00	Cd 108	107.904186	0.89
Ni 58	57.9353471	67.76	Cd 110	109.903007	12.43
Ni 60	59.9307890	26.16	Cd 111	110.904182	12.86
Ni 61	60.9310586	1.25	Cd 112	111.9027614	23.79
Ni 62	61.9283464	3.66	Cd 113	112.9044013	12.34
Ni 64	63.9279680	1.16	Cd 114	113.9033607	28.81
Cu 63	62.9295992	69.09	Cd 116	115.904758	7.66
Cu 65	64.9277924	30.91	In113	112.904056	4.16

TABLE 47.2 (CONTINUED)
Atomic Weights of the Elements Based on the Carbon 12 Standard

Symbol	Amu	Abundance	Symbol	Amu	Abundance
In 115	114.903875	95.84	Dy 160	159.925203	2.294
Sn 112	111.904823	0.95	Dy 161	160.926939	18.88
Sn 114	113.902781	0.65	Dy 162	161.926805	25.53
Sn 115	114.9033441	0.34	Dy 163	162.928737	24.97
Sn 116	115.9017435	14.24	Dy 164	163.929183	28.18
Sn 117	116.9029536	7.57	Ho 165	164.930332	100.00
Sn 118	117.9016066	24.01	Er 162	161.928787	0.136
Sn 119	118.9033102	8.58	Er 164	163.929211	1.56
Sn 120	119.9021990	32.97	Er 166	165.930305	33.41
Sn 122	121.903440	4.71	Er 167	166.932061	22.94
Sn 124	123.905271	5.98	Er 168	167.932383	27.07
Sb 121	120.9038237	57.25	Er 170	169.935476	14.88
Sb 123	122.904222	42.75	Tm 169	168.934225	100.00
Te 120	119.904021	0.089	Yb 168	167.933908	0.140
Te 122	121.903055	2.46	Yb 170	169.934774	3.03
Te 123	122.904278	0.87	Yb 171	170.936338	14.31
Te 124	123.902825	4.61	Yb 172	171.936393	21.82
Te 125	124.904435	6.99	Yb 173	172.938222	16.13
Te 126	125.903310	18.71	Yb 174	173.938873	31.84
Te 128	127.904464	31.79	Yb 176	175.942576	12.73
Te 130	129.906229	34.49	Lu 175	174.940785	97.40
I 127	126.904477	100.00	Lu 176	175.942694	2.60
Xe 124	123.90612	0.096	Hf 174	173.940065	0.199
Xe 126	125.904281	0.090	Hf 176	175.941420	5.23
Xe 128	127.9035308	1.919	Hf 177	176.943233	18.55
Xe 129	128.9047801	26.44	Hf 178	177.943710	27.23
Xe 130	129.9035095	4.08	Hf 179	178.945827	13.79
Xe 131	130.905076	21.18	Hf 180	179.946561	35.07
Xe 132	131.904148	26.89	Ta 180	179.947489	0.0123
Xe 134	133.905395	10.44	Ta 181	180.948014	99.9877
Xe 136	135.907219	8.87	W 180	179.946727	0.126
Cs 133	132.905433	100.00	W 182	181.948225	26.31
Xe 130	129.906277	0.101	W 183	182.950245	14.28
Xe 132	131.905042	0.097	W 184	183.950953	30.64
Xe 134	131.904490	2.42	W 186	185.954377	28.64
Xe 135	134.905668	6.59	Re 185	184.952977	37.07
Xe 136	135.904556	7.81	Re 187	186.955765	62.93
Xe 137	136.905816	11.32	Oa 184	183.952514	0.018
Xe 138	137.905236	71.66	Os 186	185.953852	1.59
La 138	137.907114	0.089	Os 187	186.955762	1.64
La 139	138.906355	99.911	Os 188	187.955850	13.20
Ce 136	135.90714	0.193	Os 189	188.958156	16.10
Ce 138	137.905996	0.250	Os 190	189.958455	26.40
Ce 140	139.905442	88.48	Os 192	191.961487	41.00
Ce 142	141.909249	11.07	Ir 191	190.960603	38.50
Pr 141	140.907657	100.00	Ir 193	192.962942	61.50
Nd 142	141.907731	27.09	Pt 190	189.959937	0.012
Nd 143	142.909823	12.14	Pt 192	191.961049	0.78
Nd 144	143.910096	23.83	Pt 194	193.962679	32.80
Nd 145	144.912582	8.29	Pt 195	194.964785	33.70
Nd 146	145.913126	17.26	Pt 196	195.964947	25.40
Nd 148	147.916901	5.74	Pt 198	197.967879	7.23
Nd 150	149.920900	5.63	Au 197	196.966560	100.00
Sm 144	143.912009	3.16	Hg 196	195.965812	0.146
Sm 147	146.914907	15.07	Hg 198	197.966760	10.02
Sm 148	147.914832	11.27	Hg 199	198.968269	16.84
Sm 149	148.917193	13.84	Hg 200	199.968316	23.13
Sm 150	149.917285	7.47	Hg 201	200.970293	13.22
Sm 152	151.919741	26.63	Hg 202	210.970632	29.80
Sm 154	153.922218	22.53	Hg 204	203.973481	6.85
Eu 151	150.919860	47.77	Tl 203	202.972336	29.50
Eu 153	152.921243	52.23	Tl 206	204.974410	70.50
Gd 152	151.919803	0.20	Pb 204	203.973037	1.37
Gd 154	153.920876	2.15	Pb 206	205.974455	25.15
Gd 155	154.922629	14.73	Pb 207	206.975885	21.11
Gd 156	155.922130	20.47	Pb 208	207.976641	52.38
Gd 157	156.923967	15.68	Bi 209	208.980388	100.00
Gd 158	157.924111	24.87	Th 232	232.03805381	100.00
Gd 160	159.927061	21.90	U 234	234.04094740	0.0058
Tb 159	158.925350	100.0	U 235	235.04392525	0.715
Dy 156	155.924287	0.0524	U 238	238.05078578	99.28
Dy 158	157.924412	0.0902			

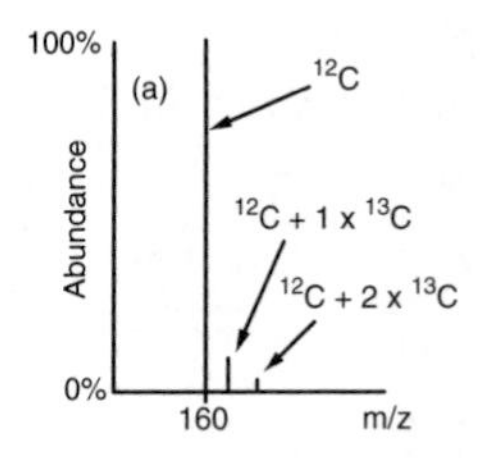

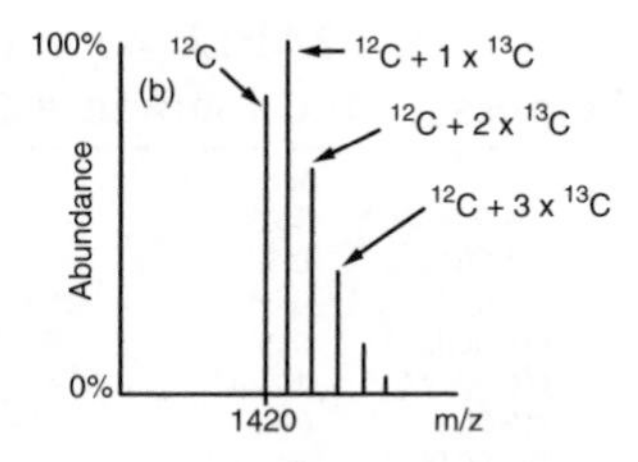

Figure 47.1
(a) A carbon compound having one fluorine and ten carbon atoms has a major molecular ion peak at m/z 160 in its mass spectrum, which corresponds to a preponderance of the ^{12}C isotope. The peaks at m/z 161 and 162, containing respectively one and two ^{13}C isotopes, are much smaller because the chances of finding one or two isotopic atoms among the total of only 10 atoms are low. (b) A carbon compound having 100 carbon atoms in the molecule has a molecular ion peak at m/z 1420, representing only ^{12}C atoms, which is smaller than the peak at m/z 1421, corresponding to those molecular ions having 99 ^{12}C atoms plus one ^{13}C atom. Peaks representing ions having successively 2 ^{13}C (plus 98 ^{12}C), 3 ^{13}C (plus 97 ^{12}C), and 4 ^{13}C (plus 96 ^{12}C atoms) in the molecule are progressively less abundant but still prominent in comparison with a compound containing only ten carbon atoms in total.

This probability distribution means that, for a carbon compound containing 100 carbon atoms as in fluorohectane, the relative abundance of those ions containing 99 ^{12}C atoms and just one ^{13}C atom becomes $100 \times 1.1\% = 110\%$, as against 100% for those ions containing only ^{12}C atoms (Figure 47.1b). Thus, in the molecular ion region of its mass spectrum, the peak corresponding to the molecular ion based on ^{12}C is smaller than the next peak representing the relative number of those ions containing one ^{13}C atom. Ions containing two, three, four, or more ^{13}C atoms also become relatively more abundant, as shown in Figure 47.1b.

If oxygen or nitrogen atoms are in the molecule, their isotopes will add to this complexity, but in practice these atoms present little complication because they are usually present in much smaller numbers than for carbon, and the relative abundances of the minor isotopes are quite small. In mass spectrometry of organic compounds containing only small numbers of carbon atoms, the molecular ion peak is simply taken to be that corresponding to the ^{12}C isotope (the biggest peak or line in Figure 47.1a), which is clearly differentiated from the other, much smaller isotopic peaks. However, as the number of carbon atoms increases, selection of the "all" ^{12}C isotope peak means that it does not constitute the biggest peak in the molecular ion region (Figure 47.1b). This effect can lead inadvertently to misidentification of molecular mass for an unknown sample containing a large number of carbon atoms if the largest peak in the molecular ion region is assumed to be due only to ^{12}C atoms.

For other elements that occur with major relative abundances of more than one isotope in the natural state, the isotope pattern becomes much more complex. For example, with chlorine and bromine, the presence of these elements is clearly apparent from the isotopes ^{35}Cl and ^{37}Cl for chlorine and ^{79}Br and ^{81}Br for bromine. Figure 47.2a shows the molecular ion region for the compound chlorodecane. Now, there are new situations in that ^{12}C, ^{13}C, ^{35}Cl, and ^{37}Cl isotopes all have probabilities of occurring together. Thus, there are molecular ion peaks for ^{12}C + ^{35}Cl, ^{13}C + ^{35}Cl, ^{12}C + ^{37}Cl, and so on. Even so, the isotopic ratio of 3:1 for ^{35}Cl to ^{37}Cl is very clear (Figure 47.2a). Similarly, the pattern for bromine, having a 1:1 ratio of the isotopes ^{79}Br and ^{81}Br is equally clear in bromodecane (Figure 47.2b). Indeed, the numbers of such atoms occurring in a molecule can be "counted" accurately from the pattern of isotopic peaks in the mass spectrum. Figure 47.2c shows the molecular ion region for a substance containing one chlorine and one bromine atom.

A common mistake for beginners in mass spectrometry is to confuse average atomic mass and isotopic mass. For example, the *average* atomic mass for chlorine is close to 35.45, but this average is of the numbers and masses of ^{35}Cl and ^{37}Cl isotopes. This average must be used for instruments that cannot differentiate isotopes (for example, gravimetric balances). Mass spectrometers do differentiate isotopes by mass, so it is important in mass spectrometry that isotopic masses be used

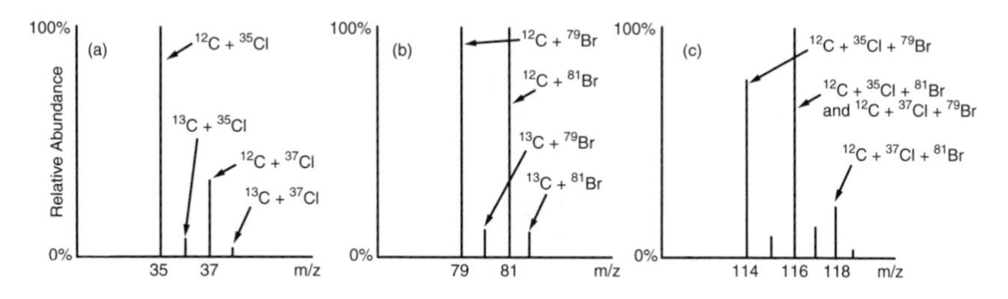

Figure 47.2
Partial mass spectra showing the isotope patterns in the molecular ion regions for ions containing carbon and (a) only one chlorine atom, (b) only one bromine atom, and (c) one chlorine and one bromine atom. The isotope patterns are quite different from each other. Note how the halogen isotope ratios appear very clearly as 3:1 for chlorine in (a), 1:1 for bromine in (b), and 3:4:1 for chlorine and bromine in (c). If the numbers of halogens were not known, the pattern could be used in a reverse sense to decide their number.

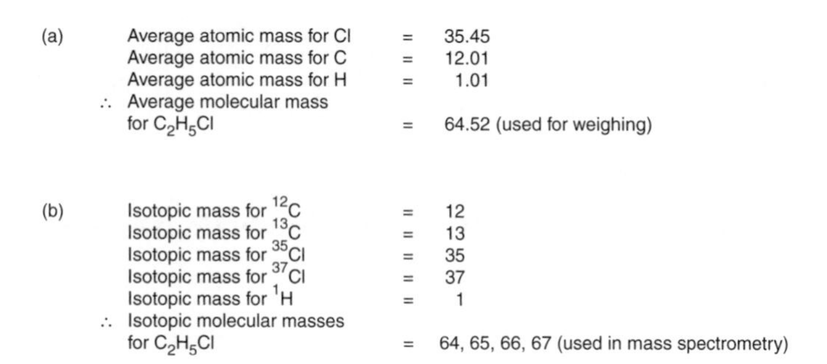

(a)	Average atomic mass for Cl	=	35.45
	Average atomic mass for C	=	12.01
	Average atomic mass for H	=	1.01
∴	Average molecular mass for C_2H_5Cl	=	64.52 (used for weighing)
(b)	Isotopic mass for ^{12}C	=	12
	Isotopic mass for ^{13}C	=	13
	Isotopic mass for ^{35}Cl	=	35
	Isotopic mass for ^{37}Cl	=	37
	Isotopic mass for ^{1}H	=	1
∴	Isotopic molecular masses for C_2H_5Cl	=	64, 65, 66, 67 (used in mass spectrometry)

Figure 47.3
An illustration of the use of average and isotopic mass. In (a), the average molecular mass for chloroethane is calculated, and in (b), the isotopic masses are calculated. An ordinary use of average mass in, for example, mass balances for chemical reactions would use the average molecular mass of 64.52. However, a mass spectrometer can distinguish the different isotopic masses easily, so it is necessary to use these isotopic masses. For example, the molecular ion region in a mass spectrum contains isotopic molecular ions, as shown in Figures 47.1 and 47.2. Only approximate isotopic and average masses are used for purposes of illustration.

in calculating mass and not average atomic masses. A simple example of this difference is illustrated in Figure 47.3.

For organometallic compounds, the situation becomes even more complicated because the presence of elements such as platinum, iron, and copper introduces more complex isotopic patterns. In a very general sense, for inorganic chemistry, as atomic number increases, the number of isotopes occurring naturally for any one element can increase considerably. An element of small atomic number, lithium, has only two natural isotopes, but tin has ten, xenon has nine, and mercury has seven isotopes. This general phenomenon should be approached with caution because, for example, yttrium of atomic mass 89 is monoisotopic, and iridium has just two natural isotopes at masses 191 and 193. Nevertheless, the occurrence and variation in patterns of multi-isotopic elements often make their mass spectrometric identification easy, as depicted for the cases of dimethylmercury and dimethylplatinum in Figure 47.4.

Accurate Determination of Isotope Ratios

Special instruments (isotope ratio mass spectrometers) are used to determine isotope ratios, when needed, to better than about 3%. Such special instruments are described in Chapters 6, 7, and 48. The methods of ionization and analysis for such precise measurements are not described here.

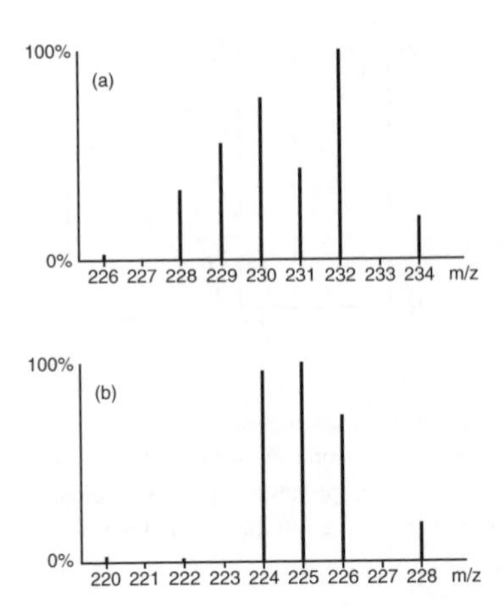

Figure 47.4
The isotope patterns for two simple organometallic compounds in the molecular ion region: (a) dimethylmercury and (b) dimethylplatinum. The seven isotopes of mercury show clearly and appear quite different from the six isotopes of platinum. Since there are only two carbon atoms, the contribution from ^{13}C is negligible.

Examples of the Use of Accurate Isotope Ratios

Archaeology

Dating of wooden objects has commonly been carried out by counting the number of tree rings (dendrochronology). As climate changes, the thicknesses of the rings change also because more wood is laid down in warmer, wetter weather than in colder, drier weather. By matching tree ring patterns, it has been possible to date wooden objects accurately for periods going back thousands of years. Although the rings are thicker when laid down in warmer weather and vice versa in cooler weather, it is not possible to gauge the actual average temperature that each ring represents from the thickness alone. However, isotopic analysis can guage temperature. The $^{12}C/^{13}C$ ratio for tree rings has been measured for recent times, for which accurate summer and winter temperatures are known. From these data, a graph can be constructed that relates average temperature to the ratio of $^{12}C/^{13}C$. Such measurements can reveal differences of only 0.2°C and can be used to form an accurate assessment of average annual temperatures for thousands of years (Figure 47.5).

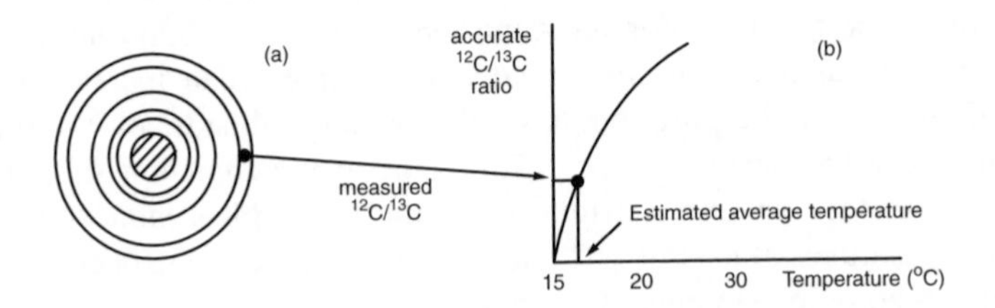

Figure 47.5
(a) A cross-section of a tree trunk reveals a number of rings for each year of growth. Comparison of the ring patterns for trees of different ages shows that the pattern is consistent for a given region. Thus, by looking at the patterns for trees or timbers from different ages, the patterns can be matched or overlapped to build a profile of tree ring thickness with age. Radioactive dating allows approximate estimation of the date when a tree was growing, but the tree ring method provides a more accurate assessment of dates for more recent times. (b) A graph showing the ratio $^{12}C/^{13}C$ for tree growth and average temperature. The graph can be used to derive an accurate assessment of average summer temperatures in the geographical region of interest. From dendrochronological tree ring dating and accurate $^{12}C/^{13}C$ isotope ratios, the average summer temperatures for a region can be gauged quite accurately.

Geology

In a similar vein, mean seawater temperatures can be estimated from the ratio of ^{16}O to ^{18}O in limestone. The latter rock is composed of calcium carbonate, laid down from shells of countless small sea creatures as they die and fall to the bottom of the ocean. The ratio of the oxygen isotopes locked up as carbon dioxide varies with the temperature of sea water. Any organisms building shells will fix the ratio in the calcium carbonate of their shells. As the limestone deposits form, the layers represent a chronological description of the mean sea temperature. To assess mean sea temperatures from thousands or millions of years ago, it is necessary only to measure accurately the $^{16}O/^{18}O$ ratio and use a precalibrated graph that relates temperatures to $^{16}O/^{18}O$ isotope ratios in sea water. A graph similar to that shown in Figure 47.5 is used.

Environmental Science

Roadside soils and vegetation contain high concentrations of lead, mostly derived from petroleum additives, but less widely known is the relatively high concentration of lead found in upland soils or soils in remote areas of Scotland. It is not clear that such contamination should be derived from petroleum additives. It has been recognized for many years that the isotopic composition of lead varies with its source, as in lead derived from natural nuclear fission reactions of uranium and thorium. The ratios of the four lead isotopes — ^{204}Pb, ^{206}Pb, ^{207}Pb, and ^{208}Pb — can be measured accurately using an isotopic ratio mass spectrometer that separates and measures simultaneously the four isotopic ion beams for the element. Comparison of lead from different sources has revealed that surface lead from highland areas remote from roads not only had an isotopic distribution different from that present in deeper soil samples, but it was also different from that obtained from lead samples taken from the sides of major roads, where contamination was expected from lead additives in petrol. Thus the lead found in the highland areas remote from roads could not have arisen from the petroleum additives. Rainwater over wide areas of the country was found to contain lead corresponding to an input from petroleum additives, and therefore it was not clear why the highland areas should have had a lead isotope composition different from those of petroleum additives.

Rare Earth Element Analysis

The separation of the rare earth elements lanthanum to lutetium provides a significant analytical challenge. Because they are difficult to separate by standard methods, ion-exchange resin systems have been developed to differentiate the elements. An initial ion-exchange separation is followed by isotopic analysis on each fraction using a ratio mass spectrometer. The separation process is time consuming. Standard mass spectrometric analyses of isotope content without any prior separation are not feasible because the rare earth elements have overlapping isotopes of equal mass (isobaric isotopes). For example, at mass 176, the ^{176}Yb, ^{176}Lu, and ^{176}Hf isotopes are isobaric. The problem has been partly resolved by a rapid ion-exchange separation of the rare earths into just two fractions, one containing the lower rare earths (LREE: La, Ce, Nd, Sm, Eu) and the other comprising the higher rare earth elements (HREE: Gd, Dy, Er, Yb). The two fractions can be examined separately using an isotope ratio mass spectrometer that has a number of ion collectors, which can simultaneously collect and measure the numbers of ions in the different ion beams. With a suitable choice of which isotopes to monitor, it is possible to measure the content of all of the rare earths in a short time. For example, although lanthanum and cerium have isobars at 138, lanthanum has an isotope at 139, which cerium does not have, and therefore this mass can be used to estimate lanthanum. Similarly, cerium has an isotope at 140 that is not present in lanthanum

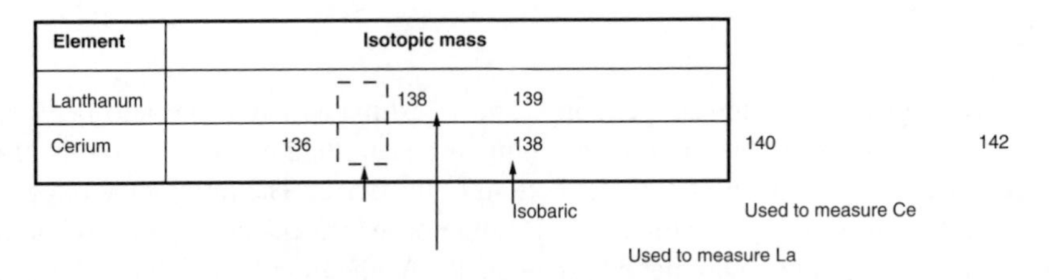

Figure 47.6
The masses of the naturally occurring isotopes for lanthanum and cerium are shown. For lanthanum, the isotope at 138 is only present in 0.09% natural abundance and is isobaric with ^{138}Ce. For this reason the isotope ^{139}La is used to measure the amount of lanthanum. Similarly, ^{136}Ce and ^{138}Ce are present in low abundance; ^{140}Ce is present in greatest abundance and is used to measure the amount of cerium. Another isotope of cerium, ^{142}C, although quite abundant, is isobaric with ^{142}Nd and is therefore not used for measurement.

(Figure 47.6). By choosing which isotope to measure, all of the rare earth elements can be analyzed accurately and quickly following their ion-exchange separation into just two fractions.

Conclusion

Elemental isotopic compositions (isotope ratios) can be used mass spectrometrically in a routine sense to monitor a substance for the presence of different kinds of elements, as with chlorine or platinum. It can also be used in a precise sense to examine tiny variations in these ratios, from which important deductions can be made in a wide variety of disciplines.

Chapter 48

Variations in Isotope Ratios

Background

The existence of isotopes of elements is readily observable by mass spectrometry (see Chapter 47). For elements that exist naturally or artificially in different isotopic forms, such as nitrogen ($^{15}N/^{14}N$), the ratio of the isotopes in any one sample can be measured by mass spectrometry. However, the absolute isotope ratios are not as important as are variations in their ratios since these reflect changes undergone by the sample in comparison with a norm. These variations are important in a diverse range of areas from geochronology to food safety. For various instrumental reasons, accurate assessment of isotope ratios cannot be made from routine mass spectra. Detailed measurement of isotope ratios requires special considerations, particularly in the construction of ion sources, inlet systems, and detectors. Instruments such as inductively coupled plasma mass spectrometers and thermal ionization mass spectrometers are widely used and have been described elsewhere in this book (see Chapters 6 and 7). Once accurate (precise) ratios have been measured, it is quickly apparent that the ratios are not constant. There are several important reasons for the inconstancy, ranging from the effects of natural radioactivity to detailed changes arising from chemical reactions.

One of the most significant sources of change in isotope ratios is caused by the small mass differences between isotopes and their effects on the physical properties of elements and compounds. For example, ordinary water (mostly $^{1}H_2{}^{16}O$) has a lower density, lower boiling point, and higher vapor pressure than does "heavy" water (mostly $^{2}H_2{}^{16}O$). Other major changes can occur through exchange processes. Such physical and kinetic differences lead to natural local fractionation of isotopes. Artificial fractionation (enrichment or depletion) of uranium isotopes is the basis for construction of atomic bombs, nuclear power reactors, and depleted uranium weapons.

The determination of small variations in isotope ratios is used in many areas of investigation, ranging from interplanetary research to the labeling of drugs in order to characterize their source and time of manufacture. Accordingly, suitable reliable instrumentation capable of measuring accurate isotope ratios is increasingly important. To make best use of these instruments it is equally important to understand how the ratios can change through natural effects. This chapter describes some of the more significant of these effects, whereby fractionation of isotopes occurs, together with criteria for making decisions about the significance of observed variations in isotope ratios.

Radioactive and Nonradioactive Isotopes

Few of the naturally occurring elements have significant amounts of radioactive isotopes, but there are many artificially produced radioactive species. Mass spectrometry can measure both radioactive and nonradioactive isotope ratios, but there are health and safety issues for the radioactive ones. However, modern isotope instruments are becoming so sensitive that only very small amounts of sample are needed. Where radioactive isotopes are a serious issue, the radioactive hazards can be minimized by using special inlet systems and ion pumps in place of rotary pumps for maintaining a vacuum. For example, mass spectrometry is now used in the analysis of $^{240}Pu/^{239}Pu$ ratios.

Standards for Isotope Ratios

Because variations in accurate isotope ratio measurements typically concern only a few parts per 1000 by mass and there are no universal absolute ratios, it is necessary to define some standards. For this purpose, samples of standard substances are produced and made available at two major centers: IAEA (International Atomic Energy Authority, U.K.) and NIST (National Institute for Standards and Technology, U.S.). Standards from other sources are also available. These primary standards can be used as such, or alternative standards can be employed if the primary ones are not available. However, any alternative standards need to be related accurately to the primary ones (see formulae below). For example, the material PDB (PeeDee belemnite), used particularly as a standard for the ratio of $^{13}C/^{12}C$ isotopes, is no longer readily available, and a new standard, VPDB, is used in its place. The composition of the isotopes in VPDB is known accurately in relation to PDB (the V in VPDB stands for "Vienna," the location of the meeting at which the change was approved). Similarly, SMOW (standard mean ocean water) is another recommended standard for oxygen ($^{18}O/^{16}O$) and hydrogen ($^{2}H/^{1}H$). SMOC (standard mean ocean chloride) and AIR (actually "atmospheric air") are used respectively as standards for chlorine ($^{37}Cl/^{35}Cl$) and nitrogen ($^{15}N/^{14}N$). The standard ratio of $^{34}S/^{32}S$ is obtained from CDT (canyon diabolo troilite) or VCDT (a preparation of silver sulfide). These standard ratios are listed in Table 48.1.

Basic Facets of Measuring Isotope Ratios

Before measurement it must be decided exactly which isotopes are to be compared. For oxygen, it is usually the ratio of ^{18}O to ^{16}O, and for hydrogen it is ^{2}H to ^{1}H. Such isotope ratios are measured by the mass spectrometer. For example, examination of a sample of a carbonaceous compound provides abundances of ions at two m/z values, one related to ^{13}C and one to ^{12}C (it could be $^{13}CO_2$ at m/z 45 and $^{12}CO_2$ at m/z 44). By convention, the heavier isotope is always compared with the lighter isotope. The ratio of isotopes is given the symbol R (Figure 48.1).

For example, if a carbonaceous sample (S) is examined mass spectrometrically, the ratio of abundances for the carbon isotopes ^{13}C, ^{12}C in the sample is R_S. This ratio by itself is of little significance and needs to be related to a reference standard of some sort. The same isotope ratio measured for a reference sample is then R_R. The reference ratio also serves to check the performance of the mass spectrometer. If two ratios are measured, it is natural to assess them against each other as, for example, the sample versus the reference material. This assessment is defined by another ratio, α (the fractionation factor; Figure 48.2).

Figure 48.2 shows that α compares the ratio of atom or ion abundances for two isotopes in each of two samples. If α is not equal to 1, then the isotopes in one sample must have a different ratio from those in the other. If isotopes behave chemically almost identically on a universal stage,

TABLE 48.1
Isotope Characteristics of Some Standard Substances

Reference Number/code	Material	δD_{VSMOW}	$\delta^{13}C_{VPDB}$	$\delta^{15}N_{AIR}$	$\delta^{18}O_{VSMOW}$	$\delta^{34}S_{CDT}$
IAEA-VSMOW	Vienna Standard Mean Ocean Water	0.00			0.00	
IAEA-SLAP	Standard Light Antarctic Precipitation	-428			-55.5	
IAEA-GISP (NIST-GISP)	Greenland Ice Sheet Precipitation (water)	-189.5 (-190)			-24.8	
IAEA-302A	Deuterium (^2H) labelled water	+508.4				
IAEA-302B	Deuterium (^2H) labelled water	+996				
IAEA-OH-1	Water	-3.9 D			-0.1 D	
IAEA-OH-2	Water	-30.8 D			-3.3 D	
IAEA-OH-3	Water	-61.3 D			-8.7 D	
IAEA-OH-4	Water	-109.4 D			-15.3 D	
IAEA-304A	O-18 labelled water				+251.7	
IAEA-304B	O-18 labelled water				+502.5	
IAEA-NBS30 (NIST-8538)	Biotite	-65.7 (-66)			+5.1	
IAEA-CH-7 (NIST-8540)	Polyethylene	-100.3 (-100)	-31.8			
IAEA-NBS22 (NIST-8539)	Oil	-118.5 (-120)	-29.7			
IAEA-NBS28 (NIST-8546)	Quartz sand				+9.58 (9.6)	
IAEA-NBS127	Barium sulphate				+9.3	
IAEA-NBS18 (NIST-8543)	Carbonatite		-5.01 (-5.0)		-23 VPDB (+7.2)	
IAEA-NBS19 (NIST-8544)	Carbonatite		+1.95 (defines VPDB scale)		-2.2 VPDB (+28.6)	
IAEA-CO-1	Marble		+2.48		-2.44 VPDB	
IAEA-CO-8	Calcite		-5.75		-22.7 VPDB	
IAEA-CO-9	Barium carbonate		-47.1		-15.3 VPDB	
IAEA-LSVEC (NIST-8545)	Lithium carbonate		-46.5 (-47)		-26.5 VPDB (+3)	
IAEA-305A	N-15 labelled ammonium sulphate			+39.8		
IAEA-305B	N-15 ammonium sulphate			+375.3		
IAEA-310A	N-15 labelled Urea			+47.2		
IAEA-310B	N-15 labelled Urea			+244.6		

D = Delta value
Data obtained from IAEA, Vienna (http://www.iaea.org/); bracketed data obtained from NIST, Gaithersburg, Md (http://ois.nist.gov/)
Data correct at 30th October 2000.

(*continued*)

TABLE 48.1 (CONTINUED)
Isotope Characteristics of Some Standard Substances

Reference Number/code	Material	δD_{VSMOW}	$\delta^{13}C_{VPDB}$	$\delta^{15}N_{AIR}$	$\delta^{18}O_{VSMOW}$	$\delta^{34}S_{CDT}$
IAEA-USGS32	Potassium sulphate			+180		
IAEA-USGS26 (NIST-8551)	Ammonium sulphate			+53.7 (+53.5)		
IAEA-N1	Ammonium sulphate			+0.4		
IAEA-N2	Ammonium sulphate			+20.3		
IAEA-311	Ammonium sulphate			+2.05 (Atom %)		
IAEA-USGS25	Ammonium sulphate			-30.4		
IAEA-NO-3	Potassium nitrate			+4.7		
IAEA-NSVEC	Air-N_2			-2.8		
IAEA-NBS127 (NIST-8557)	Barium sulphate					+20.3 (+20)
IAEA-NBS123 (NIST-8556)	Sphalerite					+17.1 (+17)
IAEA-S1	Silver sulphide					-0.3 (defines VCDT scale)
(NIST-8555)	Silver sulphide					(+21)
(NIST-8553)	Elemental sulphur					(+16)
IAEA-309A	C-13 labelled UI-glucose		+93.9			
IAEA-309B	C-13 labelled UI-glucose		+535.3			
IAEA-303A	C-13 labelled sodium bicarbonate		+93.3			
IAEA-303B	C-13 labelled sodium bicarbonate		+466			
IAEA-C2	Travertine		-8.25 PDB			
IAEA-C3	Cellulose	-	24.91 PDB			
IAEA-C4	Wood	-	23.96 PDB			
IAEA-C5	Wood	-	25.49 PDB			
IAEA-CH-6 (NIST-8542)	Sucrose		-10.4 (-10.5)			
IAEA-C6	Sucrose	-	10.8 PDB			
IAEA-C7	Oxalic acid		-14.48			
IAEA-C8	Oxalic acid		-18.3			
IAEA-USGS24 (NIST-8541)	Graphite		-16.1 (-16)			

PDB: $^{18}O/^{16}O = 0.0020672$

$\delta^{18}O_{PDB/SMOW} = 30.92$

For any one of a series of measurements, the isotope ratio (R_{AB}) is given by the formula,

$$R_{AB} = \frac{\text{the abundance of atoms of isotope A in the sample}}{\text{the abundance of atoms of isotope B in the sample}}$$

where the mass of isotope A is greater than the mass of isotope B.

For example, in the standard substance PDB (Table 1),

$$R_{PDB} = \frac{\text{abundance of the } ^{13}\text{C isotope}}{\text{abundance of the } ^{12}\text{C isotope}} = 0.112372$$

and, the alternative standard, VPDB (Table 1) gives,

$$R_{VPDB} = \frac{\text{abundance of the } ^{13}\text{C isotope}}{\text{abundance of the } ^{12}\text{C isotope}} = 0.112591$$

Thus, it can be said that VPDB is slightly enriched in ^{13}C compared with PDB or that VPDB is slightly depleted in ^{12}C compared with PDB.

Figure 48.1
The measured ratio of abundances for two isotopes (A,B) is defined and illustrated for the two standard substances PDB and VPDB.

In figure 1, the enrichment or depletion of $^{13}C/^{12}C$ was clear from the measured ratios. To quantify this change, a fractionation factor (α) is used where,

$$\alpha = \frac{R_{VPDB}}{R_{PDB}} = 1.001949$$

If a value R_S represents a $^{13}C/^{12}C$ ratio in a sample(s) as measured by mass spectrometry, then this ratio can be compared with either PDB or VPDB. Thus,

$$\alpha_{S,PDB} = \frac{R_S}{R_{PDB}} \qquad \text{or,} \quad \alpha_{S,VPDB} = \frac{R_S}{R_{VPDB}}$$

For example, from a series of measurements, the mean $^{13}C/^{12}C$ isotope ratio in a sample(s) of a carbonate was measured as $R_S = 0.112386$ so that,

$$\alpha_{S,PDB} = \frac{0.112386}{0.112372} = 1.000125$$

or $$\alpha_{S,VPDB} = \frac{0.112386}{0.112591} = 0.998179$$

These results illustrate a common finding: the α values are very often close together and near to the value "1".

Figure 48.2
Differences in measured isotope ratios (R) can be compared for two substances by using a new ratio α. The two substances may constitute part of a series of measurements on a sample, which are to be internally compared, or they may constitute results from a series that is compared with an external value from a standard substance.

any such variations in α must be due to local perturbations of the isotope ratios. These local perturbations provide valuable information in a wide range of applications. Among other sources of variation, local perturbations may arise chemically over long geological times, chemically in short reaction times, or radioactively over both long and short times. If two substances are compared and the isotope ratios are different, it is as if the isotopes have been fractionated (such as occurs during distillation of a mixture of two substances); viz., the original mix of isotopes has been

As shown in figure 2 for two substances A,B, $\alpha_{AB} = \frac{R_A}{R_B}$ and α_{AB} is usually = 1.

Therefore, to emphasise the small change from "1", this is subtracted from α and mutiplied by 1000.

$$\therefore \alpha_{AB} - 1 = \frac{RA}{RB} - 1, \quad \text{and } (\alpha_{AB} - 1) \times 1000 = \left(\frac{R_A}{R_B} - 1\right) \times 1000$$

These last values are termed per mil ($^0/_{00}$) and are given the symbol δ. They may be compared with the common percentage (%), which is frequently used to emphasise differences between values.

$$\delta_{AB} = \left(\frac{R_A}{R_B} - 1\right) \times 1000 = \left(\alpha_{AB} - 1\right) \times 1000 \ \%.$$

Note that $\delta_{AB} \neq \delta_{BA}$ since,

$$\delta_{BA} = \left(\frac{R_B}{R_A} - 1\right) \times 1000$$

from which,

$$\frac{R_B}{R_A} = \frac{\delta_{BA}}{1000} + 1 \quad \text{and} \quad \frac{R_A}{R_B} = \frac{1000}{\alpha_{BA} + 1000}$$

but by definition

$$\frac{R_A}{R_B} = \frac{\delta_{AB}}{1000} + 1$$

$$\therefore \frac{\delta_{AB}}{1000} + 1 = \frac{1000}{\alpha_{BA} + 1000} \qquad \therefore \delta_{AB} + 1000 = \frac{10^6}{\delta_{BA} + 1000}$$

$$\therefore \delta_{AB} = \frac{10^6}{\delta_{BA} + 1000} - 1000 = \frac{1000}{10^{-3}.\delta_{BA} + 1} - 1000$$

$$\therefore \delta_{AB} = \left(\frac{1}{10^{-3}.\delta_{BA} + 1} - 1\right) \times 1000$$

This expression is frequently used. Alternatively,

$$\left(\frac{1}{10^{-3}.\delta_{BA} + 1} - 1\right) \times 1000 = \left(\frac{1 - 10^{-3}\delta_{BA} - 1}{10^{-3}\alpha_{BA} + 1}\right) \times 1000 = \frac{\delta_{BA}}{10^{-3}\delta_{BA} + 1}$$

$$\therefore \delta_{AB} = \left(\frac{-1000\,\delta_{BA}}{1000 + \delta_{BA}}\right) \times 1000$$

This alternative expression can be used and it also shows that since δ is likely to be small (« 1000) then,

$$\frac{-\delta_{BA} \times 1000}{\delta_{BA} + 1000} \cong -\delta_{AB}$$

viz., $\delta_{AB} \cong -\delta_{BA}$

Figure 48.3
To emphasize small variations in fractionation factors (α), a new term (δ) is introduced, which accentuates small differences. Since the latter are very small, they are usually multiplied by 1000 (mil).

enriched or depleted in one or more isotopes. Thus, α is a fractionation factor that refers to variations found in a series of samples or between various samples and a standard reference substance.

Variations in isotope ratios are not large so the fractionation factor α does not differ much from 1. However, it is the variation itself that is important, and to accentuate this effect clearly it is usual to subtract 1 from α. Then, for convenience, it is usual to multiply the variation by 1000, and, finally, the resulting number is given the symbol δ (delta; Figure 48.3). For a sample isotope ratio R_S measured against a standard reference ratio R_R, delta is regarded as referring to the sample. For a sample (A) and a reference substance (R), a series of equations (Figure 48.3) can be applied. Similar quantitative comparisons are made in everyday life when ratios are discussed. Ratios are often written as fractions (e.g., 3/5) or decimals (0.6), and it becomes convenient to multiply them by 100, when they become percentages (60%). However, when differences in ratios need to be emphasized, a slightly different formula is used; the percentage change is used. The same type of formulation is used for isotope measurements, the δ value being a difference in ratios relative to

For a sample substance (S) for which an isotope ratio (R_S) has been measured and then compared with the same isotope ratio (R_A) for a standard substance (A),

$$\delta_{SA} = \left(\frac{R_S}{R_A} - 1\right) \times 1000 \qquad (1)$$

If the isotope ratios for the one standard (A) and another standard (B) are known then,

$$\delta_{AB} = \left(\frac{R_A}{R_B} - 1\right) \times 1000 \qquad (2)$$

From equations (1,2) by simple transposition,

$$\frac{R_S}{R_A} = \frac{\delta_{SA}}{1000} + 1 \quad \text{and} \quad \frac{R_A}{R_B} = \frac{\delta_{AB}}{1000} + 1$$

$$\text{but} \quad \frac{R_S}{R_B} = \frac{R_S}{R_A} \cdot \frac{R_A}{R_B} = \left(\frac{\delta_{SA}}{1000} + 1\right)\left(\frac{\delta_{AB}}{1000} + 1\right)$$

$$\therefore \quad \delta_{SB} = \left(\frac{R_S}{R_B} - 1\right) \times 1000 = \left[\left(\frac{\delta_{SA}}{1000} + 1\right)\left(\frac{\delta_{AB}}{1000} + 1\right) - 1\right] \times 1000 \qquad (3)$$

On expansion, equation (3) becomes,

$$\delta_{SB} = \delta_{SA} + \delta_{AB} + \delta_{SA} \cdot \delta_{AB} \cdot 10^{-3} \qquad (4)$$

Either equation (3) or (4) may be used.

For example, from figure 1.

$$\alpha_{VPDB,PDB} = \frac{R_{VPDB}}{R_{PDB}} = 1.001949$$

and, $\delta_{VPDB,PDB} = (\alpha_{VPDB,PDB} - 1) \times 1000 = 0.001949 \times 1000 = +\underline{1.949}$

Suppose a $^{13}C/^{12}C$ isotope ratio has been measured for a sample(s) using VPDB as a standard and a value of $\delta_{S,VPDB}$ of + 2.003 has been obtained. Then,

$$\begin{aligned}\delta_{S,PDB} &= \delta_{S,VPDB} + \delta_{VPDB,PDB} + \delta_{S,VPDB} \cdot \delta_{VPDB,PDB} \times 10^{-3} \\ &= 2.003 + 1.949 + 0.004 \\ &= +\underline{\underline{3.956}}\end{aligned}$$

Figure 48.4
The δ values for a sample measured against a standard substance can be changed into δ values against a second standard substance if the δ value for the two standards is known.

the absolute standard, which is emphasized (or magnified) through multiplication by 1000 (per mil) rather than by 100 (percent). Differences in two isotope ratios for the same element are very small, and multiplication by 1000 produces numbers closer to or greater than 1. It is more readily understandable and convenient for humans to compare δ values such as 5 or 30 "per mil" rather than to compare the same values written as 0.005 and 0.03.

If a sample substance (S) has been compared against one standard compound (A) to give δ_{SA} but comparison with another standard (B) is required (δ_{SB}), then this change can be effected easily if the relation between the two standards δ_{AB} is known (Figure 48.4). For delta values, the order in which the suffixes appear is important. For a sample S measured against reference substance A, delta is written as δ_{SA}. This is not the same as δ_{AS}, as can be seen in Figure 48.4.

Where it is necessary to compare a sample S against a second standard B, other simple equations can be used (Figure 48.4). If one analyst has used standard A for sample comparison but wants to compare the sample against another standard B, it is only necessary to know the relative delta values of the two standards (δ_{AB}) and to apply the equations shown in Figure 48.4.

Relationship between α and δ

For two samples A and B measured against the same reference standard X, the relationship between the α and δ values can be approximated by the expression shown in Equation 48.1 (see also Figure 48.5).

For two substances (P,Q) each measured against a standard X, the values δ_{PX}, δ_{QX} are known from α_{PX}, α_{QX},

$$\delta_{PX} = (\alpha_{PX} - 1) \times 1000 \quad \text{and} \quad \delta_{QX} = (\alpha_{QX} - 1) \times 1000$$

$$\therefore \delta_{PX} - \delta_{QX} = \Delta_{PQ} = 1000\,(\alpha_{PX} - \alpha_{QX})$$

$$\text{but,} \quad \ln(\alpha_{PQ}) = \ln(\alpha_{PX}/\alpha_{QX}) = \ln(\alpha_{PX}) - \ln(\alpha_{QX})$$

$$\approx \alpha_{PX} - \alpha_{QX}$$

$$\therefore \Delta_{PQ} \cong 1000.\ln \alpha_{PQ}$$

For example, from figure 4 for the two standard substances VPDB, PDB,

$$\delta_{VPDB} - \delta_{PDB} = 1.949 = 1.95$$

but, $\alpha_{VPDB,PDB} = 1.001949$

and $\ln \alpha_{VPD,PDB} = 1.001947$

$$\therefore 1000 \ln \alpha_{VPD,PDB} = 1.947 = 1.95$$

$$\therefore \Delta_{VPDB,\,PDB} \cong 1000 \ln \alpha_{VPDB,PDB}$$

Figure 48.5
The difference in δ values for two substances (P,Q) measured against a standard substance is approximately equal to 1000 times the natural logarithm of their fractionation factor (α_{PQ}).

Instead of the fractionation factor (α_{PQ}) for two substances (P,Q), a slightly different fractional abundance (f_{PQ}) may be defined:

$$R_{PQ} = \frac{\text{abundance of isotope P}}{\text{abundance of isotope Q}}, \text{ where mass P> mass Q}$$

$$f_{PQ} = \frac{\text{abundance of isotope P}}{\text{abundance of isotope P + abundance of isotope Q}}$$

$$= \frac{R_{PQ}}{R_{PQ} + 1}$$

If this is multiplied by 100, it becomes "atom%", a percentage that relates the proportion of heavy isotope to the sum of the heavy and light isotopes.

$$\text{Thus, Atom\%} = \left(\frac{R}{R+1}\right) \times 100.$$

The "Atom % excess" is defined as the difference between the atom% for the sample and the atom % for the standard. For example, if VPDB is compared with PDB then, for $^{13}C/^{12}C$ ratios,

$R_{VPDB} = 0.112591$ and $R_{PDB} = 0.112372$ and,

$$\text{Atom \% } (^{13}\text{C isotope}) \text{ for VPDB} = \frac{0.112591 \times 100}{1.112591} = 10.12\%$$

$$\text{Atom \% } (^{13}\text{C isotope}) \text{ for PDB} = \frac{0.112372 \times 100}{1.112372} = 10.10\%$$

$\therefore$ Atom % excess VPDB/PDB = 10.12 - 10.10 = <u>0.02%</u>

Figure 48.6
Two further expressions that are used for the excess of one isotope over another are "atom%" and "atom% excess."

$$\delta_{AX} - \delta_{BX} \approx 1000 \cdot \ln \alpha_{AB} \tag{48.1}$$

The difference between using this approximation and the true value is only minor and is often ignored. Note that both α and δ are dimensionless units.

Other Units of Measurement

Two further expressions are used in discussions on isotope ratios. These are the "atom%" and the "atom% excess," which are defined in Figure 48.6 and are related to abundance ratios R. It has been recommended that these definitions and some similar ones should be used routinely so as to conform with the system of international units (SI). While these proposals will almost certainly be accepted by mass spectrometrists, their adoption will still leave important data in the present format. Therefore, in this chapter, the current widely used methods for comparison of isotope ratios are fully described. The recommended SI-compatible units such as atom% excess are introduced where necessary.

Precision and Accuracy of Measurement

For variations in isotope ratios, considerations of precision and accuracy become very important because the measured ion abundance ratios normally vary by only very small amounts. For any one sample, its isotope ratio is usually measured ten or more times to gain an estimate of the accuracy of measurement or the precision in the method being used. Precision is a criterion that reveals how consistent a series of measurements is on one sample. By itself, it does not describe how accurate the measured result is with respect to the true result. For example, for any one sample, a series of determinations of any property that gave values of 52, 61, 59, 70, 34, and 62 could not be considered to be very precise since the results vary by ±12% (see Figure 48.7). In contrast, a series of measurements that gave values of 65, 65, 66, 65, 64, 65, and 66 would obviously be more precise (±0.69%; see Figure 48.8). However, this last level of precision does not describe the accuracy of the measurements. If, in this last precise series of Figure 48.8, the true value should have been 83, then clearly, the measurements must have been highly precise but they were all wrong! To be significant, the determination of precision needs to be done by measurements several times for any one sample. Often, each sample is measured 10 to 20 times. Precision tells the analyst how well the measurements can be reproduced, but there needs also to be some assessment of accuracy as well. The accuracy of an analytical method is best assessed by carrying out a series of measurements on one substance for which the true result is known. Generally, this substance would be one of the standards used in isotope ratio work (e.g., VPDB or SMOW). These factors are considered in greater detail in Figure 48.9.

Omission of Spurious Results

Usually, 10 to 20 measurements are made of the isotope ratio for one substance. Sometimes, one or more of these measurements appears to be sufficiently different from the mean value that the question arises as to whether or not it should be included in the set at all. Several statistical criteria are available for reaching an objective assessment of the reliability of the apparently rogue result (Figure 48.10). Such odd results are often called outliers, and ignoring them gives a more precise mean value (lower standard deviation). It is not advisable to remove such data more than once in any one set of measurements.

In this example, large numbers are used for the isotope ratios so as to make the procedure easier to follow.

Suppose for one sample that *n* measurements are made of an isotope (R); the measured ratios are $R_1, R_2 \ldots\ldots R_n$. A simple mean value ($\bar{R}$) is given by $R_1\ R_2\ R_3$ etc.

$$\bar{R} = \frac{(R_1 + R_2 + R_i \ldots\ldots R_n)}{n}$$

Statistically, the standard deviation (SD) from the mean is given by the formula,

$$SD = \left(\Sigma\,(\bar{R} - R_i)^2/(n-1)\right)^{\frac{1}{2}}$$

For the example of 6 determinations given in the main text,

Measurement	Value	$(\bar{R} - R_i)$	$(\bar{R} - R_i)^2$
R_1	52	+ 4.3	18.5
R_2	61	- 4.7	22.09
R_3	59	- 2.7	7.29
R_4	70	-13.7	187.69
R_5	34	+ 22.3	497.29
R_6	62	- 5.7	32.49
$\bar{R}$	338/6		Σ = 765.35
	$\bar{R}$ = 56.3		

$$\therefore\ SD = \sqrt{\frac{765.35}{5}} = 12.4$$

Therefore, the mean ratio = 56.3 ± 12.4

The precision of measurement does not appear to be very high. Confidence levels in the precision may be made by use of "Student *t*" Tables.

$$\text{Confidence} = \bar{R} \pm (t \times SD)/\sqrt{n}$$

From standard Student tables, the value for $t/\sqrt{n}$ = 1.049 at the 95% confidence level. Thus, mean value = 56.3 ± 13.0 (95% confidence level) and one could be confident that 95% of measured values would fall in the range 69.3 - 43.3. This is a large range and is not very precise.

Statistically, a similar indication of precision could be achieved by utilising the 95% probability level if the results fell on a "Gaussian" curve, *viz.*, the confidence would lie within two standard deviations of the mean. $\bar{R} \pm 2 \times SD$ = 56.3 ± 24.8

Figure 48.7
From a series of isotope ratio measurements, the precision of measurement can be assessed statistically, as shown here. Precision reveals the reproducibility of the measurement method, but it does not provide information on the accuracy of the measurement (see also Figures 48.8 and 48.9).

Causes of Variation in Isotope Ratios

Chemical (Kinetic) Effects

Masses of isotopes are important in bond strengths because they affect vibration force constants. Since reactions proceed through the breaking and formation of bonds, it is not surprising that isotopes affect chemical reaction rates through their effect on the force constants. However, the difference in force constants for two isotopic atoms forming similar bonds is very small; therefore the effect of isotopes on rates of reaction is also very small. However, by use of accurate isotope ratios, it is possible to investigate reaction mechanisms through the differences in the effects of isotopes. Where the rate of reaction depends primarily on one bond-forming or bond-breaking step, isotope effects can be used to reveal this step. The effect is frequently made easier to discern by artificially altering the isotope ratio, particularly when individual pure isotopes are available. For example, ^{2}H and ^{13}C isotopes can be obtained in 99–100% purity and are frequently used to

In a second series of measurements, the ratios (R_i) are listed below (main text).

Measurement	Value	$(\bar{R} - R_i)$	$(\bar{R} - R_i)^2$
R_1	65	+ 0.1	0.01
R_2	65	+ 0.1	0.01
R_3	66	- 0.9	0.81
R_4	65	+ 0.1	0.01
R_5	64	+ 1.1	1.21
R_6	65	+ 0.1	0.01
$\bar{R}$	66	- 0.9	0.81
	456/7		$\Sigma = 2.87$
	$\bar{R} = 65.1$		

$$\therefore \text{SD} = \sqrt{\frac{2.87}{6}} = 0.69$$

Therefore, the mean ratio = 65.1 ± 0.69

The precision of measurement is clearly much better than in the series given in Figure 7 since the variation from the mean is now less than 1.

At the 95% confidence level, $t/\sqrt{n} = 0.923$

$\therefore$ Confidence in the results $= \bar{R} \pm (t \times \text{SD})/\sqrt{n} = 65.1 \pm 0.65$

Therefore the results of this series of measurements show that the analyet could be confident that 95% of the results lie within ± 0.65 of the mean (within 1% of the mean).

Figure 48.8
For a second series of measurements, the precision is much better than for the series shown in Figure 48.7. Neither Figure 48.8 nor Figure 48.7 reveal any information on the accuracy of measurement.

Suppose that a true value for the ratio measured by the series in Figure 7,8 is 64.5. In the first series (Figure 7), the mean value was found to be 56.3 ± 12.4 and this certainly encompasses the true result.

The degree of accuracy can be assessed from the error in measurement. If R_a is the true ratio (64.5) and $\bar{R}$ is the mean, then

$$\% \text{ error} = \frac{\bar{R} - R_a}{R_a} \times 100$$

Thus, for the first set of measurements (Figure 7), $\bar{R} = 56.3$

$$\text{and the error} = \frac{56.3 - 64.5}{64.5} \times 100 = -12.7\%$$

The accuracy is 100 + error = 100 + 12.7 = 112.7%

This first series is not precise and also not very accurate. For the second series (Figure 8) with a mean of 65.1, the error is given by,

$$\text{error} = \frac{65.1 - 64.5}{64.5} \times 100 = +0.9\ \%$$

The accuracy is 100 - 0.9 = 99.1 %.

This second series is precise, very accurate but slightly under-estimates the true result. If this error was found to be consistent throughout several series of measurements, there would be good reason to "correct" all subsequent determinations by a factor of 100.9/100 = 1.009, *viz.*, the method involves systematic errors.

In contrast, the method of Figure 7 is not only inaccurate and imprecise but it also over-estimates the true value.

Figure 48.9
An assessment of the accuracy of any method of measurement requires comparison with a known true value. If the true value is unknown, then results from several different methods of analysis need to be assessed so as to choose the best one.

assess reaction mechanisms (Figure 48.11). The effect of isotopes on force constants is most readily observed in infrared spectra, which reveal bond vibration and rotation frequencies. Thus, a typical $^{12}C = O$ stretching vibration typically occurs near 1700 cm^{-1},while the $^{13}C = O$ is about 20–30 cm^{-1} less.

In any series of measurements, one or more results may appear to be very different from all the others. Such spurious results occur in all methods of analysis and there is a tendency to simply ignore the seemingly odd results. However a simple inspection of all the results is not sufficiently rigorous for making a decision on when to reject or retain any one result.

There are several statistical tests for reaching such a decision, the most popular probably being the "Chauvenet" criterion. Application of this criterion here uses the results shown in figure 7.

In the results given in figure 7, there are 6 measurements giving a mean ($\bar{R}$) of 56.3 and a standard deviation (SD) of 12.4. One of the measurements (number five; R_5 = 34) is suspiciously low. A factor p may be calculated:

$$p = (R_5 - \bar{R}) / SD \quad \text{(for 6 measurements)}$$
$$\therefore\ p = 1.80$$

The Chauvenet tables relate the number of measurements to p values and assess the probability that the unusual value may be real or may be ignored. In the present example, the tables show that p = 1.73 for 6 measurements. Since the calculated value (1.80) is greater than 1.73, the result may be rejected.

Now there are 5 measurements giving a new $\bar{R}$ = 60.8 and a new SD = 6.4 and the overall spread of results is 60.8 ± 6.4, clearly an improvement on 56.3 ± 12.4. The precision is better and the error from the true value (64.5) is -5.7%, giving a accuracy of 105.7%, which is a significant improvement on the original accuracy of 112.7%.

This process of rejecting values should not to be repeated on any one set of measurements, unless the number of measurements is large.

Figure 48.10
Chauvenet "t" tables can be used to decide whether one measurement in a series of measurements is a true "outlier" or can be rejected as statistically not significant.

Hydrolysis of the thiol ester **1** gives the products **2**, **3**.

$$\underset{\mathbf{1}}{CF_3COSC_2H_5} \xrightarrow{H_2O} \underset{\mathbf{2}}{CF_3CO_2H} + \underset{\mathbf{3}}{HSC_2H_5}$$

By use of ^{18}O enriched H_2O, it was found that, at pH 0-2, there was ^{18}O exchange with the starting material **1**. Therefore, it could be proposed that an intermediate **4** is first formed:

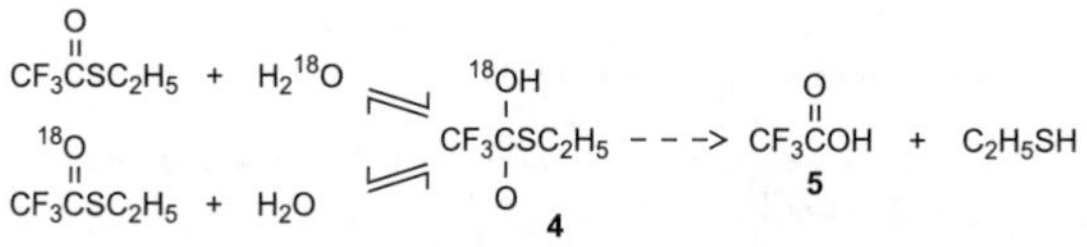

$$CF_3C(=O)SC_2H_5 + H_2{}^{18}O \rightleftharpoons \underset{\mathbf{4}}{CF_3C(^{18}OH)(OH)SC_2H_5} \dashrightarrow \underset{\mathbf{5}}{CF_3C(=O)OH} + C_2H_5SH$$
$$CF_3C(={}^{18}O)SC_2H_5 + H_2O \rightleftharpoons \mathbf{4}$$

From the intermediate **4**, loss of water simply drives the reaction back to starting material, but the water molecule that is eliminated may be $H_2{}^{16}O$ or $H_2{}^{18}O$. Therefore, there is a build-up of ^{18}O in the starting material and in the product acid **5**. This sort of exchange process was found to be common in many similar systems.

Figure 48.11
Use of the ^{18}O isotope to investigate a reaction mechanism.

These chemical effects become important in medicine because living systems operate mostly through the reactions of enzymes, which catalyze all sorts of metabolic reactions but are very sensitive to small changes in their environment. Such sensitivity can lead to preferential absorption of some deleterious isotopes in place of the more normal, beneficial ones. One example in metabolic systems can be found in the incorporation of a radioactive strontium isotope in place of calcium.

Natural radioactive processes in themselves give rise to changes of one element into another. Emission of an alpha particle reduces the atomic number of an element by two units, and emission of a beta particle increases the atomic number by one unit. Thus, for isotopes of elements near

Lead occurs naturally as a mixture of four non-radioactive isotopes, ^{204}Pb, ^{206}Pb, ^{207}Pb and ^{208}Pb, as well as the radioactive isotopes ^{202}Pb and ^{205}Pb. All but ^{204}Pb arise by radioactive decay of uranium and thorium. Such decay products are known as radiogenic isotopes.

For example, after successively emitting eight helium nuclei and six beta particles from ^{238}U the end result is the non-radioactive ^{206}Pb isotope.

Similarly, ^{235}U emits seven helium nuclei and 4 beta particles during its transition through several other elements, until ^{207}Pb is reached and ^{232}Th starts the successive elimination of six helium nuclei and 4 beta particles, which leads to ^{208}Pb.

Thus, the ratios of lead isotopes 204,206,207 and 208 can vary markedly depending on the source of the lead. One use of these ratios lies in determination of the ages of rocks from the abundances of the various isotopes and the half-lives of their precursor radioactive isotopes.

Figure 48.12
Ratios of lead isotopes depend on the source of the lead. They vary because lead is an end product of radioactive decay from elements of greater atomic number.

uranium or thorium in the periodic table (e.g., lead), their isotope ratios are affected by the presence or absence of the radioactive sources (Figure 48.12).

Chemical (Physical) Effects

The small differences in physical properties of substances containing elements with isotopes are manifested through measurement of isotope ratios. When water evaporates, the vapor is richer in its lighter isotopes ($^{1}H_2{}^{16}O$) than the heavier one ($^{1}H_2{}^{18}O$). Such differences in vapor pressures vary with temperature and have been used, for example, to estimate sea temperatures of 10,000 years ago (see Chapter 47).

These effects of differential vapor pressures on isotope ratios are important for gases and liquids at near-ambient temperatures. As temperature rises, the differences for volatile materials become less and less. However, diffusion processes are also important, and these increase in importance as temperature rises, particularly in rocks and similar natural materials. Minerals can exchange oxygen with the atmosphere, or rocks can affect each other by diffusion of ions from one type into another and vice versa. Such changes can be used to interpret the temperatures to which rocks have been subjected during or after their formation.

Instruments Used to Measure Accurate Isotope Ratios

Analyzers

Almost any type of analyzer could be used to separate isotopes, so their ratios of abundances can be measured. In practice, the type of analyzer employed will depend on the resolution needed to differentiate among a range of isotopes. When the isotopes are locked into multielement ions, it becomes difficult to separate all of the possible isotopes. For example, an ion of composition $C_6H_{12}O_2$ will actually consist of many compositions if all of the isotopes (^{12}C, ^{13}C, ^{1}H, ^{2}H, ^{16}O, ^{17}O, and ^{18}O) are considered. To resolve all of these isotopic compositions before measurement of their abundances is difficult. For low-molecular-mass ions (H_2O, CO_2) or for atomic ions (Ca, Cl), the problems are not so severe. Therefore, most accurate isotope ratio measurements are made on low-molecular-mass species, and resolution of these even with simple analyzers is not difficult. The most widely used analyzers are based on magnets, quadrupoles, ion traps, and time-of-flight instruments.

Ionization Methods

Almost any kind of ion source could be used, but, again, in practice only a few types are used routinely and are often associated with the method used for sample introduction. Thus, a plasma torch is used most frequently for materials that can be vaporized (see Chapters 14–17 and 19). Chapter 7, "Thermal Ionization," should be consulted for another popular method in accurate isotope ratio measurement.

Sample Preparation

The above-mentioned chapters all contain references to methods of sample preparation. In general, the sample will govern what kind of preparation is needed. For very intractable and complex substances, such as bone and ceramics, it is convenient to vaporize a sample from the surface by using a laser. To some extent, the laser can be used to provide a depth profile of the sample too. Other substances can be broken down through treatment with acids or alkalis or other chemical means to produce solutions of the material that can then be analyzed (see Chapter 19), with the ions being separated through ion-exchange chromatography before isotopic measurements take place. This last is often used to remove the effects of isobaric interferences between elements.

Examples of Isotope Ratio Measurements

Isotopic Dilution Analysis

Isotopic dilution analysis is widely used to determine the amounts of trace elements in a wide range of samples. The technique involves the addition to any sample of a known quantity (a spike) of an isotope of the element to be analyzed. By measuring isotope ratios in the sample before and after addition of the spike, the amount of the trace element can be determined with high accuracy. The method is described more fully in Figure 48.13.

Although isotope-dilution analysis can be very accurate, a number of precautions need to be taken. Some of these are obvious ones that any analytical procedure demands. For example, analyte preparation for both spiked and unspiked sample must be as nearly identical as possible; the spike also must be intimately mixed with the sample before analysis so there is no differential effect on the subsequent isotope ration measurements. The last requirement sometimes requires special chemical treatment to ensure that the spike element and the sample element are in the same chemical state before analysis. However, once procedures have been set in place, the highly sensitive isotope-dilution analysis gives excellent precision and accuracy for the estimation of several elements at the same time or just one element.

A major example of isotope-dilution analysis lies in the procedure itself, which does not require any quantitative isolation of the elements being investigated. The relation between the abundance of the element under investigation and the spike is such that, once the spike has been intimately mixed with the sample, any losses of sample have no effect on the result (Figure 48.14).

It is not necessary that there be two isotopes in both the sample and the spike. One isotope in the sample needs to be measured, but the spike can have one isotope of the same element that has been produced artificially. The latter is often a long-lived radioisotope. For example, ^{234}U, ^{235}U, and ^{238}U are radioactive and all occur naturally. The radioactive isotope ^{233}U does not occur naturally but is made artificially by irradiation of ^{232}Th with neutrons. Since it is commercially available, this last isotope is often used as a spike for isotope-dilution analysis of natural uranium materials by comparison with the most abundant isotope (^{238}U).

For the purposes of this illustration, suppose that the level of of a trace element (X) is to be determined in a sample and that the element has two suitable isotopes (A,B).

Then,

Number of atoms of isotope A in the sample = N'. Ab'(A)

Number of atoms of isotope B in the sample = N'. Ab'(B)

Where N' is the total number of atoms of element X in the sample and Ab'(A), Ab'(B) are the percentage abundances of the two isotopes.

As shown in diagram (a), the measured ratio of isotopes, $R' = \frac{N'.\,Ab'(A)}{N'.\,Ab'(B)} = \frac{Ab'(A)}{Ab'(B)}$

From this measured ratio, the proportional abundances are

$$Ab'(A) = \frac{R'}{R'+1} \quad \text{and} \quad Ab'(B) = \frac{1}{R'+1}$$

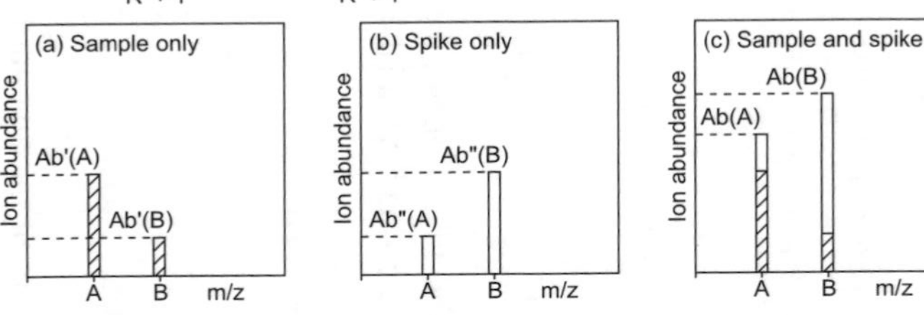

As shown in diagram (b), for N" atoms of spike, the measured ratio of isotopes in the spike, $R'' = \frac{N''.Ab''(A)}{N''.Ab''(B)} = \frac{Ab''(A)}{Ab''(B)}$

From this measured ratio, the proportional abundances are :

$$Ab''(A) = \frac{R''}{R''+1} \quad \text{and} \quad Ab''(B) = \frac{1}{R''+1}$$

After a known weight of spike has been mixed with the sample (C),

the measured ratio of isotopes in the mixture, $R_x = \frac{N'.Ab'(A) + N''.Ab''(A)}{N'.Ab'(B) + N''.Ab''(B)}$

From this last equation, by rearrangement,

$$N' = N''\left\{\frac{Ab''(A) - R_x Ab''(B)}{R_x.Ab'(B) - Ab'(A)}\right\} \qquad (1)$$

Number of atoms of A,B in the sample = N'.

Number of moles of A,B in the sample = N'/Av, where Av = Avagadro's number.

Mass of A,B in the sample (w') = (N'/Av)∗m'_{AB}, where m'_{AB} = the weighted mean atomic mass of A,B in the sample.

$$\therefore\ N' = \frac{w'.}{m'_{AB}}$$

Similarly, for the spike,

mass of A,B in the spike (w") = (N"/Av)∗m''_{AB} and $w'' = \frac{w''.\,Av}{m''_{AB}}$

Inserting these into equation (1) gives (2)

$$w' = w''.\,\frac{m'_{AB}}{m''_{AB}}\left\{\frac{Ab''(A) - R_X Ab''(B)}{R_X.Ab'(B) - Ab'(A)}\right\} \qquad (2)$$

In equation (2), the only unknown is w', since R_X is measured and w", Ab'(A), Ab'(B), Ab"(A), Ab"(B), m'_{AB} and m''_{AB} are measured as above or are known from standard listings of isotope ratios. Thus, knowing the weight (w") of spike added to the sample, the weight of the trace element (X) can be determined.

The mean atomic mass for two isotopes, $m_{AB} = \frac{Ab(A) \times m_A + Ab(B) \times m_B}{Ab(A) + Ab(B)}$

where m_A, m_B are the individual atomic masses of the isotopes. Since, m_A, m_B are standard values and Ab(A), Ab(B) are known or can be measured, m_{AB} can be calculated easily.

For the case of only the isotope (A) in the sample and B is an additional isotope, all terms (B) disappear from equation (2).

Thus, suppose the A isotope occurs only in the sample and B isotope occurs only in the spike. Then, Ab'(B) and Ab"(A) disappear from equation (2) and give equation (3).

$$w' = w''\,\frac{m_A}{m_B}\left\{\frac{-\,R_X.Ab''(B)}{-\,Ab'(A)}\right\} = w''\,\frac{m_A}{m_B}\cdot R_X\left\{\frac{Ab''(B)}{Ab'(A)}\right\} \qquad (3)$$

Figure 48.13

(a) Isotope-dilution analysis is a precise, accurate, and versatile technique for measuring trace amounts of elements in a variety of sample types. (b)

DATING ROCKS FROM THE $^{40}K/^{40}Ar$ ISOTOPE RATIO.

The rate of decay of ^{40}K is 5.543 x 10^{-10} yr. For any time difference (t), equation (1) relates the amount of ^{40}K existing now to that existing some time ago (in the present case, when the rocks were formed).

$$^{40}K_{now} / ^{40}K_{original} = \exp(-\lambda t)$$

or

$$\ln\left(\frac{^{40}K_{now}}{^{40}K_{original}}\right) = -5.543 \times 10^{-10} (t). \quad (1)$$

The amount of $^{40}K_{now}$ can be obtained by analysing the rock sample for its potassium content and then using standard tables which show the abundance of ^{40}K in any sample of potassium.

The value for $^{40}K_{original}$ is composed of $^{40}K_{now}$ plus the amount of argon (^{40}Ar) that has been formed over time.

$$\therefore \quad \frac{^{40}K_{now}}{^{40}K_{original}} = \frac{^{40}K_{now}}{^{40}K_{now} + ^{40}Ar}$$

but, radioactive decay of ^{40}K produces only 11.7% of ^{40}Ar (the remaining 88.3% is ^{40}Ca). Thus, the value for the amount of ^{40}Ar measured must be adjusted so as to give the correct amount of ^{40}K that has decayed.

$$\therefore \quad \frac{^{40}K_{now}}{^{40}K_{original}} = \frac{^{40}K_{now}}{^{40}K_{now} + ^{40}Ar/0.117} \quad (2)$$

Combining (2) with (1)

$$\ln\left(\frac{^{40}K_{now}}{^{40}K_{now} + ^{40}Ar/0.117}\right) = -5.543 \times 10^{-10} (t).$$

or

$$\ln\left\{1 + \left(\frac{^{40}Ar}{^{40}K_{now}}\right) \frac{}{/0.117}\right\} = -5.543 \times 10^{-10} (t).$$

$$\therefore \quad t = \frac{1}{5.543 \times 10^{-10}} \ln\left\{1 + \frac{^{40}Ar}{^{40}K_{now}}\left(\frac{1}{0.117}\right)\right\} \quad (3)$$

In equation (3), the amount of ^{40}Ar is obtained by isotope analysis (spiking with ^{38}Ar). Since $^{40}K_{now}$ can be measured by a variety of methods, equation (3) is easily solved. For example, if $^{40}Ar/^{40}K_{now}$ is 0.05, then the time (t) is approximately 6 x 10^8 years, *viz*, the rocks were first formed (or previously melted) some 600 million years previously.

Figure 48.14
By using K/Ar isotope ratios, potassium-containing rocks can be dated to their first formation, even through millions of years.

Potassium/Argon Dating (Chronology)

The element potassium occurs naturally as ^{39}K, ^{40}K, and ^{41}K. Of these, the ^{39}K is most abundant and ^{40}K is radioactive. The decay of ^{40}K follows two paths, one giving ^{40}Ar and the other ^{40}Ca, as shown below.

$$^{40}K \rightarrow ^{40}Ar* \rightarrow ^{40}Ar$$

and,

$$^{40}K \rightarrow ^{40}Ca$$

From the radioactive decay constants and measurement of the amount of argon in a rock sample, the length of time since formation of the rock can be estimated. Essentially, the dating method requires fusion of a rock sample under high vacuum to release the argon gas that has collected through radioactive decay of potassium. The amount of argon is determined mass spectrometrically, usually using the spiking method described above. Then, if the concentration of potassium is known from any suitable method of analysis, the ratio of argon to potassium — together with the known decay constants for potassium — gives the age of the rock sample (Figure 48.14). The spiking uses argon that has been enriched in ^{38}Ar, and the two isotopes compared are ^{40}Ar and ^{38}Ar. At the same time, the ratio $^{38}Ar/^{36}Ar$ is used to correct for any atmospheric argon that may have contaminated the sample argon during workup. There are some limitations to this dating method because of the possibility that some argon may have been lost before the rock sample was taken, through melting or fracture of the rock, through diffusion, and through recrystallization of the rock by natural processes.

Plutonium Contamination in the Environment

Plutonium (Pu) is an artificial element of atomic number 94 that has its main radioactive isotopes at ^{240}Pu and ^{239}Pu. The major sources of this element arise from the manufacture and detonation of nuclear weapons and from nuclear reactors. The fallout from detonations and discharges of nuclear waste are the major sources of plutonium contamination of the environment, where it is trapped in soils and plant or animal life. Since the contamination levels are generally very low, a sensitive technique is needed to estimate its concentration. However, not only the total amount can be estimated. Measurement of the $^{240}Pu/^{239}Pu$ isotope ratio provides information about its likely source. Manufacture of nuclear bombs or shells requires a plutonium isotope ratio of about 0.07, which increases to about 0.2 upon their detonation due to formation of more ^{240}Pu. Nuclear reactors produce an isotope ratio of about 0.4–0.8. Thus, accurate measurement of the $^{240}Pu/^{239}Pu$ ratio gives information concerning its likely origin. Such precise, accurate isotope ratio measurements on environmental samples have been obtained in the fentogram-per-milliliter range by using a double-spiking technique with a $^{236}U/^{233}U$ mix and a plasma source mass spectrometer having multiple ion collectors in conjunction with a single-ion counter.

Conclusion

Accurate, precise isotope ratio measurements are important in a wide variety of applications, including dating, examination of environmental samples, and studies on drug metabolism. The degree of accuracy and precision required necessitates the use of special isotope mass spectrometers, which mostly use thermal ionization or inductively coupled plasma ionization, often together with multiple ion collectors.

Chapter 49

Transmission of Ions through Inhomogeneous RF Fields

Background

Separation of ions according to their m/z values can be effected by magnetic and/or electric fields used as mass analyzers, which are described in Chapters 24 through 27. However, apart from measurement of m/z values, there is often a need to be able to transmit ions as efficiently as possible from one part of a mass spectrometer to another without any mass separation.

Slow ions formed in atmospheric-pressure sources need to be collected and transmitted to a mass analyzer operating under high vacuum. On the path from the ion source to the inlet of the mass analyzer, vacuum pumps reduce the background gas pressure from atmospheric to that of the high vacuum. Ions from the source must be transmitted across this varying pressure region without significant losses due to low-energy ion/molecule interactions and other effects. Ion transmission guides are used for this purpose.

Often, ions are accelerated by potential differences of several thousand volts. These fast ions are then directed from a region under high vacuum into a gas collision cell at moderate gas pressure (where the energy of collision leads to the ions fragmenting) and finally back into a high vacuum for entry into a mass analyzer. The ion/molecule collisions lead to scattering of the ion beam, and it is essential to have some means of keeping the ions on track. Ion transmission guides are used for this purpose. Therefore, any consideration of ion transmission guides needs to examine the different requirements for keeping slow or fast ions traveling through high- or low-pressure regions on trajectories that will carry them from their point of origin to a mass analyzer.

Apart from losses caused through collisional processes with background neutral gases, ions can also fail to reach the mass analyzer through space-charge effects (mutual repulsion of like-charged ions) and the influence of stray electric fields within the instrument. Careful instrument design can reduce stray fields, but the intrinsic space-charge effects of the ions cannot be removed. The space-charge and collisional effects lead to spreading of the ion beam such that, in the worst cases, the beam may be almost entirely dissipated before it reaches the inlet of the mass analyzer. Thus, collisional and space-charge phenomena lead to ion losses during their movement from one part of a mass spectrometer to another. Loss of a significant proportion of ions leads to poor sensitivity in the mass spectrometer if many of the initially formed ions are lost before they reach the mass analyzer.

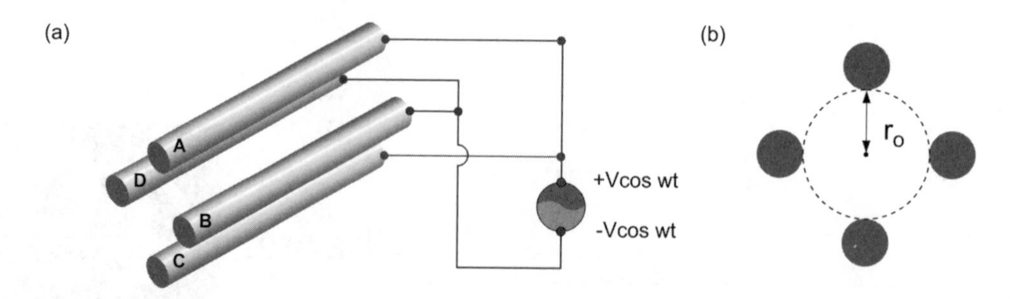

Figure 49.1
(a) Four parallel rods of a quadrupole operating in an RF-only mode. One pair of rods (A,C) is connected to one side of an RF generator (V cos ωt), and the other pair (B,D) is connected to the negative phase (–V cos ωt). The RF field has frequency ω and maximum amplitude V volts. (b) An end view of the four rods (A,B,C,D), showing the radius (r_0) of the inter-rod space.

To counteract problems, inhomogeneous RF fields have been designed to guide ions from one vacuum region to another with high efficiencies even though collisions of ions and gas molecules still occur and space charges still exist. For example, a quadrupole mass analyzer operates with both RF and DC fields applied to the rods. However, it can also be operated in an all-RF mode, there being no DC component (Figure 49.1). In the RF-only mode, all ions are transmitted irrespective of m/z value, viz., no significant mass separation occurs. Instead, the quadrupole acts as a transmission guide for either slow or fast ions. These quadrupole guides have been extended to include hexapoles (six rods), octopoles (eight rods), and even more poles (*n*-polar or multipoles), all of which give increasingly greater ion transmission efficiencies. Generally, the efficiency of ion transmission increases with the number of poles (Figure 49.2).

A different ion guide is the ion tunnel, which also uses only RF fields to transmit ions. It is not a rod device but consists of a series of concentric circular electrodes. It is perhaps best described as operating like a series of ion traps. This chapter gives details of some of the fundamental characteristics of rod-type transmission guides (multipoles).

Gas pressure and ion/molecule collisions

For efficient transmission of ions, multipolar guides provide a means of reducing ion losses during their transit from one part of a mass spectrometer to another. It is useful to understand some of the reasons for ion losses.

Space-charge effects cause an ion beam to spread apart. A beam of ions of the same electric charge (all positive or all negative) means that the electric field from one ion can be felt by all its neighbors. The repulsive effect drops relative to their distance apart (*d*), approximately as d^{-6}. Consequently, a dense beam of ions produces a large effect, while a beam containing few ions per unit volume will be subject to only a feeble effect. Ion guides help to contain ions and to offset the tendency for the beam to spread. In this way, most of a beam of ions traveling from one aperture to another across the ends of an ion guide will exit through the second aperture.

A second important need for some guidance system lies in stray electric fields. Clearly, a sufficiently large potential arranged transversely to an ion beam can serve to deflect ions away from the intended direction. Such stray fields can be produced easily by sharp edges or points on the inside of a mass spectrometer and even more so in an ion guide itself. Considerable care is needed in the construction and design of mass spectrometers to reduce these effects to a minimum.

Finally, probably the most important item affecting an ion beam is the overall gas pressure inside the instrument. Generally, a mass spectrometer operates under a high vacuum, in which

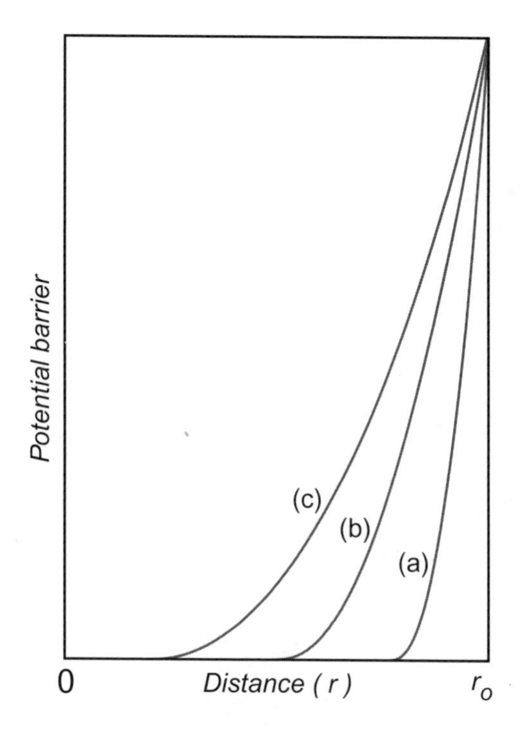

Figure 49.2
The spatial variation of electric confining potential (a barrier), which prevents ions from straying outside the cylindrical region illustrated in Figure 49.1b (radius r_0). As an ion moves from the center of the homogeneous field (o) toward any one of the rods of the multipole at a distance r_0 from the center, the potential barrier begins to increase, reaching a maximum at the rod itself. The steepness of the barrier is least for quadrupoles (c), in which it varies as r^2. It is greatest in the ion tunnel (a), in which it varies according to e^r and is intermediate for the hexapole (varies as r^4).

ion/molecule collisions are much rarer than at higher pressures. However, some parts of a spectrometer may be at a considerably higher pressure than the general vacuum, as with atmospheric-pressure inlets and gas collision cells. Therefore, it is useful to consider how ion/molecule collisions are affected by changes in background pressure and type of neutral gas.

Without some form of transmission guide, collisions between ions and neutral background molecules lead to scattering of an ion beam. Collision between an ion and a neutral gas molecule can be elastic or inelastic. Elastic in this context simply means that the total kinetic energy of the two colliding entities remains the same after collision as it was before. Alternatively, if some of the kinetic energy is turned into internal electronic, vibrational, and rotational energy in one or both of the colliding species, then the total kinetic energy is less after collision than it was before, and the collision is said to be inelastic (Figure 49.3). The subject of inelastic ion/molecule collisions forms the basis of collisionally activated decomposition (CAD) and is not discussed here in detail.

When RF-only multipole devices are used as ion transmission guides, it is necessary to inject the ions into the rod assembly. The trajectories of ions through the inhomogeneous RF fields of a multipole are complex, and the inhomogeneous fields themselves provide no forward momentum along the direction of the main axis of the multipole assembly. If the ions are to travel from one end of the space between the multipole rods to the other, then they must be given some forward momentum along the long axis of the rods. This forward momentum may refer to slow ions or fast ions. The slow ions have kinetic energies of only a few electron volts, an energy that can be gained through ion repulsion (space-charge effects) rather than applied intentionally by superimposing an electric potential. Fast ions have forward kinetic energies of hundreds or thousands of electron volts as a direct result of the application of a suitable electric accelerating potential.

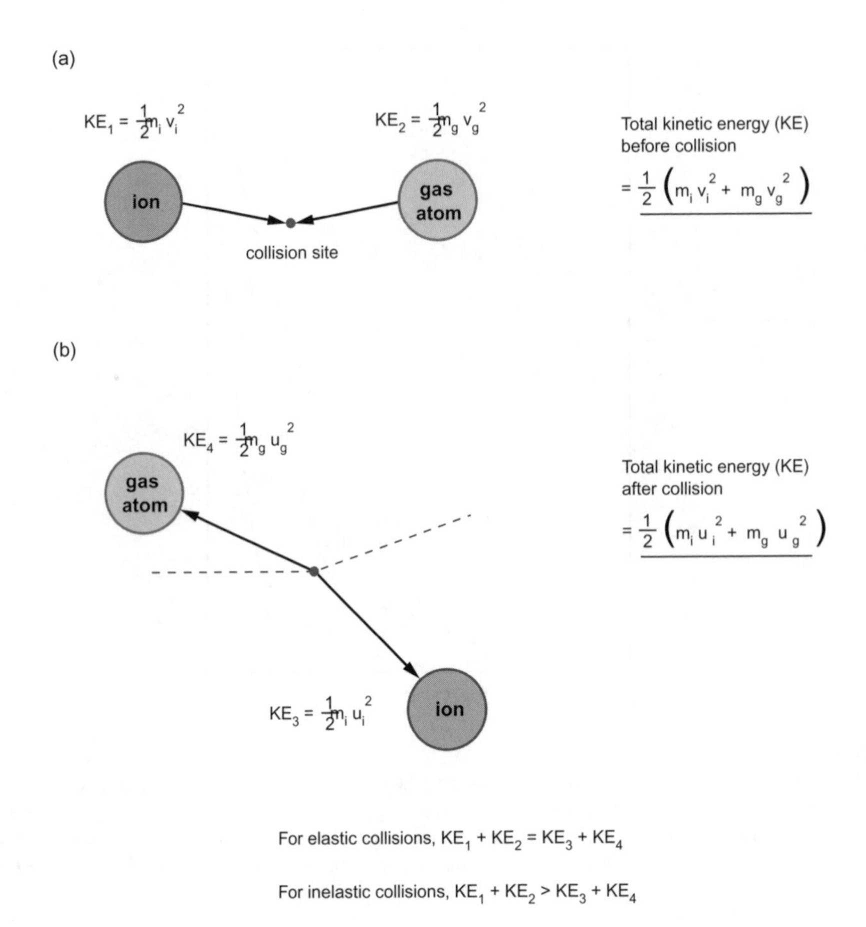

Figure 49.3
In (a), an ion and a gas atom approach each other with a total kinetic energy of $KE_1 + KE_2$. After collision (b), the atom and ion follow new trajectories. If the sum of $KE_1 + KE_2$ is equal to $KE_3 + KE_4$, the collision is elastic. In an inelastic collision (b), the sums of kinetic energies are not equal, and the difference appears as an excess of internal energy in the ion and gas molecule. If the collision gas is atomic, there can be no rotational and no vibrational energy in the atom, but there is a possibility of electronic excitation. Since most collision gases are helium or argon, almost all of the excess of internal energy appears in the ion.

To consider the effects of ion/molecule collisions on the trajectory of an ion, it is useful to consider:

1. The frequency of collisions and consequent deviations from an initial ion trajectory
2. The kinetic energies of both the ions and the background molecules
3. Elastic and inelastic collisions

These three factors are discussed briefly here and apply to helium as background gas and ions of relatively small mass. For other background gases and ions of greater mass, the effects are qualitatively similar but differ in scale.

For an ion traveling in a straight line in a complete vacuum, where there are no collisions, the velocity of the ion depends only on the electric potential difference through which it was accelerated, as shown by Equation 49.1 and Figure 49.4a.

$$v = (z \cdot e \cdot V/m)^{0.5} \qquad (49.1)$$

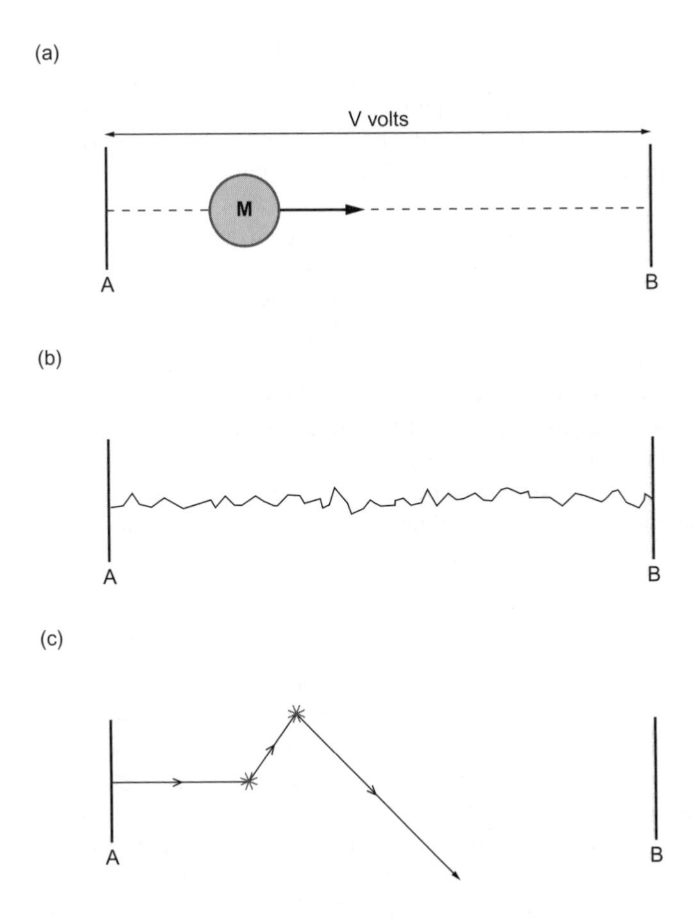

Figure 49.4
In a vacuum (a) and under the effect of a potential difference of V volts between two electrodes (A,B), an ion (mass m and charge ze) will travel in a straight line and reach a velocity v governed by the equation, $mv^2 = 2zeV$. At atmospheric pressure (b), the motion of the ion is chaotic as it suffers many collisions. There is still a driving force of *V* volts, but the ions cannot attain the full velocity gained in a vacuum. Instead, the movement (drift) of the ion between the electrodes is described by a new term, the mobility. At low pressures, the ion has a long mean free path between collisions, and these may be sufficient to deflect the ion from its initial trajectory so that it does not reach the electrode B.

Frequency of Collisions

For background atoms or molecules such as helium at 25°C and a pressure of 1 atm (about 1 bar), the mean free path is about 2×10^{-8} m (one collision every 0.15 nsec). In a high-vacuum region (typically 10^{-9} bar), the mean free path is very much longer at about 20 m, indicating a collision every 0.15 sec. An ion moving through a gas at 1-atm pressure is subjected to frequent collisions, moving only a few molecular diameters before colliding with another gas molecule (Figure 49.4b). As the background pressure decreases, an ion travels farther and farther until, in a high vacuum, it may travel several centimeters before striking a neutral molecule (Figure 49.4c).

Statistically, in a high-pressure region, an ion will be struck by neutral molecules randomly from all angles. The ion receives as many collisions from behind as in front and as many collisions from one side as from the other. Therefore, it can be expected that the overall forward motion of the ion will be maintained but that the trajectory will be chaotic and similar to Brownian motion (Figure 49.4b). Overall, the ion trajectory can be expected to be approximately along the line of its initial velocity direction, since it is still influenced by the applied potential difference V.

Under high-vacuum conditions, an ion can travel quite long distances before it meets a neutral molecule. A resulting collision causes both ion and neutral to be deflected from their original paths. However, unlike the higher pressure conditions, because collisions are so infrequent, an ion can be deflected right away from its initial velocity direction. In the absence of any forces to bring it back onto its original trajectory, the ion will eventually strike the walls of the mass spectrometer and be lost entirely (Figure 49.4c).

For pressures between atmospheric and a high vacuum, an ion trajectory is more indeterminate, ranging from major deflections at low pressures to only small changes in direction at higher pressures. In general, as the pressure of background gas diminishes, the ion trajectory becomes increasingly erratic. In the absence of some guiding electric field, this erratic trajectory can lead the ion to strike some part of the mass spectrometer before it can travel from its origin to its destination, usually a mass analyzer. Any beam of ions suffers losses through scattering due to collisions with neutral gas molecules, and the effect becomes severe as gas pressure is reduced because a deflected ion then has a long mean free path, which allows it to hit the walls of the mass spectrometer and be discharged. The guides in the multipole transmission system deflect ions back onto their original direction, thereby preventing serious losses by scattering.

Velocity Distribution

It is useful to consider the effects of collisions with neutral gas molecules on the velocities of fast or slow ions. In a gas such as helium at 25°C, the atoms have a distribution of velocities (a Boltzmann distribution) with a mean velocity of about 1.3×10^3 m/sec. In a gas, the constantly moving atoms collide with each other in such a way that their kinetic energies are continually redistributed throughout the mass of gas. Faster atoms tend to lose kinetic energy upon collision with slower atoms and vice versa. Overall, this situation results in a continuous adjustment of velocities (kinetic energies), leading to a distribution of velocities about a mean value characteristic of the temperature of the gas. From the point of view of mass spectrometry, an ion can be considered to be slow if it has a kinetic energy (KE) of only a few electron-volts (0 = KE = 5 eV) and fast if it has a kinetic energy of several hundred or thousand electron volts.

In the slow ion range, an ion of m/z ratio 100 and a kinetic energy of 1 eV will have a velocity of about 1.4×10^3 m/sec in a vacuum. This velocity is very similar to the mean velocity of background gas atoms. Therefore, upon ion/atom collision, there will be little change in the respective kinetic energies of the ion or the neutral gas atom. The ion continues on its trajectory with approximately the same energy it had before collision, although it may be deflected from its original direction in the absence of RF fields from ion transmission guides. For such low-energy collisions, there is almost no probability that some of the collisional energy will be converted into an excess of internal energy in the ion or the atom — almost all collisions are said to be elastic.

In the case of a fast ion of m/z 100 accelerated through a potential of 1000 V, its velocity in a vacuum is about 4.5×10^4 m/sec. This velocity is much faster than the mean speed of the background gas molecules. With such higher energy collisions, an ion will lose some kinetic energy, and the background atom will gain some. The result of numerous such collisions will be the slowdown of the originally fast ion as it collides with slower, neutral, background molecules. The ion also still suffers from deflections to its original trajectory. At these higher energies, there is an additional effect in that some of the collisional energy can be converted into an excess of internal energy in the ion (or the neutral); viz., the collision is no longer elastic and is described as inelastic. This increase in the internal energy of an ion appears as additional rotational, vibrational, or even electronic energy, which can be sufficient to cause the ion to fragment. Regardless if it does decompose, the ion or its charged fragments will have been deflected, and an ion transmission guide is necessary to return the ion to its original direction.

Excess of Internal Energy Caused by Inelastic Collisions

The total collisional energy between an ion and a neutral gas atom can be calculated from Equation 49.2, in which m_g, m_i are the masses of the colliding neutral gas molecule and the ion, respectively. E_{LAB} is the energy imparted to the ion, and E_{CM} is the collisional energy referred to the center-of-mass of the ion and molecule.

$$E_{CM} = E_{LAB}[m_g/(m_i + m_g)] \quad (49.2)$$

For a slow ion of 1 eV kinetic energy (E_{LAB} = 1) and mass m_i = 100 colliding with a helium atom (m_g = 4), the collisional energy E_{CM} = 0.04 eV. Only small changes in rotational energy can be expected from such low energy collisions.

For fast ions of 1000 eV energy and mass = 100 colliding with a helium atom, E_{CM} rises to 40 eV. A proportion of this energy is converted into an excess of internal energy in the ion, which can appear as increased internal rotational, vibrational, and electronic energy. Because bond breaking requires only about 3–4 eV (330–440 kJ), the excess of internal energy gained from collisions with fast ions may be sufficient to break bonds and to cause the ion to fragment. As ions become even faster, although E_{CM} continues to increase as well, the proportion of E_{CM} that is converted into excess of internal energy in the ion tends to a limit of about 10 eV.

Note that if the background gas is changed to argon (m_g = 40) for the ions considered above, E_{CM} changes from 0.04 to 0.4 eV for slow ions and from 40 to 400 eV for fast ions. For slow ions there is still almost no conversion of E_{CM} into an excess of internal energy, but for fast ions, the amount of internal energy appearing in the ion will be approximately 10 eV (nearly 1000 kJ). This amount of internal energy is certainly enough to effect bond breaking (fragmentation) in the ion and is the basis of collisionally activated decomposition (CAD).

Overall Effects of Ion/Molecule Collisions and the Use of Ion Transmission Guides

Quadrupoles or hexapoles are used as transmission guides for both slow and fast ions. In both cases, the objective is to ensure that as many ions as possible are guided from the entrance of the device to its exit. The ions are usually in transit in a straight line between an ion source and a mass analyzer. Any ions within the transmission guides that are deflected from the desired trajectory are pushed or pulled back on course by the action of the inhomogeneous RF fields applied to the poles of the guides.

For slow ions, transmission guides are frequently used to carry ions from an atmospheric-pressure ion inlet (as in electrospray [ES] or atmospheric-pressure chemical ionization [APCI] ion sources) to the low-pressure region of a mass analyzer. Often, more than one guide is used to bridge the high- and low-pressure regions. Slow ions have about the same kinetic energy after traversing the guides as they had on entering it. The guides reduce ion losses to the walls of the mass spectrometer without affecting their kinetic energies along the direction of the main (long) axes of the guides.

Equally for fast ions, transmission guides help to keep ions (and their fragments resulting from collision) on track. The kinetic energies of fast ions change after collision with gas molecules but not from the action of the guide fields themselves. The RF transmission guides are frequently used as collision cells, which redirect any deflected ions back toward their original trajectory. The collisions are used to induce fragmentation, and the transmission guides offset the scattering effects.

It is worth comparing this use of RF fields as transmission guides with those used in ion traps, where ion/molecule collisions are encouraged by the introduction of a bath gas. Once injected into an ion trap, ions follow complex trajectories with sizes dependent on the velocities of the ions. Faster ions have larger trajectories than slower ones. If such a collection of ions having different trajectories were to be sent to the ion detector, it would arrive spread out, and mass resolution would be poor, with ions of one m/z value overlapping those of their nearest neighbors. By introducing a collision gas, the faster ions are intentionally caused to lose kinetic energy through multiple collisions while retained in the RF trapping fields. Eventually all ions arrive at about the same kinetic energy, which is approximately that of the bath-gas molecules. The bath gas drains this kinetic energy to the walls of the trap. At the stage when all the ions have about the same kinetic energy, they are then ejected to the detector. Unlike the transmission guides discussed above, which are designed to reduce collisional scattering losses at pressures from atmospheric down to high vacuum, the ion trap uses a carefully regulated background gas pressure that is sufficient to slow the faster ions, thereby improving mass resolution without causing fragmentation. If collisionally induced decomposition is required from an ion trap, the trapped ions have to be first sped up enough to cause bond breaking, and then the resulting fragment ions must be slowed again to achieve reasonable mass resolution.

Although ion transmission guides and ion traps both use the same universal physical laws to achieve control over ion behavior, the ways in which the laws are used are different, as are the objectives. The guides do not retain ions to gain control over their velocities and are used simply to transmit both slow and fast ions over a very wide range of gas pressures. Ion traps retain ions over a relatively long period of time so as to adjust their kinetic energies and thereby improve mass resolution. The so-called bath gas is used at carefully controlled pressures.

Inhomogeneous RF Fields Applied to Rod Assemblies

Quadrupole

As shown in Figure 49.1, a quadrupole ion guide consists of four parallel rods spaced a fixed distance apart. An RF field is applied to one opposed pair of rods such that the positive phase ($V \cos \omega t$) is 180° out of phase with the negative phase ($-V \cos \omega t$) of the same field applied to the other pair of opposed rods. The RF field changes in amplitude cyclically so that it reaches a maximum on one pair of rods and then falls to reach a maximum of opposite polarity on the other pair of rods. The net RF field within the quadrupole rod set is inhomogeneous and has components at right angles (in the x-, y-directions) to the long axis (the z-direction) of a rod assembly but has no component along this main axis. Any ion placed in the field would simply oscillate in the x,y-plane. For this reason, if ions are to travel through the transmission guide, they must have some kinetic energy in the z-direction.

After an ion enters one end of the rod assembly, it follows a complicated path before reaching the other end (Figure 49.5). Since there is no accelerating effect of the RF fields along the central axis (z-axis) of the rod assembly, movement of an ion in this direction is not considered further, as it is not a consequence of the inhomogeneous RF fields. Generally, ions enter the rod assembly after having been accelerated sufficiently to have a kinetic energy of about 5 V or less. The drift velocity in the z-direction is given by Equation 49.3, in which an ion of mass m and z charges is either accelerated through an intentionally applied potential of V volts or through an effective V^* volts caused by natural space-charge effects from following ions.

$$v = (zeV/m)^{0.5} \text{ or } v = (zeV^*/m)^{0.5} \tag{49.3}$$

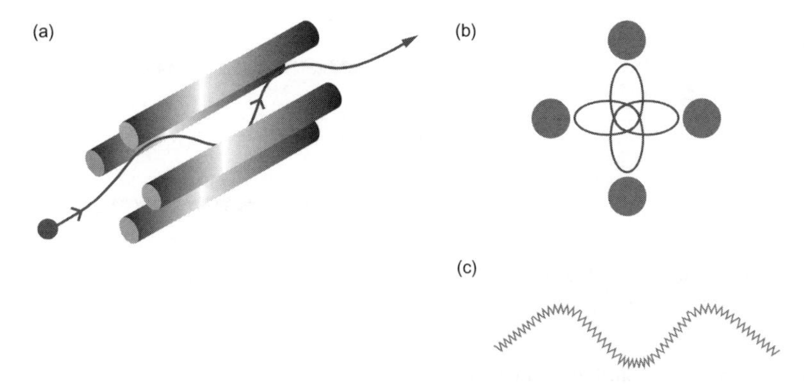

Figure 49.5
(a) An illustration of the guided motion of an ion traveling through the inhomogeneous field of a quadrupole operating under RF-only conditions. In (b), the motion is seen end-on, revealing a sort of four-leaf-clover pattern. (c) Rather than the smooth motion shown in (a, b), the trajectory has a superimposed rapid oscillatory motion as the ion attempts to follow the rapid oscillations of the RF field.

In the x- and y-directions (Figure 49.5b), an ion trajectory is more difficult to visualize. It is essentially the sum of two main effects: one a simple oscillation caused by the rapid cyclic alternations of the RF field (Figure 49.5c), the other a more complicated drift or guided motion due to the inhomogeneity of the RF field within the space between all four rods.

The simpler oscillating motion arises because as ions pass through the alternating RF field they experience an alternating electric potential, pulling the ions first one way, then the other, then back again, and so on. The ions, being electrically charged, try to follow the rapidly changing swings in potential. Accordingly, the ion trajectory describes an oscillatory motion at the frequency of the changing RF field (Figure 49.5c). An ion is a heavy species (compared with an electron, for example) and has significant molecular inertia. Consequently, it cannot respond immediately to the rapid voltage swings of the RF field. The resulting oscillatory motion of the ions can be described as damped and is slightly out of phase (lags behind) with the oscillations of the applied RF field. Nevertheless, the motion is still an oscillatory one in the x,y-plane.

As well as the motion in the z-direction and the oscillatory motion in the x,y-plane, ions in the inhomogeneous RF field experience yet another more complicated motion. This last effect is caused by the RF field, which produces an "effective" or "guiding" potential. This field is also oscillatory but of a very different frequency and amplitude from the simpler oscillatory motion just described. One way to understand this extra motion is to consider how the rapid oscillations of an ion in the RF field are affected by the inhomogeneity of the field. As an ion begins an oscillation in the inhomogeneous RF field, it passes from one region of the field at one potential to another nearby at a different electric potential. Thus, at one extreme (apogee) of the oscillation, the ion finds itself in a part of the field at a lower potential than is the case as it swings back to the other extreme (Figure 49.6). This asymmetrical effect of the inhomogeneous field tends to elongate one side of the oscillation toward the weaker part of the RF field and to truncate the oscillation toward the stronger side. The ion motion in the x,y-plane always drifts toward the weakest part of an inhomogeneous field. Equation 49.4 describes how the guiding force depends mostly on a small number of factors.

$$\text{Guiding potential (force), } V^* = (z^2/4m\Omega^2)E_0^{\,2} \qquad (49.4)$$

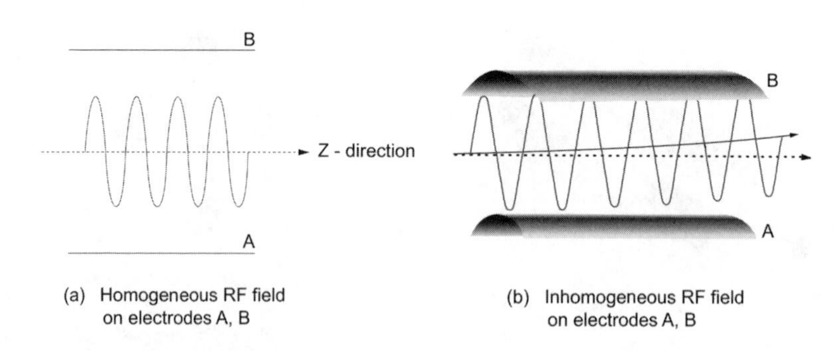

Figure 49.6
In a homogeneous RF field (a) applied to electrodes (A, B), an ion traveling in the z-direction will be subjected to equal forces on either side of its trajectory and will oscillate about a mean position. In an inhomogeneous field (b), the swings toward the weaker part of the RF field (B) are greater than those toward the stronger part of the field (electrode A). The forces about the median position are not equal, and the ion will drift toward electrode (B) as it oscillates. This drift motion is sometimes called "guided" motion and is a result of the inhomogeneous electric field.

First, the guided motion depends on the square of the charge on the ion (z^2) and is therefore independent of the sign of the charge. Regardless if it is a positive or negative ion, the guiding motion is always toward the weakest parts of the inhomogeneous RF field. As the field changes, an ion guided toward one rod of one pair will have its trajectory reversed, and it will then be guided toward one of the other pair of rods and so on. The motion always drifts toward the weakest part of the inhomogeneous RF field at any instant.

Note that the guiding effect is also directly proportional to the amplitude of the RF potential (E_0) and is inversely proportional to the mass (m) of the ion and to the square of the frequency (Ω) of the RF field. Therefore, the greater the mass of an ion or the higher the frequency of the field, the smaller is the guiding effect. Conversely, the greater the number of charges on the ion or the larger the amplitude (voltage) of the applied RF field, the greater is the guiding effect.

Generally, ions do not strike any of the rods because, as they approach one rod, the weakest part of the field shifts because the RF potentials on the two pairs of rods are 180° out of phase. This shifting field causes an ion to swing around and start heading toward one rod of another pair and so on. The motion is illustrated in Figure 49.5b for a quadrupole. The action of the inhomogeneous field is to constrain an ion within the rod assembly, which behaves like a potential well, forcing an ion toward the central axis of the transmission guide.

It is perhaps worth noting here that, if a quadrupole assembly is used in this all-RF mode, there is no significant mass separation as ions of different mass move through the guide. However, if a DC potential is applied to one pair of rods, the guiding potential changes to that shown in Equation 49.5, in which F is the applied DC potential.

$$\text{Guiding force (potential), } V^* = (z^2E_0^2/4m\Omega^2) + zF \qquad (49.5)$$

Now, in addition to a force that is proportional to z^2, there is an extra one proportional to z; viz., an ion will move continuously toward one or the other of the rod pairs depending on the sign of its charge. The interplay of these two effects largely controls whether or not the rod assembly allows an ion of given mass to pass through it or not. The system then becomes a mass analyzer or filter that can be adjusted through changes in E_0 and F to control which ions are allowed to pass through and those that are not permitted to do so (see Chapter 25).

Therefore, in the RF mode, ions transmitted through the rod guide are subjected to (1) an oscillation in step with the variations of the RF field in the x,y-plane, (2) a drift or guided motion caused by the inhomogeneity of the RF field (x,y-plane), and (3) a forward motion (z-direction) due to any initial velocity of the ions on first entering the rod assembly. The separate motions

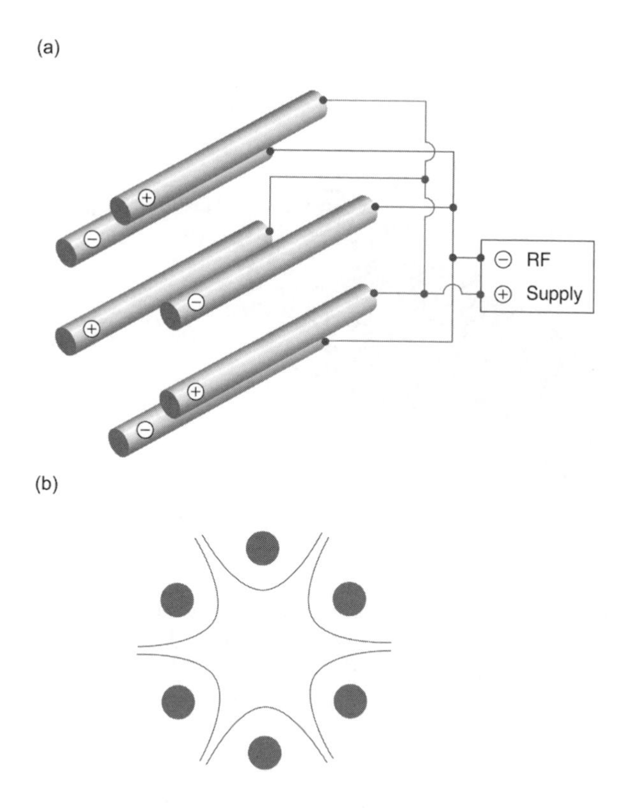

Figure 49.7
(a) In the hexapole, the six rods are connected in two pairs of three, as shown. (b) The approximate shape of the confining potential is shown end-on.

interact to provide an ion with a constant total energy in its passage through the transmission guide. However, this total energy is the sum of the potential and kinetic energies, which are not static but interchange throughout the ion motion. The specific proportion of kinetic and potential energies depends on the actual position of the ion on its journey. As an ion nears a rod, the kinetic energy falls to a minimum, and the potential energy rises to a maximum. Toward the central axis of the quadrupoles, the kinetic energy is greatest and the potential energy least.

From Figure 49.2, it can be seen that the quadrupole assembly provides a "potential well" to contain the ions in their journey along the main quadrupole axis. The potential well of the quadrupole has not very steep sides and, compared with steep-sided hexapoles or higher *n*-poles or ion tunnels, the quadrupole is not as efficient as the others in containing ions inside the rod assembly.

Hexapole

The principles applying to the quadrupoles also apply to the hexapole, except that there are now six rods arranged in two pairs of three rather than four in two pairs of two. In the hexapole ion guide, the all-RF field is arranged so that there are two pairs of three nonadjacent rods, the two pairs being 180° out of phase with each other (Figure 49.7). The resulting motion is similar to that in a quadrupole but somewhat more complex. The steepness of the walls of the confining electric potential is now much greater than that of the quadrupole (Figure 49.2). The hexapole arrangement results in a better guidance system for ions and increases the efficiency of their transmission through the assembly.

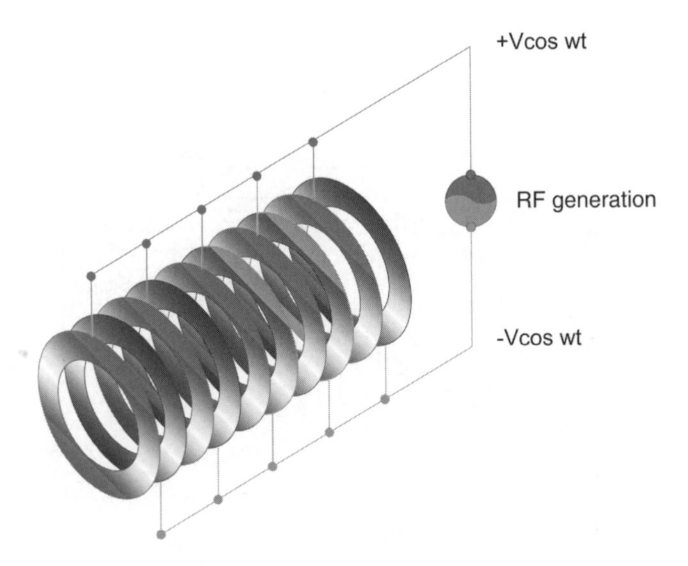

Figure 49.8
A simplified view of an ion tunnel, composed of a series of electrodes. The positive phase (+V cos ωt) from an RF generator is laced on alternate rings, and the negative phase (–V cos ωt) is placed on the remaining rings. A typical ion tunnel can have 60–80 such rings, and ions are guided along the central axis of the assembly.

Octopole

The octopole configuration is similar to the quadrupole and hexapole in affording a good ion guidance system in its all-RF mode. The poles are connected in two pairs of four nonadjacent rods, and each pair is connected to an RF voltage supply, 180° out of phase with each other. The octopole is even more efficient than the hexapole as an ion guide.

As the number of rods is increased, the rate of increase in ion efficiency falls off, and the devices become more difficult to construct. Consequently, quadrupoles and hexapoles are used most frequently in commercial instruments. More recently, the concept has been extended to a set of ring electrodes, which are the most efficient ion guides (Figure 49.8). However, although the ring set (ion tunnel) uses RF fields similar to the ones outlined here, there are sufficient differences that are not discussed here.

Conclusion

When quadrupole, hexapole, and other multipolar devices are operated in an RF-only mode, they act as guides, transmitting ions from one section of a mass spectrometer to another, offsetting the effects of space charge, stray electric fields, and collisions with neutral background molecules. The devices operate at atmospheric pressure or in high vacuum, serving as "bridges" between high- and low-pressure regions in a mass spectrometer.

Appendix A

Collection of Summaries

Chapter 1: Chemical Ionization (CI)

- This article should be read in conjunction with Chapter 3, "Electron Ionization."
- In a high vacuum (low pressure; 10^{-5} mbar), molecules and electrons interact to form ions (electron ionization, EI). These ions are usually injected into the mass spectrometer analyzer section.
- Still under vacuum but at higher pressure (typically 10^{-3} mbar), the initially formed ions collide with neutral molecules to give different kinds of ions before they are injected into the analyzer. As an example, at low pressure, methane gas (CH_4) is ionized to give molecular ions ($CH_4^{\bullet+}$) but, at higher pressures, these ions collide with other CH_4 molecules to give carbonium ions (CH_5^+).
- If a substance (M) is under investigation, its molecules can collide with CH_5^+ with transfer of a proton (H^+) to give $[M + H]^+$, which can be called quasi-molecular or protonated molecular ions.
- Because this chemical reaction occurs between the CH_5^+ and M species, the original methane (CH_4) is called a reagent gas, the CH_5^+ species are reagent gas ions, and the process is known as chemical ionization (CI).
- There are other reagent gases, e.g., hydrogen (H_2), butane (C_4H_{10}), methanol (CH_3OH), ammonia (NH_3), and so on.
- The different reagent gases produce quasi-molecular (protonated) ions $[M + H]^+$ having more or less of an excess of internal energy, which may be enough to cause some fragmentation. Some reagent gases cause no decomposition (fragmentation) of $[M + H]^+$ ions, but others do to some extent.
- Other reactions between reagent gas ions and molecules (M) can occur. As examples, ions $[M + X]^+$ can be formed, where X is Na, NO, or NH_4.
- Mass spectra from CI are not the same as those from EI.
- Mass spectra from EI usually contain many fragment ions and, often, not many molecular ions ($M^{\bullet+}$). Mass spectra from CI contain few fragment ions and, usually, many quasi-molecular ions $[M + X]^+$.
- Molecular ions ($M^{\bullet+}$) from EI have the same relative molecular mass (molecular weight) as the original molecules (M), but quasi-molecular ions $[M + X]^+$ have a different mass from that of the original molecules.
- In similar ways, negative ions $[M - H]^-$ or $[M + X]^-$ can be formed by using a negative reagent gas.
- EI and CI are complementary techniques and are usually combined into one ion source so that rapid switching between the two can take place.

Summary

Chemical ionization produces quasi-molecular or protonated molecular ions that do not fragment as readily as the molecular ions formed by electron ionization. Therefore, CI spectra are normally simpler than EI spectra in that they contain abundant quasi-molecular ions and few fragment ions. It is advantageous to run both CI and EI spectra on the same compound to obtain complementary information.

Chapter 2: Laser Desorption Ionization (LDI)

- A laser is a device for producing ultraviolet, visible, or infrared light of a definite wavelength, in contrast to most other light sources, which give out radiation over a range of wavelengths. The output of a single wavelength of light is described as being monochromatic. If the photons are all in phase (as in lasers), the light is also said to be coherent.
- Lasers can be tunable: Although only one wavelength is emitted at any one setting, the actual wavelength can be varied over a small range by changing the setting of the laser.
- Other notable characteristics of the laser are the intensity of the light emitted, its pulsed nature, and the fine focusing that is possible.
- For many lasers used in scientific work, the light is emitted in a short pulse lasting only a few nanoseconds, but the pulses can be repeated at very short intervals. Other lasers produce a continuous output of light.
- The emitted beam of coherent radiation is narrow and can be focused into a very small area. This means that the density of radiation that can be delivered for any one pulse over a small area is very high, much higher than can be delivered by conventional light sources operating with similar power inputs.
- If the target at which a laser beam is directed can absorb light of the laser wavelength, then the target will absorb a large amount of energy in a very small space in a very short time.
- The absorption of so much energy by a small number of target molecules in such a short time means that their internal energy is greatly increased rapidly, and the normal processes of energy dissipation (such as heat transfer) do not have time to occur. Much of this excess of energy is converted into kinetic energy so that the target molecules are vaporized (ablated) and leave the target zone.
- Some of the target molecules gain so much excess internal energy in a short space of time that they lose an electron and become ions. These are the molecular cation-radicals found in mass spectrometry by the direct absorption of radiation. However, these initial ions may react with accompanying neutral molecules, as in chemical ionization, to produce protonated molecules.
- The above direct process does not produce a high yield of ions, but it does form many molecules in the vapor phase. The yield of ions can be greatly increased by applying a second ionization method (e.g., electron ionization) to the vaporized molecules. Therefore, laser desorption is often used in conjunction with a second ionization step, such as electron ionization, chemical ionization, or even a second laser ionization pulse.
- Laser desorption is particularly good for producing ions from analytically difficult materials. For example, lasers can be used with bone, ceramics, high-molecular-mass natural and synthetic polymers, and rock or metal specimens. Generally, few fragment ions are formed.
- Ionization can be improved in many cases by placing the sample in a matrix formed from sinapic acid, nicotinic acid, or other materials. This variant of laser desorption is known as matrix-assisted laser desorption ionization (MALDI). The vaporized acids transfer protons to sample molecules (M) to produce protonated ions $[M + H]^+$.
- The laser can be used as a finely focused beam that, with each pulse, drills deeper and deeper into the specimen to give a depth profile. Alternatively, the beam can be defocused and moved over an area at lower power to explore only surface features of a specimen.

Summary

Lasers are used to deliver a focused, high density of monochromatic radiation to a sample target, which is vaporized and ionized. The ions are detected in the usual way by any suitable mass spectrometer to produce a mass spectrum. The yield of ions is often increased by using a secondary ion source or a matrix.

Chapter 3: Electron Ionization (EI)

- Molecules can interact with energetic electrons to give ions (electron ionization, EI), which are electrically charged entities. The interaction used to be called electron impact (also EI), although no actual collision occurs.
- Loss of an electron from a molecule (M) gives a (positive) radical cation, written $M^{\bullet+}$.
- Gain of an electron by a molecule gives a (negative) radical anion ($M^{\bullet-}$).
- The mass of an electron is very small, and, for most practical purposes, the mass of an ion is the same as that of the corresponding neutral species, viz., $M \approx M^+ \approx M^-$.
- The interacting electrons are energized by acceleration in an electric field before they interact with molecules to produce ions.
- Standard EI spectra are obtained with an electron energy of 70 eV (electrons accelerated through 70 V).
- For most compounds, it is easier to produce positive ions than negative ones, and most EI mass spectrometry is concerned with positive ions.
- Mass spectrometrically, the mass-to-charge ratio (m/z) is important. However, if $z = 1$, then, conveniently, $m/z = m$.
- Removal of one (negative) electron from a molecule leaves a single positive charge ($z = 1$). Loss of two electrons would give $z = 2$, and so on.
- For most EI mass spectrometry, $z = 1$, and higher charges can usually be neglected.
- The ion ($M^{\bullet+}$) derived from the parent molecule by loss of an electron is called a molecular ion.
- Depending on the structure of substance M and the energy of the incident electron, the resulting ion ($M^{\bullet+}$) may break up (fragment) to give ions of smaller mass (A^+, B^+, etc.).
- A mass spectrum is a chart showing on the x-axis the mass of each ion (M_m, M_a, M_b, etc.) and on the y-axis the number (abundance) of ions at each mass.
- The ion having greatest abundance is said to form the base peak in the spectrum. The base peak may or may not be the same as the molecular ion peak.

Summary

The interaction of electrons with molecules gives molecular ions, some of which can break down to give smaller fragment ions. The collection of molecular and fragment ions is separated by a mass analyzer to give a chart relating ion mass and abundance (a mass spectrum).

Chapter 4: Fast-Atom Bombardment (FAB) and Liquid-Phase Secondary Ion Mass Spectrometry (LSIMS) Ionization

- In fast-atom bombardment (FAB), an atom gun is used to project heavy, fast atoms (often argon or xenon) onto the surface of a target solution (matrix).
- The solution (or matrix) consists of the substance under investigation dissolved in a high-boiling-point solvent that does not evaporate quickly in the vacuum of a mass spectrometer.
- The impact of the fast atoms on the solution surface results in desorption of secondaries (positive ions, negative ions, and neutrals) into the low-pressure gas-phase region above the matrix surface.
- By selecting either a large positive or negative voltage on a plate with a slit in it held above the surface, the desorbed negative or positive ions can be accelerated away from the surface and into a mass analyzer.
- The mass spectrometer provides a mass spectrum of the ions coming from the matrix, some of which arise from the matrix itself and some from substances dissolved in it.

- Usually, FAB yields molecular or quasi-molecular ions, which have little excess of internal energy and therefore do not fragment. This ionization method is mild and good for obtaining molecular mass (molecular weight) information.
- Substrate molecules (M) usually give molecular radical cations ($M^{•+}$) or cationated molecular ions ($[M + X]^+$, X = H, Na, K, etc.) or negative ions $[M - H]^-$.
- FAB has been used with a wide variety of substances, including thermally labile compounds such as peptides, sugars, carbohydrates, and organometallics.
- Molecular masses up to about 2000 Da are routine; above this value, the efficiency of the process drops off, but it is still useful up to about 4000–5000 Da.
- Confusingly, FAB is sometimes called secondary ion mass spectrometry (SIMS), the "secondary" referring to the nature of the process (primary bombardment, secondary emission), but see next item.
- Historically, the term SIMS was developed for bombardment of solid surfaces with ions, so, for greater descriptive precision, the name "liquid secondary ion mass spectrometry" (LSIMS) is better and can be used synonymously with FAB.
- Instead of bombarding the matrix surface with fast atoms, fast ions can be used. Often these are cesium (Cs^+) ions.
- As with fast atoms, bombardment of the matrix with fast ions causes very similar desorption of ions and neutrals.
- In the case of fast ions, the terminology of secondary ion emission mass spectrometry (SIMS) is more obvious in that a primary incident beam of ions onto a target releases secondary ions after impact.
- Where the target is a liquid (matrix), the more descriptive term LSIMS should be used, as noted above.

Summary

The impact of a primary beam of fast atoms or ions on a target matrix (substrate and solvent) causes desorption of molecular or quasi-molecular ions characteristic of the substrate. The process is called FAB for atom bombardment or LSIMS for ion bombardment.

Chapter 5: Field Ionization (FI) and Field Desorption (FD)

- A large electric potential applied to a needle provides a very intense field at its tip, where the radius of curvature is small.
- Similarly, a sharp edge (razor blade) or a very sharp curve can also provide an intense electric field.
- For any given electric potential, as the radius of curvature of a tip or edge becomes smaller, the electric field becomes increasingly stronger.
- By growing thin whiskers along a sharp edge or thin wire, the ends of the whiskers become regions of very small radius of curvature and, consequently, provide very intense electric fields.
- A molecule (M) lying on one of these tips experiences the effects of the intense field such that its own electric fields are distorted and the normal barrier to movement of electrons away from or onto the molecule becomes smaller.
- The distortion caused by the field allows an electron to pass from the molecule to the tip if the applied potential is positive or from the tip to the molecule if the potential is negative. This is called field ionization (FI), and the electron transfer occurs through quantum tunneling. Little or no vibrational excitation occurs, and the ionization is described as mild or soft.
- If the applied potential is positive, a positive ion ($M^{•+}$) is produced, and, if negative, a negative ion ($M^{•-}$) is formed. Since there is no vibrational excitation, no fragment ions are produced.
- A positive ion formed on such a tip held at a high positive potential is repelled and flies off the tip almost immediately after formation and into the mass spectrometer, where its m/z value is measured. Similarly, negative ions can be mass measured.

- The process of field ionization presupposes that the substance under investigation has been volatilized by heat, so some molecules of vapor settle onto the tips held at high potential. In such circumstances, thermally labile substances still cannot be examined, even though the ionization process itself is mild.
- To get around this difficulty, a solution of the substance under investigation can be placed on the wire and the solvent allowed to evaporate. When an electric potential is applied, positive or negative ions are produced, but no heating is necessary to volatilize the substance. This technique is called field desorption (FD) ionization.
- Both FI and FD provide good molecular mass information, but few if any fragment ions, and allow thermally labile substances such as peptides, nucleosides, and glycerides to be examined, as well as inorganic salts.

Summary

In field ionization (or field desorption), application of a large electric potential to a surface of high curvature allows a very intense electric field to be generated. Such positive or negative fields lead to electrons being stripped from or added to molecules lying on the surface. The positive or negative molecular ions so produced are mass measured by the mass spectrometer.

Chapter 6: Coronas, Plasmas, and Arcs

- Extra energy can be added to atoms, molecules, and ions, causing them to become energetically excited.
- For atoms, molecules, and ions, the extra energy can make them move faster (an increase in kinetic or translational energy).
- For molecules and ions having more than one atom, the extra energy can make the component bonds rotate and vibrate faster (rovibrational energy). Isolated atoms, having no bonds, cannot be excited in this way.
- For atoms, molecules, and ions, the extra energy may be sufficient to cause one or more electrons to move from one orbital to another (electronic excitation).
- As excited atoms, molecules, or ions come to equilibrium with their surroundings at normal temperatures and pressures, the extra energy is dissipated to the surroundings. This dissipation causes the particles to slow as translational energy is lost, to rotate and vibrate more slowly as rovibrational energy is lost, and to emit light or x-rays as electronic energy is lost.
- The loss of energy returns the particles to their original (ground) state, viz., their energy state at normal temperatures and pressures.
- For electronically excited species, the emitted light can be used for spectroscopic purposes, as in fluorescence analysis.
- If an electric potential is applied between two electrodes in a gas, electrons are released from the cathode (negative electrode). The electrons are accelerated by the electric field and collide with atoms or molecules of gas.
- These collisions can be sufficiently energetic such that the gas molecules become electronically excited, and, as the excited atoms return to their ground state, they emit light. Thus, passage of electrons (an electric current) through a gas under the right conditions leads to the emission of light from the gas.
- The color of the emitted light depends on what type of gas is present. For example, sodium atoms glow with a yellow light (as in the familiar yellow street lights), and neon glow with a dark red light (as in the familiar neon lights).
- The appearance of light during passage of a current through a gas (called a discharge) is a manifestation of electronic excitation, but the conditions for excitation are dependent on the pressure of the gas and the voltage applied to initiate the discharge. At low voltages there is no discharge; usually, 200–800 V are needed to start the discharge process.

- As the voltage is increased, intermittent discharges occur because the discharge loses energy to its surroundings. Eventually the discharge becomes self-sustaining and, by maintaining a constant current flow, the discharge continues and light is emitted until the power is switched off.
- The exact conditions of gas pressure, current flow, and applied voltage under which the discharge occurs determine if it is of the corona, plasma, or arc type. The color of the emitted light may also change, depending not only on the type of gas used but also on whether it is a corona, plasma, or arc discharge.
- All three types of discharge involve the formation of ions as part of the process. For various reasons, most of the ions are positive. The ions can be examined by mass spectrometry. If small amounts of a sample substance are introduced into a corona or plasma or arc, ions are formed by the electrons present in the discharge or by collision with ions of the discharge gas.
- Thus, either the emitted light or the ions formed can be used to examine samples. For example, the mass spectrometric ionization technique of atmospheric-pressure chemical ionization (APCI) utilizes a corona discharge to enhance the number of ions formed. Carbon arc discharges have been used to generate ions of otherwise analytically intractable inorganic substances, with the ions being examined by mass spectrometry.
- Since a discharge is characterized by having a substantial population of charged species (electrons and ions), it responds to an applied electromagnetic field. The applied field moves electrons in one direction and positive ions in the opposite direction, in accordance with Maxwell's laws.
- If the applied electromagnetic field is an alternating one, then the electrons and ions are pushed (or pulled) backward and forward as the sign of the field changes. At high frequencies of applied fields, this motion causes multiple collisions between ions and neutral species and between electrons and ions and neutral species.
- The multiple energetic collisions cause molecules to break apart, eventually to form only atoms, both charged and neutral. Insertion of sample molecules into a plasma discharge, which has an applied high-frequency electric field, causes the molecules to be rapidly broken down into electronically excited ions for all of the original component atoms.
- This is the basic process in an inductively coupled plasma discharge (ICP). The excited ions can be examined by observing the emitted light or by mass spectrometry. Since the molecules have been broken down into their constituent atoms (as ions) including isotopes, these can be identified and quantified by mass spectrometry, as happens with isotope ratio measurements.

Summary

Depending on gas type and pressure, an electric discharge can be maintained in a gas by applying an electric potential across two electrodes in the gas to produce ions and emitted light. Whether the discharge is corona, plasma, or arc depends on the voltage and current flowing through it. The discharge contains electrons and ions, and the latter can be examined mass spectrometrically.

Chapter 7: Thermal Ionization (TI), Surface Emission of Ions

- When a metal wire (filament) is heated in a vacuum, electrons are formed as a cloud above the surface.
- The yield of electrons depends on the temperature of the filament and on the fundamental degree of difficulty in separating the electrons from the metal. The latter is measured in electron volts as a work function, ϕ.
- The electrons are accelerated from the filament by an anode, which has a positive potential with respect to the filament.
- If there is a different material (M) on the heated metal surface, M will be evaporated as the temperature increases. The ionization energy (ionization potential, I) of M is also measured in electron volts.

- As well as the evaporation of neutrals from the sample, some positive ions (M^+) are also produced, the number depending on the temperature and the energy difference, $I - \phi$. This production of positive ions is known as surface or thermal ion emission. The m/z values of the ions are characteristic of the elements in the sample.
- These positive ions can be accelerated toward a cathode, which is held at a negative potential with respect to the filament.
- With a suitable arrangement of potentials with respect to the filament, it is then possible to obtain a flow of positive ions from any substance previously deposited on the filament.
- The flow of positive ions is normally passed into a suitable ion analyzer in order to separate them according to m/z value. For this surface ion emission process, z is always equal to 1, and, therefore, m/z = m. The flow of ions at each m/z value generates an ion current that is used to measure the abundances of the ions.
- Different types of mass analyzer can be used as quadrupoles.
- This thermal ionization process requires filament temperatures of about 1000–2000°C. At these temperatures, many substances, such as most organic compounds, are quickly broken down, so the ions produced are not representative of the structure of the original sample substance placed on the filament.
- Ionization energies (I) for most organic substances are substantially greater than the filament work function (ϕ); therefore $I - \phi$ is positive (endothermic) and few positive ions are produced.
- Many inorganic substances are stable at the filament temperatures used, or they are changed to simpler substances. The ionization energies for inorganic elements are generally low so that $I - \phi$ is negative (exothermic). Inorganic samples produce good yields of positive ions characteristic of the elements present in the sample substance.
- Organics produce no useful positive ions, but the ions produced by inorganic samples are remarkably free from background interference, and the resulting mass spectra are relatively simple. The ion currents derived from the positive sample ions at each m/z value, being free from background ions, represent an accurate measure of the amount of each element.
- Thermal ionization has three distinct advantages: the ability to produce mass spectra free from background interference, the ability to regulate the flow of ions by altering the filament temperature, and the possibility of changing the filament material to obtain a work function matching ionization energies. This flexibility makes thermal ionization a useful technique for the precise measurement of isotope ratios in a variety of substrates.
- Thermal or surface emission of ions is one of the oldest ionization techniques used for isotope ratio measurements.

Summary

Heating inorganic substances to a high temperature on a metal filament yields characteristic positive ions that can be mass analyzed for m/z value and abundance to obtain accurate isotope ratios.

Chapter 8: Electrospray Ionization (ESI)

- Electrospray is both an atmospheric-pressure (API) liquid inlet system for a mass spectrometer, and, at the same time, it is an ionization source.
- Electrospray uses an electric field to produce a spray of fine droplets.
- In many applications of mass spectrometry, it is necessary to obtain a mass spectrum from a sample dissolved in a solvent. The solution cannot be passed directly into the mass spectrometer because, in the high vacuum, the rapidly vaporizing solvent would entail a large pressure increase, causing the instrument to shut down.

- Therefore, the sample solution, which may or may not come from a liquid chromatographic column, is passed along a narrow capillary tube, the end of which is maintained at a high positive or negative potential.
- The high potential and small radius of curvature at the end of the capillary tube create a strong electric field that causes the emerging liquid to leave the end of the capillary as a mist of fine droplets mixed with vapor. This process is nebulization and occurs at atmospheric pressure. Nebulization can be assisted by use of a gas flow concentric with and past the end of the capillary tube.
- The droplets, which carry positive or negative charges depending on the sign of the applied potential, pass into and along a small evaporation region. Much of the excess of solvent vapor is allowed to pass to atmosphere or can be gently exhausted to waste.
- As the droplets pass through the evaporation region, solvent evaporates, and the droplets rapidly become much smaller. At the same time, because the surface area of the droplets gets smaller and smaller, the density of electrical charge on the surface increases until a point of instability is reached.
- Eventually, not only neutral solvent molecules but also ions start to desorb from the surface of each droplet.
- Ions, residual droplets, and vapor formed by electrospray are extracted through a small hole into two evaporation chambers (evacuated) via a nozzle and a skimmer, passing from there into the analyzer of the mass spectrometer, where a mass spectrum of the original sample is obtained.
- This inlet/ion source is a simple system with no moving parts and yields many ions from the original dissolved sample. Even more attractive is the tendency for electrospray to produce multicharged ions, a benefit that makes accurate measurement of large relative molecular masses much easier.
- When multicharged ions are formed, the simple rule of thumb used widely in mass spectrometry that $m/z = m$ because, usually, $z = 1$ no longer applies; for $z > 1$ then $m/z < m$, and the apparent mass of an ion is much smaller than its true mass. Accurate mass measurement is much easier at low mass than at high, and the small m/z values, corresponding to high mass with multiple charges, yield accurate values for the high mass.
- Electrospray can be used with sector, time-of-flight, and quadrupole instruments. The technique has been used extensively to couple liquid chromatographs to mass spectrometers.

Summary

A sample to be examined by electrospray is passed as a solution in a solvent (made up separately or issuing from a liquid chromatographic column) through a capillary tube held at high electrical potential, so the solution emerges as a spray or mist of small droplets (i.e., it is nebulized). As the droplets evaporate, residual sample ions are extracted into a mass spectrometer for analysis.

Chapter 9: Atmospheric-Pressure Ionization (API)

- Under suitable conditions, a stream of solvent containing a substrate (solute) dissolved in it can be broken up into a spray of fine droplets at atmospheric pressure (nebulized).
- Some of the droplets carry an excess of positive electric charge and others an excess of negative electric charge.
- The spray or stream of droplets is passed along a tube that is usually heated.
- As the droplets proceed along the tube, solvent evaporates to yield smaller droplets.
- Because the electrically charged droplets retain their charge but get smaller, their electric field increases.
- At some point, mutual repulsion between like charges causes charged particles (ions) to leave the surface of the droplet (ion evaporation). These ions can be detected by the mass spectrometer.
- Additional ionization occurs by collision between the ions and other neutral species (ion/molecule collision; see Chapter 1). Unless special steps are taken (see Chapters 8 and 11), the ions formed do not fragment, so little or no structural information is obtained. However, the lack of fragmentation does mean that good relative molecular mass data can be obtained. The assembly of ions formed by ion

evaporation and chemical ionization passes through a small orifice (a skimmer) into the mass spectrometer for mass analysis in the usual way (see Chapters 25 through 27).

- Greater sensitivity is attained if the original solvent is polar (e.g., water or methanol).
- Greater sensitivity is attained if the solvent already contains ions through addition of an electrolyte.
- Greater sensitivity is attained if an additional ionization mode is included. This can be in the form of a radioactive source, a heated filament, or a plasma or glow discharge.
- All three of the previous items can be combined for maximum sensitivity.
- The rate of evaporation of solvent from droplets can be increased by blowing a drying gas across the stream. Nitrogen is frequently used as the drying gas.
- Practical inlet systems for attaching a high-pressure liquid chromatography (HPLC) column to a mass spectrometer utilize atmospheric-pressure ionization (see Chapters 8 and 11).

Summary

Evaporation of solvent from a spray of electrically charged droplets at atmospheric pressure eventually yields ions that can collide with neutral solvent molecules. The assemblage of ions formed by evaporation and collision is injected into the mass spectrometer for mass analysis.

Chapter 10: Z-Spray Combined Inlet/Ion Source

- Z-spray is a novel kind of electrospray that functions as a combined inlet and ion source. Chapter 8 ("Electrospray Ionization") should be consulted for comparison.
- For conventional electrospray, a solution of an analyte is sprayed from a narrow tube into a region where the solvent and other neutral molecules are pumped away and residual ions are directed into the analyzer of a mass spectrometer.
- The spraying process is aided by placing a high electric potential on the end of the narrow tube from which the solution exits and by using a nebulizing gas flowing past the end of the tube. Very small electrically charged droplets are produced, positive for a positive electric potential and vice versa.
- Solvent evaporates from the droplets as they move toward an opening (skimmer) and then into the mass analyzer.
- Before reaching the skimmer, much of the solvent has evaporated, and mostly only residual ions carry on through the skimmer opening. In conventional electrospray sources, the trajectory of the ions from the solution exit tube to the skimmer is a straight line of sight.
- Before ions actually pass into the spectrometer analyzer, there is usually a second drying stage to remove any final solvent.
- The ions that pass into the analyzer have near-ambient thermal energies and do not fragment but give excellent molecular or quasi-molecular ions. These ions can be investigated for their m/z values by almost any kind of mass analyzer.
- Ions can be induced to fragment by increasing an electric potential known as a cone voltage, which speeds them. Accelerating the ions causes them to collide more energetically with neutral molecules, a process that causes them to fragment (collision-induced decomposition).
- The Z-spray source utilizes exactly these same principles, except that the trajectory taken by the ions before entering the analyzer region is not a straight line but is approximately Z-shaped. This trajectory deflects many neutral molecules so that they diffuse away toward the vacuum pumps.
- The Z-trajectory ensures excellent separation of ions from neutral molecules at atmospheric pressure. In line-of-sight or conventional electrospray sources, the skimmer is soon blocked by ions and molecules sticking around the edges of the orifice. In Z-spray sources, the final skimmer, being set off to one side, is not subjected to this buildup of material.
- Z-spray sources require much less frequent maintenance than do conventional electrospray sources.

Summary

The Z-spray inlet/ion source is a particularly efficient adaptation of the normal in-line electrospray source and gets its name from the approximate shape of the trajectory taken by the ions between their formation and their entrance into the analyzer region of the mass spectrometer. A Z-spray source requires much less maintenance downtime for cleaning.

Chapter 11: Thermospray and Plasmaspray Interfaces

- Thermospray is both a liquid inlet system for a mass spectrometer and, at the same time, an ionization source.
- Plasmaspray, or discharge-assisted thermospray, is a modification of thermospray in which the degree of ionization has been enhanced.
- As the name implies, thermospray uses heat to produce a spray of fine droplets. Plasmaspray does not produce the spray by using a plasma but, rather, the droplets are produced in a thermospray source and a plasma or corona is used afterward to increase the number of ions produced.
- In many applications of mass spectrometry, it is necessary to obtain a mass spectrum from a sample dissolved in a solvent. The solution cannot be passed straight into the mass spectrometer because, in the high vacuum, the rapidly vaporizing solvent would entail a large pressure increase, causing the instrument to shut down.
- Therefore, the sample solution, which may or may not come from a liquid-chromatographic (LC) column, is passed along a narrow capillary tube at the end of which it is strongly heated.
- The strong localized heating causes the liquid to vaporize very rapidly, forming a supersonic jet that leaves the end of the capillary as a mist of fine droplets mixed with vapor.
- The droplets, which are electrically charged, pass into and along a small evaporation region.
- As the droplets pass through this region, they evaporate solvent and rapidly become much smaller. At the same time, because the surface area of the droplets gets increasingly smaller, the density of electrical charge on the surface increases until a point of instability is reached.
- Eventually, not only neutral solvent molecules but also ions start to desorb from the surface. With much of the solvent removed, the ions and residual solvent pass through two chambers, each under partial vacuum to remove more solvent. After passing through the two chambers, the ions are passed to the m/z analyzer.
- Although this system is simple with no moving parts, unfortunately not many ions from the original dissolved sample are produced, and the thermospray inlet/ion source is not very sensitive considering the achievable sensitivities of standard mass spectrometers.
- To increase the number of ions, a plasma or corona discharge is produced in the mist issuing from the capillary. The electrical discharge induces more ionization in the neutrals accompanying the few thermospray ions. This enhancement increases the ionization of sample molecules and makes the technique much more sensitive; to distinguish it from simple thermospray, it is called plasmaspray.
- Ions formed by thermo- or plasmaspray are extracted through a small hole into the mass spectrometer analyzer, where a mass spectrum of the original dissolved sample is obtained.
- Thermospray and plasmaspray can be used with both sector and quadrupole instruments. They have been used extensively to couple liquid chromatographs to mass spectrometers.

Summary

A sample to be examined by thermospray is passed as a solution in a solvent (made up separately or issuing from a liquid chromatographic column) through a capillary tube that is strongly heated at its end, so the solution vaporizes and emerges as a spray or mist of droplets. As the droplets evaporate, residual ions are extracted into a mass spectrometer for analysis. Plasmaspray is a modification of thermospray designed to enhance the yield of ions.

Chapter 12: Particle-Beam Interface

- The particle-beam interface is used to remove solvent from a liquid stream without, at the same time, removing the solute (or substrate).
- A flow of liquid, for example from high-performance liquid chromatography (HPLC), is treated in such a way that most of the solvent evaporates to leave solute molecules that pass into an ionization region (ion source).
- A stream of liquid issuing from a narrow tube can be broken up into a spray of small droplets by injecting helium gas just before the end of the tube. This nebulization is analogous to the action of an aerosol spray-can nozzle.
- The flow of droplets enters an evaporation chamber that is heated sufficiently to prevent condensation.
- Solvent evaporates from the droplets.
- The mix of tiny drops is formed into a particle beam on passing through the exit nozzle of the evaporation chamber.
- The beam of tiny drops passes from the exit nozzle across an evacuated space and into another small orifice (skimmer 1). In this evacuated region, about 90% of the originally injected helium and solvent is removed by vacuum pumps to leave a stream of droplets so small that they are called *clusters*.
- The clusters are composed of aggregates of solvent (S_m), substrate (M_n), and mixed substrate/solvent molecules ($S_m.M_n$), where m, n are integers (1, 2, 3, … , etc.).
- The particle stream then passes through a second evacuated region between skimmer 1 and a second orifice (skimmer 2), where more residual solvent and helium are removed.
- Finally, the beam — composed mainly of single substrate and solvent molecules and very small clusters — is passed through a heated wire grid, where the last declustering and desolvation occurs, leaving a beam of substrate molecules.
- The beam of substrate molecules then passes straight into the ion source (electron ionization, EI, or chemical ionization, CI) for ionization before entry into the mass analyzer.

Summary

A stream of a liquid solution can be broken up into a spray of fine drops from which, under the action of aligned nozzles (skimmers) and vacuum regions, the solvent is removed to leave a beam of solute molecules, ready for ionization. The collimation of the initial spray into a linearly directed assembly of droplets, which become clusters and then single molecules, gives rise to the term *particle beam interface*.

Chapter 13: Dynamic Fast-Atom Bombardment and Liquid-Phase Secondary Ion Mass Spectrometry (FAB/LSIMS) Interface

- In fast-atom bombardment (FAB), an atom gun is used to fire heavy fast atoms at the static surface of a target solution (also called a matrix).
- In dynamic FAB, this solution is not stationary but flows steadily over the target area. Usually, the liquid flow is the eluant from a liquid chromatography column, but it need not be.

- The solution or matrix consists of the substance under investigation (the solute) dissolved in a high-boiling-point solvent that evaporates only slowly in the vacuum of the mass spectrometer at an operating temperature of about 20 to 30°C.
- On leaving the chromatographic column, the liquid flow passes along a narrow tube, into the FAB ion source, and then into the target zone of the fast atoms.
- The impact of fast atoms on the solution surface results in desorption of secondaries (positive ions, negative ions, and neutrals) into the low-pressure gas phase above the target matrix surface.
- By selecting either a large positive or negative electrical potential on a plate, with a slit in it, held above the target area, the desorbed negative or positive ions are extracted into the analyzer of the mass spectrometer.
- The spectrometer provides a mass spectrum of the ions, some of which come from anything dissolved in the solution or matrix (solute ions) and some from the matrix solvent itself.
- Components of a mixture emerging from a liquid chromatographic column are dissolved in the eluting solvent, and this solution is the one directed across the target, as described above. Thus, as the components reach the target, they produce ions. These ions are recorded by the spectrometer as an ion current.
- The passage of a component of a mixture over the atom gun target area is accompanied by first a rise and then a fall in the ion current, and a graph of ion yield against time is an approximately triangular-shaped peak.
- A graph or chart of ion current (y-axis) vs. time (x-axis) is therefore a succession of peaks corresponding to components eluting from the chromatographic column. This chart is called a *total ion current* (TIC) *chromatogram.*
- The area under each peak represents the amount of substance eluting from the column, and the time at which it emerges is called the *retention time* for that component.
- Dynamic FAB is an interface between a liquid chromatograph and a mass spectrometer and is, at the same time, an ion source. As an inlet/ion source, this technique fulfils a similar function to plasmaspray and electrospray, both of which are combined inlet/ion sources.
- Instead of bombarding the matrix with fast atoms, fast ions (FIB) can be used. Often these are cesium ions (Cs^+). As with fast atoms, fast ions cause desorption of ions and neutrals from the surface of a bombarded matrix.
- The term *liquid secondary ion mass spectrometry* (LSIMS) is sometimes used synonymously with FAB and is preferred by some as being more descriptive, since FAB could apply to bombardment of solid or liquid surfaces and does not indicate the types of secondaries investigated. In practice, little confusion is likely to result from using either term. Strictly, LSIMS can refer to the use of fast ions (FIB).

Summary

By allowing any solution, but particularly the eluant from a liquid chromatographic column, to flow continuously (dynamically) across a target area under bombardment from fast atoms or ions (FAB or FIB), any eluted components of a mixture are ionized and ejected from the surface. The resulting ions are detected and recorded by a mass spectrometer. The technique is called *dynamic FAB* or *dynamic LSIMS.*

Chapter 14: Plasma Torches

- Plasma consists of a gaseous mixture of neutral species, ions, and electrons. The charged species are in approximately equal concentrations.
- For mass spectrometric purposes, the plasma is normally created in argon, a monatomic gas. The plasma then consists of electrons, positive argon ions, and neutral argon atoms.

- There are different conditions for producing a plasma, which can be started in gases at low or high (atmospheric) pressures. In a plasma torch, a flow of argon gas is used at atmospheric pressure.
- All methods of plasma production require some electrons to be present as electric-discharge initiators. For a plasma torch, the initiating electrons are introduced from a piezoelectric spark directed into argon gas flowing in the interval between two concentric quartz tubes.
- Near the outlet from the torch, at the end of the concentric tubes, a radio high-frequency coil produces a rapidly oscillating electromagnetic field in the flowing gas. The applied high-frequency field couples inductively with the electric fields of the electrons and ions in the plasma, hence the name inductively coupled plasma or ICP.
- Electrons from a spark are accelerated backward and forward rapidly in the oscillating electromagnetic field and collide with neutral atoms. At atmospheric pressure, the high collision frequency of electrons with atoms induces chaotic electron motion. The electrons gain rapidly in kinetic energy until they have sufficient energy to cause ionization of some gas atoms.
- When ionization occurs, an incident electron collides with an atom, producing a positive ion and second electron. Thus two electrons appear after collision where there was only one initially.
- The two electrons emerging from the collision are again speeded until each produces another electron by collisional ionization of another atom of argon. The process continues so the first incident electron becomes two, the two become four, and so on. This cascade increases the number of electrons and ions in the gas to form a plasma within a few milliseconds.
- The density of ions and electrons increases quickly in the argon gas, at the same time increasing their kinetic energies as they are pulled back and forth in the applied electromagnetic field and undergo frequent collisions with neutral gas atoms. Some recombination of ions and electrons also occurs to form neutrals.
- An approximate equilibrium is set up in the plasma, with the electrons, ions, and atoms having velocity distributions similar to those of a gas that has been heated to temperatures of 7,000 to 10,000°C.
- Since the plasma is ignited toward the end of the concentric tubes from which argon gas is issuing, the plasma appears as a pale-blue-to-lilac flame coming out of the end of the tube, which is why the system is referred to as a *torch*, as in a welding torch.
- If a sample is introduced as a solution into the middle of the start of the flame, the combination of high temperatures, energetic electrons, and ions breaks down the sample molecules into constituent atoms and their ions. These elemental ions and atoms emerge from the end of the flame.
- By using a sampling device, the ions are siphoned from the end of the plasma flame and led into an ion mass analyzer, such as a quadrupole instrument, where the abundances of the ions and their m/z values are recorded.
- For several reasons — including the complete breakdown of sample into its substituent elements in the plasma and the use of an unreactive monatomic plasma gas (argon) — background interferences in the resulting mass spectra are of little importance. Since there are no or very few background overlaps with sample ions, very precise measurements of sample ion abundances can be made, which facilitate the determination of precise isotope ratios.
- Inductively coupled plasmas are used to obtain the ions needed to measure either relative concentrations (amounts) of the various elements in a sample or to obtain accurate elemental isotope ratios.

Summary

A discharge ignited in argon and coupled inductively to an external high-frequency electromagnetic field produces a plasma of ions, neutrals, and electrons with a temperature of about 7000 to 10,000°C. Samples introduced into the plasma under these extremely energetic conditions are fragmented into atoms and ions of their constituent elements. These ions are examined by a mass analyzer, frequently a quadrupole instrument.

Chapter 15: Sample Inlets for Plasma Torches, Part A: Gases

- When an argon plasma torch is used to identify the elements present in a sample or to measure isotope ratios, the sample must be introduced into the center of the plasma flame through an inlet tube.
- Once inside the hot plasma, which is at a temperature of about 8000 K and contains large numbers of energetic electrons and ions, the sample molecules are broken down into their constituent elements, which appear as ions. The ions are transported into a mass analyzer such as a quadrupole or a time-of-flight instrument for measurement of m/z values and ion abundances.
- The m/z values of the ions provide identification of the elements present, and the abundances of the ions provide accurate isotope ratios.
- For substances that are gases or are very volatile at ambient temperatures, it is relatively easy to introduce them into the flame. Liquids and solids are more difficult to introduce and are discussed in Parts B and C of this discussion (Chapters 16 and 17, respectively).
- The gas or vapor to be examined is mixed with argon gas to constitute the needed flow of gas into the plasma flame, and the sample vapor or gas is swept along with this argon makeup gas.
- If samples are introduced batchwise, then the sample enters the flame as a plug and the elements are measured transiently. If the samples are introduced continuously, then the measurement of isotope ratios can also be continuous as long as sample is flowing into the flame.
- The flame can become unstable if too large an amount of sample is introduced or if the sample contains substances that can interfere with the basic operation of the plasma. For example, water vapor, air, and hydrogen all lead to instability of the plasma flame if their concentrations are too high.
- In some instances, the plasma flame can go out altogether if the amounts of sample or other contaminants rise too high. This possibility has led to the development of a wide variety of gas/liquid separators that treat the sample before it is introduced to the flame.
- In some inlet devices, the volatile sample materials are first separated from entrained hydrogen gas or air by condensing them in a coolant bath. Subsequently, when all of the volatile sample components have been condensed and the hydrogen or air has been swept away, the sample is reheated and sent to the plasma flame.
- Instability in the flame leads to varying efficiencies in ion formation within the plasma (varying plasma temperature) and, therefore, to variations in measured isotope ratios (lack of accuracy).
- Some elements (S, Se, Te, P, As, Sb, Bi, Ge, Sn, Pb) are conveniently converted into their volatile hydrides before passed into the plasma. The formation of the hydrides by use of sodium tetrahydroborate (sodium borohydride) can be batchwise or continuous.
- The effluent from a GC column is already in the gas phase and needs only to be mixed with argon makeup gas before passage into the flame. Precautions need to be taken to divert temporarily the GC flow when the first solvent peak emerges because it contains far too much material for the plasma to withstand.
- Other volatile compounds of elements can be used to transport samples into the plasma flame. For example, hydride reduction of mercury compounds gives the element (Hg), which is very volatile. Osmium can be oxidized to its volatile tetroxide (OsO_4), and some elements can be measured as their volatile acetylacetonate (acac) derivatives, as with $Zn(acac)_2$.

Summary

Gases and volatile materials can be swept into the center of an argon plasma flame, where they are fragmented into ions of their constituent elements. The m/z values of ions give important information for identification of the elemental composition of a sample, and precise measurement of ion abundances is used to provide accurate isotope ratios.

Chapter 16: Sample Inlets for Plasma Torches, Part B: Liquid Inlets

- When an argon plasma torch is used to identify the elements present in a sample or to measure isotope ratios, the sample must be introduced into the center of the plasma flame through an inlet tube.
- Once inside the hot plasma, which is at a temperature of about 8000 K and contains large numbers of energetic electrons and ions, the sample molecules are broken down into their constituent elements, which appear as ions. The ions are transported into a mass analyzer such as a quadrupole or a time-of-flight instrument for measurement of m/z values and ion abundances.
- The m/z values of the ions provide identification of the elements present, and the abundances of the ions provide accurate isotope ratios.
- For substances that are gases or are very volatile at ambient temperatures, it is relatively easy to introduce them into the flame. Gases and vapors are discussed in Part A (Chapter 15). Solids are more difficult to handle and are discussed in Part C (Chapter 17).
- A liquid sample must be vaporized to a gas or, more likely, to a vapor consisting of an aerosol of gas, small droplets, and even small particles of solid matter. To be examined, the aerosol is mixed with argon gas to make up the needed flow of gas into the plasma and is then swept into the flame.
- If samples are introduced batchwise, then each one enters the flame as a plug, and the elements are measured transiently. If more than one m/z ration must be examined, the analyzer needs to be a quadrupole or time-of-flight instrument.
- If samples are introduced continuously, then the measurement of isotope ratios can also be continuous as long as sample is flowing into the flame, and several m/z ratios can be examined with almost any kind of mass spectrometer.
- The flame can become unstable if too large an amount of vaporized liquid is introduced or if the sample contains substances that can interfere with the basic operation of the plasma. For example, water vapor, organic solvents, air, and hydrogen all lead to instability of the plasma flame if their concentrations become too high.
- Instability in the flame leads to varying efficiencies in ion formation within the plasma (varying plasma temperature) and therefore to variations in measured isotope ratios (lack of accuracy).
- In some instances, the plasma flame can go out altogether if the levels of sample or other contaminants rise too high. This problem has led to the development of a wide variety of gas/liquid separators and/or desolvation chambers that condition the sample before it is introduced to the flame. These separators and chambers reduce the amount of solvent flowing into the flame.
- Some elements (S, Se, Te, P, As, Sb, Bi, Ge, Sn, Pb) in liquid samples are conveniently converted into their volatile hydrides before being passed into the plasma, as discussed in Part A (Chapter 15). For some samples, any volatile solvent is first evaporated in a sample holder, which is then heated strongly to vaporize the resulting solid residue, as discussed in Part C (Chapter 17).

Summary

Solutions can be examined by inductively coupled plasma mass spectrometry (ICP/MS) by either evaporating the solvent first and then volatilizing the solid residue or by nebulizing the solution and desolvating the resulting spray of fine droplets. After vaporization, residual sample (solute) constituents are swept into the center of an argon plasma flame, where they are fragmented into ions of their constituent elements. The m/z values of ions give important information for identification of the elemental composition of the sample, and measurement of ion abundances is used to provide accurate isotope ratios.

Chapter 17: Sample Inlets for Plasma Torches, Part C: Solid Inlets

- When an argon plasma torch is used to identify the elements present in a sample or to measure isotope ratios, the sample must be introduced into the center of the plasma flame through an inlet tube.
- Once inside the hot plasma, which is at a temperature of about 8000 K and contains large numbers of energetic electrons and ions, the sample molecules are broken down into their constituent elements, which appear as ions. The ions are transported into a mass analyzer such as a quadrupole or a time-of-flight instrument for measurement of m/z values and ion abundances.
- The m/z values provide identification of the elements present, and the abundances of the ions give accurate isotope ratios.
- For substances that are gases or are very volatile at ambient temperatures, it is relatively easy to introduce them into the flame. Gases and vapors are discussed in Part A (Chapter 15). Liquids are more difficult to handle and are discussed in Part B (Chapter 16).
- The solid to be examined must be vaporized in some way. This vaporization can be done by using the heat of the plasma flame or, more usually, the solid is ablated separately and the resulting aerosol is mixed with argon gas and swept into the center of the flame.
- If solid samples are vaporized quickly, then the sample enters the flame as a small plug and the elements must be measured over a short period of time. This mode is useful for high sensitivity because the entire sample passes through the flame in a short time. (The abundances of ions appear as a sharp peak on the output.) If samples are introduced continuously, then ultimate sensitivity may be reduced, but isotope ratios can be determined continuously to provide high accuracy.
- The flame can become unstable if too much sample is introduced or if the sample contains substances that can interfere with the basic generation of electrons and ions in the plasma. For example, water vapor, air, and hydrogen all lead to instability of the plasma flame if their concentration is too high.
- In some instances, the plasma flame can go out altogether if the levels of sample or other contaminants rise too high.
- Instability in the flame leads to varying efficiencies in ion formation within the plasma (varying plasma temperature) and, therefore, to variations in measured isotope ratios (lack of accuracy).
- Some solid samples can be vaporized easily, but others require very high temperatures. The inlet systems must be able to cover a vaporization range of about 100 to 2000°C.
- Some solids' inlet systems are also suitable for liquids (solutions) if the sample is first evaporated at low temperatures to leave a residual solid analyte, which must then be vaporized at higher temperatures.
- The major methods used for vaporization (ablation) include lasers, electrically heated wires, or sample holders and electrical discharges (arcs, sparks).

Summary

After vaporization, solid samples are swept into the center of an argon plasma flame, where they are fragmented into ions of their constituent elements. The m/z values of ions give important information for identification of the elemental composition of the sample, and measurement of ion abundances is used to provide accurate isotope ratios.

Chapter 18: Lasers and Other Light Sources

- Lasers are sources of highly collimated, coherent, and intense beams of light that may be obtained commercially from the ultraviolet into the far infrared.
- The small cross-sectional area covered by a laser light beam coupled with the energy density in the beam leads to power levels reaching from milliwatts to many hundreds of kilowatts per square meter.

- The weaker lasers are used in such systems as CD players and recorders and in communications and distance-measuring devices. The more-intense laser beams are used for welding and cutting of metals, cloth, skin, etc. and have even been examined as a means of inducing thermonuclear fusion reactions.
- The high intensity of a laser beam (many photons per unit area in unit time) means that, upon irradiation, each molecule of a photon-absorbing substance can absorb one, two, three, or more photons in rapid succession before the molecule can relax and return to its ground state. So much energy may be absorbed in a short space of time that the irradiated material is vaporized, often as a plasma containing ions.
- This behavior is unlike the more usual experience of light sources that are capable only of providing one photon per absorbing event in the time taken for the absorbing molecule to return to its ground state.
- Multiphoton absorption leads to electrons in the irradiated molecules raised to highly excited states. The excitation may be such that the sample is ionized directly, as in the plasma which is the basis of laser desorption ionization (LDI). Generally, there is only a small excess of ions formed in this way, and a secondary ionization is necessary to obtain a better yield of charged species.
- A laser beam is capable of putting so much energy into a substance in a very short space of time that the substance rapidly expands and volatilizes. The resulting explosive shock wave travels through the sample, subjecting it to high temperatures and pressures for short times. This process is also known as ablation.
- The ablated vapors constitute an aerosol that can be examined using a secondary ionization source. Thus, passing the aerosol into a plasma torch provides an excellent means of ionization, and by such methods isotope patterns or ratios are readily measurable from otherwise intractable materials such as bone or ceramics.
- If the sample examined is dissolved as a solid solution in a matrix, the rapid expansion of the matrix, often an organic acid, covolatilizes the entrained sample. Proton transfer from the matrix occurs to give protonated molecular ions of the sample. Normally thermally unstable, polar biomolecules such as proteins give good yields of protonated ions. This is the basis of matrix-assisted laser desorption ionization (MALDI).
- The three techniques — laser desorption ionization, laser ablation with secondary ionization, and matrix-assisted laser desorption — are all used for mass spectrometry of a wide variety of substances from rock, ceramics, and bone to proteins, peptides, and oligonucleotides.

Summary

The energy density in a tightly collimated laser beam is particularly useful for injecting large amounts of energy into sample molecules within a very short space of time. After absorption of this energy, the molecules relax by converting it into rotational, vibrational, and kinetic energy within a few picoseconds, and the sample is vaporized. Direct ionization can occur during this process, but the level of ionization usually needs to be assisted by having a secondary ion source (as with plasma torches used in isotope work) or by proton transfer from a matrix to give quasi-molecular ions of thermally sensitive molecules such as peptides or proteins.

Chapter 19: Nebulizers

- Samples to be examined by inductively coupled plasma and mass spectrometry (ICP/MS) are frequently in the form of a solution of an analyte in a solvent that may be aqueous or organic.
- For mass spectrometric ionization and introduction into a plasma flame, the analyte needs to be separated from most of the accompanying solvent. One way to accomplish this separation is to break the solution down into small droplets using a nebulizer.
- Nebulizers convert bulk liquid into an aerosol, consisting of a mix of small droplets of various sizes and solvent vapor. Such devices are used to transfer analyte solutions into the flame of a plasma torch.

- The aerosol is swept to the torch in a stream of argon gas. During passage from the nebulizer to the plasma flame, the droplets rapidly become smaller, as solvent evaporates, and eventually become very small. In many cases, almost all of the solvent evaporates to leave dry particulate matter of residual analyte.
- To assist evaporation of solvent, the argon stream carrying the aerosol can be passed through a heated tube called a desolvation chamber, operated at temperatures up to about 150°C.
- The large quantities of solvent vapor produced from the evaporating droplets must be removed before reaching the plasma flame, which is done by having cooling tubes sited after the heated desolvation chamber to condense the vapor into liquid. This condensed liquid is run to waste.
- After desolvation, the remaining fine particulate matter and residual droplets are swept by the argon carrier into the plasma flame, where fragmentation and ionization occur.
- Depending on the type of nebulizer used and its efficiency, there may be initially a significant proportion of large droplets in the aerosol. Heavier than the very fine droplets, the larger droplets are affected by gravity and by turbulent flow in the argon sweep gas, which cause them to deposit onto the walls of the transfer tube.
- To assist in the deposition of these larger droplets, nebulizer inlet systems frequently incorporate a spray chamber sited immediately after the nebulizer and before the desolvation chamber. Any liquid deposited in the spray chamber is wasted analyte solution, which can be run off to waste or recycled.
- A nebulizer inlet may consist of (a) only a nebulizer, (b) a nebulizer and a spray chamber, or (c) a nebulizer, a spray chamber, and a desolvation chamber. Whichever arrangement is used, the object is to transfer analyte to the plasma flame in as fine a particulate consistency as possible, with as high an efficiency as possible.
- The transfer efficiencies of analyte solution from the nebulizer to the plasma flame depend on nebulizer design and vary widely from about 5–20% up to nearly 100%.
- There is a very wide range of designs for nebulizers, but most are based on some form of gas/liquid sprayer or on ultrasonics.
- In the gas/liquid spray form of nebulizer, a stream of gas interacts with a stream of liquid. Depending on the relative velocity of the two streams and their relative orientation, the liquid flow is broken down into a spray of droplets, as in the common hair sprays.
- For ultrasonic nebulizers, the liquid is fragmented into droplets by an acoustic standing wave, usually produced by a piezoelectric transducer.

Summary

For mass spectrometric analysis of an analyte solution using a plasma torch, it is necessary to break down the solution into a fine droplet form that can be swept into the flame by a stream of argon gas. On the way to the flame, the droplets become even smaller and can eventually lose all solvent to leave dry analyte particulate matter. This fine residual matter can be fragmented and ionized in the plasma flame without disturbing its operation.

Chapter 20: Hybrid Orthogonal Time-of-Flight (oa-TOF) Instruments

- Time-of-flight (TOF) instruments utilize the times taken by ions to pass (fly) along an evacuated tube as a means of measuring m/z values and therefore of obtaining a mass spectrum. Often a reflectron is used to direct the ions back along the TOF tube.
- TOF instruments can be operated as stand-alone mass spectrometers (in-line with an ion source), but they frequently are used in combination with other techniques to give hybrid instruments. In such

hybrids, the TOF analyzer is usually placed at right angles (orthogonal) to a beam of ions, and hence the term orthogonal TOF hybrids.

- Orthogonal TOF is the name commonly given to what should properly be called orthogonally accelerated TOF mass spectrometry. Therefore, it is sometimes referred to by the acronym oa-TOF, especially in official publications, but it is more usual to hear it referred to simply as orthogonal TOF.
- For stand-alone or hybrid TOF mass spectrometry, the ions examined must all start from some point at the same instant. From this zero time, the ions are accelerated through a short region by applying a short pulse of electric potential of several kilovolts. The acceleration gives the ions velocities that vary in proportion to the square root of their m/z values.
- The process is rather like a sprint race, with all ions leaving the starting line at the same time. However, unlike a normal race, the result is always the same; viz., the ions arrive at the finish (collector or detector) in procession and strictly in the order of increasing m/z values. Ions of the smallest m/z values arrive first, followed successively by others of increasing m/z value.
- Flight times are extremely short (microseconds) for all of the ions, and therefore the scanning of the total mass spectrum from m/z 1 to about m/z 2000–3000 appears to be instantaneous on a human time scale. The arrival of ions at the finishing point is determined by a time-to-digital (TDC) microchannel plate collector (detector).
- TOF instruments have been hybridized with sector and quadrupole analyzers. These hybrid arrangements produce decided advantages over each technique alone.
- Because TOF spectra are obtained in such a short time frame, additional spectra can be accumulated rapidly. Thus, in the span of one second, it is possible to accumulate several thousand mass spectra.
- This attribute of rapid accumulation of spectra leads to excellent reproducibility and better signal-to-noise characteristics, and it makes full use of small quantities of sample.
- In the orthogonal mode, an ion beam is directed into one end of a TOF tube at right angles to the main TOF axis (orthogonally). A pusher electrode placed alongside and parallel to a positive ion beam and carrying a pulsed positive electric potential deflects sections of the beam away from the main beam and approximately at right angles to it. The ions pass along the TOF tube to reach the ion collector.
- Advantages of hybrid TOF instruments vary with the actual hybrid and are discussed in the relevant sections.

Summary

Hybrid time-of-flight (TOF) mass spectrometers make use of a TOF analyzer placed at right angles to a main ion beam. Ions are deflected from this beam by a pulsed electric field at right angles to the ion beam direction. The deflected ions travel down the TOF tube for analysis. Hybrid TOF mass spectrometers have many advantages arising from the combination of two techniques, neither of which alone would be as useful.

Chapter 21: Hybrid Magnetic-Sector Time-of-Flight (Sector/TOF) Instruments

- Magnetic/electric-sector instruments are used to manipulate ion beams by making use of the deflection of charged species (ions) in magnetic or electric fields.
- For such instruments, the ions are accelerated from an ion source by using an electric potential of several thousand volts.

- By using the property that ions are deflected in a magnetic field in proportion to both the square root of their m/z values and the potential through which they have been accelerated, it is possible to measure the m/z values very accurately.
- The electric fields in such instruments are used to focus the fast-moving ion beam according to the kinetic energies of the ions contained in it. This property allows ions of individual m/z values to be focused sharply before or after deflection in the magnetic field.
- This focusing action gives an ion beam, in which the m/z values can be measured so accurately that the resolution of a magnetic/electric-sector instrument (separation of ions of different m/z values) is measured as a few parts per million, compared to the more modest few parts per thousand in, say, a quadrupole or ion-trap instrument.
- The ions are detected by an electron multiplier placed in line with the beam.
- Thus, ions are produced, deflected in a magnetic field, then focused in an electric field, and finally detected by an electron multiplier or other ion detector.
- If, just before the ion beam reaches the ion detector, a pusher electrode is used alongside it to deflect the beam at right angles (orthogonal) to its original direction into the flight tube of a time-of-flight sector (TOF analyzer), the m/z values can be measured by the TOF section.
- The pusher electrode must be operated by placing very short pulses of electric potential on it. The short pulses are required to ensure that all the ions are started at the same time along the TOF analyzer, since the latter must time the flights of the ions very accurately in order to measure m/z values.
- A magnetic-sector/TOF hybrid has two means of measuring m/z values, one very accurately in a conventional magnetic/electric-sector sense, and the other somewhat less accurately in a time-of-flight sense.
- This hybrid is used in one form to measure highly accurate m/z values to obtain excellent elemental compositions of ions and therefore molecular formulae from molecular ions; in the other form, it is used to obtain MS/MS data at high resolution.
- Mixtures of substances can be examined without the need for initial chemical or analytical separation.
- The hybrid has other advantages of sensitivity, low signal-to-noise ratio, fast switching between MS and MS/MS modes, use with continuous or pulsed ion sources, and use with high- or low-energy collision-induced ion decomposition.

Summary

The use of a magnetic/electric-sector instrument in conjunction with an orthogonal time-of-flight instrument produces a hybrid mass spectrometer. The m/z values of ions can be measured in MS mode at high resolution to give elemental compositions of ions or at somewhat lower resolution in MS/MS mode to give structural information about the ions. Mixtures of substances can be investigated without the need for initial separation of the individual components by, for example, liquid or gas chromatography.

Chapter 22: Hybrid Hexapole Time-of-Flight (Hexapole/TOF) Instruments

- This system is very similar to that of a hybrid quadrupole time-of-flight (Q/TOF) instrument but without the initial quadrupole section.
- An ion beam can be produced from a number of different sources, but for this instrument — used for biochemical examination of thermally unstable, large molecules — an atmospheric-pressure inlet such as APCI or ES would generally be used. These can be operated with liquid inlets from chromatographic columns or simply from static solutions.

- The atmospheric-pressure ionization inlets produce a stream of cations or anions that have been formed at about room temperature and have little excess of thermal energy. Consequently, these ions are very stable and exhibit little or no fragmentation (see Chapter 10).
- In the hybrid hexapole/TOF instrument, these ions are collimated or guided through two consecutive hexapole sections, each operating at different pressures.
- A hexapole assembly is incapable of separating ions according to their m/z values. However, it is capable of accepting an ion beam and ensuring that the beam is kept as narrow as possible and remains on a straight-line track.
- The gas pressure in the first hexapole section is influenced by gas leaking in from the inlet system and, at about 10^{-4} mbar, is higher than desirable if excessive ion/neutral collisions are to be prevented. Therefore, the first hexapole is separated from the second hexapole section by a small orifice, which allows ions to pass through.
- Differential pumping of the two hexapole sections keeps the second at a pressure of about 10^{-5} mbar. The two consecutive hexapole sections are sometimes described as bridges between the pressure in the inlet and the following vacuum in the TOF mass analyzer.
- After passing through the hexapoles, the ion beam emerges in front of a pusher electrode built into the end of the TOF analyzer.
- The TOF analyzer is positioned at right angles (orthogonal) to the incoming ion beam.
- Application of a pulse of high electric potential (about 1kV) to the pusher electrode over a period of about 3 μsec causes a short section of the ion beam to be detached and accelerated into a TOF analyzer. A positive potential is used to accelerate positively charged ions and vice versa for negative ions.
- The detached section of ions sets off along the TOF analyzer, with the ions having velocities proportional to the square roots of their m/z values.
- A reflectron is a special device designed to reverse the direction of travel of ions as they near the end of the time-of-flight tube by applying an opposing electric-field gradient. At some point in the reflectron, the ions are stopped and then accelerated back out, returning through the flight tube along a slightly different trajectory.
- The total trajectory of the ions is approximately V-shaped, the top of one leg of the V being the position of the pusher electrode and the top of the other being the position of the ion collector (a microchannel plate detector).
- The reflectron increases the spatial separation of the ions of different m/z values by making them travel up and down the flight tube, so the distance traveled is twice what it would be if the ions simply passed once along the tube from one end to the other. The reflectron also narrows the energy spread for individual m/z values, thus improving mass resolution. TOF analyzers are not necessarily equipped with a reflectron.
- The TOF analyzer provides the full mass spectrum of all the ions in the main ion beam at the time the pulse of electric potential was applied, m/z values being derived from the flight times of the ions along the TOF analyzer.
- The upper limit of the mass range is about 10,000 mass units (Daltons).

Summary

The hybrid system is a modified Q/TOF and is ideal for examining mass spectra of substrates dissolved in solvents, as occurs with substances emerging from a liquid chromatographic column and passing into an atmospheric-pressure ionization source. By operating two hexapole sections as a bridge between the atmospheric-pressure inlet/ion source and an orthogonal TOF analyzer, a full mass spectrum can be obtained of all ions from the ion source and therefore of substrates eluting from an LC column. The solution used can also come from a static supply and not just from an LC column.

Chapter 23: Hybrid Quadrupole Time-of-Flight (Q/TOF) Instruments

- A quadrupole analyzer can be operated in either RF mode only or in RF/DC mode.
- In its RF-only configuration (wide band-pass mode), all ions produced in an ion source and accelerated gently into the quadrupole analyzer will pass through it, whatever their m/z values. In contrast, the RF/DC mode is selective for chosen m/z values.
- Upon emerging from the quadrupole, the ions are accelerated through about 40 V and focused into the time-of-flight (TOF) analyzer. A pusher electrode is sited alongside this focused ion beam.
- Application of a pulse of high electric potential (about 1 kV) to the pusher electrode over a period of about 3 μs causes a short section of the ion beam to be detached and accelerated into the TOF analyzer. A positive potential is used to accelerate positively charged ions and vice versa.
- The TOF analyzer is placed at right angles (orthogonal) to the main ion beam, and therefore the pusher electrode accelerates a short section of this beam at right angles to its original direction.
- The detached section of ions sets off along the TOF analyzer, the ions having velocities proportional to the square roots of their m/z values. Thus ions of smaller m/z value arrive first and those of larger value arrive last.
- TOF flight tubes often include a reflectron.
- The reflectron is a device that uses an opposing electric-field gradient to reverse the direction of travel of ions as they near the end of the flight tube. At some point in the reflectron, the ions are stopped and then accelerated back out, returning through the flight tube or along a slightly different trajectory.
- Where the return path is different, the trajectory of the ions is approximately V-shaped; the top of one leg of the V is the position of the pusher electrode, and the top of the other is the position of the ion collector (a microchannel plate detector).
- The reflectron increases the spatial separation of the ions of different m/z values by making them travel up and down the flight tube, so the distance traveled is twice what it would be if the ions simply passed along the tube from one end to the other. The reflectron also narrows the energy spread for individual m/z values, thus improving mass resolution.
- The TOF analyzer provides the full mass spectrum of all the ions in the main ion beam at the time the pulse of electric potential was applied, m/z values being derived from the flight times of the ions along their trajectory in the TOF analyzer.
- If the initial quadrupole analyzer is operated in its RF/DC mode (narrow band-pass mode), any one m/z value can be selected to pass right through the quadrupole analyzer. Ions of all other m/z values are shut out.
- The main ion beam emerging from the quadrupole now has only ions of one selected m/z value. If the TOF is operating, the mass spectrum would consist of only one main peak (with peaks for isotopes).
- However, if the selected ions can be collided with a neutral gas in a hexapole device before reaching the pusher electrode, the collisions will fragment some of the selected ions. Thus the ions reaching the pusher electrode will consist of ions of the originally selected m/z value plus ions of m/z values resulting from fragmentation. These ions travel into the TOF analyzer and give an MS/MS product ion spectrum.

Summary

By operating a quadrupole analyzer as a gate together with an orthogonal TOF analyzer, a full mass spectrum can be obtained of all ions from an ion source if the gate is open. Alternatively, by selectively opening and closing the gate, ions of selected m/z values can be chosen for MS/MS studies. In either case, the TOF analyzer is used to obtain the mass spectrum.

Chapter 24: Ion Optics of Magnetic/Electric-Sector Mass Spectrometers

- Substances are converted into species having positive or negative charges (ions) in the ion source.
- For an ion of mass (m) and a number (z) of positive or negative charges, the value m/z is an important mass spectrometric observable quantity.
- A stream of ions (an ion beam) is directed out of the ion source toward a collector that records its arrival.
- As with a light beam and glass lenses, an ion beam can be directed and focused using electric and magnetic fields, often called lenses by analogy with their optical counterparts.
- The system of electric and magnetic fields or lenses is called the ion optics of the mass spectrometer.
- Electric lenses correct aberrations in the shape of the ion beam.
- Electric and magnetic fields can be used sequentially to focus the beam of ions. The use of crossed electromagnetic fields is described in the discussion of quadrupoles (Chapter 25).
- Another important property of electric and magnetic fields is their ability to separate ions according to their individual masses (m_1, m_2, … , m_n) or, more strictly, their mass-to-charge ratio (m_1/z, m_2/z, … , m_n/z).
- After the ion source, the ion optics split the ion beam into its component m/z values (compare splitting white light into a spectrum of colors).
- By changing the strengths of the electric and magnetic fields, ions of different m/z values can be focused at just one spot (the collector).
- From the strengths of the electric and magnetic fields, m/z values are obtained.
- A chart showing the number of ions (abundance) arriving at the collector and their respective m/z values is a mass spectrum.

Summary

The ion optics of a magnetic-sector mass spectrometer cause the ion beam leaving the ion source to arrive at a collector after being separated into individual m/z values and focused.

Chapter 25: Quadrupole Ion Optics

- In a quadrupole mass analyzer, four parallel rods are arranged equidistantly from a central (imaginary) axis.
- Static and alternating (radio frequency) electric potentials are applied to opposite pairs of rods to give a resultant fluctuating electric field.
- Positive or negative ions (electrically charged species) from a source are injected along the central axis of the quadrupole assembly.
- For particular magnitudes and frequencies of the electric fields, only ions of selected mass can pass (filter) through the assembly to reach an ion detector.
- Those ions with masses too large or too small to pass through the quadrupole strike the rods and are lost.
- The selection of which ions can pass through the quadrupoles to reach the detector is made by varying the electric potentials and/or their frequency; usually it is easier to keep the frequency constant and alter the voltage.
- By altering the electric fields in a consistent manner, the masses of all ions formed in the source can be scanned sequentially from low mass to high or vice-versa to give a mass spectrum.
- The appearance of the mass spectrum is closely similar to that provided by a magnetic-sector instrument.

- If no DC (static) voltage is used, the remaining all-RF field guides all ions through the quadrupole assembly. There is no separation by m/z, and the quadrupole in this mode is often used as an ion/gas collision cell.

Summary

Through the application of DC and RF voltages to an assembly of four parallel rods, ions can be filtered along their central axis and mass measured to give a mass spectrum. In the all-RF mode, the assembly is used as a guide for ions of all m/z values.

Chapter 26: Time-of-Flight (TOF) Ion Optics

- In a time-of-flight (TOF) mass spectrometer, ions formed in an ion source are extracted and accelerated to a high velocity by an electric field in an analyzer consisting of a long, straight drift tube. The ions pass along the tube until they reach a detector.
- After the initial acceleration phase, the velocity reached by an ion is inversely proportional to its mass (strictly, inversely proportional to the square root of its m/z value).
- Since the distance from the source to the detector is fixed, the time taken for an ion to traverse the analyzer in a straight line is proportional to its velocity and hence its mass (strictly, proportional to the square root of its m/z value). Thus each m/z value has its characteristic time of flight from the source to the detector.
- In effect, the ions race each other along the drift tube, but the winners are always the ions of smallest m/z value, since these have the shortest flight times. The last to arrive at the detector are always those of greatest mass, which have the longest flight times. However, as in a race, for there to be a separation at the finish line (the detector), the ions must all start from the ion source at the same time (no handicapping allowed!).
- The times taken for ions of differing m/z values to reach the detector are of the order of a few microseconds, and the separation in times of arrival for ions of differing m/z value is less than this. Thus, if ions of different mass are to be separated adequately in a time domain (good resolution), they should all start from the ion source at exactly the same time or, more practically, within a few nanoseconds of each other.
- Ions for TOF mass spectrometry must be extracted from the ion source in instantaneous pulses. Therefore, either ions are produced continuously but are extracted from the source in pulses, or ions are produced directly in pulses.
- TOF mass spectrometry is ideally suited to those ionization methods that inherently produce ions in pulses, as with pulsed laser desorption or Cf-radionuclide ionization.
- There is no theoretical upper limit on m/z that can be examined, and TOF mass spectrometry is useful for substances having very high molecular mass. In practice, the current upper limit is about 350,000.
- Unfortunately, ions even of the same m/z value do have a spread of velocities after acceleration, so the resolution achievable with TOF is not very high because bunches of ions of one m/z value overlap those at the next m/z value.
- By use of an electrostatic ion mirror called a reflectron, arrival times of ions of the same m/z value at the detector can be made more nearly equal. The reflectron improves resolution of m/z values.
- After reflection in the reflectron, the ions must again pass along the length of the analyzer to reach the detector.
- The improvement in resolution with the reflectron is achieved at the expense of some loss in overall sensitivity due to loss of ions in the reflectron and in the second length of analyzer.
- For very high mass, when sensitivity is frequently critical, the reflectron is not used and lower resolution is accepted.

- The mass spectrum gives the abundances of ions for different times of arrival at the detector. Since the times are proportional to the square root of the m/z values, it is simple to convert the arrival times into m/z values.

Summary

After acceleration through an electric field, ions pass (drift) along a straight length of analyzer under vacuum and reach a detector after a time that depends on the square root of their m/z values. The mass spectrum is a record of the abundances of ions and the times (converted to m/z) they have taken to traverse the analyzer. TOF mass spectrometry is valuable for its fast response time, especially for substances of high mass that have been ionized or selected in pulses.

Chapter 27: Orthogonal Time-of-Flight (oa-TOF) Ion Optics

- Orthogonal TOF is the name commonly given to orthogonally accelerated time-of-flight mass spectrometry. It is sometimes referred to by the acronym oa-TOF, especially in official publications, but it is more usual for it to be referred to simply as orthogonal TOF; this abbreviation is used here.
- Orthogonal TOF optics are compared with those from magnetic-sector instruments.
- In conventional mass spectrometry with electric and magnetic sectors arranged in-line (see Chapter 26), an ion beam consists of a stream of ions of all m/z values, which is separated into individual m/z values by the magnetic sector before being collected by single-point or multipoint detectors (see Chapters 28 and 29).
- In TOF mass spectrometry, ions of different m/z values are detected as a function of their velocities along a flight tube (see Chapters 20 through 23).
- Thus, it can be said that conventional magnetic sectors separate ions into individual m/z values by dispersion in space (spatially) and not according to their flight times. Contrarily, TOF analyzers separate ions of different m/z values according to their velocities (temporally) but not spatially.
- These two types of analyzer are frequently used alone but can be used in tandem, with ions from a first magnetic analyzer passing through a region in which there is applied an electric field at right angles to the direction of the ion beam. This orthogonal electric field is pulsed at very short time intervals.
- Ions accelerated from the first analyzer have a velocity that is proportional to the initial accelerating voltage in the ion source. Upon reaching the orthogonal zone, the pulsed electric field gives these ions a further velocity but now in a direction at right angles to their original velocity. The resultant velocity is given by the vector sum of the initial and second velocities.
- After the pulsed electric field has been applied, a pulse of ions travels along a TOF analyzer placed at a right angle to the original ion beam. When the pulse is off, the ions have only their original velocities and continue into a different ion collector.
- The pulsed ions start their journeys down the TOF flight tube all at the same time; they separate in the TOF analyzer according to their velocities and arrive at the TOF ion collector at different times (temporally separated).
- Therefore, the orthogonal TOF mass spectrum is a snapshot of all the ions in the sampled ion beam at any one moment in time. The arrangement has advantages over magnetic sectors alone and TOF instruments alone (see Chapter 20 for further discussion).
- An orthogonal acceleration TOF mass spectrometer can be used with continuous ion sources with a high sampling efficiency (typically 20 to 30%). Consequently, the orthogonal TOF has a much higher duty cycle than a scanning instrument, which may have a duty cycle of only 0.1 to 1% when used to record a mass spectrum. Consequently, the sensitivity will be much higher for the orthogonal TOF mass spectrometer.
- Pulses of ions can be directed into the TOF analyzer at the rate of about 30 kHz, and, therefore, more than 30,000 spectra per second can be collected and summed. There are significant improvements in signal-to-noise ratios and speed of acquisition of data.

Summary

In combined sector/TOF analyzers, a beam of ions accelerated from an ion source by an electric field and sent into a sector instrument is further subjected to a second pulsed electric field applied at right angles to its initial direction. The resultant pulse of ions sets off along the flight tube of a TOF analyzer, where the ions separate into m/z values and are recorded (along with their respective abundances) as a mass spectrum. The combined sector/TOF analyzers have several significant advantages, not least for MS/MS studies and improved signal-to-noise ratios.

Chapter 28: Point Ion Collectors (Detectors)

- A mass spectrum is a plot of mass-to-charge (m/z) values for ions versus their abundances.
- A mass spectrometer analyzer disperses ions according to their various m/z values.
- Recording of the dispersed ion beams can take place simultaneously across a plane, as in an array detector, or, as described here, by being brought to a focus at one point sequentially.
- By placing a suitable detector at the focus (a point detector), the arrival of ions can be recorded.
- Point detectors are usually a Faraday cup (a relatively insensitive device) or, more likely, an electron multiplier (a very sensitive device) or, less likely, a scintillator (another sensitive device).
- Arrival of ions, which have a positive or negative charge, causes an electric current to flow either directly (Faraday cup) or indirectly (electron multiplier and scintillator detectors).
- The flow of electric current marks the arrival of ions, and its magnitude marks the abundance of ions arriving at a given m/z value.
- The electric current is used to drive a recorder of some kind, which can be an oscilloscope, pen recorder, or galvanometer UV recorder.
- More likely now, the analog electrical signal will be digitized and processed by a computer (data) system. Alternatively, a time-to-digital converter can be used as an ion collector (see Chapter 31).
- A data system stores the mass spectrum until required, when it can be printed at leisure or viewed immediately on the computer screen.

Summary

Having been separated according to their m/z values, ions can be focused sequentially at a point where there is a detector — usually a Faraday cup or an electron multiplier —– that generates an electrical current proportional to the number of ions arriving.

Chapter 29: Array Collectors (Detectors)

- A mass spectrum is a plot of mass-to-charge (m/z) values for ions versus their abundances.
- A mass spectrometer analyzer disperses ions according to their various m/z values either in space or in time.
- Recording of a dispersed ion beam can take place either at a point (see Chapter 28, "Point Ion Collectors") or across a plane, as in the array collector described here.
- An array collector is a collection of point collectors (elements) assembled in a plane.
- Ions of a given m/z value are collected at one of the small point ion detectors; ions of larger or smaller m/z values are collected at other point collectors placed on either side.

- By having a large number of point ion collectors in a line in a plane, many different m/z values can be recorded at the same time (concurrently rather than sequentially, as with a single-point ion collector).
- Depending on the dispersion achieved at the array plane, m/z values can be separated by fractions of a unit mass, by unit mass, or by only tens of mass units.
- Each element of an array detector is essentially a small electron multiplier, as with the point ion collector, but much smaller and often shaped either as a narrow linear tube or as somewhat like a snail shell.
- The arrival of ions at the opening of one of the array elements causes a shower of electrons to pass to the end of the collector, where they are recorded as a current flow, which is usually amplified.
- The magnitude of the current flow is proportional to the number of ions arriving at the array element per unit time.
- The amplified signal can be recorded directly by analog means but, more likely, the signal is digitized and processed by a computer (data system).
- The resolution achievable by an array assembly depends critically on the number of elements in the array, the separation of one element from another, and the degree of dispersion of the ions in the ion beam.
- By collecting all ions at the same time, a mass spectrum can be obtained instantaneously instead of over a period of time as with a point ion collector.
- Array detectors are particularly useful for detecting ions from either a very small amount of a substance or when ionization is not continuous but intermittent.

Summary

The ions in a beam that has been dispersed in space according to their various m/z values can be collected simultaneously by a planar assembly of small electron multipliers. All ions within a specified mass range are detected at the same time, giving the array detector an advantage for analysis of very small quantities of any one substance or where ions are produced intermittently during short time intervals.

Chapter 30: Comparison of Multipoint Collectors (Detectors) of Ions: Arrays and Microchannel Plates

- A mass spectrum is a plot of mass-to-charge (m/z) values for ions versus their abundances.
- A mass spectrometer analyzer disperses ions either in space or in time according to their various m/z values.
- Recording of a dispersed ion beam can take place either at a point (see Chapter 28, "Point Ion Collectors") or across a plane, as in the multipoint collectors discussed here.
- A multipoint collector is an assembly (array) of single-point collectors (elements), packed closely together in a plane.
- Ions arrive at one end of each element of a multipoint collector and trigger a cascade of electrons, which moves toward the opposite end and is detected electronically. The resulting electric current corresponds to the ion current.
- The strength of the ion current relates to the number of ions per second arriving at the collector plate, and a mass spectrum can be regarded as a snapshot of the current taken over a definite period of time. Because of the finite time taken to produce a mass spectrum, it is a record of the abundances of ions (often mistakenly called intensities of ions).
- Thus, a mass spectrum records ion abundances in one dimension. In the second dimension, it records m/z ratios. The mass spectrum is a record of m/z values of ions and their abundances.

- In a mass spectrometer, ions can arrive at a multipoint collector as a spatially dispersed beam. This means that all ions of different m/z values arrive simultaneously but separated in space according to each m/z value. Each element of the array, depending on its position in space, detects one particular m/z value (see Chapter 29, “Array Collectors”).
- Alternatively, the ions in a mass spectrometer can also arrive at a multipoint collector as a temporally dispersed beam. Therefore, at any point in time, all ions of the same m/z value arrive simultaneously, and different m/z values arrive at other times. All elements of this collector detect the arrival of ions of one m/z value at any one instant of time. This type of detector, which is also an array, is called a microchannel plate collector of ions.
- To differentiate their functions and modes of operation, the array collector of spatially dispersed m/z values is still called an array collector for historical reasons, but the other multipoint detector of a temporally dispersed range of m/z values is called a microchannel plate (typically used in time-of-flight instruments).
- Each element of an array or a microchannel plate ion collector is essentially an electron multiplier, similar in operation to the type used for a point ion collector but very much smaller.
- Unlike the array collector, with a microchannel plate all ions of only one m/z value are detected simultaneously, and instrument resolution does not depend on the number of elements in the microchannel array or on the separation of one element from another. For a microchannel plate, resolution of m/z values in an ion beam depends on their being separated in time by the analyzer so that their times of arrival at the plate differ.
- In a beam of ions separated in time according to m/z value, the total time taken for ions of different m/z values to arrive at a microchannel plate is so short (about 30 μsec) that the spectrum appears to have been obtained instantaneously. Thus, for practical purposes, the array and microchannel plate collectors produce an instantaneous mass spectrum, even though the first detects a spatially dispersed set of m/z values and the second detects a temporally dispersed set.
- Both types of multipoint array are particularly useful for detecting ions produced either from a very small amount of a substance or when ionization is not continuous but intermittent.

Summary

After the analyzer of a mass spectrometer has dispersed a beam of ions in space or in time according to their various m/z values, they can be collected by a planar assembly of small electron multipliers. There are two types of multipoint planar collectors: an array is used in the case of spatial separation, and a microchannel plate is used in the case of temporal separation. With both multipoint assemblies, all ions over a specified mass range are detected at the same time, or apparently at the same time, giving these assemblies distinct advantages over the single-point collector in the analysis of very small quantities of a substance or where ions are produced intermittently during short time intervals.

Chapter 31: Time-to-Digital Converters (TDC)

- In time-of-flight (TOF) mass spectrometers, a pulse of ions is accelerated electrically at zero time. Having attained a maximum velocity, the ions drift along the flight tube of the analyzer. The times of arrival of ions at a detector are noted.
- From the flight times, it is easy to deduce the m/z values for the ions and then to produce a mass spectrum.
- The times taken for the ions to drift along a typical analyzer tube are only a few microseconds, and a very accurate clock is needed if the derived m/z values are to be accurate.
- This timing is done electronically by using a microchannel array ion collector and time bins. The microchannel array sends an electrical pulse to a time bin. Since a pulse is recorded, the ion arrival times are already digitized, hence the name time-to-digital converter (TDC).

- A series of consecutive time bins covers a length of time of a few milliseconds, with each bin representing a time of only a fraction of a nanosecond. When an ion arrives at the microchannel array detector, one time bin notes the resulting electronic pulse.
- By observing which bin in the series has been affected by the pulse from an ion arrival event and knowing how much time each bin represents, it is an easy matter to calculate the time taken for the ion to drift the length of the analyzer tube.
- For example, if each bin represents 0.3 nsec and bin number 200 has been affected by an ion arrival, then the flight time must have been $200 \times 0.3 = 60$ nsec. Knowing the length of the drift tube, the ion drift velocity can be calculated, and from that calculation its m/z value can be deduced.
- Time bins allow very small time intervals to be measured accurately and therefore provide accurate m/z values.
- One drawback of time bins relates to the events they record. If two or more ions arrive at the array detector at the same instant, the resulting electrical pulse is the same as if only one ion had arrived. The bins are blind to multiple concurrent events.
- This problem is known as dead time. To offset this effect, an algorithm is used to adjust the actual number of events into a true number of events. Since the numbers of ions represent ion abundances, the correction adjusts only abundances of ions before a mass spectrum is printed.
- Ion arrival times are not adjusted, and the observed m/z values are just those originally measured.

Summary

Accurate m/z values can be obtained on time-of-flight mass spectrometers by electronically measuring the timing of ion impacts using very small time intervals. Ion arrival times represent pulses of electricity (digits) and do not need to be converted from analog signals into digitized ones, hence the name time-to-digital converter.

Chapter 32: Origin and Uses of Metastable Ions

- Metastable ions are most readily detected following electron ionization.
- In the few microseconds that molecular ions (M^+) spend in an ion source following electron ionization, many have sufficient energy to decompose to give fragment ions (F_1^+, F_2^+, … , F_n^+).
- This assembly of ions (M^+, F_1^+, … , F_n^+) in the source is analyzed by the mass spectrometer to give a mass (m/z) versus ion abundance chart, i.e., a mass spectrum.
- Ions formed in the source in this way compose the majority and are called normal ions.
- A small proportion of the ions formed in the source have insufficient energy to fragment there but have just sufficient energy to decompose in the few microseconds of flight between the source and the detector. These are the metastable ions.
- Modern mass spectrometers are set to transmit and measure normal ions but, under normal circumstances, the metastable ones are not recorded.
- By adjusting magnetic and electric fields in the ion optics (see Chapters 33 and 34 on linked scanning), metastable ions can be investigated.
- Normal ions (M^+, F_1^+, … , F_n^+) in a spectrum can provide a molecular structure for substance M if the fragments can be theoretically reassembled. The problem is rather like deducing an original jigsaw puzzle by putting the pieces together correctly. For most molecules containing more than a few atoms, this reassembly exercise is difficult and often problematic.
- To carry out the reassembly exercise, connectivities between the ions must be determined. For example, does the ion F_3^+ come from M^+ or F_1^+ or F_2^+? This process can be likened to completing a jigsaw by finding where and how the pieces fit together.

- In mass spectrometry, the required connectivities (links) can be obtained from observations on metastable ions. For example, finding a metastable ion for a supposed process, $M^+ F_3^+$, proves the process exists.
- By measuring a mass spectrum of normal ions and then finding the links between ions from the metastable ions, it becomes easier to deduce the molecular structure of the substance that was ionized originally.

Summary

Metastable ions are useful for determining the paths by which molecular ions of an unknown substance have decomposed to give fragment ions. By retracing these fragmentation routes, it is often possible to deduce some or all of the molecular structure of the unknown.

Chapter 33: Linked Scanning and Metastable Ions in Quadrupole Mass Spectrometry

- Ions formed in an ion source can be described as normal or metastable.
- Normal ions are readily and easily observed by quadrupole mass spectrometers.
- Metastable ions cannot be detected with simple quadrupole instruments.
- Knowledge of which normal ions in a mass spectrum fragment to which others is important and can be obtained from observations on metastable ions.
- Metastable ions can be detected efficiently by using three quadrupoles in tandem, a QQQ instrument.
- In use, the second quadrupole is set with its fields to pass all ions through to the third quadrupole. This setup is known as QqQ.
- By linking the way in which the first and third quadrupoles are scanned, specific metastable fragments can be detected, which is one form of linked scanning.
- Rather than looking at just the low-abundance metastable ion processes occurring in the second quadrupole, extra fragmentation can be induced by having a neutral collision gas present in this quadrupole.
- Collision of normal ions from the first quadrupole with gas molecules in the second quadrupole increases fragmentation, a process known as either collisionally induced dissociation (CID) or collisionally activated decomposition (CAD).
- Again, as for metastable ions, linked scanning of the first and third quadrupoles reveals important information on fragmentation processes, viz., which normal ions fragment to give which product ions.
- Linked scanning provides important information about molecular structure and the complexities of mixtures, and it facilitates the detection of trace components of mixtures.
- Linked scanning is particularly easy with a triple quadrupole instrument.

Summary

Triple quadrupole instruments can be used to detect metastable ions or can be used for linked scanning to obtain information about molecular structure.

Chapter 34: Linked Scanning and Metastable Ions in Magnetic-Sector Mass Spectrometry

- A normal, routine electron ionization mass spectrum represents the m/z values and abundances of molecular and fragment ions derived from one or more substances.
- The molecular ions yield molecular mass information and molecular formulae. These in themselves are very useful but are usually insufficient to fully elucidate the structure of an unknown substance.

- The fragment ions are characteristic of any given substance in that the way a compound fragments can be regarded as a fingerprint. For the fragment ions to be useful, it is necessary to know how they relate to each other. In other words, it is necessary to understand which fragment should be connected to which others in the overall chemical structure.
- The required connections between fragment ions cannot be obtained from a normal, routine mass spectrum, which records only those ions formed in the ion source and that find their way to an ion collector.
- Ions that fragment in flight between the ion source and the ion collector are called metastable ions and can give information about connections between fragment ions.
- By obtaining the metastable ion connections, it becomes easier to build up a likely structure for an unknown substance. This process is similar to reassembling a jigsaw puzzle by putting together all the pieces (fragments) with the right connections and ending with a complete picture (chemical structure).
- To examine metastable ions in electric/magnetic-sector instruments it is necessary to manipulate one or more of the electric or magnetic fields.
- There are three important fields, two electric and one magnetic.
- The accelerating voltage V for ions leaving the source and the electric-sector voltage E for energy focusing are the two electric ones. The field B is the magnetic one,
- Each of the fields V, E, and B can be varied by itself to examine metastable ions and their connections, but with modern automated techniques it is better to use linked scanning.
- Linking requires that two of the fields be changed simultaneously and automatically under computer control. Fields V and E or E and B are generally linked.
- By automating the linked scanning under computer control, a complete mass spectrum can be scanned for metastable ions in just a few seconds.
- At such a rate of scanning, it is even possible to examine eluants from capillary GC (gas chromatography) columns during GC/MS operations.
- Along with the normal, routine mass spectrum, the resulting metastable ion connections often supply enough information to allow a structure of an unknown substance to be deduced or to confirm the identity of one substance with another by comparison with their metastable ion behavior.
- Scanning techniques are carried out differently with such hybrid instruments as the triple quadrupole analyzer, the Q/TOF (quadrupole and time-of-flight), and double magnetic-sector instruments.

Summary

Automated linked scanning of metastable ions is valuable for deducing a whole or partial molecular structure of an unknown substance.

Chapter 35: Gas Chromatography (GC) and Liquid Chromatography (LC)

- Chromatography is a method for separating mixtures of substances into their individual components.
- Substances can be loosely categorized as volatile (including gaseous) and nonvolatile. The terms are not exact.
- At normal pressures (around atmospheric) and up to about 250°C (approaching the limit of thermal stability for most organic compounds), a volatile substance can be defined as one that can be vaporized by heat between ambient temperature (10 to 30°C) and 200 to 250°C. All other substances are nonvolatile.

- Gas chromatography (GC) deals with volatile substances that can be vaporized into a gas stream.
- Liquid chromatography (LC) concerns mostly nonvolatile substances dissolved in a liquid stream.
- LC can be used for both volatile and nonvolatile substances, but GC can handle only volatile substances.
- Chromatography was originally a method for separating and displaying mixtures of colored substances on a colorless column of solid material. The word *chromatography* is derived from chroma (color) and graph (writing).
- It was rapidly realized that this method could also be used to separate colorless substances, but the name stuck, and it is now simply a description of a process for separating any mixture, colored or colorless, into its component parts.
- Analogous to a race, in chromatography, a mixture of substances (the runners) is placed at the beginning of a column (the track) and then made to move along the column (race) to the end, where the faster ones arrive first and the slower ones arrive last.
- Unlike an ordinary race, in chromatography the runners (substances) are forced along the track (column) by either a stream of gas (GC) or a stream of liquid (LC).
- A detector is needed to sense when the separated substances are emerging from the end of the column.
- A mass spectrometer (MS) makes a very good, sensitive detector and can be coupled to either GC or LC to give the combined techniques of GC/MS or LC/MS, respectively.
- High-pressure liquid chromatography (HPLC) is simply a variant on LC in which the moving liquid stream is forced along under high pressure to obtain greater efficiency of separation.
- GC and LC (or HPLC) are two of the most widely used separation techniques in chemistry, biochemistry, pharmacology, and medical and environmental sciences.
- The coupled methods, GC/MS and LC/MS, form very powerful combinations for simultaneous separation and identification of components of mixtures. Hence, these techniques have been used in such widely disparate enterprises as looking for evidence of life forms on Mars and for testing racehorses or athletes for the presence of banned drugs.

Summary

Mixtures of substances can be separated into their individual components by passage through special (chromatographic) columns in the gas phase or liquid phase.

Chapter 36: Gas Chromatography/Mass Spectrometry (GC/MS)

- This chapter is best read in conjunction with Chapter 35, "Gas Chromatography and Liquid Chromatography," and any other chapters appropriate to the operation of a mass spectrometer.
- Gas chromatography/mass spectrometry (GC/MS) is an analytical technique combining the advantages of a GC instrument with those of a mass spectrometer.
- GC is a means of separating components of mixtures by passing them through a chromatographic column so that they emerge sequentially.
- With highly efficient capillary chromatographic columns, very small amounts of complex mixtures can be separated in the gas phase. Generally, the separated components cannot be positively identified by GC alone.
- MS is a means of examining a compound, also in the gas phase, so that its structure or identity can be deduced from its mass spectrum. MS alone is not good for examining mixtures because the mass spectrum of a mixture is actually a complex of overlapping spectra from the individual components in the mixture.

- Because a GC and an MS both operate in the gas phase, it is a simple matter to connect the two so that separated components of a mixture are passed sequentially from the GC into the MS, where their mass spectra are obtained. This combined GC/MS is a very powerful analytical technique, the two instruments complementing each other perfectly.
- The combined GC/MS system provides more information than is obvious from the simple sum of the two separate instruments.
- GC can be combined with all kinds of mass spectrometers, but, for practical and economic reasons, only quadrupolar and magnetic/electric-sector instruments are in wide use. Ion traps can also be used.
- A good GC/MS instrument routinely provides a means for obtaining the identities and amounts of mixture components rapidly and efficiently. It is not unusual to examine micrograms or less of material.
- GC/MS is used in a wide range of applications, including environmental, archaeological, medical, forensic, and space sciences, chemistry, biochemistry, and control boards for athletics and horse racing.

Summary

By combining a GC instrument with MS, the powerful combination of GC/MS can be used to analyze, both qualitatively and quantitatively, complex mixtures arising from a wide variety of sources.

Chapter 37: Liquid Chromatography/Mass Spectrometry (LC/MS)

- This chapter is best read in conjunction with Chapter 35, "Gas Chromatography and Liquid Chromatography," and any other chapters concerning the operation of a mass spectrometer.
- Liquid chromatography/mass spectrometry (LC/MS) is an analytical technique combining the advantages of an LC instrument with those of a mass spectrometer.
- LC, or sometimes HPLC (high-pressure liquid chromatography), is a means of separating components of mixtures by passing them in a solvent through a chromatographic column so that they emerge sequentially.
- With highly efficient chromatographic columns, very small amounts of complex mixtures can be separated in the liquid phase. Generally, the separated components cannot be positively identified by LC alone.
- MS is a means of examining a compound, in the gas phase, so that its structure or identity can be deduced from its mass spectrum. MS alone is not good for examining mixtures because the mass spectrum of a mixture is actually a complex of overlapping spectra from the individual components in the mixture.
- LC operates in the liquid phase, while MS is a gas-phase method, so it is not a simple matter to connect the two. An interface is needed to pass separated components of a mixture from the LC to the MS. With an effective interface, LC/MS becomes a very powerful analytical technique.
- The combined LC/MS system provides more information than is obvious from the simple sum of the two separate instruments.
- LC can be combined with all kinds of mass spectrometers, but for practical reasons only quadrupolar, magnetic/electric-sector, and TOF instruments are in wide use. A variety of interfaces are used, including thermospray, plasmaspray, electrospray, dynamic fast-atom bombardment (FAB), particle beam, and moving belt.
- A good LC/MS instrument routinely provides a means for obtaining the identities and amounts of mixture components rapidly and efficiently. It is not unusual to examine micrograms or less of material.
- LC/MS is used in a wide range of applications, including environmental, archaeological, medical, forensic, and space sciences, chemistry, biochemistry, and control boards for athletics and horse racing.

Summary

By combining an LC instrument to an MS through an interface, the powerful combination of LC/MS can be used to analyze, both qualitatively and quantitatively, complex mixtures arising from a wide variety of sources.

Chapter 38: High-Resolution, Accurate Mass Measurement: Elemental Compositions

- Atoms have relative masses close to integer numbers: for example, hydrogen (H), is near 1, helium (He) is near 4, and nitrogen (N) is near 14.
- For molecules, the integer molecular mass is obtained by using the molecular formula and adding up the relevant individual masses. Ammonia (NH_3) has an integer mass of 17, made up of 1×14 for N and 3×1 for hydrogen.
- A mass spectrometer measures mass-to-charge ratio (m/z) and, often, the charge on the ion is unity, so that m/z = m/1 = m. Thus, a mass spectrometer can be used to measure mass.
- For an ion of NH_3, the measured integer mass would be 17, viz., m/z = 17/1 = 17, for z = 1.
- In theory, this process can be reversed in that any measured mass leads to an elemental composition. For example, a measured value of 17 would imply the composition, NH_3.
- In practice, other elemental compositions could add up to 17. For example, OH (oxygen = 16, hydrogen = 1), CDH_3 (carbon = 12, deuterium = 2).
- For larger masses, the possibilities increase enormously. At mass 100, there would be literally thousands of possible elemental compositions so that, although integer mass can be measured mass spectrometrically, attempts to obtain elemental compositions will not lead to a definite answer.
- In fact, atomic masses are not integers. On the atomic scale, carbon is given a value of 12.0000. On this accurate mass scale, oxygen is 15.9949, nitrogen is 14.0031, hydrogen is 1.0078, and so on.
- The accurate mass for ammonia (NH_3) is $14.0031 + 3 \times 1.0078 = 17.0265$, and for OH the mass is $15.9949 + 1.0078 = 17.0027$, giving a mass difference of 0.0238 units between NH_3 and OH, which are potentially separable.
- A mass spectrometer that can measure mass accurately to several decimal places (rather than just to the nearest integer) can be used to measure such differences.
- In the example given above, a measured mass of 17.0265 would indicate the definite composition NH_3 and eliminate the other possibility of OH.
- Even for large molecules, the ability to measure accurate mass means that elemental compositions can be obtained from the accurately measured molecular mass.
- A simple mass spectrometer of low resolution (many quadrupoles, magnetic sectors, time-of-flight) cannot easily be used for accurate mass measurement and, usually, a double-focusing magnetic/electric-sector or Fourier-transform ion cyclotron resonance instrument is needed.
- Accurate mass measurement on a molecular ion of any substance gives directly the molecular formula; for fragment ions, similar measurement gives their elemental compositions.
- The double-focusing mass spectrometer is arguably the finest instrument for obtaining molecular and elemental compositions.

Summary

A double-focusing mass spectrometer can measure mass accurately to several decimal places, thus enabling the determination of molecular formulae and elemental compositions of fragment ions.

Chapter 39: Choice of Mass Spectrometer

Summary

The choice of a particular mass spectrometer to perform a given task must take into account the nature of the substances to be examined, the degree of separation required for mixtures, the types of ion source and inlet systems, and the types of mass analyzer. Once these individual requirements have been defined, it is much easier to discriminate among the large number of instruments that are commercially available. Once suitable mass spectrometers have been identified, the final choice is often a case of balancing capital and running costs, reliability, ease of routine use, after-sales service, and the reputation of the manufacturer.

Chapter 40: Analysis of Peptides and Proteins by Mass Spectrometry

- Amino acids are the molecular building blocks of peptides and proteins. About 20 common amino acids are known.
- Peptides and proteins are formed by linking successive amino acids into chains or rings. The order (sequence) and types of amino acids determine the chemical and physical properties of peptides and proteins.
- An enzyme is a special protein that acts as a catalyst for biochemical reactions.
- An enzyme digest is the term applied to a process whereby a peptide or protein is mixed with a selected enzyme under favorable conditions to allow reaction to occur. The enzyme splits the peptide or protein into smaller units that are easier to identify.
- Post-translational modifications to proteins are biochemical in origin and alter the measured molecular mass relative to that calculated for an untranslated sequence.
- Fast-atom bombardment (FAB) is an ionization technique that produces a protonated or deprotonated molecular ion, hence a molecular mass for the sample. It can be used for analysis of peptides up to m/z about 5000.
- Dynamic/continuous-flow FAB allows a continuous stream of liquid into the FAB source; hence it constitutes an LC/MS interface for analyses of peptide mixtures.
- Laser-desorption mass spectrometry (LDMS) or matrix-assisted laser desorption ionization (MALDI) coupled to a time-of-flight analyzer produces protonated or deprotonated molecular ion clusters for peptides and proteins up to masses of several thousand.
- Electrospray ionization (ESI) produces a series of multicharged ions that can be transformed into an accurate molecular mass for proteins with masses of tens of thousands.
- Peptides and proteins can be analyzed by mass spectrometry. Molecular mass information can be obtained particularly well by MALDI and ESI.
- Tandem mass spectrometry (MS/MS) produces precise structural or sequence information by selective and specific induced fragmentation on samples up to several thousand Daltons. For samples of greater molecular mass than this, an enzyme digest will usually produce several peptides of molecular mass suitable for sequencing by mass spectrometry. The smaller sequences can be used to deduce the sequence of the whole protein.
- Samples containing mixtures of peptides can be analyzed directly by electrospray. Alternatively, the peptides can be separated and analyzed by LC/MS coupling techniques such as electrospray or atmospheric pressure chemical ionization (APCI).

Summary

The use of mass spectrometry for the analysis of peptides, proteins, and enzymes has been summarized. This chapter should be read in conjunction with others, including Chapter 45, "An Introduction to Biotechnology," and Chapters 1 through 5, which describe specific ionization techniques in detail.

Chapter 41: Environmental Protection Agency Protocols

- The Environmental Protection Agency (EPA) authorizes control over hazardous and potentially hazardous substances and validates appropriate methods of analysis, many of which require GC/MS.
- The so-called matrix — the predominant material of which the sample is composed — is the subject of any analysis. The matrix is frequently water, soil, or sediment.
- The analysis programs of the Comprehensive Environmental Response, Compensation, and Liability Act (CERCLA) are managed through the EPA's Contract Laboratory Program (CLP).
- A laboratory wishing to register under the CLP must perform certain analytical test procedures to the EPA's satisfaction.
- The EPA's protocols, published in the official Series Methods, describe the exact procedures that must be followed when handling, preparing, and analyzing samples and reporting the results.
- Target compounds are specified for each Series Method. Volatile compounds that need to be analyzed can be extracted from the matrix by a purge-and-trap device.
- Base, neutral, and acid compounds, which may be less volatile, are extracted from the matrix with organic solvents.
- Only mass spectrometer-based analyses are discussed in this guide.
- Calibration and tuning of the mass spectrometer are achieved using either bromofluorobenzene (BFB) or decafluorotriphenylphosphine (DFTPP).
- Initial calibration for a Series Method is achieved by analyzing a set of standards made up to specified concentrations.
- Continuing calibration for a Series Method is performed using calibration check compounds.
- Surrogate compounds are added to the matrix before sample preparation to evaluate recovery levels.
- To check GC retention times, internal standards are added to a sample after its preparation for analysis.
- The National Institutes of Health–EPA mass spectral library is used to identify analyzed components of a sample by comparing their mass spectra with those of authentic specimens held in the library.
- To produce a quantitative result, chromatographic peak areas of identified target compounds are compared with peak areas of the internal standards, which are of known concentration.

Summary

The Environmental Protection Agency lays down strict guidelines for the analysis of a range of environmentally hazardous substances. Many of the analyses utilize GC/MS.

Chapter 42: Computers and Transputers in Mass Spectrometers, Part A

- Binary numbers are composed of strings of zeros and ones. A typical binary number is 1110, which is equivalent to the decimal number 14.
- Binary numbers (just two digits) are useful in computer construction because the zero can be represented by an electronic switch in the off position and the one by a switch in the on position. Movement (change) from zero to one or vice versa is then simply the change in a switch from off to on or vice versa.

- All information into or out of a computer is therefore not continuous (analog) but flows as bits switched on and off (digital), like a series of electronic pulses.
- The special electronic switches are called bits and are arranged in sets, sometimes referred to as registers or memory locations, depending on usage. The sets of bits are 8 in the simplest computers, 16 in more advanced ones, and 32 or 64 in the latest.
- Whether there are 8, 16, 32, or 64 bits in each set, the set is called a byte.
- The capacity of a computer to carry out various tasks is partly governed by the number of bytes it has. Thus, a one-megabyte memory means there are 1 million locations with 8, 16, 32, or 64 bits in each.
- A computer must communicate with a variety of peripheral devices (keyboard, mouse, printer, mass spectrometer). A central processing unit (CPU) controls the flow of information to each, rather like a choreographer directing complicated dance routines.
- Some of this control is completely automatic and is called the operating system. Generally, it cannot be altered by the user, and it uses a low-level machine language (machine code), which enables the fastest response to commands.
- The instructions for the operating system reside within a memory unit that can be read but not changed in any way — the read-only memory (ROM).
- The user can input commands through a computer program (software), which the central processor will carry out. These programs reside in memory, usually on a disk (either hard or floppy) rather than the processor's own memory banks.
- The memory units within the computer — known as random access memory (RAM) — must be capable of change as needed to run software programs.
- Most computers have far more RAM than ROM, and the more they have, the more tasks can be carried out.
- Software programs are usually written in a more user-friendly high-level language such as Fortran, Pascal, or C, which facilitates the tedious and labor-intensive task of writing a computer program.
- Although much easier to assemble, a software program written in a high-level language requires more time for the computer to execute, since all the instructions must be translated into machine code before the computer can understand them. Even a simple statement like "start" in a high-level language requires several machine-code moves to execute.
- The working of the CPU is controlled by a crystal clock having a frequency, generally, of 16 to 25 MHz, depending on the type of computer. All electronic moves are controlled by the clock and operate in sequence to its ticking.

Summary

Digital computers operate with a binary system whereby all operations consist of a series of on/off electronic switching controlled by a crystal clock. The smallest on/off unit is the bit, and these are assembled into larger units known as bytes. The various functions of a computer are controlled through a processor, the CPU, which deals with incoming and outgoing signals and the execution of instructions (software) dealing with the signals.

Chapter 43: Computers and Transputers in Mass Spectrometers, Part B

- Movement of information (operations) in a computer is controlled by a microprocessor.
- In conventional processors, the various operations are carried out sequentially, viz., one after the other in a strictly controlled succession of movements (serially).
- The total time taken to carry out a sequence of operations (or instructions) is the average time for one operation multiplied by the number of operations. To carry out 100 instructions, each taking 100 nsec, requires a total of 10,000 nsec.

- An alternative to serial execution of operations is to split the total work into smaller groups, with each group carrying out its function simultaneously with the others (in a parallel fashion), viz., flows of instructions run in parallel (at the same time as each other).
- Now, if the average time needed to complete one instruction is 100 nsec and there are 100 such operations but handled by 10 groups of processors working in parallel, then the time taken is only 1000 nsec: $(100/10) \times 100$.
- Parallel processing is inherently faster than serial processing, but special processors are needed, and these are called transputers.
- Parallel processing requires that each transputer be able to communicate efficiently with others (up to four immediate neighbors with current transputers) if the final result is not to be garbled.
- Each transputer is a microprocessor with its own memory banks and its own built-in operating mode similar to a conventional microprocessor, but a transputer has additional input and output channels enabling it to communicate with other transputers. For example, in one simple mode, five transputers could be coupled so that four of them were carrying out operations at the same time (in parallel) but controlled by the fifth.
- A special computer language (Occam) is needed to enable transputers to be programmed in this cooperative mode, yielding true parallel processing of information with all its advantages in speed.
- Transputers also operate using an instruction set with fewer options than a conventional CPU (reduced instruction set for computing, RISC). Consequently, any sequence of operations is carried out more quickly because the processor does not have to search through a large instruction set to find out exactly what to do (analogous to reading this summary instead of searching through a textbook on computers).
- Parallel processing and the RISC set give transputers a considerable speed advantage over conventional serial processors for handling information or flows of data.
- Given the huge investments in time and resources in the development of serial processors and their software programs, it is unlikely that serial processors will be dropped in favor of transputers in the near future. However, in situations where large amounts of information must be handled in a very short time, transputers are used because they hold a distinct speed advantage over conventional processors.
- A mass spectrometer, with its need for high-speed data acquisition and processing while simultaneously handling background operations (instrument control, printing, library searching), is one such application where transputers can have a revolutionary impact. It is no surprise to find them being used increasingly in mass spectrometer systems.

Summary

By carrying out a set of instructions in parallel rather than sequentially (serially), any total operation (a set of computer programs) can be carried out much faster. The transputer is a special microprocessor designed to work in parallel with other transputers. Processing speed is further enhanced by using a special programming language (Occam) and a reduced set of basic operating instructions.

Chapter 44: Computers and Transputers in Mass Spectrometers, Part C

- Most electrical signals flowing between a mass spectrometer and an attached computer are of the analog type; viz., the associated voltage varies continuously with time.
- Inside a computer, electrical signals consist of series of pulses in which voltage rises from zero to some maximum and back to zero in a very short space of time (a pulse). These are digital voltages.
- For a mass spectrometer and a computer to operate together, the analog and digital voltages must be converted by using either an analog-to-digital converter (ADC) or a digital-to-analog converter (DAC).

- The mass spectrometer provides a mass spectrum that is actually an analog voltage varying in amplitude with time as ions of different m/z values arrive at the ion collector within a period of a few seconds. An important exception to this generalization occurs with ion collectors, called time-to-digital converters because their output is already digitized.
- After passing through an ADC, the resulting large number of digits or bits of information must be reduced to a more manageable level before being passed on to the computer storage area.
- This reduction in information is achieved by a preprocessor, which uses the digital voltages corresponding to an ion peak to estimate the peak area (ion abundance) and centroid (mean arrival time of peak, equivalent to m/z value); these two pieces of information — plus a flag to identify the peak — are stored.
- By working with a preprocessor, data-storage requirements are reduced from approximately a million bits of information to about 300.
- Once the mass spectral information has been acquired, various software programs can be employed to print out a complete or partial spectrum, a raw or normalized spectrum, a total ion current (TIC) chromatogram, a mass chromatogram, accurate mass data, and metastable or MS/MS spectra.
- Apart from acquiring and manipulating data, a computer can be used to tune the mass spectrometer and to check that all systems are operating correctly. This working mode is carried out by sequentially examining various input voltages from different parts of the instrument, checking these against normal readings, and altering any that need to be adjusted.
- Thus, a computer attached to a mass spectrometer must operate on two levels. When mass spectral information is arriving, this must be acquired in real time. When the computer has spare time, it controls the operation of the instrument. Both operations are carried out at such a high speed that the dual level of computer tasks is not obvious.
- However, the two levels may become obvious if the instrument operator tries, for example, to conduct a library search while the computer is trying to acquire input from another mass spectrum; the library search has to wait. Acquiring the data is a foreground task. Other functions such as library searching are background tasks.
- Powerful mass spectrometer/computer systems can achieve simultaneous foreground/background operation, especially if transputers are used to provide the advantage of parallel processing.

Summary

A computer attached to a mass spectrometer is used both to acquire data and to control the operation of the spectrometer. Powerful transputer systems can be used to ensure that both modes of operation can be carried out almost simultaneously.

Chapter 45: Introduction to Biotechnology

- Chromosomes are extremely complex chemicals that are assembled from simple repeating units and contain all the chemical information needed to reproduce animate species. Each living organism has its own complete set of chromosomes, called the genome.
- Genes are segments of chromosomes. Some of the genes are coded to give each animate species its characteristics (e.g., color and number of eyes, type of hair, muscle), and others are coded to produce the chemicals required for the organism to live (metabolism).
- Genes are constructed from sets of deoxyribonucleic acids (DNA), which in turn consist of chains of nucleotides. These chains occur in matched pairs, twisted around each other (a double helix).
- Any one nucleotide, the basic building block of a nucleic acid, is derived from a molecule of phosphoric acid, a molecule of a sugar (either deoxyribose or ribose), and a molecule of one of five nitrogen compounds (bases): cytosine (C), thymine (T), adenine (A), guanine (G), uracil (U).

- A chain of nucleotides containing only deoxyribose as the sugar is a DNA. Similarly, RNA possesses chains nucleotides having only ribose as the sugar and is therefore a ribonucleic acid.
- The information needed to reproduce and support an animate species is given by the order in which the nitrogen bases occur along the DNA or RNA chains (-C-T-T-A-G-, for example). A sequence of three such bases (a codon) provides the fundamental unit of information.
- α-Amino acids are the molecular building blocks of peptides and proteins. About 20 amino acids are known, each corresponding to one or more codons.
- Peptides and proteins are formed by linking successive amino acids into chains or rings. The order (sequence) and types of amino acids (read from the sequence of codons) determine the chemical and physical properties of peptides and proteins (and enzymes).
- An enzyme is a special protein that acts as a catalyst for biochemical reactions (metabolism).
- The genome, through its constituent DNAs, provides all of the codes needed for building a wide range of peptides, proteins, and enzymes, which in turn utilize raw materials (food) to form an animate body and keep it going. These multiple reactions work together as a unit within a water-filled cell.
- Simple life forms, such as bacteria, consist of single cells, whereas, at the other extreme, complex life forms such as animals, contain many types of cell, each having a specific function (cells in eyes, limbs, stomach, etc.).
- Changes in genes (mutation) — and therefore in genetic information — can occur naturally (mistakes in the billions of metabolic reactions) or accidentally through damage (cosmic particles, radioactivity, smoking, etc.), or they can be effected intentionally and specifically by genetic engineering.
- Genetic engineering uses special chemicals (restriction enzymes) to snip DNA in specific places and other chemicals (ligases) to stitch cut ends. By this recombinant technology, gene DNA can be cut and a new piece inserted; the new segment may be natural (from another DNA) or synthetic (made automatically by a gene machine that produces lengths of DNA from nucleotides).
- Gene cloning is a method that uses recombinant technology to insert a gene into a vector DNA (plasmid; obtained from a bacterium). The modified vector is put back into the bacterium, which then reproduces endlessly (clones) the new gene as well as the others in the vector.
- A virus is a species containing DNA and RNA that can reproduce itself, but to do this, it needs to hijack the metabolism (cells) of a host organism, since it has no information itself with which to build cells.

Summary

Life forms are based on coded chemicals that, in the right environment, can reproduce themselves and make other chemicals needed to break down and utilize food. Within an organism, these biochemical reactions constitute normal metabolism. Biotechnology is the manipulation of these biochemical reactions at either the cellular or the molecular level.

Chapter 46: Isotopes and Mass Spectrometry

- At the sorts of temperatures that exist normally on earth, all matter is made up from about 90 elements.
- Most of these elements are familiar, such as solid iron, liquid mercury, and gaseous helium.
- Most elements can combine in a variety of ways to create millions of substances such as chalk (a combination of three elements: calcium, carbon, and oxygen) and water (a combination of hydrogen and oxygen).
- If the elements exist as such, they are still called elements, but when they combine with each other, the different combinations are called compounds. Thus mercury is an element but water is a compound.

- Whether existing separately as elements or combined as compounds, elements are composed of atoms, which are the smallest part of an element that can exist naturally.
- Atoms of mercury cling together to form the familiar liquid, atoms of iron hold together to form the solid metal, and atoms of hydrogen and oxygen combine to form molecules that hold together as water.
- All matter is composed of atoms, sometimes all of one sort (as with iron), and sometimes a combination of atoms (as with rust, which is a combination of atoms of the element iron and atoms of the element oxygen).
- Each atom comprises a small dense nucleus surrounded by electrons. Most of the mass of each atom resides in the nucleus, and for most purposes, the mass of the electrons in an atom can be ignored.
- A nucleus is composed of protons and neutrons, each of which has unit atomic mass. The number of protons characterizes each element. In going from one element to the next, the total number of protons increases by one. Thus the simplest element, hydrogen, has atoms having only one proton in the nucleus, and the next simplest, helium, has two protons in the nucleus.
- While the number of protons in an atomic nucleus characterizes each element, the mass of the nucleus comprises the total number of protons and neutrons.
- An atom of one of the simplest elements, helium, has a mass of 4 atomic units (Daltons) comprising 2 protons and 2 neutrons, whereas each atom of a heavier element, phosphorus, has 15 protons and 16 neutrons, giving it a mass of 31 Da.
- The ratio of the number of protons to neutrons is not fixed. In some elements, as with phosphorus, the ratio is 15 to 16 (1:1.07), but in the precious metal rhodium, the ratio is 45 to 58 (1:1.29).
- Phosphorus and rhodium are unusual among the elements in that they consist of atoms that naturally contain only one ratio of protons to neutrons and therefore have only one mass: 31 (15 protons plus 16 neutrons) for phosphorus and 103 (45 protons and 58 neutrons) for rhodium. Such elements are called monoisotopic — each of their atoms has one (and only one) mass in each case.
- Other elements have atoms that can have different ratios of protons to neutrons. Indeed, hydrogen actually consists of three types of atoms. All hydrogen atoms have the same number of protons (one for hydrogen), giving each a mass of 1 Dalton, but some atoms of hydrogen also contain one neutron in the nucleus as well as the proton (mass of 2 Da), while yet others have two neutrons with each proton (mass of 3 Da). Thus hydrogen has three naturally occurring isotopes of mass 1, 2, and 3 Da.
- Chemically, there are only small differences between the reactivities of the different isotopes for any one element. Thus isotopes of palladium all react in the same way but react differently from all isotopes of platinum.
- One instrument that can reveal the presence of isotopes is a mass spectrometer, which can be regarded as a very accurate weighing machine!
- The three isotopes of hydrogen are almost indistinguishable for most chemical purposes, but a mass spectrometer can see them as three different entities of mass 1, 2, and 3 Da. Isotopes of other elements can also be distinguished. Mass spectrometry is important for its ability to separate the isotopes of elements.
- A few natural isotopes are radioactive. Of the three isotopes of hydrogen, only that of mass 3 (tritium) is radioactive. Radioactive isotopes can be examined by other instrumental means than mass spectrometry, but these other means cannot see the nonradioactive isotopes and are not as versatile as a mass spectrometer.
- Many artificial (likely radioactive) isotopes can be created through nuclear reactions. Radioactive isotopes of iodine are used in medicine, while isotopes of plutonium are used in making atomic bombs.
- In many analytical applications, the ratio of occurrence of the isotopes is important. For example, it may be important to know the exact ratio of the abundances (relative amounts) of the isotopes 1, 2, and 3 in hydrogen. Such knowledge can be obtained through a mass spectrometric measurement of the isotope abundance ratio.
- All mass spectrometers can measure abundance ratios to some degree of accuracy, but special mass spectrometers have been designed to measure isotope ratios very accurately. These specialized devices are used in a wide range of applications, such as dating the antiquity of objects, unraveling details of reaction mechanisms, and testing athletes for illegal use of bodybuilding drugs.

Summary

Atoms of elements are composed of isotopes. The ratio of natural abundance of the isotopes is characteristic of an element and is important in analysis. A mass spectrometer is normally the best general instrument for measuring isotope ratios.

Chapter 47: Uses of Isotope Ratios

- An element is characterized by its atomic number (Z), in which Z = 1, 2, 3, etc. The first element (hydrogen) has Z = 1, the next (helium) has Z = 2, and so on up to the heaviest natural element (uranium) with Z = 92.
- An atom contains a nucleus, surrounded by electrons. Most of the mass of the atom is centered in the nucleus.
- The nucleus consists of protons and neutrons; the number of protons (P) is equal to the atomic number (P = Z).
- The nucleus also contains neutrons. The number of neutrons (N) for any one element is similar to but not necessarily equal to the number of protons.
- Each proton or neutron has an atomic mass close to 1 Da. Neglecting the small electron mass and other factors, the total atomic mass of an element is given by the sum (P + N).
- For each element, the number of protons is fixed. Thus, for hydrogen (Z = 1) there is just one proton (P = 1); for the next element, helium (Z = 2), there are just two protons (P = 2); and so on up to the heaviest natural element, uranium, which has atomic number 92 and therefore has Z = P = 92.
- Since the total integer atomic mass (M) is given by the number of protons and neutrons, then M = P + N. Because of the masses of the electrons in an atom and a packing fraction of mass in each nucleus, the actual atomic mass is not an integer.
- Some elements contain a fixed number of neutrons, as with fluorine (P = 9, N = 10) and phosphorus (P = 15, N = 16). For their natural occurrences, atoms of any one such element all have the same mass (F = 19; P = 31).
- Atoms of many other elements contain nuclei that have different numbers of neutrons. For example, carbon (Z = 6) can have six neutrons (M = 6 + 6 = 12), seven neutrons (M = 13), or eight neutrons (M = 14). Atoms of the same atomic number but having different numbers of neutrons (and different atomic masses) are called isotopes. Thus, naturally occurring carbon has three isotopes, for which Z = P = 6 and N = 6 or 7 or 8. These are written ${}^{12}_{6}C$, ${}^{13}_{6}C$, ${}^{14}_{6}C$.
- For any one element, the abundances (relative amounts) of isotopes can be described in percentage terms. Thus, fluorine is monoisotopic; viz., it contains only nuclei of atomic mass 19, and phosphorus has 100% abundance of atoms with atomic mass 31. For carbon, the first two isotopes occur in the proportions of 98.882 to 1.108.
- These isotope masses and their ratio of abundances are characteristic of carbon. Similarly, the isotopes of other elements that occur naturally have "fixed" ratios of isotopes, as given in Tables 47.1 and 47.2 at the end of the accompanying full text.
- In a mass spectrum, the ratios of isotopes give a pattern of isotopic peaks that is characteristic of a given element. For example, the mass spectrum of any compound containing carbon, hydrogen, nitrogen, and oxygen will show patterns of peaks due to the ${}^{12}_{6}C$, ${}^{13}_{6}C$, ${}^{14}_{7}C$, ${}^{15}_{7}N$, ${}^{16}_{8}O$, ${}^{17}_{8}O$, and ${}^{18}_{8}O$ isotopes. The ratios of isotope abundances can be estimated directly from a routine mass spectrum and are frequently useful for identifying the presence of specific elements. Such uses are discussed in the main text.
- When measured carefully, isotope ratios are found not to be fixed but to vary slightly, depending on several factors. This variation is often very small and can be difficult to detect.

- Special isotope ratio mass spectrometers are needed to measure the small variations, which are too small to be read off from a spectrum obtained on a routine mass spectrometer. Ratios of isotopes measured very accurately (usually as 0/00, i.e., as parts per 1000 [mil] rather than parts per 100 [percent]) give information on, for example, reaction mechanisms, dating of historic samples, or testing for drugs in metabolic systems. Such uses are illustrated in the main text.

Summary

Isotope ratios are very useful for (a) identifying elements from their pattern of isotopes in a spectrum obtained on an ordinary mass spectrometer or (b) obtaining detailed information after accurate measurement of isotope ratios from special isotope ratio instruments.

Chapter 48: Variations in Isotope Ratios

- Many elements exist as isotopes, viz., atoms of the same atomic number containing different ratios of protons to neutrons.
- A simple example occurs with hydrogen, which occurs naturally as three isotopes (hydrogen, deuterium, tritium), all of atomic number 1 but having atomic masses of 1, 2, and 3 respectively.
- Some elements are monoisotopic (^{19}F, ^{31}P), some have two or three isotopes (^{14}N, ^{15}N; ^{16}O, ^{17}O, ^{18}O), and a few have as many as 7–11 isotopes (^{196}Hg, ^{198}Hg, ^{199}Hg, ^{200}Hg, ^{201}Hg, ^{202}Hg, ^{204}Hg).
- Most naturally occurring isotopes are not radioactive, but there are a few notable exceptions such as ^{14}C, ^{40}K, and ^{235}U.
- Many artificially made isotopes are known, and most have very short half-lives. For example, ^{203}Au has a half-life of 53 sec.
- Approximate ratios of isotope abundance ratios are important in identifying elements. For example, the naturally occurring ^{35}Cl, ^{37}Cl isotopes exist in an abundance ratio of about 3:1, and ^{12}C, ^{13}C exist in a ratio of about 99:1.
- Routine mass spectrometry can be used to identify many elements from their approximate ratios of isotope abundances. For example, mercury-containing compounds give ions having the seven isotopes in an approximate ratio of 0.2:10.1:17.0:23.1:13.2:29.7:6.8.
- Other important areas of mass spectrometric investigation of isotope ratios need accurate, not approximate values. For example, for some investigations in archaeology, pharmaceuticals, and chemistry, very accurate precise ratios of isotope abundances are needed.
- Variations in such accurate ratios can give valuable information.
- Isotope ratios are commonly reported as relative abundance (R), and comparison of two ratios (R_1, R_2) represent the α-value (known as a fractionation factor).
- A δ-value is used to compare a measured isotope ratio in a sample with that for a standard substance containing the same isotopes but in known abundance ratio.
- δ-Values are reported as parts per 1000 (mil) rather than parts per 100 (percent).
- Standard substances are available from such agencies as the International Atomic Energy Authority (U.K.) and the National Institute for Standards and Technology (U.S.).
- Special instruments (isotope ratio mass spectrometers) are needed to measure the required very accurate, precise ratios of abundances.
- With such mass spectrometers, plasma torches and thermal ionization are the most widely used means for ionizing samples for ratio measurements.
- Almost any kind of mass analyzer can be used to measure the isotope m/z values and abundances, but the usual ones are based on magnetic sectors, quadrupoles, and time-of-flight.

- Apart from the need for a specialized instrument, sample preparation before analysis is extremely important for reliable results. The sample preparation must not itself alter the ratios of isotopes in the samples under investigation.
- Usually, during sample preparation or during ionization, samples are broken down to simpler substances (e.g., CO_2) or to their elements.
- An isotope ratio is frequently measured 10 to 20 times for each sample to obtain high accuracy and precision.
- Special sample inlet devices such as nebulizers, furnaces, and gas inlets are commonly used to avoid cross-contamination and accidental fractionation of isotopes.

Summary

Accurate, precise isotope ratio measurements are used in a variety of applications including dating of artifacts or rocks, studies on drug metabolism, and investigations of environmental issues. Special mass spectrometers are needed for such accuracy and precision.

Chapter 49: Transmission of Ions through Inhomogeneous RF Fields

- In an inhomogeneous radio frequency (RF) field, ions can be guided from one part of a mass spectrometer to another.
- The RF fields are applied to electrodes called *poles*.
- The most common ion guide is the quadrupole, in which there are four parallel metal rods spaced evenly about a central axis.
- The RF field is applied in such a way that one pair of opposed rods has the positive phase applied and the other pair has the negative phase applied.
- The inhomogeneous field produces a trapping effect for ions so that, if an ion moves close to a pole, the field pushes it away and back toward the central axis.
- The RF field has electric components only at right angles (x-, y-directions) to the main central axis (z-direction).
- If an ion has to travel from one end of a quadrupole to the other, it must have some kinetic energy in the z-direction. This kinetic energy can be induced by application of an accelerating potential to the ions before they enter the quadrupole field.
- Alternatively, ions moving toward a quadrupole can gain sufficient kinetic energy in the z-direction by space-charge repulsion from following ions.
- More poles can be used, as with the hexapole and octopole.
- In the hexapole, one phase of the RF field is applied to three evenly spaced rods and the other phase to the remaining triplet of rods.
- Similarly, in the octopole, the rods are arranged as two sets of four.
- The efficiency of the ion guiding effect increases from quadrupole to hexapole to octopole.
- The ion guides are frequently used to transmit ions from an atmospheric-pressure inlet/source system (electrospray ionization, atmospheric-pressure chemical ionization) into the vacuum region of an m/z analyzer.
- The ion guides are also used as gas collision cells. When ions collide with neutral gas atoms in such a cell, it is important that ion losses due to deflection or collision should be minimized. Ion guides perform this task.

- The ion guides can also be constructed as a series of flat rings, placed parallel to each other, the centers of the rings lying along a long linear axis.
- An RF field is applied to the rings. The positive phase is applied to every other ring and the negative phase to the remaining intervening rings. Such an ion guide is called an ion tunnel.
- Ion tunnels are even more efficient in their guiding motion on ions than are the multipoles.
- Again, ion tunnels have no electric-field component in the z-direction, and ions must be injected with some initial kinetic energy if they are to pass through the device.

Summary

By application of an RF field to a system of rods or rings, an inhomogeneous field is produced. Ions entering this field are guided along a central axis and emerge from the other end. These ion guides reduce the loss of ions that would normally result from space-charge effects, stray electric fields, and ion/molecule collisions.

Appendix B

Glossary of Mass Spectrometry Definitions and Terms

Analyzers

Average mass. The mass of an ion for a given empirical formula calculated using the atomic weight of each element, e.g., C = 12.01115, H = 1.00797, O = 15.9994.

Double-focusing analyzer. A magnetic analyzer and an electrostatic analyzer combined in either sequence to effect direction and velocity focusing.

Dynamic-field mass spectrometer. A mass spectrometer in which the separation of an ion beam depends essentially on the use of a field or fields that vary with time. These fields are generally electrostatic or magnetic.

Electrostatic analyzer. A velocity-focusing device for producing an electrostatic field perpendicular to the direction of ion travel (usually used in combination with a magnetic analyzer for mass analysis). The effect is to bring to a common focus all ions of a given kinetic energy.

Ion cyclotron resonance analyzer. A device to determine the mass-to-charge ratio (m/z) of an ion in the presence of a magnetic field by measuring its cyclotron frequency.

Ion trap analyzer. A mass-resonance analyzer that produces a three-dimensional rotationally symmetric quadrupole field capable of storing ions at selected mass-to-charge (m/z) ratios.

Magnetic analyzer. A direction-focusing device that produces a magnetic field perpendicular to the direction of ion travel. The effect is to bring to a common focus all ions of a given momentum with the same mass-to-charge (m/z) ratio.

Mass analysis. A process by which a mixture of ionic (or neutral) species is separated according to the mass-to-charge (m/z) ratios (for ions) or their aggregate atomic masses (for neutrals). The analysis can be qualitative or quantitative.

Mass resonant analyzer. A mass analyzer for mass-dependent resonant-energy transfer and measurement of the resonance frequency, power, or ion current of the resonant ions.

Mass spectrograph. An instrument in which beams of ions are separated according to their mass-to-charge ratio (m/z) and in which the deflection and intensity of the beams are recorded directly on a photographic plate or film.

Mass spectrometer. An instrument in which ions are analyzed according to their mass-to-charge ratio (m/z) and in which the number of ions is determined electrically (or via scintillator, vidicon, etc.).

Mass spectrometer configuration. Multianalyzer instruments should be named for the analyzers in the sequence in which they are traversed by the ion beam, where B is a magnetic analyzer, E is an electrostatic analyzer, Q is a quadrupole analyzer, TOF is a time-of-flight analyzer, and ICR is an ion cyclotron resonance analyzer. For example: BE mass spectrometer (reversed-geometry double-focusing instrument), BQ mass spectrometer (hybrid sector and quadrupole instrument), EBQ (double-focusing instrument followed by a quadrupole).

Mattauch-Herzog geometry. An arrangement for a double-focusing mass spectrometer in which a deflection of $\pi/4\sqrt{2}$ radians in a radial electrostatic field is followed by a magnetic deflection of $\pi/2$ radians.

Monoisotopic ion mass. The mass of an ion for a given empirical formula calculated using the exact mass of the most abundant isotope of each element, e.g., C = 12.000000, H = 1.007825, O = 15.994915.

m/z. An abbreviation used to denote the dimensionless quantity formed by dividing the mass of an ion by the number of charges carried by the ion. It has long been called the mass-to-charge ratio, although m is not the ionic mass nor is z a multiple of the electronic charge, e–. The abbreviation m/e, therefore, is not recommended. Thus, for example, for the ion $C_7H_7^{2+}$, m/z = 45.5.

Neir-Johnson geometry. An arrangement for a double-focusing mass spectrometer in which a deflection of $\pi/2$ radians in a radial electrostatic-field analyzer is followed by a magnetic deflection of $\pi/3$ radians. The electrostatic analyzer uses a symmetrical object-image arrangement, while the magnetic analyzer is used asymmetrically.

Nominal ion mass. The mass of an ion with a given empirical formula calculated using the integer mass numbers of the most abundant isotope of each element, e.g., C = 12, H = 1, O = 16.

Quadrupole analyzer. A mass filter that creates a quadrupole field with a DC component and an RF (radio frequency) component in such a manner as to allow transmission only of ions having a selected mass-to-charge (m/z) ratio.

Single-focusing mass spectrometer. An instrument in which ions with a given mass-to-charge ratio (m/z) are brought to a focus although the initial directions of the ions diverge.

Static fields mass spectrometer. A mass spectrometer that can separate ion beams with fields that do not vary with time. These fields are generally electrostatic or magnetic.

Time-of-flight analyzer. A device that measures the flight time of ions with an equivalent kinetic energy over a fixed distance.

Vacuum system. Components associated with lowering the pressure within a mass spectrometer. A vacuum system includes not only the various pumping components but also valves, gauges, and associated electronic or other control devices; the chamber in which ions are formed and detected; and the vacuum envelope.

Wien analyzer. A velocity filter with crossed homogeneous electric and magnetic fields for transmitting only ions of a fixed velocity.

Data System

Amplifier bandwidth. The range of signal frequencies over which an amplifier is capable of undistorted or unattenuated transmission. An operational amplifier should transmit DC voltage accurately; the upper (bandwidth) limit is defined as the 3-dB point (attenuation factor of two). Because bandwidth can vary with gain, the product of gain × bandwidth can be a more useful parameter.

Amplifier complex. A number of operational amplifiers configured for a specific function, packaged and used as a single unit.

Amplifier noise. Can be of two kinds: white noise results from random fluctuations of signal over a power spectrum that contains all frequencies equally over a specified bandwidth; pink noise results when the frequencies diminish in a specified fashion over a specified range.

Analog signal. A signal that can be expressed as a continuously variable mathematical function of time.

Data acquisition. The process of transforming spectrometer signals from their original form into suitable representations, with or without modification, with or without a computer system.

Data logging. Implies data collection with storage for later data processing.

Data processing. Once information is obtained with an appropriate data system, the information must be interpreted appropriately for the end use. Data processing involves the steps leading to this end use; data processing does not necessarily imply application of modern computer techniques.

Data reduction. The process of transforming the initial digital or analog representation of output from a spectrometer into a form that is amenable to interpretation, e.g., a bar graph, a table of masses versus intensities.

Data system. Components used to record and process information during the analysis of a sample. The system includes a computer and an analog-to-digital conversion module as well as other control devices for data recording, storage, and manipulation.

Differential amplifier. An operational amplifier with two inputs of opposite-gain polarity with respect to its output. Differential-output amplifiers can also have two opposite-sense outputs.

Digital signal. A signal that represents information in a computer-compatible form as a sequence of (binary) numbers that can describe discrete samples of an analog signal.

Firmware. Computer programs stored in a semipermanent form, usually semiconductor memory, and used repeatedly without modification. Firmware can be changed only by replacing or removing hardware.

Hardware. The physical components of a computer system.

Hard wired. A preprocessor that is capable of performing only certain defined tasks and no others without major physical modification.

Off-line. A data-acquisition method in which the mass spectra are produced some time after the original experiment.

Operational amplifier. A linear, high-gain DC voltage or current amplifier with high input impedance, low output impedance, and the capability of producing a bipolar output from a bipolar input.

Preprocessor. A device in a data-acquisition system that performs a significant amount of data reduction by extracting specific information from raw signal representations in advance of the main processing operation. A preprocessor can constitute the whole of a data-acquisition interface, in which case it must also perform the data-acquisition task (conversion of spectrometer signal to computer representation), or it can specialize solely in data treatment.

Preprogrammed. A preprocessor that incorporates specific but readily alterable instructions to perform a particular task.

Real time. A data-acquisition method in which the mass spectra are generated within the same time frame as the original experiment.

Signal conditioning. The process of altering the relationship of a transducer (ion or neutral detector) output with respect to time or other parameters (frequency, voltage, or current).

Signal processing. The mechanisms involved in analyzing, routing, sampling, or changing the representation of a signal.

Single-ended amplifier. An operational amplifier with a single input (or output).

Software. Computer programs, whether inside or outside a computer.

Sample Introduction

Batch inlet. The historic term for a reservoir inlet. The term *reservoir inlet* is preferred because a direct-inlet probe is also a form of batch inlet. Batch gas inlet or batch vapor inlet are, however, completely descriptive terms.

Continuous inlet. An inlet in which sample passes continuously into the mass spectrometer ion source, as distinguished from a reservoir inlet or a direct-inlet probe.

Crucible direct-inlet probe. Holds the sample in a cup-shaped device (the crucible) rather than on an exposed surface. A direct-inlet probe is assumed to be a crucible type unless otherwise specified.

Direct-exposure probe. Provides for insertion of a sample on an exposed surface, such as a flat surface or a wire, into (rather than up to the entrance of) the ion source of a mass spectrometer.

Direct GC/MS interface. An interface in which all the effluent from a gas chromatograph passes into the mass spectrometer ion source during an analysis, without any splitting of the effluent.

Direct-inlet probe. A shaft or tube having a sample holder at one end that is inserted into the vacuum system of a mass spectrometer through a vacuum lock to place the sample near to, at the entrance of, or within the ion source. The sample is vaporized by heat from the ion source, by heat applied from an external source, or by exposure to ion or atom bombardment. Direct-inlet probe, direct-introduction probe, and direct-insertion probe are synonymous terms. The use of DIP as an abbreviation for these terms is not recommended.

Direct liquid introduction interface. An interface that continuously passes all, or a part of, the effluent from a liquid chromatograph to the mass spectrometer; the solvent usually functions as a chemical ionization agent for ionization of the solute.

Dual viscous-flow reservoir inlet. An inlet having two reservoirs, used alternately, each having a leak that provides viscous flow. This inlet is used to obtain precise comparisons of isotope ratios in two samples.

Dynamic headspace GC/MS. The distillation of volatile and semivolatile compounds into a continuously flowing stream of carrier gas and into a device for trapping sample components. Contents of the trap are then introduced onto a gas chromatographic column. This is followed by mass spectrometric analysis of compounds eluting from the gas chromatograph.

Effusion separator (or effusion enricher). An interface in which carrier gas is preferentially removed from the gas entering the mass spectrometer by effusive flow (e.g., through a porous tube or through a slit). This flow is usually molecular flow, such that the mean free path is much greater than the largest dimension of a traverse section of the channel. The flow characteristics are determined by collisions of the gas molecules with surfaces; flow effects from molecular collisions are insignificant.

GC/MS interface. An interface between a gas chromatograph and a mass spectrometer that provides continuous introduction of effluent gas from a gas chromatograph to a mass spectrometer ion source.

Jet separator. An interface in which carrier gas is preferentially removed by diffusion out of a gas jet flowing from a nozzle. Jet separator, jet-orifice separator, jet enricher, and jet orifice are synonymous terms.

Liquid chromatograph/mass spectrometer (LC/MS) interface. An interface between a liquid chromatograph and a mass spectrometer that provides continuous introduction of the effluent from a liquid chromatograph to a mass spectrometer ion source.

Membrane separator. A separator that passes gas or vapor to the mass spectrometer through a semipermeable (e.g., silicon) membrane that selectively transmits organic compounds in preference to carrier gas. Membrane separator, membrane enricher, semipermeable membrane separator, and semipermeable membrane enricher are synonymous terms.

Moving-belt (ribbon or wire) interface. An interface that continuously applies all, or a part of, the effluent from a liquid chromatograph to a belt (ribbon or wire) that passes through two or more orifices, with differential pumping into the mass spectrometer's vacuum system. Heat is applied to remove the solvent and to evaporate the solute into the ion source.

Nonfractionating continuous inlet. An inlet in which gas flows from a gas stream being analyzed to the mass spectrometer ion source without any change in the conditions of flow through the inlet or by the conditions of flow through the ion source. This flow is usually viscous flow, such that the mean free path is very small in comparison with the smallest dimension of a traverse section of the channel. The flow characteristics are determined mainly by collisions between gas molecules, i.e., the viscosity of the gas. The flow can be laminar or turbulent.

Sample introduction. The transfer of material to be analyzed into the ion source of a mass spectrometer before or during analysis.

Sample introduction system. A system used to introduce sample to a mass spectrometer ion source. Sample introduction system, introduction system, sample inlet system, inlet system, and inlet are synonymous terms.

Separator GC/MS interface. An interface in which the effluent from the gas chromatograph is enriched in the ratio of sample to carrier gas. Separator, molecular separator, and enricher are synonymous terms. A separator should generally be defined as an effusion separator, a jet separator, or a membrane separator.

Solvent-divert system. Used in conjunction with an interface, it permits temporary interruption of the flow from a chromatograph to a mass spectrometer by briefly opening a valve to a pumping line. Thus effluent present at a high concentration (usually solvent) does not enter the mass spectrometer ion source.

Splitter GC/MS interface. An interface in which the effluent from the gas chromatograph is divided before admission to the mass spectrometer without enrichment of sample with respect to carrier gas.

Static headspace GC/MS. The partitioning of volatile and semivolatile compounds between two phases in a sealed container. An aliquot of the headspace gas generated is injected onto a gas chromatographic column. This is followed by mass spectrometric analysis of compounds eluting from the gas chromatograph.

Thermal desorption. The vaporization of ionic or neutral species from the condensed state by the input of thermal energy. The energy input mechanism must be specified.

Thermospray interface. Provides liquid chromatographic effluent continuously through a heated capillary vaporizer tube to the mass spectrometer. Solvent molecules evaporate away from the partially vaporized liquid, and analyte ions are transmitted to the mass spectrometer's ion optics. The ionization technique must be specified, e.g., preexisting ions, salt buffer, filament, or electrical discharge.

Vacuum-lock inlet. An inlet through which a sample is first placed in a chamber; the chamber is then pumped out, and a valve is opened so that the sample can be introduced to the mass spectrometer ion source. A vacuum-lock inlet commonly uses a direct-inlet probe, which passes through one or more sliding seals, although other kinds of vacuum-lock inlets are also used.

Scanning of Spectra

Accelerating voltage (high voltage) scan. An alternative method of producing a momentum (mass) spectrum in magnetic-deflection instruments. This scan can also be used, in conjunction with a fixed radial electrical field, to produce an ion kinetic energy spectrum.

Ion kinetic energy spectrum. A spectrum obtained when a beam of ions is separated according to the translational energy-to-charge ratios of the ionic species contained within it. A radial electric field achieves separation of the various ionic species in this way.

Linked scan. A scan, in an instrument with two or more analyzers, in which two or more of the analyzer fields are scanned simultaneously to preserve a predetermined relationship between parameters that characterize these fields. Often these parameters are the field strengths, but they can be the frequencies in the case of analyzers that use alternating fields.

Linked scan at constant B/E. A linked scan at constant B/E can be performed on a sector instrument that incorporates at least one magnetic sector plus one electric sector. It involves scanning the magnetic-sector field strength (B) and the electric-sector field strength (E) simultaneously, holding the accelerating voltage (V) constant, to maintain the ratio of the two field strengths that transmit main-beam ions of predetermined mass-to-charge ratio. These preselected main-beam ions are the precursor (parent) ions whose fragment-ion spectrum is required. The observed fragmentation reactions occur in a field-free region traversed before the two sectors scanned in this way. The term *B/E linked scan* is not recommended, since it might suggest that the ratio B/E varies during the scan.

Linked scan at constant B^2/E. A linked scan at constant B^2/E can be performed on a sector instrument that incorporates at least one electric sector plus one magnetic sector. It involves holding the accelerating voltage fixed and scanning the magnetic field (B) and the electric field (E) simultaneously to maintain the ratio B^2/E at a constant value. This constant value corresponds to the ratio of the two fields that transmit main-beam ions of predetermined mass-to-charge ratio; these preselected main-beam ions are the fragment ions whose precursor ion spectrum is required. The observed fragmentation reactions occur in a field-free region traversed before the two sectors scanned in this way. This term should not be used without prior explanation of the meanings of B and E. The term B^2/E linked scan is not recommended.

Linked scan at constant $(B/E)(1 - E)^{1/2}$. A linked scan at constant $(B/E)(1 - E)^{1/2}$ can be performed on a sector instrument that incorporates at least one electric sector plus one magnetic sector. It involves holding the accelerating voltage fixed and scanning the magnetic field (B) and electric field (E) simultaneously to maintain the quantity $(B/E)(1 - E)^{1/2}$ at a constant value. This constant value is equal to B_3/E_0, where E_0 and B_3 are, respectively, the electric-sector field and magnetic-sector field required to transmit m_3 ions in the main ion beam; m_3 represents the mass $(m_1 - m_2)$ of the selected neutral fragment whose precursor (parent) ion spectrum is required. The observed fragmentation reactions occur in a field-free region traversed before the two sectors scanned in this way. This term should not be used without prior explanation of the meanings of B, E, and E_0. The term $(B/E)(1 - E)^{1/2}$ linked scan is not recommended. The above three definitions are merely examples of the types of linked scan that might be used. Other linked scans can readily be defined in a similar manner.

Linked scan at constant E^2/V. A linked scan at constant E^2/V can be performed on a sector instrument that incorporates at least one electric sector plus one magnetic sector. The electric-sector field (E) and the accelerating voltage (V) are scanned simultaneously to maintain the ratio E^2/V at a constant value equal to the value of this ratio that transmits the main beam of ions through the electric sector. The magnetic-sector field is set at a fixed value such that main-beam ions of a predetermined m/z are transmitted by the magnet. The preselected main-fragmentation reactions that are observed occur in a field-free region traversed before the two sectors scanned in this way. This term should not be used without prior explanation of the meaning of E and V. The term E^2/V linked scan is not recommended.

Magnetic-field scan. The usual method of producing a momentum (mass) spectrum in instruments.

Mass spectrum. A spectrum obtained when ions (usually in a beam) are separated according to the mass-to-charge (m/z) ratios of the ionic species present. The mass-spectrum plot is a graphical representation of m/z versus measured abundance information.

Fixed-precursor ion scans (sector instruments). (a) Mass selection followed by ion kinetic energy analysis: If a precursor (parent) ion is selected by a magnetic sector, all product (daughter) ions formed from it in the field-free region between the magnetic sector, and a following electric sector can be identified by scanning an ion kinetic energy spectrum. (b) Linked scan at constant B/E or at constant E^2/V: Both of these linked scans give a spectrum of all product (daughter) ions formed from a preselected precursor (parent) ion.

Fixed-product ion scans (sector instruments). High-voltage scan or linked scan at constant B^2/E. Both techniques give a spectrum of all precursor (parent) ions that fragment to yield a preselected product (daughter) ion.

Constant neutral loss (or fixed neutral fragment) scans. The linked scan at constant $B[1 - (E/E_0)]^{1/2}/E$ gives a spectrum of all product (daughter) ions that have been formed by loss of a preselected neutral fragment from any precursor (parent) ions.

2E mass spectrum. Processes of the partial charge-transfer type:

$$m^{2+} + N \rightarrow m^+ + N^+$$

that occur in a collision cell (containing a gas, N) located in a field-free region preceding a magnetic- and electric-sector combination (placed in either order) can be detected as follows: If the instrument slits are wide, and if the electric-sector field E is set to twice the value required to transmit the main ion beam, the only ions to be transmitted will be those with a kinetic energy-to-charge ratio twice, or almost exactly twice, that of the main ion beam. The product ions of the process shown fulfill this condition. If the magnetic field B is scanned, a mass spectrum of such singly charged product ions, and thus of their doubly charged precursors, is obtained. Such a spectrum is called a 2E mass spectrum.

E/2 mass spectrum. Processes of the charge-stripping type:

$$m^+ + N \rightarrow m^{2+} + N + e^-$$

that occur in a collision cell (containing a gas, N) located in a field-free region preceding a magnetic- and electric-sector combination (placed in either order) can be detected as follows: If the instrument slits are wide, and if the electric-sector field E is set to half the value required to transmit the main ion beam, the only ions to be transmitted will be those with a kinetic energy-to-charge ratio half that of the main ion beam. The product (daughter) ions of the charge-stripping process fulfill this condition. If the magnetic field B is scanned, a mass spectrum of such doubly charged product ions, and thus their singly charged precursors, is obtained. Such a spectrum is called an E/2 mass spectrum. Interference from product ions from processes of the type:

$$m_1^+ + N \rightarrow m_2^+ + N + (m_1 - m_2)$$

where m_2 = 0.5 ml, can arise in E/2 mass spectra.

Charge-inversion mass spectrum. Charge-inversion processes of the type:

$$m^+ + N \rightarrow m^- + N^{2+}$$

or

$$m- + N \rightarrow m^+ + N + 2e^-$$

that occur in a collision cell (containing a gas, N) located in a field-free region preceding a magnetic- and electric-sector combination (placed in either order) can be detected as follows: If the instrument slits are wide, and if the connections to the two sectors, appropriate to transmission of either positive or negative main-beam ions, are simply reversed, the negative or positive product ions of the two processes, respectively, will be transmitted. If the magnetic field is scanned, a spectrum of such product ions will be obtained, and this spectrum is called a charge-inversion mass spectrum. These spectra are sometimes referred to as charge reversal, or as –E and +E spectra, respectively. The terms 2E, E/2, –E, or +E mass spectrum should not be used without prior explanation of the meaning of 2E, E, +E, and –E.

Momentum spectrum. A spectrum obtained when a beam of ions is separated according to the momentum-to-charge (m/z) ratios of the ionic species present. A magnetic-sector analyzer achieves separation of the various ionic species in this way. If the ion beam is homogeneous in translational energy, as is the case with sector instruments, separation according to the m/z ratios is also achieved.

Scanning method. The sequence of control over operating parameters of a mass spectrometer that results in a spectrum of masses, velocities, momenta, or energies.

Selected-ion monitoring (SIM). Describes the operation of a mass spectrometer in which the ion currents at one (or several) selected m/z values are recorded, rather than the entire mass spectrum. The use of the terms multiple-ion detection (MID), multiple-ion (peak) monitoring (MPM), and mass fragmentography are not recommended.

Ion Detection and Sensitivity

Base peak. The peak in a mass spectrum corresponding to the m/z value that has the greatest intensity. This term can be applied to the spectra of a pure substance or mixtures.

Detection limit. The detection limit of an instrument should be differentiated from its sensitivity. The detection limit reflects the smallest flow of sample or the lowest partial pressure that gives a signal that can be distinguished from the background noise. One must specify the experimental conditions used and give the value of signal-to-noise ratio corresponding to the detection limit.

Detection of ions. The observation of electrical signals due to particular ionic species by a detector under conditions that minimize ambiguities from interferences. Ions can be detected by photographic or suitable electrical means.

Electron multiplier. A device to multiply current in an electron beam (or in a photon or particle beam after conversion to electrons) by incidence of accelerated electrons upon the surface of an electrode. This collision yields a number of secondary electrons greater than the number of incident electrons. These electrons are then accelerated to another electrode (or another part of the same electrode), which in turn emits secondary electrons, continuing the process.

Faraday cup (or cylinder) collector. A hollow collector, open at one end and closed at the other, used to measure the ion current associated with an ion beam.

Intensity relative to base peak. The ratio of intensity of a particular peak in a mass spectrum to the intensity of the mass peak of the greatest intensity. This ratio is generally equated to the normalized ratio of the heights of the respective peaks in the mass spectrum, with the height of the base peak being taken as 100.

Photographic plate recording. The recording of ion currents (usually associated with ion beams that have been spatially separated by m/z values) by allowing them to strike a photographic plate, which is subsequently developed.

Resolution: 10% valley definition, m/Δm. Let two peaks of equal height in a mass spectrum at masses m and m/Δm be separated by a valley that at its lowest point is just 10% of the height of either peak. For similar peaks at a mass exceeding m, let the height of the valley at its lowest point be more (by any amount) than 10% of either peak height. Then the resolution (10% valley definition) is m/Δm. It is usually a function of m; therefore, m/Δm should be given for a number of values of m.

Resolution: peak width definition, m/Δm. For a single peak made up of singly charged ions at mass m in a mass spectrum, the resolution can be expressed as m/Δm, where Δm is the width of the peak at a height that is a specified fraction of the maximum peak height. It is recommended that one of three values be used: 50%, 5%, or 0.5%. For an isolated symmetrical peak, recorded with a system that is linear in the range between 5% and 10% levels of the peak, the 5% peak-width definition is technically equivalent to the 10% valley definition. A common standard is the definition of resolution based upon Δm being full width of the peak at half its maximum height, sometimes abbreviated FWHM.

Resolution energy. A value derived from a peak showing the number of ions as a function of their translational energy.

Resolving power (mass). The ability to distinguish between ions differing slightly in mass-to-charge ratio. It can be characterized by giving the peak width, measured in mass units, expressed as a function of mass, for at least two points on the peak, specifically for 50% and for 5% of the maximum peak height.

Sensitivity. Different measures of sensitivity are recommended, depending on the nature of the sample and the required inlet system. The first, which is suitable for nonvolatile materials as well as gases, depends upon the observed change in ion current for a particular change of flow rate of sample through the ion source. The recommended unit is coulomb per microgram. A second method of stating sensitivity, most suitable for gases, depends upon the change of ion current relative to the change of partial pressure of the sample in the ion source. The recommended unit is amperes per pascal. It is important that the relevant experimental conditions corresponding to sensitivity measurement always be stated. These typically include details of the bombarding electron current, slit dimensions, angular collimation, gain of the detector, scan speed, mass range scanned, and whether the measured signal corresponds to a single mass peak or to the ion beam integrated over a specified mass range. Sample flow into the ion source per unit time should be noted; indication of the time involved in the determination should also be given, i.e., counting time or bandwidth. The sensitivity should be differentiated from the detection limit.

Total ion current (TIC). (a) After mass analysis: the sum of all the separate ion currents carried by the different ions contributing to the spectrum. (b) Before mass analysis: the sum of all the separate ion currents for ions of the same sign.

Ionization Nomenclature

Adiabatic ionization. A process whereby an electron is removed from the ground state of an atom or molecule, producing an ion in its ground state.

Appearance energy. The minimum energy that must be imparted to an atom, molecule, or molecular moiety in order to produce a specified ion. The use of the alternative term appearance potential is not recommended.

Associative ionization. Occurs when two excited gaseous atoms or molecular moieties interact and the sum of their internal energies is sufficient to produce a single, additive ionic product.

Atmospheric-pressure ionization. Chemical ionization performed at atmospheric pressure.

Auto ionization. Occurs when an internally supraexcited atom or molecular moiety loses an electron spontaneously without further interaction with an energy source. (The state of the atom or molecular moiety is known as a pre-ionization state.)

Charge-exchange (charge transfer) ionization. Occurs when an ion/atom or ion/molecule reaction takes place in which the charge on the ion is transferred to the neutral species without any dissociation of either.

Chemical ionization. The formation of new ionized species when gaseous molecules interact with ions. The process involves the transfer of an electron, a proton, or other charged species between the reactants. When a positive ion results from chemical ionization (CI), the term can be used without qualification. When a negative ion results, the term negative ion chemical ionization can be substituted. Specifics relating to the ionization should be given; e.g., it should be specified if negative ions are formed from sample molecules via resonance capture of thermal electrons generated in a CI source.

Chemi-ionization. A process:

$$A^* + M \rightarrow AM^+ + e^-$$

whereby gaseous molecules are ionized when they interact with other internally excited gaseous molecules or molecular moieties. The terms *chemi-ionization* and *chemical ionization* must not be used interchangeably.

Desorption ionization (DI). General term to encompass the various procedures (e.g., secondary ion mass spectrometry, fast-atom bombardment, californium fission fragment desorption, thermal desorption) in which ions are generated directly from a solid or liquid sample by energy input. Experimental conditions must be clearly stated.

Dissociative charge transfer. Occurs when an ion/molecule reaction takes place in which the charge on the ion is transferred to the neutral species. The new ion then dissociates to one or more fragment ions.

Dissociative ionization. A process in which a gaseous molecule decomposes to form products, one of which is an ion.

Electron attachment. A resonance process whereby an electron is incorporated into an atomic or molecular orbital of an atom or molecule.

Electron energy. The potential difference through which electrons are accelerated before they are used to bring about electron ionization. The term *ionizing voltage* is sometimes used in place of electron energy.

Electron ionization. Ionization of any species by electrons. The process can be written for atoms or molecules as:

$$M + e^- \rightarrow M^{+\bullet} + 2e^-$$

and for radicals as:

$$M^\bullet + e^- \rightarrow M^+ + 2e^-$$

Field desorption. The formation of ions in the gas phase from a material deposited on a solid surface (known as an emitter) that is placed in a high electrical field. Field desorption is an ambiguous term because it implies that the electric field desorbs a material as an ion from some kind of emitter on which the material is deposited. There is growing evidence that some of the ions formed are due to thermal ionization and some to field ionization of material vaporized from the emitter. Because there is little or no ionization unless the emitter is heated by an electric current, field desorption is a misnomer. However, the term is firmly implanted in the literature, and most users (by no means all) understand what is meant regardless of the implications of the term. Because no better simple term has been suggested to take its place, it is recommended with reluctance that it be retained.

Field ionization. The removal of electrons from any species by interaction with a high electrical field.

Ionic dissociation. Decomposition of an ion into another ion of lower formula weight, plus one or more neutral species.

Ionization. A process that produces an ion from a neutral atom or molecule.

Ionization cross-section. A measure of the probability that a given ionization process will occur when an atom or molecule interacts with an electron or a photon.

Ionization efficiency. The ratio of the number of ions formed to the number of electrons, photons, or particles that are used to produce ionization

Ionization efficiency curve. Shows the number of ions produced as a function of energy of the electrons, photons, or particles used to produce ionization.

Ionization energy. The minimum energy of excitation of an atom, a molecule, or a molecular moiety that is required to remove an electron in order to produce a positive ion.

Ion-pair formation. An ionization process in which a positive fragment ion and a negative fragment ion are the only products.

Laser ionization. Occurs when a sample is irradiated with a laser beam. In the irradiation of gaseous samples, ionization occurs via a single- or multiphoton process. In the case of solid samples, ionization occurs via a thermal process.

Multiphoton ionization. Occurs when an atom or molecule and its associated ions have energy states in which they can absorb the energy in two or more photons.

Penning ionization. Occurs through the interaction of two or more neutral gaseous species, at least one of which is internally excited:

$$A^* + M \rightarrow A + M^+ + e^-$$

Photo-ionization. Ionization of any species by photons:

$$M + h\nu \rightarrow M^{+\bullet} + e^-$$

Electrons and photons do not "impact" molecules or atoms. They interact with them in ways that result in various electronic excitations, including ionization. For this reason it is recommended that the terms *electron impact* and *photon impact* be avoided.

Spark (source) ionization. Occurs when a solid sample is vaporized and partially ionized by an intermittent electric discharge. Further ionization occurs in the discharge when gaseous atoms and small molecular moieties interact with energetic electrons in the intermittent discharge. It is recommended that the word *source* be dropped from this term.

Surface ionization. Takes place when an atom or molecule is ionized when it interacts with a solid surface. Ionization occurs only when the work function of the surface, the temperature of the surface, and the ionization energy of the atom or molecule have an appropriate relationship.

Thermal ionization. Takes place when an atom or molecule interacts with a heated surface or is in a gaseous environment at high temperatures. Examples of the latter include a capillary arc plasma, a microwave plasma, or an inductively coupled plasma.

Vertical ionization. A process whereby an electron is removed from a molecule, in its ground state or an excited state, so rapidly that a positive ion is produced without change in the positions or momenta of the atoms. The resultant ion is often in an excited state.

Types of Ions, Ion Structures

Adduct ion. An ion formed by interaction of two species, usually an ion and a molecule, and often within the ion source, to form an ion containing all the constituent atoms of one species as well as an additional atom or atoms.

Bond fission. Confusion can arise when a hyphen is used in the symbolism. Thus, D(R – X) has been used to mean the dissociation energy of the bond between R and X while $(X - CH_2)^+$ might mean the next higher homologue of X^+ or the ion formed from X^+ by removal of a CH_2 group. Thus, it is recommended that a dash should not be used to indicate a bond, except in a conventional structural formula such as that for the acetone molecular ion shown here. In other cases, the next higher homologue of X^+ should be written $(XCH_2)^+$, without any dash. In the event that it is necessary to emphasize that a bond is breaking, it should be represented by two dots and a wavy line (no other bond than the breaking bond being illustrated):

$$[R_1 \cdot\wr\cdot CH_2CH_2R_2]^{+\bullet}$$

When in addition to indicating fragmentation of the bond, it is necessary to emphasize the mass number of the fragments formed, this is done by writing the mass number at the top (right-hand fragment) or the bottom (left-hand fragment) as shown:

$$[CH_3CH_{2\,29}\cdot\wr\cdot CH_2CH_{2\,57}\cdot\wr\cdot{}^{47}CH_2SH]^{+\bullet}$$

Loss of a particular group should be indicated by the use of a minus sign located outside the parentheses or to the right of the "+•" sign. Spaces should be left on either side of the minus sign to reduce any confusion as to its meaning. Thus one would write:

$$(M)^{+\bullet} - CH_2 \text{ or } (C_6H_5CH_3)^{+\bullet} - H$$

It is recommended that the convention used by Budzikiewicz, Djerassi, and Williams (*Mass Spectrometry of Organic Compounds*, Holden-Day, 1967, p. 2) be followed in referring to α-cleavage as: "fission of a bond originating at an atom which is adjacent to the one assumed to bear the charge; the definition of β-, γ-, then follows automatically." The process:

$$R_1R_2C{=}O^{+\bullet} \longrightarrow \overset{+}{O}{\equiv}C{-}R_2 + R_1$$

would thus be described as "α-fission of a ketone with expulsion of a radical $R_1^\bullet$." The carbon atoms of the radical R_1 are called the α-, β-, γ-carbons, starting with the atom nearest the functional group. The symbol '⌒▲' is recommended for indicating the movement of two electrons (heterolysis). The symbol '⌒▲' is recommended for indicating the movement of one electron (homolysis).

Cluster ion. An ion formed by the combination of two or more molecules of a chemical species, often in association, with a second species. For example, $((H_2O)_nH)^+$ is a cluster ion.

Daughter ion. An electrically charged product of a reaction of a particular parent ion. In general, such ions have a direct relationship to a particular precursor ion and, indeed, may relate to a unique state of the precursor ion. The reaction need not necessarily involve fragmentation. It could, for example, involve a change in the number of charges carried. Thus, all fragment ions are daughter ions but not all daughter ions are necessarily fragment ions.

Dimeric ion. An ion formed either when a chemical species exists in the vapor phase as a dimer and can be detected as such, or when a molecular ion can attach to a neutral molecule within the ion source to form an ion such as $(2M)^{+\bullet}$, where M represents the molecule.

Even-electron ion. An ion containing no unpaired electrons, e.g., CH_3^+ in its ground state.

Fragment ion. An electrically charged dissociation product of an ionic fragmentation. Such an ion may dissociate further to form other electrically charged molecular or atomic moieties of successively lower formula weight.

Ion structures. Capital letters enclosed in parentheses should be used to represent an ion that contains different structural features; the charge should be located outside the parentheses. Thus, one should write $(ABC)^+$ to refer to an even-electron species, or $(ABC)^{+\bullet}$ to emphasize that one is dealing with an odd-electron species. An exception is that an ion may be written (A^+BC) or $(A^{+\bullet}BC)$ if it is necessary to discuss structures in which the charge and the odd electron are localized. The empirical formula of a molecular ion can be written in four ways: The molecular ion of aniline, for example, can be written $(C_6H_7N)^+$, or it can be written $(C_6H_5NH_2)^{+\bullet}$ to emphasize that one is discussing an ion formed from a molecule with the structure $C_6H_5NH_2$ by the removal of a single electron, the upper half of a bracket indicating that the structure of the ion may or may not be the same as that of the molecule from which it was formed. If one wishes to discuss a particular structure in which an odd electron is localized, then one can write $(C_6H_5N^+H_2^\bullet)$. The electron ionization process of aniline, for example, can be written:

$$C_6H_5HN_2 + e^- \rightarrow (C_6H_5NH_2)^{+\bullet} + 2e^-$$

$$\rightarrow (C_6H_7N)^{+\bullet} + 2e^-$$

Structural formulae of ions in which it is desired to indicate the localization of an odd electron can, alternatively, be written without the parentheses. Thus,

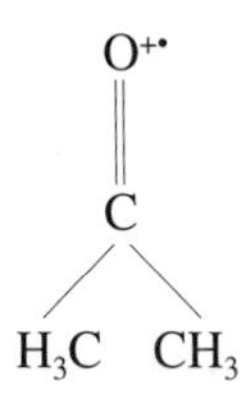

Isotopic ion. Any ion containing one or more of the less abundant naturally occurring isotopes of the elements that make up its structure.

Isotopic molecular ion. A molecular ion containing one or more of the less abundant naturally occurring isotopes of the atoms that make up the molecular structure. Thus, for ethyl bromide there exist molecular isotope ions such as $^{13}CCH_5Br^{+\bullet}$, $C_2H_4DBr^{+\bullet}$, $C_2H_5{}^{81}Br^{+\bullet}$, $^{13}C_2H_5{}^{81}Br^{+\bullet}$, etc.

Isotopically enriched ion. Any ion containing an abundance of a particular isotope above the level at which it occurs in nature.

Metastable ion. An ion that is sufficiently excited to dissociate into a particular daughter ion and neutral species during the flight from the ion source to the detector. The dissociation is most readily observed when it takes place in one of the field-free regions in a mass spectrometer.

Molecular ion. An ion formed by the removal (positive ions) or addition (negative ions) of one or more electrons from a molecule without fragmentation of the molecular structure. The mass of this ion corresponds to the sum of the masses of the most abundant naturally occurring isotopes of the various atoms that make up the molecule (with a correction for the masses of the electrons lost or gained). For example, the mass of the molecular ion of the ethyl bromide C_2H_5Br will be 2×12 plus 5×1.0078246 plus 78.91839 minus the mass of the electron (m_e). This is equal to $107.95751\mu - m_e$, the unit of atomic mass based on the standard that the mass of the isotope $^{12}C = 12.000000$ exactly.

Negative ion. An atom, radical, molecule, or molecular moiety that has gained one or more electrons, acquiring an electrically negative charge. The use of the term anion as an alternative is not recommended, except in the context of chemical reactions or structures.

Odd-electron ion. Synonymous with radical ion.

Parent ion. An electrically charged molecular moiety that may dissociate to form fragments, of which one or more may be electrically charged, and one or more are neutral species. A parent ion can be a molecular ion or an electrically charged fragment of a molecular ion.

Positive ion. An atom, radical, molecule, or molecular moiety that has lost one or more electrons, acquiring an electrically positive charge. The use of the term *cation* as an alternative is not recommended, except in the context of chemical reactions or structures. The use of mass ion is not recommended.

Precursor ion. Synonymous with parent ion.

Principal ion. A molecular or fragment ion that is made up of the most abundant isotopes of each of its atomic constituents. In the case of compounds that have been artificially isotopically enriched in one or more positions (such as $CH_3{}^{13}CH_3$ or CH_2D_2), the principal ion can be defined by treating the heavy isotopes as new atomic species. Thus, in the above two examples, the principal ions would be of masses 31 and 18, respectively.

Product ion. Synonymous with daughter ion.

Progeny fragment ions. Includes daughters, granddaughters, great-granddaughters, etc. It is a more generic term than daughter or product ion. Given the sequential fragmentation scheme:

$$M_1^+ \rightarrow M_2^+ \rightarrow M_3^+ \rightarrow M_4^+ \rightarrow M_5^+$$

M_4^+ = parent to M_5^+

= daughter and granddaughter of M_2^+

= daughter and great-granddaughter of M_1^+

M_3^+ = parent to M_4^+

= daughter and granddaughter of M_1^+

= daughter of M_2^+

Protonated molecule. An ion formed by interaction of a molecule with a proton abstracted from an ion, as often occurs in chemical ionization according to the reaction:

$$M + XH^+ \rightarrow MH^+ + X$$

The symbolism $(M + H)^+$ can also be used to represent the protonated molecule. The widely used term *protonated molecular* ion to describe the MH^+ ion is not recommended because it suggests an association product of a proton with a molecular ion.

Radical ion. An ion containing an unpaired electron that is thus both an ion and a free radical. The presence of the odd electron is denoted by placing a dot alongside the symbol for the charge. Thus, $C_2H_6^{+\bullet}$ and $SF_6^{-\bullet}$ are radical ions.

Rearrangement ion. An electrically charged dissociation product, of a molecular or parent ion, in which atoms or groups of atoms have transferred from one portion of the molecule or molecular moiety to another during the fragmentation process.

Rearrangement reactions. These reactions may be skeletal, in which case the symbol —▭→ is employed:

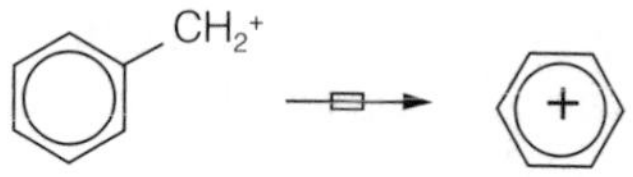

They may also involve the specific transfer to atoms within a molecule together with bond fission. An example of this is the McLafferty rearrangement, which is defined as β-cleavage with concomitant specific transfer of a γ-hydrogen atom in a six-member transition state in mono-unsaturated systems, regardless of whether the rearrangement is formulated by a radical or by an ionic mechanism, and irrespective of which fragment retains the charge.

Singly, doubly, triply, etc. charged ion. An atom, molecule, or molecular moiety that has gained or lost one, two, three, or more electrons. The term multiply charged ion is used to refer to ions that have gained or lost more than one electron where the number of electrons lost or gained is not designated.

Stable ion. An ion that is not sufficiently excited to dissociate into a daughter ion and associated neutral fragments, or to react further in the time frame of the mass spectrometric analysis under stated experimental conditions.

Unstable ion. An ion that is sufficiently excited to dissociate within the ion source, under stated experimental conditions.

Ion/Molecule Reactions

Association reaction (associative combination). The reaction of a (slow moving) ion with a neutral species, wherein the reactants combine to form a single ionized species.

Charge-exchange reaction. Synonymous with charge-transfer reaction.

Charge-inversion reaction. An ion/neutral reaction wherein the charge on the reactant ion is reversed in sign.

Charge-permutation reaction. An ion/neutral reaction involving a change in the magnitude and/or sign of the charges on the reactants. Some of the possible reactions of ions M^{2+}, M^+, and M– with a neutral species N would be categorized on the basis of the above definitions as follows (all are ion/neutral reactions and also charge-permutation reactions):

$$M^{2+} + N \rightarrow M^+ + N^+ \text{ (particle charge transfer)}$$

$$M^+ + N \rightarrow M^{2+} + N + e^- \text{ (charge stripping)}$$

$$M- + N \rightarrow M^+ + N + 2e^- \text{ (charge stripping and charge inversion)}$$

Charge-stripping reaction. An ion/neutral reaction that increases the positive charge on the reactant ion.

Charge-transfer reaction. An ion/neutral reaction wherein the charge on the reactant ion is transferred to the reactant neutral species so that the reactant ion becomes a neutral entity.

Collisional activation. An ion/neutral process wherein excitation of a (fast) projectile ion is brought about by the same mechanism as in collision-induced dissociation. (The ion may decompose subsequently).

Collisional excitation. An ion/neutral process wherein the (slow) reactant ion's internal energy increases at the expense of the translational energy of either (or both) of the reacting species. The scattering angle can be large.

Collision-induced dissociation (or decomposition), abbreviated CID. An ion/neutral process wherein the (fast) projectile ion is dissociated as a result of interaction with a target neutral species. This is brought about by conversion during the collision of part of the translational energy of the ion to internal energy in the ion. The term collisional-activated dissociation (or decomposition), abbreviated CAD, is also used.

Elastic collision. A collision resulting in elastic scattering.

Elastic scattering. An ion/neutral interaction wherein the direction of motion of the ion is changed, but the total translational energy or internal energy of the collision partners remains the same.

Impact parameter. The distance between two particles at their closest approach, had they continued in this original direction at their original speeds.

Inelastic collision. A collision resulting in inelastic scattering.

Inelastic scattering. An ion/neutral interaction wherein the direction of motion of the ion is changed, and the total translational energy of the collision partners is reduced.

Interaction distance. The greatest distance between two particles at which it is discernible that they will not pass at the impact parameter.

Ion energy loss spectrum. A spectrum that shows the loss of translation energy among ions involved in ion/neutral reactions.

Ionizing collision. An ion/neutral reaction in which an electron or electrons are stripped from the ion and/or the neutral species in the collision. Generally, this term describes collisions of fast-moving ions or atoms with a neutral species in which the neutral species is ionized. Care should be taken to emphasize if charge stripping of the ion has taken place.

Ion/molecule reaction. An ion/neutral reaction in which the neutral species is a molecule.

Ion/neutral exchange reaction. An association reaction that subsequently or simultaneously liberates a different neutral species.

Ion/neutral reaction. Interaction of a charged species with a neutral reactant to produce either chemically different species or changes in the internal energy of one or both of the reactants.

Partial charge-exchange reaction. Synonymous with partial charge-transfer reaction.

Partial charge-transfer reaction. An ion/neutral reaction that reduces the charge on a multiply charged reaction ion.

Superelastic collision. A collision that increases the translational energy of the fast-moving collision partner.

Translational spectroscopy. A technique to investigate the distribution of the velocities of product ions from ion/neutral reactions.

Common Abbreviations

ADC. analog-to-digital converter

AIR. (atmospheric) air, a standard for nitrogen and chlorine isotopes

APCI. atmospheric-pressure chemical ionization, also called plasmaspray

API. atmospheric-pressure ionization

***B*.** a term meaning the strength of a magnetic field, often used to refer to a magnetic field or sector

B/E. a method of linked scanning for metastable ions, which uses combined magnetic and electric fields

B^2/E. a method of linked scanning for metastable ions, which uses combined magnetic and electric fields

CAD. collisionally activated decomposition

CDT. canyon diabolo troilite (a standard for sulfur isotopes; see VCDT)

CE. (sometimes CZE), capillary electrophoresis (or capillary zone electrophoresis)

CEM. channel electron multiplier

CE/MS. capillary electrophoresis and mass spectrometry used as a combined technique

CI. chemical ionization

CIA. collisionally induced activation

CID. collisionally induced dissociation (or decomposition)

δ. a value representing a comparison of isotope ratios

DAC. digital-to-analog converter

DCI. desorption (or direct) chemical ionization

DEI. desorption (or direct) electron ionization

DSI. direct sample insertion

e. elementary charge on an electron ($\cong 1.602 \times 10^{-19}$ C)

E. a term meaning the strength of an electric field, often used to refer to an electric field or sector

E^2/V. a method of linked scanning for metastable ions, which uses electric fields (see V)

E_{CM}. energy (of collision) referred to the center of mass of the colliding particles

E_{LAB}. energy (of collision) referred to a laboratory frame

EI. electron ionization

EPA. Environmental Protection Agency (U.S. agency responsible for many methods of analysis used by mass spectrometrists)

ESA. electrostatic analyzer

ESI. electrospray ionization (should not be confused with EI)

ETV. electrothermal vaporization

FAB. fast-atom bombardment (ionization or technique)

FD. field desorption (ionization)

FI. field ionization

FIB. fast-ion bombardment (ionization or technique)

FTICR. Fourier-transform ion cyclotron resonance

GC/IRMS. gas chromatography isotope ratio mass spectrometry

GC/MS. gas chromatography and mass spectrometry used as a combined technique

GD. glow discharge

GD/IRMS. glow discharge isotope ratio mass spectrometry

HPLC. high-pressure liquid chromatography

ICP/AES. inductively coupled plasma and atomic-emission spectroscopy used as a combined technique

ICP/MS. inductively coupled plasma and mass spectrometry used as a combined technique

ICR. ion cyclotron resonance (spectroscopy)

IKES. ion kinetic energy spectroscopy

IRMS. isotope ratio mass spectrometry

ISDMS. isotope dilution mass spectrometry

ITMS. ion trap mass spectrometry

LA. laser ablation

LASER. light amplification by stimulated emission of radiation

LC/MS. liquid chromatography and mass spectrometry used as a combined technique

LDI. laser-desorption ionization

LDMS. laser-desorption mass spectrometry

LSIMS. liquid-phase secondary ion mass spectrometry

MALDI. matrix-assisted laser desorption ionization

MIKES. mass-analyzed ion kinetic energy spectroscopy

MPI. multiphoton ionization

MS/MS. two mass spectrometers used in tandem (sometimes written as MS2)

MS^n. application of successive mass spectrometric measurements (*n* of them), particularly in linked scanning of m/z, which is the ratio of the mass (m) of an ion and the number of charges (z) on it. Older publications used m/e, but as e is the actual charge on an electron and not the number of charges on the ion, the use of m/e was abandoned.

m/z. mass-to-charge ratio, a measure of molecular mass

PDB. PeeDee Belemnite (a carbon isotope standard; see VPDB)

PID. photon-induced dissociation (or decomposition)

PPNICI. pulsed positive ion/negative ion chemical ionization

Py/GC/MS. pyrolysis, gas chromatography, and mass spectrometry used as a combined technique

Py/MS. pyrolysis and mass spectrometry used as a combined technique

oa-TOF. orthogonally accelerated time of flight

Q. quadrupole field or instrument

QET. quasi-equilibrium theory (of mass spectrometric fragmentation)

Q/TOF. used for two mass analyzers (quadrupole and time-of-flight) used in combination

QQQ (or QqQ). a triple quadrupole analyzer (if q is used, it means the central quadrupole is also a collision cell)

RA (%RA). relative abundance (percent relative abundance)

REMPI. resonance enhanced multiphoton ionization

RI. resonance ionization

RIC. reconstructed ion current

RISC. reduced instruction set for computing

SFC/MS. supercritical fluid chromatography and mass spectrometry used as a combined technique

SID. surface-induced dissociation (or decomposition)

SIM. selected (or single) ion monitoring

SIMS. secondary ion mass spectrometry

SIR. selected (or single) ion recording

SMOC. standard mean ocean chloride (a standard for chlorine isotopes)

SMOW. standard mean ocean water (a standard for oxygen and hydrogen isotopes)

SRM. selected reaction monitoring

SSMS. spark source mass spectrometry

TDC. time-to-digital converter

TIC. total ion current

TG/MS. thermogravimetry and mass spectrometry used as a combined technique

TI. thermal (surface) ionization

TIMS. thermal ionization mass spectrometry

TOF. time of flight

V. a term used to describe a voltage difference between one electrode and another

VCDT. Vienna canyon diabolo troilite (actually silver sulfide used as a replacement standard for CDT [sulfur isotopes])

VPDB. Vienna PeeDee Belemnite (a replacement for the PDB isotope standard)

VDU. visual display unit (computer screen)

z. the number of charges on an ion (see **m/z**)

Z-spray. "Z" refers to the approximate shape of the trajectory of particles formed by electrospray ionization

Appendix C

Books on Mass Spectrometry and Related Topics

The following list is not intended to be exhaustive. The titles of the books, which cover a wide range of topics, are largely self-explanatory. A rapidly developing area such as mass spectrometry requires a steady stream of new books to encapsulate explanations and applications of the advances and to describe the reasons for their importance in mass spectrometry. In assembling such books, any writer builds on previous publications and very often has to regard many facts or even whole areas as understood by the reader. While this approach is fine for regular longtime users of mass spectrometry, it is unfortunate for newcomers to the field, who may well find that a perusal of older books will explain some things better than more recent books. Certainly such older books will contain facts and information that are important, but they may not be directly relevant to recent applications. For these reasons, a good sprinkling of useful older books is included in the following list:

Adams, F., Gijbels, R., and Van Grieken, R., *Inorganic Mass Spectrometry,* Wiley Interscience, New York, 1988.

Adams, R.P., *Identification of Essential Oils by Ion Trap Mass Spectrometry,* Academic Press, San Diego, CA 1989.

Asamoto, B. and Dunbar, R.C., *Analytical Applications of Fourier Transform Ion Cyclotron Resonance Spectroscopy,* VCH, New York, 1991.

Baillie, T.A., *Stable Isotopes: Application in Pharmacology, Toxicology and Clinical Research,* Macmillan, London, 1978.

Barker, J. and Ando, D.J., *Mass Spectrometry: Analytical Chemistry by Open Learning,* Wiley, Chichester, U.K., 1999.

Beckey, H.D., *Principles of Field Ionisation and Field Desorption Mass Spectrometry,* Pergamon Press, Oxford, 1977.

Benninghoven, A., Evans, C.A., McKeegan, K., Storms, H.A., and Werner, H.W., *Secondary Ion Mass Spectrometry,* Wiley, New York, 1990.

Benninghoven, A., Rudenauer, F.G., and Werner, H.W., *Secondary Ion Mass Spectrometry: Basic Concepts, Instrumental Aspects, Applications and Trends,* Wiley, New York, 1987.

Beynon, J.H., *Mass Spectrometry and Its Application to Organic Chemistry,* Elsevier, Amsterdam, 1960.

Beynon, J.H., Saunders, R.A., and Williams, A.E., *The Mass Spectra of Organic Molecules,* Elsevier, Amsterdam, 1963.

Biemann, K., *Mass Spectrometry; Organic Chemical Applications,* McGraw-Hill, New York, 1962.

Brown, M.A., *Liquid Chromatography/Mass Spectrometry Applications in Agricultural, Pharmaceutical and Environmental Chemistry,* Oxford University Press, Oxford, 1998.

Budde, W.L., *Analytical Mass Spectrometry: Strategies for Environmental and Related Applications,* American Chemical Society, Washington, D.C., 2001.

Busch, K.L., Glish, G.L., and McLuckey, S.A., *Mass Spectrometry/Mass Spectrometry: Techniques and Applications of Tandem Mass Spectrometry,* VCH, New York, 1988.

Caprioli, R.M., *Continuous-Flow Fast Atom Bombardment Mass Spectrometry,* Wiley, New York, 1990.

Chapman, J.R., *Computers in Mass Spectrometry,* Academic Press, London, 1978.

Chapman, J.R., *Practical Organic Mass Spectrometry,* Wiley, Chichester, U.K., 1993.

Chapman, J.R., *Practical Mass Spectrometry: A Guide for Chemical and Biochemical Analysis,* Wiley, Chichester, U.K., 1995.

Cole, R.B., *Electrospray Ionization Mass Spectrometry: Fundamentals, Instrumentation and Applications,* Wiley, Chichester, U.K., 1997.

Cooks, R.G., *Collision Spectroscopy,* Plenum Press, New York, 1978.

Cooks, R.G., Beynon, J.H., Caprioli, R.M., and Lester, G.R., *Metastable Ions,* Elsevier, Amsterdam, 1973.

Cotter, R.J., *Time-of-Flight Mass Spectrometry,* ACS Symposium Series, Vol. 549, American Chemical Society, Washington, DC, 1994.

Cotter, R.J., *Time-of-Flight Mass Spectrometry: Instrumentation and Applications in Biomedical Research*, American Chemical Society, Washington, D.C., 1997.

Dass, C., *Principles and Practice of Biological Mass Spectrometry*, Wiley, Chichester, U.K., 2000.

Dawson, P.H., *Quadrupole Mass Spectrometry*, Elsevier, Amsterdam, 1976.

Field, F.H. and Franklin, J.L., *Electron Impact Phenomena*, Academic Press, New York, 1957.

Fox, A., Morgan, S.L., Larsson, L., and Oldham, G., *Analytical Microbiology Methods: Chromatography and Mass Spectrometry*, Plenum Press, New York, 1990.

Franklin, J.L., *Ion-Molecule Reactions, Parts I and II*, Hutchinson & Ross, Stroudsburg, PA, 1979.

Gerhatds, P., Bons, U., Sawazki, J., Szigan, J., and Wertman, A., *GC/MS in Clinical Chemistry*, Wiley, New York, 1999.

Grob, R.L., *Modern Practice of Gas Chromatography*, Wiley, New York, 1985.

Halker, J.M. and Rose, M.E., *Introduction to Bench-Top GC/MS*, HD Science, Nottingham, U.K., 1990.

Harrison, A.G., *Chemical Ionization Mass Spectrometry*, CRC Press, Boca Raton, FL, 1992.

Hill, H.C., *Introduction to Mass Spectrometry*, Heyden, London, 1972.

Hites, R.A., *Handbook of Mass Spectra of Environmental Contaminants*, Lewis Publishers, Boca Raton, FL, 1992.

Hoffmann, E. de, Charette, J., and Stroobant, V., *Mass Spectrometry: Principles and Applications*, Wiley, Chichester, U.K., 1996.

Hoffmann, E. de and Stroobant, V., *Mass Spectrometry: Principles and Applications*, Wiley, Chichester, U.K., 2001.

Holland, J.G. and Eaton, A., *Applications of Plasma Source Mass Spectrometry*, The Royal Society of Chemistry, Cambridge, 1991.

Jaeger, H., *Capillary Gas Chromatography Mass Spectrometry in Medicine and Pharmacology*, Huetig, New York, 1987.

James, P., *Proteome Research: Mass Spectrometry (Principles and Practice),* Springer-Verlag, Heidelberg, 2000.

Jarvis, K.E., Gray, A.L., Williams, J.G., and Jarvis, I., *Plasma Source Mass Spectrometry*, The Royal Society of Chemistry, Cambridge, 1990.

Jinno, K., *Hyphenated Techniques in Supercritical Fluid Chromatography and Extraction*, Elsevier, Amsterdam, 1992.

Johnstone, R.A.W. and Rose, M.E., *Mass Spectrometry for Chemists and Biochemists*, Cambridge University Press, Cambridge, 1996.

Keith, L.H., *Identification and Analysis of Organic Pollutants in Water*, Ann Arbor Science, Ann Arbor, MI, 1976.

Kinter, M. and Sherman, N.E., *Protein Sequencing and Identification Using Tandem Mass Spectrometry*, Wiley, Chichester, U.K., 2000.

Kiser, R.W., *Introduction to Mass Spectrometry and Its Applications*, Prentice Hall, Englewood Cliffs, NJ, 1965.

Kitson, F.G., Larsen, B.S., and McEwen, C.N., *Gas Chromatography and Mass Spectrometry*, Academic Press, New York, 1996.

Knapp, D.R., *Handbook of Analytical Derivatization Reactions*, Wiley, New York, 1979.

Knewstubb, P.F., *Mass Spectrometry and Ion-Molecule Reactions*, Cambridge University Press, London, 1969.

Laeter, J.R. de, *Applications of Inorganic Chemistry*, Wiley, New York, 2001.

Lee, T.A., *A Beginner's Guide to Mass Spectral Interpretation*, Wiley, Chichester, U.K., 1998.

Lehman, T.A. and Bursey, M.M., *Ion Cyclotron Resonance Spectroscopy*, Wiley, New York, 1976.

Lindon, J.C., Tranter, G.E., and Holmes, J.L., *Encyclopedia of Spectroscopy and Spectrometry*, Academic Press, New York, 2000.

Litzow, M.R. and Spalding, T.R., *Mass Spectrometry of Inorganic and Organometallic Compounds*, Elsevier, Amsterdam, 1973.

Lubman, D.M., *Lasers and Mass Spectrometry*, Oxford University Press, Oxford, 1990.

March, R.E. and Hughes, R.J., *Quadrupole Mass Spectrometry*, Wiley, New York, 1989.

March, R.E. and Todd, J.F.J., *Practical Aspects of Ion Trap Mass Spectrometry*, CRC Press, Boca Raton, FL, 1995.

Massey, H.S.W., *Negative Ions*, Cambridge University Press, London, 1976.

Matsuo, T., Caprioli, R.M., Gross, M.L., and Seyama, Y., *Biological Mass Spectrometry: Present and Future*, Wiley, New York, 1994.

McFadden, W.H., *Techniques of Combined Gas Chromatography/Mass Spectrometry*, Wiley, New York, 1973.

McLafferty, F.W., *Tandem Mass Spectrometry*, Wiley, New York, 1983.

McLafferty, F.W., *Interpretation of Mass Spectra,* University Science Books, 1996.

McMaster, M.C. and McMaster, C., *GC/MS: A Practical User's Guide,* Wiley, Chichester, U.K., 1998.

Meisel, W.S., *Computer Orientated Approaches to Pattern Recognition,* Academic Press, New York, 1972.

Mellon, F.A., Selh, R., and Startin, J.R., *Mass Spectrometry of Natural Substances*, Royal Society of Chemistry, London, 2000.

Message, G.M., *Practical Aspects of GC/MS*, Wiley, New York, 1984.

Middleditch, B.S., *Practical Mass Spectrometry*, Plenum Press, New York, 1979.

Middleditch, B.S., *Analytical Artifacts*, Elsevier, Amsterdam, 1989.

Millar, J.M., *Chromatography Concepts and Contrasts*, Wiley, New York, 1988.

Millard, B.J., *Quantitative Mass Spectrometry*, Heyden, London, 1978.

Montaudo, G. and Lattimer, R.P., *Mass Spectrometry of Polymers*, CRC Press, Boca Raton, FL, 2001.

Montaser, A., *Inductively Coupled Plasma Mass Spectrometry*, Wiley, Chichester, U.K., 1998.

Murphy, R.C., *Mass Spectrometry of Lipids*, Plenum Press, New York, 1993.

Niessen, W.M.A. and van der Greef, J., *Liquid Chromatography-Mass Spectrometry*, Marcel Dekker, New York, 1992.

Niessen, W.M.A., *Liquid Chromatography-Mass Spectrometry*, Marcel Dekker, New York, 1998.

Pfleger, K., Maurer, H.H., and Weber, A., *Mass Spectral and GC Data of Drugs, Poisons, Pesticides, Pollutants and Their Metabolites,* VCH, Weinheim, Germany, 1992.

Platzner, I.T., *Modern Isotope Ratio Mass Spectrometry*, Wiley, Chichester, U.K., 1997.

Prokai, L., *Field Desorption Mass Spectrometry*, Marcel Dekker, New York, 1990.

Reed, R.I., *Ion Production by Electron Impact*, Academic Press, New York, 1962.

Reed, R.I., *Applications of Mass Spectrometry to Organic Chemistry*, Academic Press, New York, 1966.

Schlag, E.W., *ZEKE Spectroscopy*, Cambridge University Press, London, 1998.

Silberring, J., Ekman, R., Desiderio, D.M., and Nibbering, N. M., *Mass Spectrometry and Hyphenated Techniques in Neuropeptide Research*, Wiley Interscience, New York, 2002.

Siuzdak, G., *Mass Spectrometry for Biotechnology*, Academic Press, New York, 1996.

Smith, R.M., *Gas and Liquid Chromatography in Analytical Chemistry*, Wiley, Chichester, U.K., 1988.

Smith, R.M. and Busch, K.L., *Understanding Mass Spectra: A Basic Approach*, Wiley, Chichester, U.K., 1998.

Snyder, A.P., *Biochemical and Biotechnological Applications of Electrospray Ionization Mass Spectrometry*, Oxford University Press, Oxford, 1998.

Snyder, A.P., *Interpreting Protein Mass Spectra: A Comprehensive Resource*, American Chemical Society, Washington, D.C., 2000.

Sparkman, O.D., *Mass Spectrometry Desk Reference,* Global View Publishing, Pittsburgh, PA, 2000.

Standing, K.G. and Ens, W., *Methods and Mechanisms for Producing Ions from Large Molecules*, Plenum Press, New York, 1991.

Stauffer, D.B. and McLafferty, F.W., *The Wiley/NBS Registry of Mass Spectral Data*, Wiley Interscience, New York, 1989.

Stemmler, E.A. and Hites, R.A., *Electron Capture Negative Ion Mass Spectra of Environmental Contaminants and Related Compounds*, VCH, Weinheim, Germany, 1988.

Suelter, C.H. and Watson, J.T., *Biomedical Applications of Mass Spectrometry*, Wiley Interscience, New York, 1990.

Taylor, H.E., *Inductively Coupled Plasma-Mass Spectroscopy*, Academic Press, New York, 2000.

Tuniz, C., *Accelerator Mass Spectrometry: Ultrasensitive Analysis for Global Science*, CRC Press, Boca Raton, FL, 1998.

Vickerman, J.C., Brown, A., and Reed, N.M., *Secondary Ion Mass Spectrometry: Principles and Applications*,

Wangzhao, Z., *Advanced Inductively Coupled Plasma*: *Mass Spectrometry Analysis of Rare Elements*, Balkema Publishers, 1999.

Watson, J.T., *Introduction to Mass Spectrometry*: *Biomedical, Environmental and Forensic Applications*, Raven Press, New York, 1976.

Wilson, R.G., Stevie, F.A., and Magee, C.W., *Secondary Ion Mass Spectrometry: A Practical Handbook for Depth Profiling and Bulk Impurity Analysis*, Wiley, Chichester, U.K., 1989.

Willoughby, R., Sheehan, E., and Mitrovitch, S., *A Global View of LC/MS: How to Solve Your Most Challenging Analytical Problems,* Global View Publishing, Pittsburgh, PA, 1998.

Yergey, A.L., Edmonds, C.G., Lewis, I.A.S., and Vestal, M.L., *Liquid Chromatography/Mass Spectrometry: Techniques and Applications,* Plenum Press, New York, 1990.

Appendix D

Regular Publications in Mass Spectrometry

The following journals are either devoted entirely to articles on mass spectrometry or contain significant numbers of papers on the subject. They are obtainable through direct subscription or through library systems.

Analytical Biochemistry
Analytical Chemistry
Analytical Instrumentation
Applied Spectroscopy Reviews
Biological Mass Spectrometry
(formerly *Biomedical and Environmental Mass Spectrometry*)
Chromatographic Science
European Spectroscopy News
European Mass Spectrometry
International Journal of Mass Spectrometry
International Journal of Mass Spectrometry and Ion Processes
(formerly *Journal of Mass Spectrometry and Ion Physics*)
Journal of Analysis and Applied Pyrolysis
Journal of Chromatographic Science
Journal of Chromatography
Journal of Environmental Monitoring
Journal of Liquid Chromatography
Journal of Mass Spectrometry
(now incorporates *Organic Mass Spectrometry and Biological Mass Spectrometry*)
Journal of the American Society for Mass Spectrometry
Mass Spectrometry Reviews
Organic Mass Spectrometry
Rapid Communications in Mass Spectrometry
Spectroscopy Europe
Trends in Analytical Chemistry

Appendix C

Regular Publications in Mass Spectrometry

Appendix E

Publications Containing Occasional Papers Related to Mass Spectrometry

Apart from the well-known journals covering aspects of chemistry, physics, biology, medicine, geology, environmental science, electrical engineering, and forensic science, which all have occasional articles that use mass spectrometry for analytical purposes, the following journals frequently contain papers in which mass spectrometry plays a major role:

Angewandte Chemie

Applied Spectroscopy

Journal of Chemical Physics

Journal of Organic Chemistry

Journal of Physical Chemistry

Journal of the American Chemical Society

Journal of the Chemical Society (Perkin Transactions 1 and Perkin Transactions 2)

Mass Spectrometry Reviews

Nature

Physics Reviews

Index

A

B

C

F

G

J

K

L

M

N

Q

R

S

T

U

V

W

X

Y

Z